The Duckfoot Site

Volume 1
Descriptive Archaeology

Edited by
Ricky R. Lightfoot and Mary C. Etzkorn

OCCASIONAL PAPER NO. 3
CROW CANYON ARCHAEOLOGICAL CENTER
Cortez, Colorado
1993

©1993 Crow Canyon Archaeological Center
ISBN 0-9624640-2-3
Library of Congress Catalog Card Number: 92-75597

Copyedited by Fiorella Ljunggren
Line drawings by Thomas May
Production coordinated by Lynn L. Udick
Production by Louise M. Schmidlap and Mary C. Etzkorn
Artifact photography by Rick Bell
Photographic prints by Lou Swenson, Gallery 145, Dolores, Colorado
Cover design by Nancy Leach, Graphic Interpretations, Durango, Colorado
Printed by Thomson-Shore, Inc., Dexter, Michigan

This book is printed on recycled acid-free paper.

The Duckfoot Site

Volume 1

Descriptive Archaeology

OCCASIONAL PAPERS OF THE CROW CANYON ARCHAEOLOGICAL CENTER

William D. Lipe, General Editor

CROW CANYON ARCHAEOLOGICAL CENTER

Crow Canyon Archaeological Center is a private, independent, not-for-profit organization committed to excellence in sustained interdisciplinary archaeological research integrated with experiential education programs. The Center seeks to broaden public involvement in, awareness of, and support for the conservation of our cultural heritage and resources, and to increase public knowledge of prehistoric and contemporary Native American cultures.

Crow Canyon Archaeological Center
23390 County Road K
Cortez, Colorado 81321

Telephone: 303-565-8975
FAX: 303-565-4859
Bitnet: alr@csn.org

Contents

Illustrations

Tables

Preface

The Duckfoot site investigation was the first major research project to be undertaken by the Crow Canyon Archaeological Center when, in 1983, the center was transformed from a supplemental education program (Crow Canyon School) into a research and education institution with archaeology as its primary focus. With a small professional staff, a limited budget, and the help of an enthusiastic group of pioneering students, the first Crow Canyon field season began at Duckfoot on May 1, 1983. The archaeological staff that year consisted of E. Charles (Chuck) Adams and Bruce A. Bradley, who jointly oversaw both field and laboratory operations—Adams as Research Director and Bradley as Associate Director. Work crews that first year, and in all succeeding years, were composed of program participants ranging from junior high students to senior citizens. Most participated in one-week educational programs, although some stayed longer, and a dedicated handful returned year after year to form a Duckfoot "core" field and laboratory crew.

The next four years saw many changes in field and laboratory procedures, the result of accumulated experience in working with the public and gradual increases in budget and staffing. In 1984, Bruce Bradley and Ricky R. Lightfoot codirected the Duckfoot excavations after Chuck Adams began work on another project within the organization. In 1985, Lightfoot assumed the lead role and remained Duckfoot director for the duration of the investigation. Also in 1985, Crow Canyon hired a full-time Laboratory Director, Angela L. Schwab; until that time, various field personnel had overseen lab operations on a rotating basis. Fieldwork on the Duckfoot project ended in November 1987.

The variable levels of funding and staffing affected the consistency of the data used in reporting. Field and laboratory recording systems during the early years were prototypes of more rigorous procedures gradually implemented throughout the course of the project. The reader will notice some "unevenness" in reporting, particularly in discussions of stratigraphy and presentations of artifact data. Stratigraphic observations were recorded in less detail during the early days of the project, and profiles were not consistently mapped to scale. Detailed profiles exist for all of the pit structures but for only some of the rooms; because postabandonment stratigraphy throughout the roomblock was similar, the reader may extrapolate from rooms for which detailed observations were made to those for which less information was recorded. Similarly, artifacts recovered from living and use surfaces of structures excavated early in the investigation were not systematically point-located (mapped as individual items or groups of related items), as eventually became standard practice. Therefore, much of the artifact information for these units is simply summarized in text, and maps do not necessarily show specific artifact locations. Structures for which artifact information was not recorded systematically are identified in Chapter 2.

Laboratory analysis procedures were also refined over the years. Certain changes in the system were imposed retroactively on the Duckfoot assemblage, but others were too time-consuming to implement as a matter of course. Therefore, there are some inconsistencies in the analytic information reported for the Duckfoot assemblage, particularly for stone artifacts, and the Duckfoot data are not strictly comparable to other data generated for subsequent Crow Canyon research projects. Analytic systems used specifically for the Duckfoot project are described in detail in Chapters 3 through 9 of this volume.

The history of reporting for the Duckfoot project extends as far back as the first year of excavation. At the close of every field season, a brief descriptive report was prepared to fulfill our most immediate obligations to state and federal government agencies, the professional community, and the program participants who had contributed their time and labor to the investigation. The first two reports took the form of Crow Canyon newsletter articles (Center for American Archeology 1983a, 1983b). The remainder were short manuscripts that collectively constituted an informal annual report series intended for limited distribution (Adams 1984; Lightfoot 1985, 1987; Lightfoot and Van West 1986; Lightfoot and Varien 1988). All consisted primarily of discussions of architecture; because it was not possible to analyze all artifacts recovered during a given field season by the time the annual report was being prepared, artifact descriptions were limited to observations made in the field. The purpose of this book is to draw together architectural information presented piecemeal in the individual annual reports; to present new data heretofore unavailable, primarily the results of laboratory analyses; and to use this information to offer a comprehensive interpretation of life at the Duckfoot site during the late Pueblo I period. The length and internal organization of the Duckfoot monograph necessitate its publication as two separate, but related, volumes. Volume 1 describes in detail site architecture, stratigraphy, and chronology and presents the results of basic laboratory analyses. Volume 2 draws on information presented in Volume 1 to support interpretations of activity-area distribution, social organization, and site-formation processes; it is a revised version of Lightfoot's (1992a) Ph.D. dissertation. Although each volume is designed to stand on its own, together the two form a complementary set that, it is hoped, will provide readers with a greater understanding of the development of early southwestern Pueblo culture.

Acknowledgments

Many individuals and institutions contributed to the success of the Duckfoot Project. We especially wish to acknowledge the approximately 4,000 participants in Crow Canyon's research programs who served as field and laboratory crew from 1983 through 1987. They paid their way, they paid *our* way, and they paid in the flesh with their accumulated sweat, blisters, gnat bites, and sunburns. To each and every one of them we extend our sincerest thanks for their contribution of labor and moral support.

The concept of an institution that combined long-term, high-quality archaeological research with a broad-based public education program was pioneered by Stuart Struever at the Center for American Archaeology in Illinois. As the Crow Canyon Archaeological Center's first President, Struever extended this concept to southwestern Colorado; the Duckfoot Project was the fledgling organization's first field research, initiated in 1983. Struever's vision and leadership thus made the project possible in the first place, and the laboratory and editorial support systems that the Center developed during his years as President have made this publication possible.

Prior to the development of the Crow Canyon Archaeological Center, the campus near Cortez was occupied by the Crow Canyon School, which was founded by Edward and Joanne Berger, who developed innovative interdisciplinary education programs in this setting. Ed Berger then became the first Executive Director of the Crow Canyon Archaeological Center in 1983, and Joanne Berger filled several key staff roles in administration, education, and research. The Bergers were already familiar with the Duckfoot site and recommended it as a potential research location in 1983. They also provided essential administrative guidance and collegial support in the early years of the project.

The initial research design for the Duckfoot Project was developed by E. Charles Adams, and the first year of field and laboratory work was directed by Adams and Bruce Bradley. The leadership and vision of Ian (Sandy) Thompson, Executive Director of Crow Canyon from 1985 to 1991, provided the focus necessary to sustain a research effort of such long duration, and those same qualities in his successor, Gerald L. Vincent, ensured the successful completion of the project. As informal advisor and later Research Director, William D. Lipe provided intellectual stimulus and guidance from the early days of excavation through the publication of this book.

The Duckfoot site is located on a 40-acre parcel of land that was jointly owned by Carol L. Claycomb, Eugene P. Crawford, James P. Crawford, and Wesley Dunlap. We thank these individuals for allowing us to excavate on their land and for generously donating all the recovered materials to the Bureau of Land Management's Anasazi Heritage Center. We would also like to thank Archie E. Hanson, who recently bought an interest in the property that includes Duckfoot. He helped us gain permission from the previous landowners to donate the Duckfoot collection to the Anasazi Heritage Center.

Crow Canyon staff members and volunteers Jenny Adams, Allen Denoyer, Jane Ferguson, Carrie Lipe, John Moore, Maureen Stoll, Russ Stoll, Carla Van West, and Mark Varien at different times served as field and/or laboratory assistants, and numerous other individuals donated their services in various capacities throughout the investigation. Bill and Mickie Thurston, long-time friends and supporters of Crow Canyon, provided labor and logistical support, including making the arrangements for an aerial-photography flyover of the site; Figures 1.2 and 2.11 in this volume were photographed by Bill in 1983 from a helicopter provided by then Chairman of the Board Ray Duncan. George and Betty Havers volunteered as field and lab assistants for up to three months every year from 1985 through 1987. When Pit Structure 4 was discovered only four weeks before the end of the final field season at Duckfoot, a host of colleagues came to the rescue and helped us finish the excavations. Those who volunteered in the cold, windy, rainy, and generally inhospitable weather of late October and early November 1987 include Cindy Bradley, Jerry Fetterman, Melissa Gould, Jim Hampson, Michelle Hegmon, Linda Honeycutt, Wayne Howell, and Jamie Merewether. John A. Campbell, Professor of Geology at Fort Lewis College, donated his time to assist us in identifying an assortment of rocks and minerals that had languished in the "to-be-analyzed" box for up to eight years. Volunteer Carole Graham conducted a mano-metate matching analysis that contributed to addressing questions of household organization. Louise Schmidlap supervised a pottery-refitting analysis conducted by research interns Katina Owen, Koranee Sangruchi, Lisa Shifrin, and Gary Wood. We appreciate the assistance of, and the opportunity to know, four individuals from the local Community Connections program—Terry Foy, Shelly Runck, Marsha Sperry, and Charlie Yelenik—who washed hundreds of bags of artifacts during the winter of 1987.

A number of individuals provided valuable assistance in preparing the manuscript for publication. Bruce Bradley and Angela Schwab reviewed portions of the text, especially discussions of material culture, for accuracy and clarity. Fiorella Ljunggren copyedited text and helped establish clear editorial standards to be used not only for this report, but for future Crow Canyon publications as well. Rex Adams put in many hours at the University of Arizona libraries tracking down obscure references and checking them for accuracy. Rick Bell took time out of his busy Crow Canyon schedule to photograph the Duckfoot artifacts and advise us in matters of design and layout. Tom May drafted all maps and line drawings appearing in this volume and pasted up the final camera-ready copy. The On Target scheduling software used to organize and coordinate the entire publication effort was donated by Semantic Corporation.

Assistance for refitting studies was provided by the Wenner-Gren Foundation for Anthropological Research. Support for faunal and botanical analyses was granted by the Colorado Historical Society, and for manuscript preparation, by the Colorado Endowment for Humanities. Funding from the Karen S. Greiner Endowment for Colorado Archaeology assisted Michelle Hegmon's analysis of white ware pottery temper.

Because Crow Canyon's archaeological program is so large, the "downstream" research costs—the cost of specialized laboratory studies of field data as well as the writing, editing, and publication of finished research manuscripts—represent a major challenge for the Center. It is not too much to say that the overall success of the research program depends on Crow Canyon's ability to meet these costs.

In response to this situation, Peggy and Steve Fossett established the **Fossett Research Fellowship** in 1989. The purpose of this special fund was not only to provide the salary for the Director of Research, Dr. William D. Lipe, but also to meet the costs of the many laboratory and manuscript preparation activities. By establishing this specialized research fund, the Fossetts highlighted a crucial need at Crow Canyon. To put their plan into motion, they created a multiyear challenge grant. Their challenge was taken up by a group of individuals who shared the Fossetts' perception that Crow Canyon's important research efforts could not come to fruition without significant, targeted funding. Contributors to the Fossett Research Fellowship include Richard Angell and Nancy Tani; Patrick H. Arbor; Rogers Aston; Judith Ann Calder and Christopher B. Cohen; Mr. and Mrs. Nance G. Creager; Gaylord Freeman; Mr. and Mrs. J. H. Frisbie; Richard H. Goss; Mr. and Mrs. Robin E. Hernreich; Timothy E. Kemper; John D. and Catherine T. MacArthur Foundation; Martha K. Maytnier; Maurice Perlstein, Jr.; Mr. and Mrs. James R. Porter; Joseph J. Ritchie; Sally Rose; Abraham J. Stern; Harris Suzuki; Mr. and Mrs. Preston Thompson; Byron Vasiliou; Christopher Wegren; Col. and Mrs. Fraser E. West.

Meanwhile, a corollary effort has been underway to meet the specific costs associated with the publication and distribution of research volumes. Crow Canyon is indebted to the following donors to the Research Publication Fund: Amsted Industries Foundation; Anonymous; Mr. and Mrs. Albert G. Boyce, Jr.; Stefan Brecht; Robert W. Cox; Marjorie Y. Crosby; Wesley M. Dixon, Jr.; George F. Feldman; Dr. and Mrs. David M. Gibson; John M. Hopkins; Clayton R. Jackson; Mrs. Harold B. James; Mark O. L. Lynton; Nick G. Maggos; Mr. and Mrs. R. A. McClevey, Jr.; William H. McLaughlin; Robert K. Mohrman; Mr. and Mrs. Kenneth F. Montgomery; George A. Ranney, Sr.; Rex Rice; Scott Family Foundation; Mr. and Mrs. Robert L. Seiffert; O. J. Sopranos; Carole B. Ely and Robert F. Wickham; Gordon P. Wilson.

List of Contributors

KAREN R. ADAMS (Ph.D., University of Arizona, 1988) is Director of the Environmental Archaeology program at the Crow Canyon Archaeological Center. She formerly served as an independent archaeobotanical consultant on the Duckfoot project and currently consults on other southwestern U.S. and northern Mexico projects.

MARY C. ETZKORN (B.A., University of Illinois, Champaign-Urbana, 1977) is an Editor and former Laboratory Assistant at the Crow Canyon Archaeological Center.

JANNIFER W. GISH (M.A., Arizona State University, 1975) is Director of Quaternary Palynology Research, Rio Rancho, New Mexico, and a Research Associate of the Crow Canyon Archaeological Center.

G. TIMOTHY GROSS (Ph.D., Washington State University, 1987) is Principal Archaeologist at Affinis, an environmental consulting firm in El Cajon, California, and an Adjunct Professor of Anthropology at San Diego State University, San Diego.

MICHELLE HEGMON (Ph.D., University of Michigan, 1990) is an Assistant Professor in the Department of Sociology and Anthropology at New Mexico State University, Las Cruces, and a Research Associate of the Crow Canyon Archaeological Center.

J. MICHAEL HOFFMAN (Ph.D., University of Colorado, Boulder, 1973; M.D., University of Maryland School of Medicine, 1970) is a Professor of Anthropology at Colorado College, Colorado Springs, and a board-certified forensic anthropologist consulting for a variety of agencies in Colorado.

RICKY R. LIGHTFOOT (Ph.D., Washington State University, 1992) is a Research Archaeologist at the Crow Canyon Archaeological Center.

LISA K. SHIFRIN (B.A., Northwestern University, 1984) is a graduate student in the Department of Anthropology at Washington State University, Pullman.

MARK D. VARIEN (M.A., University of Texas, Austin, 1984) is a Research Archaeologist at the Crow Canyon Archaeological Center and a student in the Ph.D. program in anthropology at Arizona State University, Tempe.

DANNY N. WALKER (Ph.D., University of Wyoming, 1986) is the Assistant State Archaeologist for the State of Wyoming Department of Commerce and an Adjunct Assistant Professor of Anthropology at the University of Wyoming, Laramie.

1

Introduction

Ricky R. Lightfoot, Mary C. Etzkorn, Karen R. Adams, and Danny N. Walker

The Duckfoot site, located on private land approximately 5 km west of Cortez, Colorado (Figure 1.1), is a late Pueblo I Anasazi habitation site consisting of 19 contiguous surface rooms, 4 pit structures, and an extensive midden. Site layout is typical for this period: the surface rooms form a slightly curved, double-row roomblock to the north of the pit structures, which are arranged in a straight southwest-northeast line (Figure 1.2). The midden is located south of the pit structures. An isolated surface room, which may or may not be associated with the main occupation of the site, was also recorded west of the roomblock. The site takes its name from a pottery object resembling a duck's foot that was found on the first day of excavation; its mate was found four years later, a month before excavations were completed (Figure 1.3). The zoomorphic feet and legs are believed to have supported a pottery vessel, although the vessel itself was never recognized in the large sherd assemblage from the site.

On the basis of architectural style, pottery types, and tree-ring dating results, the site is believed to have been built in the mid- to late A.D. 850s and occupied for a relatively short period of time, perhaps 20 to 25 years. With the possible exceptions of two intrusive features in post-abandonment structure fill and the isolated surface room of unknown date, site deposits appear to represent a single occupation. Duckfoot was remarkably well preserved. Abandonment was abrupt, and many structures burned with their contents still inside, presenting researchers with an ideal laboratory for the study of early Pueblo development.

From 1983 through 1987, the Crow Canyon Archaeological Center, a not-for-profit research and educational institution, conducted intensive excavations at the Duckfoot site under State of Colorado Archaeological Permit numbers 83-9, 84-15, 85-22, 86-22, and 87-26. Field and laboratory studies were designed to increase our understanding of the prehistoric occupation of the area during the mid- to late ninth century. An equally important educational goal was to increase public awareness, knowledge, and appreciation of the archaeological research process by providing a hands-on learning experience in a carefully controlled and supervised setting. With the assistance of lay participants enrolled in the center's various research and educational programs, Crow Canyon excavated over 90 percent of the site and completed basic preliminary analyses of the materials recovered. The results of field and laboratory studies are the subject of this descriptive report.

Environmental Setting

An understanding of basic environmental conditions, both past and present, is prerequisite to understanding how any group of people adapts to the constraints imposed by the physical environment. Although a detailed study of human adaptation to the environment is not a focus of this report, a brief description of the physiography, geology, soils, climate, flora, and fauna of the region will serve to familiarize the reader with existing conditions in the Four Corners area and, given certain qualifications, with conditions that may have existed when the Anasazi built, occupied, and abandoned the Duckfoot site over 1,100 years ago. In addition, the information presented in this section provides the background for discussions of resource use presented in later chapters.

Physiography, Geology, and Soils

The Duckfoot site is located on top of a ridge between Crow Canyon to the east and Alkali Canyon to the west, at an elevation of approximately 1945 m (6380 ft) (Figures 1.1

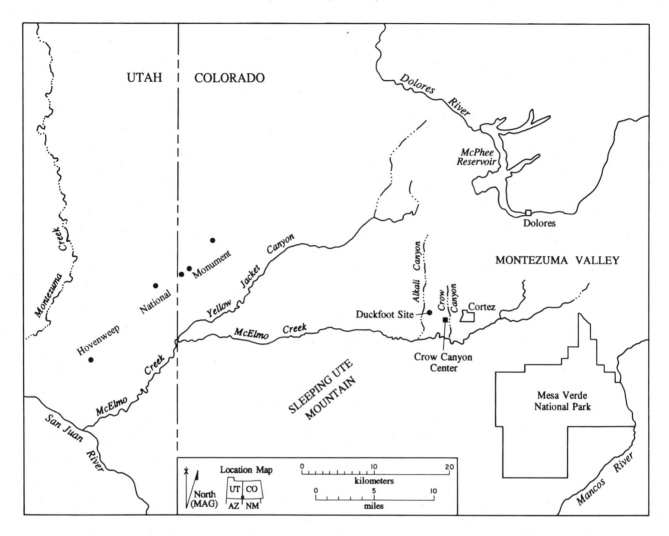

Figure 1.1. Location of the Duckfoot site, southwestern Colorado.

and 1.4). The area lies within the northeasternmost portion of the Colorado Plateau, an area of geologic uplift characterized by broad tablelands, mesas, and deeply entrenched canyons. Visible to the north and east of the site are the San Juan Mountains, with peaks rising to over 4268 m (14,000 ft); to the south, the Mesa Verde escarpment; and to the southwest, Ute Mountain, with a maximum elevation of 3042 m (9977 ft). The area immediately surrounding the site is characterized by rolling uplands dissected at intervals by medium-size canyons. Alkali and Crow canyons drain to the south into McElmo Creek, which in turn flows into the San Juan River, all part of the Colorado River drainage system. The closest reliable water supply to the site today is Alkali Canyon, slightly less than 1 km west of the site; however, water levels in this drainage are greatly affected by irrigation runoff, and prehistorically this stream may well have been intermittent. Crow Canyon, located 2.2 km east of Duckfoot, is spring-fed and possibly was the closest year-round water source when the site was occupied.

Bedrock geology in the immediate vicinity of the Duckfoot site consists of a variety of sedimentary formations overlain by Quaternary eolian sand and silt deposits (Ekren and Houser 1965:Plate 1). The uppermost formation, the Dakota Sandstone, is composed largely of sandstones and mudstones (Ekren and Houser 1965:20–21) and is of archaeological interest because it yields a number of materials used prehistorically for construction and tool manufacture. A tabular Dakota sandstone was favored for construction throughout the prehistoric Pueblo period, and a variety of sandstones and orthoquartzites (silica-cemented sandstones) from this formation were used for a number of stone tool types (Leonhardy and Clay 1985:134). Beneath the Dakota Sandstone lies the Burro Canyon Formation, composed primarily of mudstones, conglomerates, and conglomeratic sandstones; the conglomerates contain chert, siltstone, and limestone inclusions in a sand matrix (Ekren and Houser 1965:18–19, Plate 1). Prehistorically, materials from the Burro Canyon Formation were used to make

Figure 1.2.   Aerial view of the Duckfoot site during excavation. Pit Structure 1 (*center*) is completely excavated; Pit Structure 2 (*top center*), the roomblock (*left*), and the midden (*right*) are partly excavated. North is to the left.

stone tools, and Leonhardy and Clay (1985:133) note that Burro Canyon cherts and orthoquartzites are particularly well suited for chipped-stone tool manufacture. Underlying the Burro Canyon Formation is the Morrison Formation and, more specifically, in the immediate site vicinity, the Brushy Basin Member of the Morrison Formation (Ekren and Houser 1965:Plate 1). The Brushy Basin Member is composed of a variety of claystones, mudstones, siltstones, and sandstones, many of which are indurated with silica (Ekren and Houser 1965:15–16), making them excellent raw materials for chipped-stone tools. Although some clay sources are present in the claystone beds of the Morrison Formation, other clays, including some suitable for pottery manufacture, are found in greater quantities in the Mancos Shale Formation (Leonhardy and Clay 1985:131–134). The Mancos Shale is not present in the immediate vicinity of the site, but when it occurs, it overlies the Dakota Sandstone. Two Mancos Shale outcrops are found at some distance from Duckfoot: one approximately 7 km to the south, the other 6 km to the northeast (Ekren and Houser 1965:Plate 1; Haynes et al. 1976:map I-629). The Dakota Sandstone, Burro Canyon, and Morrison formations are

exposed in the eroded walls of Alkali Canyon, within short walking distance of the site.

Sources of igneous rock closest to Duckfoot are 10 to 12 km distant, at the base of Ute Mountain, along McElmo Creek. Igneous bedrock outcrops and gravel deposits on the south side of the creek (Ekren and Houser 1965:26, Plate 1) are derived from the mountain, which is an intrusive igneous formation. Igneous rock was used as temper in locally made Anasazi pottery and occasionally in the manufacture of stone tools.

Soils in the area are predominantly Inceptisols that formed in silty eolian parent material deposited during the late Pleistocene and Holocene (Clay et al. 1987; Arrhenius and Bonatti 1965). The eolian silts have been described as Mesa Verde loess (Arrhenius and Bonatti 1965) and Sage Plain loess (Shawe et al. 1968). The loess overlies sandstone and shale bedrock, and the soils that formed in these deposits are the arable mesa-top soils that are farmed to this day. The natural soil profile was exposed in the walls of the excavated pit structures at Duckfoot. These exposures revealed a 1.3-m solum consisting of an A-Bk1-Bk2-Btk1-Btk2-Ck-R horizon sequence. Calcium carbonate

Figure 1.3. Pottery objects resembling duck feet are believed to have supported a pottery vessel. The foot on the left was found on the first day of excavation and gives the site its name. Height (*left*) = 8.1 cm.

(caliche) was abundant throughout the profile below the A horizon and was most abundant toward the bottom of the profile. Alumino-silicate clays increased in the lower soil profile probably as a result of both the translocation of clay downward in the profile and the weathering of shale that underlies the soil. The Anasazi builders excavated the pit structures through the entire soil sequence, continuing through approximately 65 cm of shale and caliche and about 20 cm into sandstone bedrock. The A horizon is dark brown (7.5YR3/4), with the darkness resulting from organic humus. The B horizon has a strong subangular blocky ped structure and ranges from reddish brown (5YR4/3) on top to strong brown (7.5YR4/6) in the middle and brown (7.5YR4/4) at the bottom. The redder colors toward the top probably resulted from the oxidation of iron-rich minerals toward the surface. The C horizon is dark gray clay with calcium carbonate. It probably formed as a result of the weathering of the underlying shale deposits, although the gray color suggests that little oxidation has occurred at this depth. The characteristics of this deposit indicate that its development was influenced by a mixture of two different parent materials, Pleistocene eolian silts and a shale that may be a remnant of the Cretaceous Mancos Shale, which overlies the Dakota Sandstone (Ekren and Houser 1965). There is no apparent lithologic discontinuity marking the change in dominance of these two parent materials, but the change is signaled by the sharp increase in clay (Btk) at about 70 cm below modern ground surface.

Climate

Climatic factors that have a direct bearing on the success of dry-land farming include temperature, length of growing season, and amount and seasonal distribution of precipita-

tion. The climate records for the 30-year period ending in 1980 in nearby Cortez (elevation 1893 m [6210 ft]) indicate that the mean annual precipitation for the area around Duckfoot is approximately 32 cm (12.7 in). The precipitation is fairly evenly distributed throughout the year (National Climate Center 1983), with about half occurring in the form of winter snows and the other half deriving from summer monsoon rains. The two seasonal moisture patterns usually change in late May through June and again in November, and these months generally are drier than other months of the year. Adequate winter precipitation (snowpack) is particularly important to dry-land farmers because it provides the spring soil moisture needed for seed germination and the deep soil moisture needed to sustain growth during the dry period in late May and June (Erdman et al. 1969:19). The mean annual temperature at Cortez is 9.3°C (48.8°F), with a mean January temperature of 2.8°C (26.9°F) and a mean July temperature of 22.2°C (71.9°F) (National Climate Center 1983). Between 1951 and 1973, the average length of the growing season (consecutive frost-free days) in Cortez was 131 days (Petersen and Clay 1987:188).

Dry-land farming success in a given location cannot be predicted by treating temperature and average annual precipitation as isolated variables. Petersen (1987:219–225), following a method outlined by Siemer (1977:20), uses a measure of summer warmth that, considered in conjunction with annual precipitation, better defines the range of conditions under which corn may be successfully grown in the Mesa Verde area. The measure of summer warmth used is called "growing degree days" (GDD), a term that refers simply to the "daily accumulation of degrees over 50°F achieved during the growing season" (Petersen 1987:219). On the basis of data gathered at several weather stations in southeastern Utah and southwestern Colorado, Petersen argues that corn may be successfully grown in relatively hot areas (2500 GDD) if annual precipitation is relatively high (46 cm) and that, conversely, corn may be produced with as little as 33 cm of annual precipitation if the area is relatively cool (1600 GDD) (Petersen 1987:225). Cortez, with annual precipitation of 32 cm and an estimated GDD of 2240, today falls below the limits for successful dry-land farming as calculated by Petersen (1987:Figure 14.4). On the basis of pollen and tree-ring data, Petersen (1986:316) believes that the period from A.D. 800 to 1000 had higher summer precipitation but lower winter precipitation than is true today. Surface topography (for example, aspect and slope) and edaphic factors (for example, soil depth and moisture-holding capacity) also affect the productivity of land in marginal agricultural settings. North- and east-facing slopes tend to be cooler and have moister soil than south- or west-facing slopes. Many east-facing slopes with deep, loamy soils are present to the east of Duckfoot, and these have been agriculturally productive in recent decades without the use of irrigation. Because the ridge on which the Duckfoot site is located is approximately 50 m higher

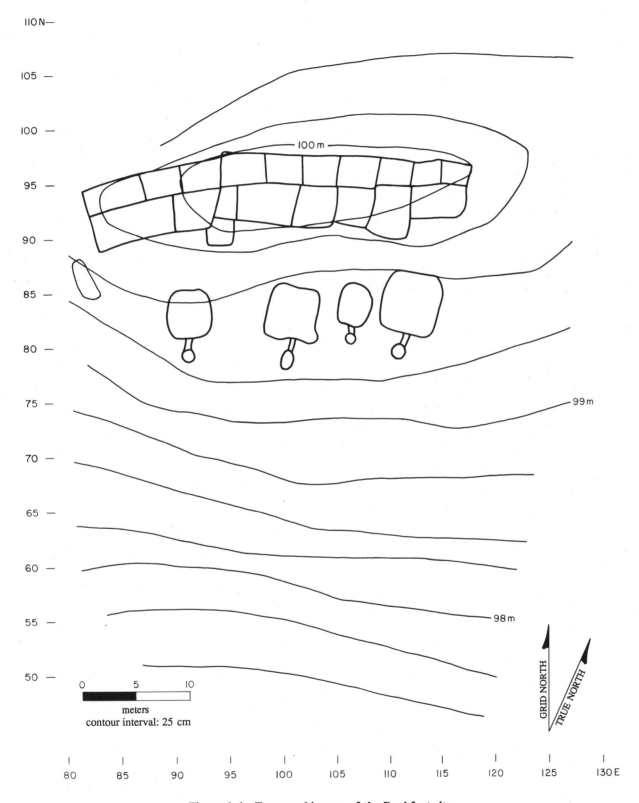

Figure 1.4. Topographic map of the Duckfoot site.

than Cortez, it possibly receives slightly more precipitation than Cortez. These topographic and edaphic factors lead Petersen (1986 and personal communication 1991) to include the area around Duckfoot in the A.D. 800–1000 dry-farming belt. In Petersen's model, the dry-farming belt was narrowest during this period and mostly restricted to areas between 2010 m (6600 ft) and 2380 m (7800 ft) in elevation.

Flora

Although extensive dry-land farming and wood cutting in historic times have affected the current vegetation distribution, a general discussion of the modern flora provides some basis for understanding the Pueblo I landscape. The suite of floristic elements present in the area today accumulated sometime after the Pleistocene, although major families, genera, and species of southwestern plant communities of essentially modern aspect were in place long before (Lowe and Brown 1982:12). Therefore, the differences in species composition in the Duckfoot area between Pueblo I times and the present are assumed to be relatively small. Vegetation in the vicinity of the Duckfoot site represents two major biotic communities, the Great Basin Desertscrub (Turner 1982) and the Great Basin Conifer Woodland (Brown 1982). Vegetation immediately surrounding the site is dominated by widely spaced pinyon pine (*Pinus edulis*) and Utah juniper (*Juniperus osteosperma*) interspersed with a variety of shrubs, most notably big sagebrush (*Artemisia tridentata*). Other woody shrubs in the nearby area include mountain mahogany (*Cercocarpus montanus*), Utah service berry (*Amelanchier utahensis*), squaw apple (*Peraphyllum ramosissimum*), bitterbrush (*Purshia tridentata*), Gambel oak (*Quercus gambelii*), and Mormon tea (*Ephedra viridis*). Herbaceous plants are important understory elements. In the early summer, one can find numerous grasses (Gramineae), legumes (Leguminosae), composites (Compositae), mustards (Cruciferae), mallows (Malvaceae), and borages (Boraginaceae). Some plants are restricted in their ranges to either relatively moist (mesic) or very dry habitats. Cattail (*Typha*) and reed (*Phragmites*), for example, are confined to mesic locations. Narrow-leaf cottonwood (*Populus angustifolia*), Fremont cottonwood (*Populus fremontii*), coyote willow (*Salix exigua*), and squawbush (*Rhus aromatica*) also do well in damp soil. In very dry locations, a few members of the cactus family are present, including cholla and prickly pear (*Opuntia*) and hedgehog (*Echinocereus*) (Turner 1982:145). A more comprehensive list of plants observed within 1.5 km of the Duckfoot site is provided in Table 1.1.

Fauna

Animals currently existing in southwestern Colorado have been discussed by Anderson (1961), Armstrong (1972),

Table 1.1. Plants Observed within 1.5 km of the Duckfoot Site

Taxon	Environment
Amelanchier utahensis	pinyon-juniper woodland
Arabis fendleri	disturbed habitat
*Artemisia tridentata**	sagebrush
Astragalus wingatanus	pinyon-juniper woodland
Balsamhoriza sagittata	mesic habitat (Alkali Canyon)
Bromus tectorum	disturbed habitat
Chenopodium sp.	disturbed habitat
Cercocarpus montanus	pinyon-juniper woodland
Cryptantha fulvocanescens	disturbed habitat
Descurainia pinnata	disturbed habitat
Ephedra viridis	pinyon-juniper woodland
Erodium cicutarium	disturbed habitat
Gilia aggregata	mesic habitat (Alkali Canyon)
Gutierrezia sp.	disturbed habitat
Helianthus annuus	disturbed habitat
*Juniperus osteosperma**	pinyon-juniper woodland
Lactuca serriola	disturbed habitat
Lappula redowskii	disturbed habitat
Opuntia spp. (prickly pear)	pinyon-juniper woodland
Peraphyllum ramosissimum	pinyon-juniper woodland
Physalis longifolia	disturbed habitat
Physaria australis	pinyon-juniper woodland
*Pinus edulis**	pinyon-juniper woodland
Populus sp.	mesic habitat (Alkali Canyon)
Purshia tridentata	pinyon-juniper woodland
Quercus gambelii	mesic habitat (Alkali Canyon)
Ranunculus sp.	mesic habitat (Alkali Canyon)
Rhus aromatica	mesic habitat (Alkali Canyon)
Rumex sp.	mesic habitat (Alkali Canyon)
Salix exigua	mesic habitat (Alkali Canyon)
Scirpus acutus	mesic habitat (Alkali Canyon)
Sisymbrium altissimum	disturbed habitat
Sitanion hystrix	disturbed habitat
Sphaeralcea digitata	sagebrush
Stipa comata	sagebrush
Stipa hymenoides	sagebrush
Streptanthus cordatus	disturbed habitat
Typha latifolia	mesic habitat (Alkali Canyon)
Yucca baccata	pinyon-juniper woodland

*Dominant member of the local vegetation.

Chase et al. (1982), Bissell and Dillon (1982), Hammerson and Langlois (1981), and Neusius (1985b, 1985c). Armstrong (1972) places southwestern Colorado in the Dolores–San Juan Faunal District of the Colorado Plateau Faunal Area—a district whose ecological complexity and diversity are reflected in the variety of mammalian species recorded (Table 1.2). There are also similarities between this faunal district and the Rocky Mountain Faunal Area to the east (Armstrong 1972), as evidenced by several relict populations of montane species. However, the Dolores–San Juan Faunal District is characterized primarily by species indigenous to the Desert Southwest. Mammalian species ob-

Table 1.2. Extant Mammalian Species in Southwestern Colorado (Montezuma County), Exclusive of Introduced or Domestic Species

Scientific Name	Common Name	Anderson (1961)	Armstrong (1972)	Bissell and Dillon (1982)
Order Insectivora				
Sorex monticolus	dusky shrew			+
Sorex nanus	dwarf shrew			+
Sorex palustris	water shrew			+
Sorex vagrans	wandering shrew	+	+	
Sorex merriami	Merriam's shrew	+	+	+
Notiosorex crawfordi	desert shrew		+	+
Order Chiroptera				
Myotis yumanensis	Yuma myotis		+	+
Myotis evotis	long-eared myotis	+	+	+
Myotis thysanodes	fringed myotis	+	+	+
Myotis volans	long-legged myotis	+	+	+
Myotis californicus	California myotis	+	+	+
Myotis leibii	small-footed myotis	+	+	+
Lasionycteris noctivagans	silver-haired bat			+
Pipistrellus hesperus	western pipistrelle		+	+
Eptesicus fuscus	big brown bat	+	+	+
Lasiurus cinereus	hoary bat		+	+
Plecotus townsendii	Townsend's big-eared bat	+	+	+
Antrozous pallidus	pallid bat		+	+
Tadarida brasiliensis	Brazilian free-tailed bat	+	+	+
Order Lagomorpha				
Sylvilagus nuttallii	Nuttall's cottontail	+	+	
Sylvilagus audubonii	desert cottontail	+	+	
Lepus americanus	snowshoe hare			+
Lepus californicus	black-tailed jackrabbit	+	+	
Order Rodentia				
Eutamias minimus	least chipmunk	+	+	+
Eutamias quadrivittatus	Colorado chipmunk	+	+	+
Marmota flaviventris	yellow-bellied marmot	+	?	+
Ammospermophilus leucurus	white-tailed antelope squirrel		+	+
Spermophilus spilosoma	spotted ground squirrel		+	+
Spermophilus variegatus	rock squirrel	+	+	+
Spermophilus lateralis	golden-mantled ground squirrel	+	+	+
Cynomys gunnisoni	Gunnison's prairie dog	+	+	+
Sciurus aberti	Abert's squirrel	+	+	+
Tamiasciurus hudsonicus	red squirrel	+	+	+
Thomomys bottae	Botta's pocket gopher		+	+
Perognathus apache	Apache pocket mouse		+	+
Perognathus flavus	silky pocket mouse		+	+
Dipodomys ordii	Ord's kangaroo rat	+	+	+
Castor canadensis	beaver	+	+	+
Reithrodontomys megalotis	western harvest mouse	+	+	+
Peromyscus crinitus	canyon mouse	+	+	+
Peromyscus maniculatus	deer mouse	+	+	+
Peromyscus boylii	brush mouse	+	+	+
Peromyscus truei	pinyon mouse	+	+	+
Peromyscus difficilus	rock mouse	?		
Onychomys leucogaster	northern grasshopper mouse		+	+
Neotoma albigula	white-throated woodrat		+	+
Neotoma mexicana	Mexican woodrat	+	+	+
Neotoma cinerea	bushy-tailed woodrat	+	+	+

Table 1.2. Extant Mammalian Species in Southwestern Colorado (Montezuma County),
Exclusive of Introduced or Domestic Species *(continued)*

Scientific Name	Common Name	Source		
		Anderson (1961)	Armstrong (1972)	Bissell and Dillon (1982)
Microtus montanus	montane vole	+	+	+
Microtus longicaudus	long-tailed vole	+	+	+
Microtus mexicanus	Mexican vole	+	+	+
Ondatra zibethicus	muskrat	+	+	+
Erethizon dorsatum	porcupine	+	+	+
Order Carnivora				
Canis latrans	coyote	+	+	+
Canis lupus	gray wolf		+	
Vulpes vulpes	red fox	+	+	+
Vulpes macrotis	kit fox		+	+
Urocyon cinereoargenteus	gray fox	+	+	+
Bassariscus astutus	ringtail	+	+	+
Procyon lotor	raccoon	+	+	+
Ursus americanus	black bear	+	+	+
Martes americana	pine martin			+
Mustela erminea	ermine			+
Mustela frenata	long-tailed weasel	+	+	+
Mustela nigripes	black-footed ferret		+	
Mustela vison	mink	+	+	+
Taxidea taxus	badger	+	+	+
Spilogale putorius	spotted skunk	+	+	+
Mephitis mephitis	striped skunk	+	+	+
Felis concolor	mountain lion	+	+	+
Felis rufus	bobcat		+	+
Order Artiodactyla				
Cervus elaphus	wapiti	+	+	+
Odocoileus hemionus	mule deer	+	+	+
Ovis canadensis	bighorn	+	+	+

served in the vicinity of the site include mule deer (*Odocoileus hemionus*), cottontail (*Sylvilagus* spp.), black-tailed jackrabbit (*Lepus californicus*), Gunnison's prairie dog (*Cynomys gunnisoni*), coyote (*Canis latrans*), and an occasional mountain lion (*Felis concolor*). Neusius (1985a, 1985b) documents the prehistoric exploitation of many of the species listed in Table 1.2 and summarizes ethnographic literature on historic faunal use in general. Historically, animal remains (meat, bone, hide, fur, feathers) have been used by aboriginal southwestern groups for food, tools, and medicinal/ceremonial purposes (Neusius 1985b:Table 16).

Cultural Setting

Before Crow Canyon's work in the area, Pueblo I sites were not recorded in the lower portions of the Montezuma Valley, the broad basin in which the Duckfoot site is located and which is loosely defined by the surrounding mesas and mountains. On the basis of environmental models and known site locations, many researchers (see, for example, Eddy et al. 1984:37–38) believed that populations during the Pueblo I period (A.D. 750–900) consolidated into communities located at relatively high elevations between 2012 and 2256 m (6600 and 7400 ft). This was thought to have happened in response to warmer, drier climatic conditions that made dry-land farming riskier at lower elevations.

In the early 1980s, however, Crow Canyon Archaeological Center surveyors recorded 10 Pueblo I sites between Crow Canyon and Alkali Canyon at elevations between 1878 and 1945 m (6160 and 6380 ft) and an additional 15 sites with a recognizable Pueblo I component (Van West 1986:12). Even more recently, a 486-ha (1,200-acre) tract surrounding and including the Duckfoot site was intensively surveyed as part of a real estate development project. This survey, conducted by Woods Canyon Archaeological Consultants and reported in Honeycutt and Fetterman (1991), resulted in the documentation of 8 Pueblo I sites, including Duckfoot, and an additional 13 sites with a known or suspected Pueblo I component (based on pottery types). These Pueblo I sites, although greatly outnumbered by earlier Basketmaker (A.D. 50–750) and later Pueblo II (A.D. 900–1150) sites in the development survey area, clearly document the Pueblo I presence in this portion of the valley. Five of the eight single-component Pueblo I sites are located along the same broad ridge on which Duckfoot is situated; the closest is approximately 100 m south of Duckfoot, and the most distant, roughly 600 m southwest. Although it is not known if, or how many, of these Pueblo I sites were occupied at the same time, it seems likely that at least some of them formed a loose community that included Duckfoot and its inhabitants.

Over 80 Basketmaker sites or sites with a possible Basketmaker component were recorded in the development survey area (Honeycutt and Fetterman 1991). The presence of these sites not only suggests continuity of prehistoric occupation of the area but has implications for the interpretation of Duckfoot artifact data as well. With so many earlier sites in the immediate vicinity, including one located within 30 m and six others within 300 m of Duckfoot, it is possible that materials were deliberately transported or inadvertently mixed between sites.

Research Objectives

The Duckfoot site presents an ideal laboratory for the study of late Pueblo I adaptations. Site occupation was relatively brief, there was little architectural remodeling, and most of the structures apparently were used simultaneously. Perhaps most important, many of the structures burned when the site was abandoned in the late ninth century. These circumstances resulted in exceptional preservation of living surfaces and artifact assemblages; on many structure floors, artifacts were found in the locations in which they were last used or stored. The most general goal of the Duckfoot project was to document and describe the occupation history of the site. Specific research objectives included testing and further developing current theories on site-formation processes, with an emphasis on abandonment behavior, and expanding our knowledge of late Pueblo I economy and social organization. The question of economy is addressed largely in this volume in Chapters 6, 7, 8, and 10. Social organization and site-formation processes are touched on briefly in this volume but are treated in greater detail in Volume 2.

The importance of Duckfoot as a research site is enhanced by earlier work conducted by the Dolores Archaeological Program (DAP), a large, federally funded mitigation project that produced the most extensive Pueblo I database in the history of archaeological research in the Four Corners area (Breternitz et al. 1986). The sites were located in the Dolores River valley, approximately 24 km northeast of Duckfoot, and the patterns that emerged from studies of these sites define the "typical" Pueblo I hamlet against which other sites, such as Duckfoot, can be measured. When appropriate, architectural, material culture, and environmental data from Duckfoot are examined against the backdrop of DAP research.

Site-Formation Processes

Site-formation processes are those processes—both cultural and natural—that produce the archaeological record. In a paper that stimulated controversy and discussion, Schiffer (1985) accused southwestern archaeologists of tacitly assuming that all structure floor assemblages could be interpreted as being the result of Pompeii-like abandonment scenarios. Schiffer (1985:38) concluded that "the real Pompeii premise . . . is that one can analyze house-floor assemblages *as if* they were systemic inventories—unmodified by formation processes." However, both Schiffer and his critics agree that archaeologists need methods to determine what the systemic inventory of prehistoric sites should include (Cordell et al. 1987:573–574; Schiffer 1985:21). In general, the processes by which sites form can be grouped into preabandonment, abandonment, and postabandonment categories (cf. Deal 1985). This simplified sequence emphasizes that at different periods in a site's history, different processes dominate. In this volume, discussions relevant to the study of formation processes focus on structure use and abandonment; the topic is addressed again in greater detail in Volume 2.

Economy

An objective of most archaeological research is to understand the relationship between society and the natural environment. As discussed earlier, a detailed study of human adaptation to the environment is beyond the scope of this book; however, several basic questions relating to resource availability and use, and to subsistence in particular, are addressed using Duckfoot macrobotanical, pollen, and faunal data. Although it would be ideal to describe the diet of the Duckfoot inhabitants over the entire year cycle and from year to year, the fact that structures and hearths apparently were cleaned out periodically restricts our view to the last few weeks (or at most, months) of occupation. The advantage of studying such an assemblage, of course, is that it provides a clear glimpse, relatively uncluttered by subsequent mixing of deposits, of subsistence practices during a specific period of time (abandonment) that is of primary interest. It also allows comparisons between different areas and structures within the site, which may shed light on intrasite social relationships. Nonsubsistence uses of plant and animal resources, seasonality of structure and site use, and tool technology are also examined under the broad heading of prehistoric economy.

Social Organization

One objective of the Duckfoot project was to define social groups within the site and determine how they interacted with one another. Because most of the site was excavated, including all architectural units, Crow Canyon researchers were in a better position to study social organization than archaeologists whose work is restricted to small samples of partly excavated sites. Archaeologists typically consider the household, rather than the family, to be the most basic identifiable social unit of a population (Wilk and Rathje 1982; Netting et al. 1984; Kane 1986; Schlanger 1985, 1987; Wilshusen 1988b). The term *household* is used to describe a group of people who interact on a regular and continuing basis and does not necessarily imply biological or kin relationships between the members of the group. Wilk and Netting (1984) propose that social scientists view households as activity groups, which are observable groups of people interacting in patterned ways. Archaeologically, we may be able to use material culture, including architecture, to recognize and measure the patterns that define prehistoric households. If households tend to confine many of their activities to specific structures and immediately adjacent areas, then, given adequate preservation, architectural remains should provide us with the evidence necessary to recognize the number and relative sizes of the social groups that inhabited a site. In the Duckfoot project, architectural models of household and suprahousehold organization, based partly on earlier work by the DAP (Kane 1986:355–358), were tested using Duckfoot architectural and artifactual evidence. Results of activity-area and artifact analyses were used to evaluate the nature and density of interactions between groups.

Success in characterizing social organization depends partly on our ability to interpret how structures were used. The criteria used to distinguish habitation (or domestic) structures from structures used for storage or some other nonhabitation purpose include structure size, location (front row versus back row), and feature- and artifact-assemblage characteristics. Surface habitation rooms tend to be large, located in the front (southern) row of the roomblock, and contain a variety of features and artifacts suggestive of multiple domestic activities. In particular, the presence of a hearth or other thermal feature and the recovery of food-preparation tools and equipment (metates, manos, storage and cooking vessels) provide convincing evidence for domestic use of a room regardless of its location. Conversely, storage or other nonhabitation rooms are typically smaller than domestic rooms, they are often located in the back (northern) row of the roomblock, and they contain few features or artifacts. In Pueblo I sites, each front-row habitation room is usually interpreted as being associated with the back storage room or rooms with which it shares a wall. Interpretations of pit structure use are based largely on feature- and artifact-assemblage characteristics that suggest that ritual or ceremonial activities, in addition to those of a domestic nature, took place in these structures. Ritual or ceremonial activities are inferred largely on the basis of the presence of certain features such as sipapus, paho marks, and sand-filled pits (Wilshusen 1989; Varien and Lightfoot 1989).

Investigative Strategy and Methods

From the outset, Crow Canyon's goal was to excavate as close to 100 percent of the site as possible. It is estimated that over 90 percent of the total site area, including all architectural units, was excavated; the northern periphery beyond the roomblock and the southwestern corner of the midden (roughly 4 percent of the total trash deposit) were not completely excavated (Figure 1.5). A 2-×-2-m grid was laid out to encompass the entire site. A portion of the grid in the northern end of the midden became skewed, and several 1-×-2-m units were outlined to compensate. A systematic surface collection was not conducted; however, selected artifacts were collected from modern ground surface in the roomblock area, near the pit structure depressions, and throughout the midden. Later, as excavation of specific units began, artifacts on modern ground surface, as well as those in the first few centimeters of unconsolidated fill, were recovered separately from underlying buried materials.

Before excavation started, vegetation was cleared from the areas of the site where, from surface evidence, the roomblock and pit structures appeared to be located. The approximate location, size, and shape of the roomblock were recognized on the basis of topography and numerous protruding sandstone slabs indicating wall alignments. Three of the four pit structures were visible as broad, shallow depressions south of the roomblock. It was not until the end of the final field season, when routine excavation in the courtyard area led to the detection of a very large "pit feature" nestled between Pit Structures 2 and 3, that the fourth pit structure was discovered. A relatively small portion of the site, estimated at less than 5 percent, had been vandalized by pot hunters. Damage was confined to the roomblock, primarily the upper fill deposits. Only in Room 10 had a small area of floor been disturbed.

Once approximate locations of major architectural units were established, two north-south exploratory transects were excavated across the roomblock, the open courtyard area, and Pit Structures 1 and 2. The purpose of these initial excavations was to better ascertain spatial and stratigraphic relationships between structures and to form initial impressions as to depths of buried surfaces. Materials recovered from the upper deposits in these trenches were grouped into several "arbitrary unit" proveniences because it was not possible in the higher levels to assign them with confidence to specific cultural units.

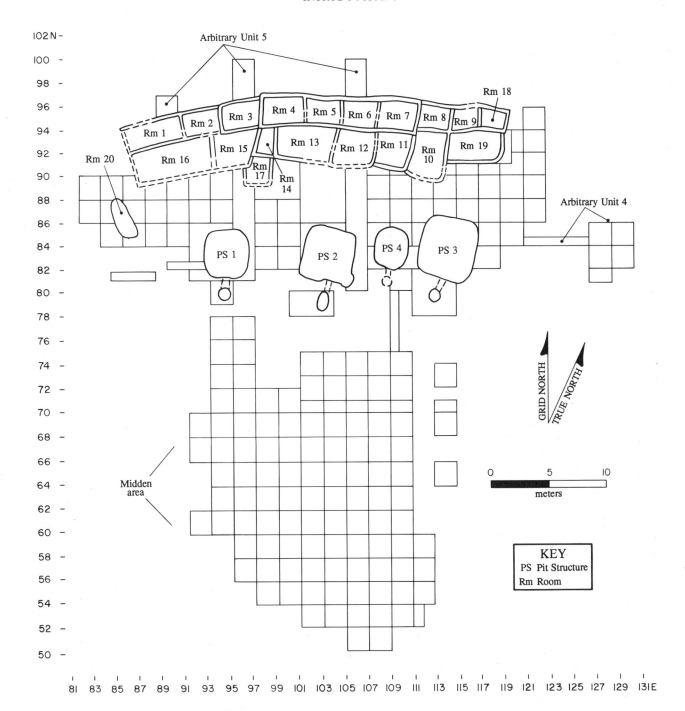

Figure 1.5. Plan of excavations, the Duckfoot site. Arbitrary Units 1, 2, and 3 are not mapped; see Chapter 2 for descriptions of these areas.

Many of the surface-room walls could be seen on modern ground surface, and others were found with minimal clearing of loose sediment from the surface. Some exploratory excavation was necessary to locate the uppermost portions of the pit structure walls. The sediments above the tops of existing structure walls were lumped into arbitrary provenience groups, whereas sediments contained within the extant walls of a structure were pro-venienced with that structure. Fill within surface rooms and pit structures was excavated in natural or cultural strata, when such stratification was discernible. When deposits had no clear stratigraphic breaks, arbitrary levels (usually 20 cm thick) were used instead. With only a few exceptions, surface rooms were excavated as whole study units. The methods used to subdivide the upper fill of pit structures were more variable but in each case involved initial excavation of a portion of the structure to expose the postabandonment fill stratigraphy. Pit structure excavation strategies and variations in excavation methods for rooms are discussed in the individual study unit descriptions in Chapter 2.

Most of the prehistoric living surfaces identified at the site were structure floors. However, other kinds of surfaces were recognized, including (1) constructional surfaces that were deliberately created but never actually used as living or use surfaces; (2) ephemeral, or indistinct, use surfaces in postabandonment fill; and (3) deposits that were placed, or allowed to accumulate, on structure floors, effectively creating secondary use/abandonment surfaces. The term *surface* was used to describe all of these contexts, and when different types of surfaces were found within a structure, a single consecutive numbering sequence was used for all. In Room 16, for instance, "Surface 1" was the upper boundary of a layer of burned materials deliberately placed on top of "Surface 2," the actual floor of the room.

In the Duckfoot assemblage, surface-associated artifacts are those that were found in contact with a prehistoric living surface or within an arbitrary 5-cm layer of fill above the surface. Not every item that was used or abandoned on a surface will be found in direct contact with the floor in an archaeological context: various processes can result in the vertical displacement of items after their initial deposition. Yet the more above-floor fill one includes with the surface, the more likely one is to include materials that were deposited after abandonment or that were incorporated in roof- and wall-construction materials. Defining living surfaces to include materials in surface contact and within 5 cm of the surface was a compromise that could be applied consistently from one context to the next. Additional provenience control was gained by mapping items that were in contact with the surface or very near it. Such items were assigned point-location (PL) numbers and recovered individually or as clusters of related artifacts. Therefore, for most surfaces, two artifact data sets are considered: the first includes all materials recovered from the 5-cm-thick surface "zone"; the second is a subset of the first and includes only those items that were mapped and assigned PL numbers. In addition, selected artifacts found in roof fall strata (strata composed entirely or primarily of collapsed roof debris) were often assigned PL numbers. Generally only formal tools, large or diagnostic sherds, and items interpreted to be in situ were included in roof fall PL sequences. Roof fall and floor sequences within a given structure were kept separate. In Chapter 2, descriptions of individual structures focus on point-located artifacts; in Chapters 3 through 5 and in Appendix A, all artifacts are included in summary presentations of material culture.

The midden and courtyard areas were excavated primarily in two separate series of contiguous 2-×-2-m squares. The first few midden units were excavated in arbitrary 20-cm levels; when no cultural or natural strata were discerned in profile, and when no appreciable differences were noted in pottery types between the different levels, it was decided to excavate the remaining midden squares in single cultural strata, from just below modern ground surface to undisturbed sterile sediments below. As with all excavation units at the site, artifacts from modern ground surface (and the first few centimeters of unconsolidated fill) were collected separately from artifacts found in subsurface deposits.

The extramural area surrounding the pit structures and extending from the pit structures to the roomblock was defined as courtyard. The courtyard area was divided into three parts that correspond to the three main subdivisions within the roomblock (described in more detail in Chapter 2) and the original three pit structures, Pit Structures 1, 2, and 3. The rationale for subdividing the courtyard was to facilitate studies of social organization. Because there were no physical boundaries separating the different areas of the courtyard, divisions were made arbitrarily along grid lines between the original three pit structures. Just how arbitrary these divisions are was made clear when, at the end of the last field season, Pit Structure 4 was found to straddle the line separating Courtyards 2 and 3. Attempts to identify the prehistoric use surface in the courtyard areas were unsuccessful because soil-formation processes and perhaps erosion had obliterated the prehistoric surface. Therefore, the courtyard grid squares, like the midden squares, were excavated in single cultural strata, exclusive of modern ground surface.

Features were treated much like miniature structures in terms of excavation strategy. Depending on size and presence or absence of internal stratigraphy, the field archaeologist had the option of subdividing feature contents horizontally and vertically. Larger features and those with obvious stratification, such as hearths, tended to be subdivided more than smaller, less stratigraphically complex features, such as post holes. Features were numbered sequentially within each major cultural unit. Feature numbers not reported in the tables in this volume were assigned to stains or disturbed areas later discovered to be of natural, rather than cultural, origin.

Because field crews were composed primarily of program participants who were not professional archaeologists, extreme care was exercised during all field operations. Trowels and whisk brooms were used for most excavations. Shovel excavation was employed in some contexts containing natural, postabandonment deposits and few artifacts (for example, upper pit structure fill), but usually by staff or adult participants under close staff supervision. Toward the end of the project, some mechanized equipment was used in selected areas of the site. The fill above roof fall in the east half of Pit Structure 3 was excavated by backhoe after the west half had been excavated with hand tools. In Pit Structure 4, naturally deposited fill above roof fall was also removed by backhoe after the top of the collapsed roof was identified in a hand-excavated exploratory trench.

Several backhoe trenches were excavated in the court-yard and trash areas after all hand excavations were completed. The purpose in digging these last few trenches was to ensure that no buried cultural deposits represent-ing earlier occupations lay beneath the site and to verify that no additional pit structures existed around the site periphery.

All deposits, except for those removed from the upper fill of pit structures and those excavated by backhoe, were sieved through ¼-in-mesh screens (infrequently, sediments were water-screened through 1⁄16-in mesh, if they contained many small items, such as tiny flakes or bone fragments, that otherwise would not have been recovered). After mid-1984, items associated with structure floors were assigned PL numbers and mapped; before then, floor artifacts had sometimes, but not always, been assigned PL numbers. Complete or temporally diagnostic artifacts found in roof fall strata frequently were point-located as well, because they potentially represented use or abandonment assemblages.

Special samples geared primarily toward gathering en-vironmental, subsistence, and dating information were collected when deemed appropriate by the field archaeolo-gist. Flotation samples were collected routinely from hearths, storage pits and bins, and any other features containing primary refuse, because these contexts were likely to yield macrobotanical remains related to the occupation of the site. Pollen samples were collected from similar prove-niences, although emphasis was placed on contexts least likely to have been contaminated by postabandonment pollen rain (for example, under slabs in direct contact with surfaces). Large pieces of burned wood, particularly frag-ments of construction beams, were routinely collected as tree-ring dating samples. Smaller pieces of wood and other plant materials were recovered on a less systematic basis as vegetal samples, depending on their size and the context in which they were found. Archaeomagnetic samples were collected from only one hearth. Sediment samples were collected from selected cultural and natural deposits for use in future grain-size analyses that might shed light on origins

of various sediments, including earthen materials used in construction. "Adobe" (construction mud or sediment) was collected, although not systematically, for the same reason. Because relatively few of the special samples collected were actually processed, only those that were analyzed and yielded interpretable results are discussed in this volume. For the sake of readability, flotation, pollen, archaeomag-netic, and sediment samples are excluded from PL tables and structure plan maps in Chapter 2. Complete records detailing provenience and, if applicable, analytic infor-mation for all materials recovered from the Duckfoot site are on file at the Anasazi Heritage Center in Dolores, Colorado.

Mapping procedures used in the field follow standard archaeological conventions. Grid north was established 25 degrees west of true north; magnetic north for the site area is approximately 13.5 degrees east of true north (U.S. Geological Survey topographic map, Mud Creek Quadran-gle, 1979). Structures, features, and point-located artifacts were mapped in relation to the site grid and a vertical datum. Artifacts mano-size or larger generally were drawn to scale; smaller items most often were mapped using geometric symbols keyed as specific material types (for example, pottery, stone, animal bone). The reader may notice "gaps" in the PL numbering sequences in some structures. PL numbers not listed were assigned to (1) un-modified items that were later discarded, (2) artifacts reassigned to proveniences other than the one being de-scribed, or (3) special samples, as described above. Infre-quently, items that were neither point-located nor collected were included on maps, if their depiction seemed essential to understanding a particular context (for instance, the sandstone rocks covering Burial 3 in Pit Structure 2).

Standard excavation procedures are described in the Crow Canyon field manual (Lightfoot and Bradley 1986). Although not all of the methods outlined in that document were in place at the beginning of the Duckfoot project, most had been adopted by 1985; in fact, many of the procedures described in the manual were instituted as a direct result of our field experience at Duckfoot.

2

Excavations

Ricky R. Lightfoot, Mary C. Etzkorn, and Mark D. Varien

Introduction

The Duckfoot excavations were structured primarily around the tripartite division inherent in the architectural layout of the site, and the highest priority was afforded investigation of the surface rooms and pit structures, which form the site "core." However, extensive work was also undertaken in nonarchitectural areas such as the courtyard and midden, and limited excavation was conducted around the site periphery. The excavation descriptions provided in this chapter focus on stratigraphy and architecture. Selected artifact information is also presented, with an emphasis on materials recovered from floor and roof fall contexts. Complete artifact information, organized by vertical provenience within major cultural units, is presented in tabular form in Appendix A. When appropriate, results of flotation, pollen, shell, animal bone, human bone, and tree-ring sample analyses are reported for specific contexts in this chapter. However, Chapters 5 through 9 and Appendix B provide more comprehensive descriptions of these materials and the analytic systems used in their study.

Architectural Suite Definition

The term *room suite* is used here to indicate a group of surface rooms that are spatially and architecturally associated with one another and with a pit structure. An *architectural suite* is made up of a pit structure, its corresponding room suite, and the associated courtyard. In Volume 2, it is argued that an architectural suite is the basic set of facilities used by a household. On the basis of architectural details, differences in construction techniques, and location of room suites in relation to pit structures, three architectural suites were identified at

the Duckfoot site (Figure 2.1). The strongest evidence supporting the three-part division of the Duckfoot pueblo is the configuration of the north wall of the roomblock. The most substantially constructed wall at the site is the one surrounding Rooms 4 through 7. It consists of massive elongate blocks, some of which exceed a meter in length and each of which was laid horizontally to form a wall that is preserved three to four courses high. The stones are interlocked in a way that suggests that the wall represents a single construction event; the uniformity of style and workmanship, too, supports the interpretation of a single episode of construction.

The back rooms west of Room 4 and east of Room 7 are more variable in terms of style and construction technique. Many of the back rooms at each end of the roomblock have thin vertical slabs at the bases of their outside walls, providing a sharp visual break from the horizontal-block wall surrounding Rooms 4 through 7. Rooms 1, 3, 8, 9, and 18 incorporate horizontally coursed block masonry, but the stones are smaller than those used in the construction of Rooms 4 through 7. The south wall of Room 8 does not line up with the south wall of Room 7, which suggests that Rooms 7 and 11 were completed first and that the dividing wall between them was no longer visible when Rooms 8 and 10 were added.

The architectural suites described above are believed to represent site organization at the time of initial construction and probably throughout much of the site's occupation. However, architectural evidence also suggests that there were at least one later episode of construction and several possible changes in the use of some existing structures. On the basis of structure size, structure placement, and the presence of hearths or other thermal features, front rooms 10, 11, 12, 13, 15, 16, and 19 are interpreted to have originally been habitation rooms. The remaining surface

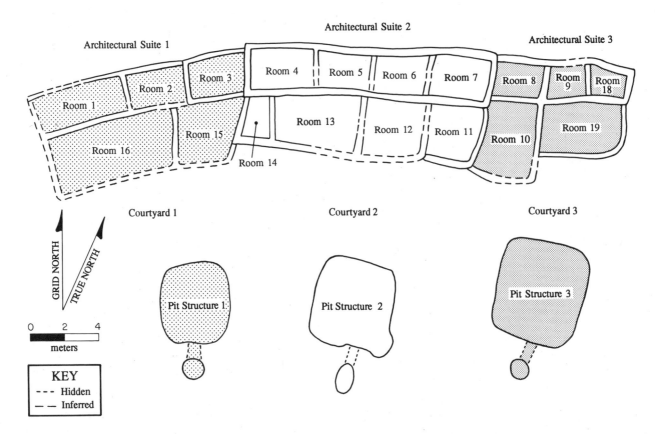

Figure 2.1. Architectural suites at the time of initial site construction. Pit Structure 4 and Room 17 were added later as part of Architectural Suite 2 (see Figure 1.5).

structures are believed to have originally served as storage rooms. Sometime during the occupation of the site, the hearth in Room 10 was capped, or deliberately sealed, with sediment, and some features in Room 15 were remodeled or removed; these two rooms appear to have changed from habitation to nonhabitation facilities. Conversely, two back rooms, Rooms 4 and 5, at some point during the occupation acquired thermal features, which may signal a change from storage to limited domestic activities. Room 17 was added to the south of Rooms 14 and 15, essentially creating a third tier in the roomblock, and Pit Structure 4 was built between Pit Structures 2 and 3 (Figure 1.5). In this volume, Room 17 and Pit Structure 4 are treated as part of Architectural Suite 2.

Architectural Suite 1

Architectural Suite 1, at the western end of the site, is made up of two large front rooms (Rooms 15 and 16), three back rooms (Rooms 1, 2, and 3), Pit Structure 1, and Courtyard 1. Tree-ring dates from Room Suite 1 indicate construction of the rooms in the mid- to late A.D. 850s. Pit Structure 1 samples yielded a cluster of dates from this

period but an even stronger cluster of A.D. 865 and 866 dates. Although this could indicate that Pit Structure 1 was built approximately 10 years after the completion of Room Suite 1, the two date clusters are interpreted as indicating that construction took place in the mid- to late A.D. 850s and extensive remodeling was carried out in the mid-A.D. 860s. Room wall junctures (Figure 2.1) indicate that Room Suite 1 was built after Room Suite 2, which is the core unit in the center of the roomblock. On the basis of tree-ring dates, Room Suite 2 is interpreted as having been built around A.D. 857, which would place the construction of Room Suite 1 that year or later, possibly in A.D. 858.

Also included in the discussion of Architectural Suite 1 is Room 20, a small jacal structure located at the west end of the courtyard (Figure 1.5). Room 20 is detached from the remainder of the roomblock, which makes its architectural-suite affiliation ambiguous. It is included in this discussion because it falls within the spatial confines of Architectural Suite 1. A single noncutting tree-ring date of A.D. 874 was obtained from a piece of charcoal in the fill of Room 20. This is among the latest dates on the site, and it indicates that the structure filled near the time the entire site was abandoned. Very little is known about the construction and use of this structure.

Room 15

Length: 3.55 m
Width: 3.10 m
Area: 11.30 m²

Room 15 (Figure 2.2) is a small front room located between Room 16 to the west and Rooms 14 and 17 to the east; it is spatially associated with at least Room 3, and possibly Room 2, in the back row of the roomblock (Figure 1.5). The structure did not burn at abandonment. The presence of an intrusive slab-lined bin in fill indicates that this location was used at least briefly after the main occupation of the site ended. Although the use surface that must have been associated with this feature was too indistinct to be detected during excavation, the designation "Surface 2" was assigned retroactively as a convenient proveniencing tool. The actual living floor of the room is designated "Surface 1."

Excavations in Room 15 began with a 2-m-wide, north-south test trench, part of a longer transect that crosscut the entire architectural suite. In Room 15, this trench was eventually expanded to include the entire structure.

Stratigraphy

The fill of Room 15 consisted of collapsed structural debris and cultural refuse mixed with postabandonment wind- and water-deposited sediments. No natural or cultural strata were defined during excavation. A distinct roof fall layer was not discernible, but sandstone rocks, the remnants of the fallen upper walls, and flecks of charcoal were found throughout fill. The sandstone slabs that formed the walls and floor of the intrusive bin were imbedded in these deposits as well. The structure appears to have deteriorated gradually as the result of natural processes. Artifacts recovered from the upper fill (Level 1) include a metate, two manos, an abrader, an obsidian projectile point, pottery sherds, chipped-stone debris, animal bones, and a *Glycymeris* shell bracelet fragment. Several fragments of isolated human bone were also recovered. Broken pottery was abundant in the fill immediately above the floor. Artifacts recovered from the lower fill (Level 2) are discussed in detail with Surface 1 artifacts later in the chapter.

Construction

The north and east walls of Room 15 are a combination of vertical sandstone slabs and horizontal sandstone blocks, with the former predominating (Figure 2.2). The extant portion of the west wall, which in places stands two courses high, is made entirely of horizontal blocks. The presence of the south wall is inferred solely on the basis of the configuration of the surrounding rooms. Sandstone rocks

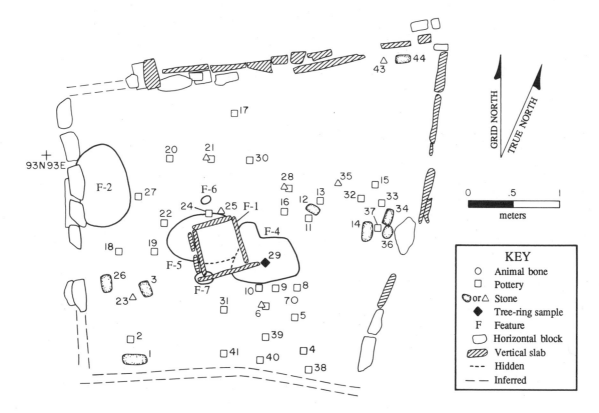

Figure 2.2. Room 15, plan. Point-located artifacts from both the floor (Surface 1) and the lower fill are described in Table 2.1. Feature 1 is associated with Surface 2 and postdates the main occupation of the site (see Table 2.2).

Table 2.1. Point-Located Artifacts from Room 15, Surface 1 and Lower Fill

No.	Description	No.	Description
Pottery Sherds		32	gray ware jar (35)
2*	gray ware jar (2)		white ware other (1)
4*	gray ware jar (1)		red ware bowl (1); Vessel 35 (1)
5	gray ware jar (16)	33	missing
	gray ware bowl (1)	37	gray ware jar (6)
6	gray ware jar (20)	38	gray ware jar (13)
	red ware bowl (1)		red ware bowl (1)
8	gray ware jar (18)	39	gray ware jar (59)
9	gray ware jar (4)	40	missing
10	gray ware jar (3)	41	gray ware jar (5)
11	gray ware jar (3)	42*	gray ware jar (1); not mapped
	white ware bowl (1)	Shaped Sherd	
13	gray ware jar (5); Vessel 127 (1)	30	complete, gray ware
	gray ware bowl (1); Vessel 117 (1)	Chipped-Stone Debris	
15	gray ware jar (35)	6	flake (1)
	gray ware other (1)	21	flake (1)
	red ware bowl (1); Vessel 35 (1)	23	core (1)
16	red ware bowl (1); Vessel 36 (1)	28	flakes (2)
17	red ware bowl (1); Vessel 36 (1)	Mano	
18*	gray ware jar (5); Vessel 117 (4)	44*	missing
	gray ware bowl (1); Vessel 117 (1)	Two-Hand Mano	
	white ware bowl (2)	1*	complete, sandstone
19*	gray ware jar (20)	3*	complete, sandstone
20*	gray ware jar (2)	26	complete, igneous
	white ware jar (1)	34	complete, sandstone
	white ware bowl (2); Vessel 119 (2)	Peckingstone	
21	gray ware jar (17)	25	complete, Dakota orthoquartzite
	white ware jar (1)	35	complete, Morrison orthoquartzite
	red ware bowl (6); Vessel 35 (2), Vessel 36 (4)	Double-Bitted Axe	
22	gray ware jar (10)	43*	complete, igneous
	white ware jar (1)	Modified Cobble	
24	gray ware jar (7); Vessel 127 (3)	12	fragmentary, igneous
27	missing	Stone Disk	
28	gray ware jar (59)	36	complete, sandstone
	gray ware bowl (2)	Dating Samples	
	white ware jar (3)	29	tree-ring sample; missing
	white ware bowl (1)	Miscellaneous	
	red ware bowl (2); Vessel 35 (1)	14*	ground stone; missing
31	gray ware jar (1)	7	animal bone

NOTES: Item counts appear in parentheses. For pottery, sherds belonging to vessels are also included in total item count. Refer to Figure 2.2 for artifact locations.

* Lower fill (Level 2) point-locations.

in fill undoubtedly were collapsed wall materials, but given the number of rocks removed during excavation, it is unlikely that any of the walls were completely coursed from top to bottom. The upper walls may have been of mud-and-stone construction, although, aside from the presence of some hardened constructional mud in fill, there is no direct evidence to support this supposition. No roof-support post holes were found, which leads to the inference that a flat roof must have been supported by the structure walls. The floor of Room 15 had been excavated prehistorically to the top of the B horizon and compacted through use.

Surface 1

Surface 1, the floor of Room 15, was an indistinct zone between sterile sediments below (B horizon) and postabandonment fill above. Artifacts, especially pottery sherds, were abundant on the floor of Room 15, and in some cases clusters of artifacts extended several centimeters into the postabandonment fill above the floor. Therefore, some artifacts assigned a Level 2 provenience were included in the Surface 1 point-location sequence. Subsequent analysis revealed that the large number of sherds on or near the floor did not originate from whole vessels in the room; rather, the sherds were probably deposited as secondary refuse near the time of abandonment. The point-located artifacts found on or near the floor are discussed here as part of the floor assemblage.

Artifacts. Point-located items are mapped in Figure 2.2 and described in Table 2.1. Of the 344 point-located sherds, 322 (93.6 percent) are gray ware, 14 (4.1 percent) are red ware, and 8 (2.3 percent) are white ware. The majority of gray ware sherds are classified as Indeterminate Plain Gray, although 4 Chapin Gray, 22 Moccasin Gray, and 1 Mancos Gray sherds were identified as well. Decorated types include Bluff Black-on-red (11 sherds) and Chapin Black-on-white (1 sherd). As with gray wares, most of the red and white wares are typed only to more general categories, such as Indeterminate San Juan Red and Indeterminate White. Both jar and bowl forms are recognized in the sherd assemblage, and one shaped sherd was recovered. (A shaped sherd is one that was modified, usually by grinding and/or flaking, on all of its edges after the original vessel broke; shaped sherds are regarded as "recycled" tools.) A special refitting study revealed that several point-located sherds from both Level 2 and the floor proper belonged to five partial or nearly complete gray, white, and red ware bowls (Chapin Gray, Bluff Black-on-red, and Indeterminate White).[1]

Stone artifacts assigned PL numbers include several two-hand manos, two peckingstones, a stone disk, and an unusual axe with a double bit (one of only two found at Duckfoot). Relatively little chipped-stone debris was recovered, although a core is among the pieces that were.

Features. Five features were associated with Surface 1 (Table 2.2 and Figure 2.2). A shallow pit (Feature 4) in the center of the room showed no signs of in situ burning, but it was filled with ash. The presence of this feature constitutes indirect evidence for the possible earlier existence of a hearth—ash receptacles are often found adjacent to hearths and fire pits in Anasazi sites. If a hearth had indeed existed, it may have been destroyed when Feature 5, a large cist adjacent to Feature 4, was constructed. Feature 5 was filled with postabandonment deposits mixed with cultural refuse. A slab-lined bin (Feature 1) eventually was constructed on top of both features, but it is interpreted to be a later, intrusive feature not associated with the main occupation of the site.

Table 2.2. Feature Summary, Room 15

No.	Type	Shape	Length (cm)	Width (cm)	Depth (cm)	Status
1[a]	bin	other	60	52	30	in use
2	cist	basin	85	69	36	in use
4	pit	basin	75	60	5	in use
5	cist	other	70	52	59	in use
6	pit	cylindrical	8	8	13	unknown
7	pit	other	20	12	11	unknown

NOTE: Feature locations are shown in Figure 2.2.
[a]Feature associated with Surface 2.

The remaining floor features consist of a second cist (Feature 2) and two shallow pits (Features 6 and 7). Feature 2, which slightly undercut the west wall of the room, was filled with postabandonment deposits. It was not clear whether Features 6 and 7 had been capped or left open before abandonment.

Numerous sherds found in Features 2 and 5 are from reconstructible vessels, most of whose other pieces are scattered across the floor of Room 15. This indicates that both features were open at abandonment and lends credence to the argument that Feature 1, which overlies Features 4 and 5, represents postabandonment reuse of the site. Enough time must have elapsed between abandonment and the construction of Feature 1 to allow Feature 5 to fill with sediment.

Four of six flotation samples collected from Features 4 and 5 were analyzed. Plant remains recovered from these two features provide evidence of both fuel (pinyon, juniper, sagebrush, four-wing saltbush, and corncobs) and food (corn, cheno-am, and groundcherry) resources.

Surface 2

Surface 2 was defined retroactively to accommodate Feature 1, the slab-lined bin built into postabandonment deposits above Features 4 and 5 (Table 2.2 and Figure 2.2). It is assumed that a use surface must have existed when the site was reused, although such a surface was never defined in the field. Sometime after the abandonment of the site, when enough time had elapsed for sediment to fill the features left open on the floor and at least partly fill the room itself, several thin sandstone slabs were placed

---

1. Selected vessel information is presented in this chapter to provide the reader with a general understanding of Duckfoot floor assemblages; additional descriptive information is provided in Chapter 3. However, an explanation of the key distinction between sherd and vessel analyses is offered here as an aid to understanding the immediate discussion: During sherd analysis, ware, type, and vessel form assignments are made for individual sherds, without regard for the vessels from which they derive. Therefore, assignments made during sherd analysis may be "contradicted" later during vessel analysis, when decisions regarding ware, type, and form are made on the basis of the complete or nearly complete vessel and not its constituent pieces. Thus, for example, a sherd typed as "Indeterminate White jar" during sherd analysis eventually may be incorporated into a vessel typed as a "Piedra Black-on-white pitcher" during vessel analysis.

upright to form a nearly square box, and another large slab was placed flush with the floor of Room 15 to form a solid base. Feature 1, like the room itself, was filled with postabandonment deposits; neither artifact nor flotation sample analysis shed light on its function. On the basis of size and shape alone, it is inferred that Feature 1 might have served as a storage bin. No artifacts of later Anasazi styles were found in the fill of the bin or in the upper fill of Room 15, so the reuse of the site might have taken place within a decade or two after abandonment. A slab-lined feature in the fill of Pit Structure 3 also represents postabandonment reuse of the site, and the two features may be contemporaneous.

Dating

Four tree-ring samples were collected from Room 15, and two of these yielded dates: A.D. 857 and 858 (Appendix B). Although neither is a cutting date, the two are only one year apart, and they are within a date cluster for Room Suite 1 that may represent the period of construction. If the two samples are from beams used in the initial construction of Room 15, they indicate construction after A.D. 857. The construction of the slab-lined bin in postabandonment deposits cannot be dated.

Interpretations

Inferring the function of Room 15 is difficult, especially because this structure appears to have been used for more than one purpose during the site occupation. The presence of an unburned pit filled with ash may indicate that a hearth once existed in the room, which in turn suggests that Room 15 functioned as a small habitation structure after initial construction. By the end of the occupation, however, a hearth was not part of the feature assemblage, which indicates that Room 15 had ceased to serve primarily as a habitation structure and instead had assumed some non-domestic function, such as storage. Exactly when this change took place is unknown, but it is tempting to view it as part of the organizational changes occurring in the mid-A.D. 860s, when Pit Structure 1 was remodeled, or in the early A.D. 870s, when Pit Structure 4 was built. In the final stages of site occupation, Room 15 apparently served as a refuse dump: some of the artifacts found in Level 2 appeared to have been discarded near the time of abandonment. Finally, sometime after abandonment, the deteriorating structure was reused by an unknown individual or group, as evidenced by the construction of a slab-lined bin in postabandonment deposits. This reuse of the site cannot be dated, but apparently it was brief. (A similar bin feature was constructed in postabandonment deposits in Pit Structure 3 but, again, could not be dated. Although it is possible that the two features represent the same reuse of the site, an association between the two cannot be demonstrated.)

Room 16

Length: 7.55 m
Width: 3.00 m
Area: 22.95 m^2

Room 16 (Figures 2.3 and 2.4) is a large front room that burned at abandonment. It is adjacent to Room 15 in the front row of the roomblock and associated with Rooms 1 and 2 in the back row (Figure 2.1). Room 16 is substantially larger than the other front rooms at Duckfoot and larger than most front rooms dating to the same time period elsewhere in the region (Kane 1986:415). Because of its size, the excavators made every attempt to find even the most subtle traces of a dividing wall within the room, but they found none. The entire room was excavated as a single horizontal unit.

Stratigraphy

Room 16 fill consisted primarily of a layer of postabandonment wind- and water-deposited sediments mixed with cultural debris, Stratum 1. This stratum rested directly on top of a layer containing numerous small (less than 1 cm in diameter), burned twigs and branches, Stratum 2. The absence of roof beams and posts in Stratum 2 suggests that the burned plant material may have been not collapsed roofing material but, rather, the remains of fuel used to intentionally burn the room at abandonment. In most areas, Stratum 2 was about 5 cm thick, but it gradually thinned toward the east end of the room. The upper boundary of Stratum 2 was designated Surface 1. Directly below the burned layer was the true floor of the room, designated Surface 2. Artifacts recovered from the postabandonment deposits (Stratum 1) in this room include pottery sherds, chipped-stone debris, animal bone, two peckingstones, a mano, a stone disk, a biface, and a nearly complete turquoise pendant.

Construction

The north and east walls of Room 16 are the best preserved; the existence of the south and west walls is inferred largely on the basis of a few remaining sandstone rocks and the presence of several post holes (Figure 2.4). The west end of the north wall is made of small- to medium-size sandstone blocks that form a single horizontal course. The eastern section of this same wall consists of a row of vertical slabs, and the middle portion consists of an earthen footing with scattered sandstone rubble. The east wall of Room 16, which in places stands two courses high, is constructed of horizontal blocks. As in Room 15, there was not enough sandstone rubble in Room 16 to account for walls constructed predominantly of coursed masonry, and upper walls may have been made of mud and stone. Eight post holes located in two of the four corners and along several edges of the room could have supported a flat roof, although the arrangement of the post holes suggests that the north and east walls may have supported a portion of the roof as well. The floor of the room, Surface 2, was a use-compacted surface that had been excavated prehistorically into sterile B-horizon sediments.

Surface 1

Surface 1 is not the floor of the room but, rather, the upper boundary of a 5-cm-thick layer of burned twigs and sediment that rested directly on the floor (Surface 2). Although it is possible that this layer consisted of burned closing material from the structure roof, it differed from roof fall found in other structures in that no large pieces of wood were included in the debris. Therefore, this burned layer is believed to be the remains of brush and other materials placed on the floor and ignited at abandonment.

Artifacts. The point-located artifacts on Surface 1 (Table 2.3 and Figure 2.3) are believed to be related to the abandonment of Room 16, although it is also possible that these items were on the roof before abandonment. Surface 1 artifacts include a nearly complete metate, several manos, one peckingstone, and numerous sherds and pieces of chipped-stone debris, including two edge-damaged flakes. Of the 33 sherds point-located on Surface 1, 29 (87.9 percent) are gray ware and 4 (12.1 percent) are red ware. Most are of indeterminate types, but two Moccasin Gray and four Bluff Black-on-red sherds are present as well. Both jars and bowls are represented in the sherd assemblage. Three point-located sherds belong to a partial Bluff Black-on-red bowl, most of whose pieces were found on the floor of Room 16 (Surface 2).

Surface 2

Surface 2, defined as the stratigraphic boundary between the burned layer above and sterile B-horizon sediments below, is the true floor of Room 16. Point-located materials are believed to have been in situ and the result of activities that took place during the use of the room.

Artifacts and Samples. The artifacts point-located on the floor of Room 16 represent a variety of domestic activities. Included in the assemblage are sherds, chipped-stone debris, three metates, four manos, and two peckingstones. A single edge-damaged flake was identified among the pieces of chipped-stone debris. A number of animal bones were also recovered but were not point-located. The sherd assemblage (273 sherds) is made up of 252 gray ware jar sherds (92.3 percent) and 21 red ware bowl sherds (7.7 percent), including 1 Chapin Gray, 21 Moccasin Gray, 32 Mancos Gray, and 13 Bluff Black-on-red sherds. During

Table 2.3. Point-Located Artifacts from Room 16, Surface 1

No.	Description
Pottery Sherds	
14	gray ware jar (4)
18	gray ware jar (6)
19	gray ware other (3)
	red ware bowl (4); Vessel 37 (3)
21	gray ware jar (2)
22	gray ware jar (9)
24	gray ware jar (5)
Chipped-Stone Debris	
11	edge-damaged flake (1)
12	flake (1)
13	edge-damaged flake (1)
23	flake (1)
Trough Metate	
20	incomplete, sandstone
Mano	
1	missing
Two-Hand Mano	
4	complete, conglomerate
5	complete, igneous
9	complete, sandstone
Peckingstone	
16	missing
Modified Cobble	
8	incomplete, sandstone
Other Modified Stone/Mineral	
7	complete, unknown orthoquartzite
Unmodified Cobble	
3	missing
Awl	
6	fragmentary, *Odocoileus hemionus*

NOTES: Item counts appear in parentheses. For pottery, sherds belonging to vessels are also included in total item count. Refer to Figure 2.3 for artifact locations.

refitting analysis, many of the point-located sherds recovered from Surface 2 were found to belong to three partial or nearly complete jars/ollas (Chapin, Moccasin, and Mancos gray) and one partial Bluff Black-on-red bowl. Artifact locations are shown in Figure 2.4, and descriptions are provided in Table 2.4.

Two pollen samples were collected and analyzed, one from a metate trough and one from the floor. The fact that sagebrush pollen dominates the floor sample could be related to the type of fuel used to burn the structure. Pinyon-type pollen predominates in the metate sample.

Features. Eighteen floor-associated features were excavated: two hearths, eight post holes, one cist, and seven miscellaneous pits (Table 2.5 and Figure 2.4). Features recorded when the floor was first exposed are believed to have been available for use when the structure was abandoned and are identified as "in use" in Table 2.5. When all visible features had been excavated, the floor was scraped

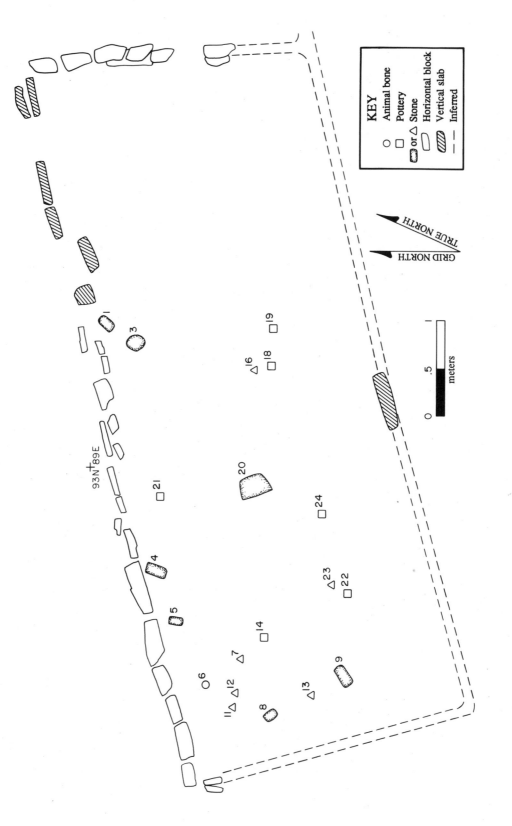

KEY

○	Animal bone
□	Pottery
⬭ or △	Stone
▨	Horizontal block
▨	Vertical slab
- - -	Inferred

GRID NORTH

TRUE NORTH

0   .5   1
meters

93N 89E

Figure 2.3. Room 16, Surface 1. Point-located artifacts are described in Table 2.3.

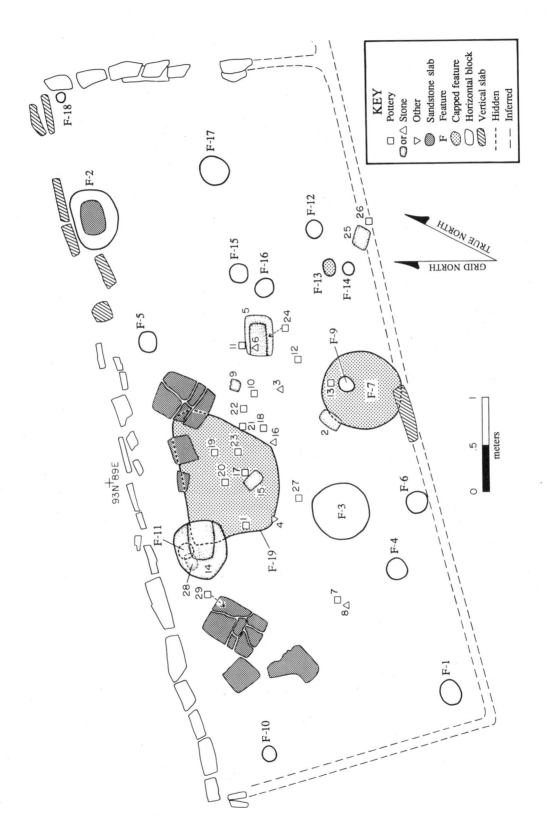

Figure 2.4. Room 16, Surface 2. Point-located artifacts are described in Table 2.4; features are described in Table 2.5.

KEY

□	Pottery
⬭ or △ Stone	
▽	Other
F	Feature
	Capped feature
	Horizontal block
	Vertical slab
---	Hidden
— —	Inferred

Sandstone slab

TRUE NORTH

GRID NORTH

93N 89E

meters

0   .5   1

Table 2.4. Point-Located Artifacts from Room 16, Surface 2

No.	Description
Pottery Sherds	
1	gray ware jar (207); Vessel 116 (77), Vessel 122 (104)
7	gray ware jar (2)
10	missing
11	gray ware jar (1); Vessel 25 (1)
12	red ware bowl (5); Vessel 37 (5)
13	gray ware jar (1)
	red ware bowl (1); Vessel 37 (1)
17	gray ware jar (14); Vessel 116 (5)
18	red ware bowl (2); Vessel 37 (2)
19	gray ware jar (4)
20	gray ware jar (4)
21	gray ware bowl (4)
22	gray ware jar (1)
23	gray ware bowl (1)
24	red ware bowl (13); Vessel 37 (12)
26	gray ware jar (10)
27	gray ware jar (2); Vessel 116 (2)
29	gray ware jar (1)
Chipped-Stone Debris	
8	flakes (2)
	edge-damaged flake (1)
16	missing
Trough Metate	
5	complete, sandstone
14	complete, sandstone
25	fragmentary, sandstone
Two-Hand Mano	
2	complete, sandstone
9	fragmentary, conglomerate
15	complete, sandstone
28	complete, conglomerate
Peckingstone	
3	complete, Morrison orthoquartzite
6	complete, Dakota orthoquartzite
Unmodified Stone/Mineral	
4	carnotite?

NOTES: Item counts appear in parentheses. For pottery, sherds belonging to vessels are also included in total item count. Refer to Figure 2.4 for artifact locations.

Table 2.5. Feature Summary, Room 16

No.	Type	Shape	Length (cm)	Width (cm)	Depth (cm)	Status
1	post hole	cylindrical	27	22	50	in use
2	pit	basin	~60	~50	~10	unknown
3	hearth	basin	65	61	12	in use
4	post hole	cylindrical	21	21	28	in use
5	post hole	cylindrical	18	17	27	in use
6	post hole	cylindrical	26	24	34	in use
7	hearth	basin	88	82	20	capped
9	post hole	cylindrical	15	13	33	in use
10	post hole	other	19	17	30	in use
11	post hole	cylindrical	20	18	23	unknown
12	pit	cylindrical	19	19	22	unknown
13	pit	cylindrical	18	13	14	capped
14	post hole	cylindrical	15	12	18	in use
15	pit	basin	22	20	8	in use
16	pit	cylindrical	22	18	12	in use
17	cist	cylindrical	32	32	23	in use
18	pit	nr	10	10	5	in use
19	pit	basin	140	95	12	capped

NOTE: Feature locations are shown in Figure 2.4.
nr = no record.

holes are believed to have held roof-support posts, although only Feature 9 still retained a post remnant. Feature 17, a small, cylindrical cist, apparently had been left open at abandonment and was filled with collapsed roofing material. Of the remaining features, all pits of unknown function, three are believed to have been left open at abandonment, two were capped, and the condition of two could not be determined because the evidence was too ambiguous.

Flotation samples collected from both hearths yielded both food (cheno-am, prickly pear, groundcherry, yucca, Indian rice grass, and corn) and fuel (sagebrush, four-wing saltbush, juniper, pinyon, and corn cupule) remains. In addition, charred wood identified as cottonwood was recovered from the lower fill of Feature 7. This type of tree grows in relatively moist habitats, and its presence in Feature 7 is one of only four occurrences in analyzed flotation samples from Duckfoot.

Dating

The one tree-ring sample collected from Room 16 yielded a cutting date of A.D. 828 (Appendix B). This date is 25 to 40 years earlier than most of the other cutting dates obtained for Architectural Suite 1, and it may indicate the use of dead or salvaged wood in the construction of this particular room (see Chapter 1 for a discussion of nearby sites, many of them possibly predating Duckfoot). Location relative to the other rooms suggests that Room 16 could not have been built any earlier than the rest of Room Suite 1, which would place its construction at no earlier than the mid- to late A.D. 850s.

with trowels, and three capped features were revealed. Capped features were used at some point during the occupation of the structure but were not in use at the time the structure was abandoned. One hearth in Room 16 (Feature 3) was still in use at abandonment; it was filled with charcoal and ash and covered by the layer of burned material blanketing Surface 2. The second hearth (Feature 7) had charcoal and ash in the lower fill, but the pit had been deliberately capped with sterile B-horizon sediments. Feature 9, a post hole with an unburned remnant of post still intact, cut through the fill of Feature 7 and into underlying sterile deposits, providing additional evidence that Feature 7 was not in use at abandonment. All eight post

Interpretations

On the basis of its size, placement in the front row of the roomblock, and feature and artifact assemblages, Room 16 is interpreted as a habitation structure. The wide variety of food-preparation tools and equipment (manos, metates, cooking jars, and serving bowls) and the presence of features related to temporary storage (cist) and cooking (hearths) clearly reflect the range of activities expected for a domestic structure. Although there is evidence of minor remodeling in the form of capped features, the remodeling does not appear to signal a functional change. In fact, one capped feature, a hearth, was replaced by a second hearth, which would have allowed the continuation of the same domestic activities that had occurred before remodeling. The burning of Room 16 at abandonment, inferred on the basis of a layer of burned vegetal material in direct contact with the living surface of the room, is believed to be related to the same abandonment activity that resulted in the destruction of six other rooms and at least three of the pit structures.

Room 1

Length: 5.25 m
Width: 2.10 m
Area: 11.46 m²

Room 1 (Figure 2.5) is an unusually large back room at the far west end of the roomblock, spatially associated with front room 16 (Figure 2.1). The entire structure was excavated as a single unit without horizontal subdivisions. A floor was never defined in this room, because the fill consisted of the redeposited B-horizon sediments from which the walls were built, and these sediments could not be distinguished from the undisturbed B-horizon sediments into which the floor was excavated. In addition, the floor lacked the artifacts, features, and signs of heavy use (compaction, staining) that typically allow recognition of floors. The room was not burned at abandonment, and apparently it deteriorated naturally over time. The gradual filling and weathering perhaps contributed to the difficulty in recognizing a distinct floor.

Stratigraphy

The fill of Room 1 was excavated in two strata. Stratum 1 consisted of a 30-cm-thick layer of collapsed wall rock mixed with postabandonment sediments. Stratum 2 had fewer rocks in a matrix of strong brown silt loam. Artifacts recovered from Stratum 1 include pottery sherds, chipped-stone debris, a maul, a biface, several peckingstones, numerous manos, and several isolated human bones. A large portion of a Moccasin Gray jar is part of the assemblage. Items recovered from Stratum 2, including a bone awl, sherds, and chipped-stone debris, could have been deposited during the use of the room or at abandonment. These items were deposited before the walls collapsed, but because a floor was not identified, the context of the items could not be adequately assessed. The number of complete artifacts in Stratum 1 also suggests the possibility that items

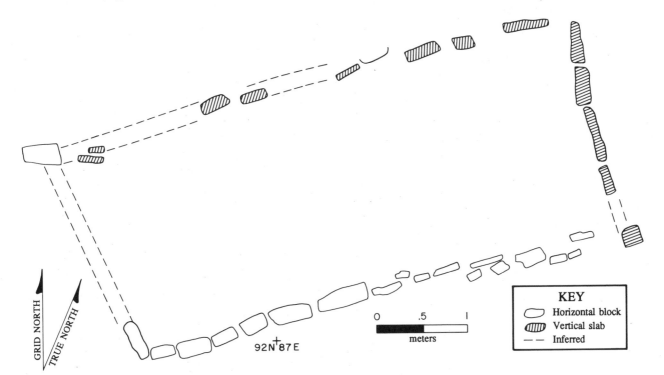

GRID NORTH
TRUE NORTH

92N 87E

0   .5   1

meters

KEY
⬭ Horizontal block
▥ Vertical slab
- - Inferred

Figure 2.5. Room 1, plan. A living surface was not identified in this room.

were stored in the room but not on the floor (for example, on a shelf or in a basket suspended from the roof beams).

Construction

The walls of Room 1 are made of a combination of vertical sandstone slabs and horizontal blocks (Figure 2.5). The north wall was defined primarily on the basis of a discontinuous alignment of vertical slabs; the east wall, on an almost uninterrupted alignment of similar stones. The south wall consists of a combination of medium-size horizontal blocks at the west end and a scattering of smaller sandstone chunks in a rough alignment at the east end. Only the basal course preserved. The west wall is inferred as a straight line between two stones in the north and south walls of the room. The amount of rubble found in fill suggests that the upper walls may have been made of coursed stone and mud. A distinct roof fall layer was not identified during excavation, but it is assumed that Room 1 had a flat roof supported by the structure walls (no post holes were present). Because the room deteriorated gradually after abandonment, it is likely that portions of collapsed roof were incorporated in both postabandonment strata but did not preserve in recognizable form.

Dating

Although no tree-ring samples were collected from Room 1, the construction style and location of this room indicate that it was built at the same time as the other surface structures in Room Suite 1, probably during the late A.D. 850s.

Interpretations

The typical back room in a Pueblo I roomblock is small, relatively devoid of artifacts and features, and believed to have served as a storage facility for food or other materials (Gross 1987). Except for its size, Room 1 conforms to the definition of a Pueblo I storage room. The lack of a definable floor diminishes the interpretability of this room; however, one reason a floor was not recognized was the lack of directly associated artifacts and features, which itself is consistent with a storage function. Many of the Stratum 1 artifacts, including the sherds belonging to a Moccasin Gray jar, were clustered in the southwest corner of the room. These items may have been stored in the structure (but not on the floor) or on the roof.

Room 2

Length: 3.40 m
Width: 2.20 m
Area: 7.25 m²

Room 2 (Figure 2.6) is located in the back row of the roomblock, between Room 1 to the west and Room 3 to the east (Figure 2.1). Room 2 is immediately north of the dividing wall between Rooms 15 and 16 and therefore could have been used by either or both front rooms. A looter's pit was apparent in Room 2 when Crow Canyon began its investigation of the Duckfoot site, but damage apparently was confined to the wall fall layer in upper fill. The room is not believed to have burned; charred wood in fill most likely originated in adjacent Room 3, which *did* burn at abandonment.

Crow Canyon excavators first investigated the spoils in the looter's pit, then expanded excavations into a test trench to define stratigraphy in the room. Finally, excavations in the trench were expanded to include the entire room.

Stratigraphy

Room 2 fill consisted of two strata, similar to those described for Room 1. The top layer, approximately 40 cm thick, consisted of collapsed wall rubble, other cultural debris, and postabandonment sediments. Below this layer, in direct contact with the floor, was a 5-cm-thick deposit containing sediments and artifacts that accumulated after abandonment but before structural collapse. The wall fall layer contained many artifacts, including manos, sherds, chipped-stone debris, animal bone, and an isolated human bone that may be from the same individual represented by Burial 3 on the floor of Pit Structure 2. Three partial or nearly complete gray ware jars/ollas (one Chapin and two Moccasin) are present in the sherd assemblage from the wall fall layer; a single sherd fits with a partial Piedra Black-on-white bowl whose pieces were scattered throughout Rooms 2 and 3. A flotation sample collected from one of the gray ware jars yielded scant evidence of food (cheno-am) and fuel and/or construction (juniper, pine) resources. Artifacts recovered from lower room fill include sherds, chipped-stone debris, and animal bone. Only one sherd from this group refit with a reconstructible vessel—in this case, the same Piedra Black-on-white bowl mentioned above.

Construction

The lower portions of all four walls are largely intact and therefore easily defined. The north, south, and west walls are made entirely of large, vertical sandstone slabs, and the east wall consists of two courses of horizontal sandstone blocks, which abut the vertical slabs that form the west wall of Room 3 (Figure 2.6). On the basis of the abundance of wall rubble in fill, the upper walls are inferred to have been a combination of coursed masonry and mud. No post holes were identified, leading to the inference that the roof must have been supported by the structure walls. Collapsed roofing materials were scattered throughout the room's fill, particularly in the upper wall fall layer, but a distinct roof fall zone was not recognized during excavation. The floor

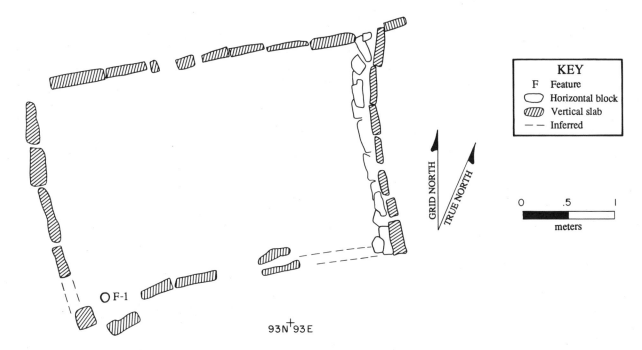

Figure 2.6. Room 2, surface. Artifacts were not mapped; refer to text for a description of floor-associated items and Feature 1.

of Room 2 had been excavated prehistorically to sterile B-horizon sediments and then compacted through use.

Surface

The living surface, or floor, was defined stratigraphically as the break between postabandonment deposits above and sterile B-horizon sediments below. It consisted of a hard-packed surface covered with flecks of charcoal and chunks of hardened constructional mud. Seven sherds, one complete hammerstone, and three pieces of chipped-stone debris were found in association with the floor but were not assigned PL numbers or mapped. The seven sherds are classified as Indeterminate Plain Gray; six are jar body sherds, and one is a rim sherd from a small dipper.

One feature (Feature 1) was identified on the floor of Room 2 (Figure 2.6). This feature is a small, circular pit, 10 cm in diameter, whose function is unknown. Its location in the southwest corner of the room suggests that it might have been a post hole, but the characteristics of its fill do not strongly support this interpretation.

Dating

Six tree-ring samples were collected from the fill of Room 2—four from the wall fall stratum and two from the lower fill immediately overlying the floor. Five noncutting dates were obtained: A.D. 822, 828, 830, 847, and 852 (Appendix B). Tree-ring dates for Room Suite 1 as a whole suggest that Room 2 was probably constructed in the late A.D. 850s.

Interpretations

On the basis of its location in the back row of rooms, its small size, and its limited artifact and feature assemblages, Room 2 is interpreted as a storage room associated with front room 15 or 16 or both. Apparently it was left empty or nearly empty when abandoned. Reconstructible vessels and other refuse in fill may have been on the structure roof when it collapsed. Because the room did not burn, the roof would have deteriorated gradually, leaving few traces except for the artifacts deposited below.

Room 3

Length: 3.30 m
Width: 2.20 m
Area: 7.00 m²

Room 3 (Figures 2.7 and 2.8) is a small back room that burned at abandonment. It is located between Room 2 to the west and Room 4 to the east and is associated with Room 15 to the south (Figure 2.1). Room 3 was excavated as a single horizontal unit without internal subdivisions. Vertical subdivisions were based on the stratigraphy described below.

Stratigraphy

Three strata were identified during excavation. The upper two consisted of collapsed wall debris, and the bottom layer consisted of burned roofing materials. The uppermost wall

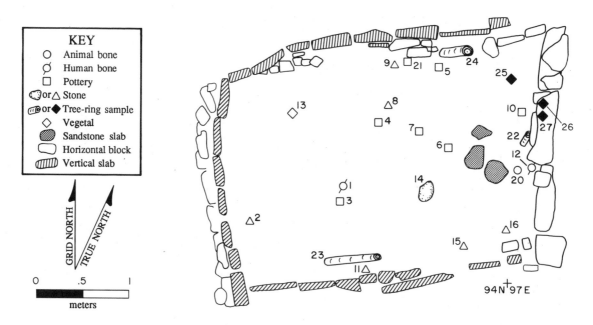

Figure 2.7. Room 3, roof fall. Point-located artifacts are described in Table 2.6.

Table 2.6. Point-Located Artifacts from Room 3, Roof Fall

No.	Description
Pottery Sherds	
3	gray ware jar (13)
4	gray ware jar (1)
5	gray ware jar (1)
6	gray ware jar (2)
7	gray ware jar (12)
10	gray ware jar (12)
21	missing
Chipped-Stone Debris	
8	cores (2)
	flakes (2)
	other (1)
Abrader	
16	fragmentary, conglomerate
Peckingstone	
2	fragmentary, Morrison orthoquartzite
9	complete, Dakota orthoquartzite
15	complete, Morrison orthoquartzite
Modified Cobble	
11	complete, igneous
14	condition unknown, igneous
Dating Samples	
22–27	tree-ring samples
28	tree-ring sample; not mapped
Miscellaneous	
1	human bone
12	human bone
20	animal bone
13	vegetal

NOTES: Item counts appear in parentheses. Refer to Figure 2.7 for artifact locations.

fall stratum was marked by abundant constructional mud and caliche inclusions, whereas the underlying wall fall layer had fewer inclusions. Directly below these two strata, in direct contact with the floor of the room, was a very distinct layer of burned vegetal material and sediment interpreted as the remains of the structure roof. Artifacts recovered from the two wall fall strata include one metate, two manos, one projectile point, sherds, chipped-stone debris, animal bones, and one human bone fragment.

Point-located artifacts recovered from the roof fall stratum are mapped in Figure 2.7 and described in Table 2.6. Some of these items may have been used, stored, or abandoned on the structure roof and then fallen into the room when the roof collapsed. Their distribution may relate to room use or abandonment activities. Point-located artifacts include sherds, chipped-stone debris, several peckingstones, animal bones, and a few human bones. All point-located sherds are from gray ware jars, and most are typed as Indeterminate Plain Gray. Six are classified as Moccasin Gray. Other artifacts were found in the roof fall stratum but were not assigned PL numbers or mapped. These include additional sherds and chipped-stone debris, several manos, and a small piece of twine. Gray, white, and red wares are present in this collection, and several of the red ware sherds belong to a partial Bluff Black-on-red bowl recovered primarily from Room 15.

Construction

The bases of the walls of Room 3 are very well preserved (Figure 2.8). The north wall consists primarily of vertical sandstone slabs, but portions of the inner wall are lined

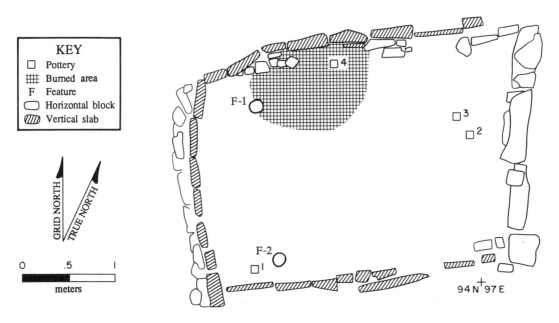

Figure 2.8. Room 3, surface. Point-located artifacts and Features 1 and 2 are described in text.

with small- to medium-size horizontal blocks. The east wall is defined by two courses of horizontal masonry. The lower portion of the south wall consists of a continuous alignment of vertical slabs. The lower portion of the west wall is also built of vertical slabs, but these are paralleled by two courses of horizontal-block masonry on the side that is within adjacent Room 2. The upper walls for the entire room were probably made predominantly of mud and stone. Shallow post holes in the northwest and southwest corners may have held roof-support posts, but given the small size of these holes, the roof's weight must have been borne at least partly by the structure walls themselves. The floor had been prehistorically excavated to sterile B-horizon sediments and, at least in some areas, compacted through use.

Surface

Stratigraphically, the floor of Room 3 was defined as the contact between sterile sediments and overlying fill. The floor was use-compacted only in the south-central portion of the room. Along the north wall, the floor sediments were fire-reddened and -blackened to a depth of 1 to 2 cm below the floor, probably as a result of contact with the burning roof.

Artifacts and Samples. Few artifacts were left on the Room 3 floor. Five Indeterminate Plain Gray jar sherds and one Indeterminate San Juan Red bowl sherd were point-located (Figure 2.8). The red ware sherd belongs to a partial Bluff Black-on-red bowl whose pieces were found mostly in Room 15. A small cluster of sherds along the north wall was also point-located, but the items were lost before analysis. No other artifacts were found in

direct association with the floor. A pollen sample collected from the floor zone yielded evidence of cholla, prickly pear, and corn pollen; a flotation sample collected from an ashy area on the floor yielded a single corn kernel fragment and remains of possible construction materials (mountain mahogany/sagebrush).

Features. Two small, basin-shaped pits (Features 1 and 2), one in the northwest corner of the room and one in the southwest corner, were identified as post holes (Figure 2.8). Feature 1 is 15 cm in diameter and 8 cm deep; Feature 2 is 15 cm long by 13 cm wide by 13 cm deep. Given their sizes and shapes, these pits ordinarily would not be identified as post holes, but Feature 1 contained the remains of a charred post. The two pits are similar in size, shape, and position with respect to a corner, and therefore both are inferred to have held roof-support posts. As there are only two post holes in the room and both are very shallow, it is likely that the roof was also supported by the structure walls.

Dating

Eight tree-ring samples were collected from Room 3 (one from wall fall, the remainder from roof fall). Four noncutting (A.D. 826, 842, 843, and 856) and two cutting (A.D. 854 and 858) dates were obtained (Appendix B). The two cutting dates and the A.D. 856 noncutting date indicate that Room 3 was built in the mid- to late A.D. 850s, and they figure prominently in the evidence used to date Room Suite 1 as a whole. The mid- to late A.D. 850s dates correspond well with a cluster of cutting dates from Pit Structure 1 and noncutting dates from other structures in Room Suite 1. If the A.D. 858 date is from a structural timber, it suggests that Room 3 was constructed after A.D. 857. This interpretation is consistent with architectural evidence indicating that construction took place after the rooms in adjacent Room Suite 2 had been built.

Interpretations

On the basis of its small size, location in the back row of the roomblock, and lack of domestic features, such as a hearth, Room 3 is interpreted as a storage facility. The paucity of artifacts on a floor that was rapidly sealed by the collapsed roof supports the inference that the room was empty at abandonment. In contrast, artifacts were abundant in the roof fall stratum. Roof fall artifacts might have been incorporated into construction mud when the room was built, or they may have been left on the roof at abandonment. The burning of Room 3 may have been deliberate, since evidence of burning was found in most pit structures and domestic rooms but not in any other storage rooms.

Pit Structure 1

Length: 4.10 m
Width: 4.05 m
Area: 14.92 m^2

Pit Structure 1 (Figures 2.9–2.13) is the westernmost of the four pit structures at Duckfoot (Figure 1.5). Its shape—rectangular with rounded corners—is typical for a Pueblo I pit structure, as are the wing wall dividing the main chamber into two distinct areas and the ventilation system consisting of a vertical shaft and horizontal tunnel. Pit Structure 1 burned at abandonment with its contents in place. Fill was excavated in arbitrary levels, and the fill above roof fall was not screened. The upper fill was excavated in several horizontal subdivisions. First, a trench approximately 1 m wide was excavated along the east wall north of the wing wall. This trench was a segment of an exploratory trench that continued across the courtyard and the roomblock. Next, the fill in the west half of the structure was excavated to about 20 cm above the floor, leaving a central balk approximately 1 m wide, whose western face bisected the structure along its north-south axis. This central profile was drawn and is described in the section on stratigraphy below. Next, the balk was excavated to about 20 cm above the floor in levels comparable to those defined in the earlier stages of excavation. The west half of the structure was then excavated to floor, and the stratigraphic profile was continued to floor. Finally, the lower fill of the balk was removed to complete excavation of the floor. Fill south of the wing wall was provenienced separately from the remainder of fill once excavations reached the level below the top of the wing wall.

Stratigraphy

The stratigraphic profile of Pit Structure 1 is presented in Figure 2.9. This cross section depicts seven major stratigraphic episodes. The pit was excavated prehistorically through approximately 1.3 m of natural deposits to a layer of bedrock; excavation continued roughly 15 cm into the bedrock substrate. The bedrock surface sloped gradually downward toward the south, but the presence of a natural "step" that extended more or less along the north-south axis resulted in the deepest portion of the structure being just east of center. Sterile sediments typical of the soil B horizon into which the pit was excavated were then placed over bedrock to level the floor. South of the wing wall these floor-construction sediments consisted of a dark brown sandy loam/strong brown silt loam intermixed with laminated sediments (Stratum 7) that apparently washed in from the ventilator shaft and tunnel. The top of Stratum 7 was designated Surface 1 (south of the wing wall only). North of the wing wall there were two constructed surfaces. A layer of strong brown silt loam (Stratum 6) had been added on top of bedrock, and the upper boundary of this layer was designated Surface 2. Internal features such as the wing wall, ash pit, and hearth were also constructed at this time. At some point during the occupation, and possibly as part of initial construction, a thin layer of clean, coarse sand and another layer of silt loam (Stratum 5) were added on top of Stratum 6 to produce the final floor (Surface 1) of the pit structure north of the wing wall. Stratum 5 was not continuous over the entire area, and underlying constructional fill (Stratum 6) and bedrock showed through in places. The structure was then used for a number of years. At abandonment, complete vessels and usable artifacts were left on the floor, ashes were left in the hearth and ash pit, and two human bodies were placed within the structure. The next stratigraphic event reflected in profile is the burning and collapse of the roof (Stratum 4). Charred beams, burned roof sediments, and ashes covered the floor, sealing the contents of the pit structure in place. The remaining strata reflect the natural filling of the pit structure depression after the structure was abandoned and the roof burned. A thin layer of gray clay (Stratum 3) washed in on top of the burned roof fall layer. The clay layer was then buried by a 1-m-thick deposit of laminated, wind- and water-deposited sediments (Stratum 2). Near the top of these laminated sediments was some type of intrusive unconformity, the size and depth of which suggest a tree fall. The entire basin eventually filled. The shallower the pit became, the slower it filled, so that the laminations caused by sorting became less and less distinct. The last stratigraphic episode reflected in profile is the result of soil-formation processes (weathering, translocation of clays, bioturbation, and accumulation of humified organic matter), which produced the dark brown loam A horizon in the upper 20 to 80 cm of fill (Stratum 1).

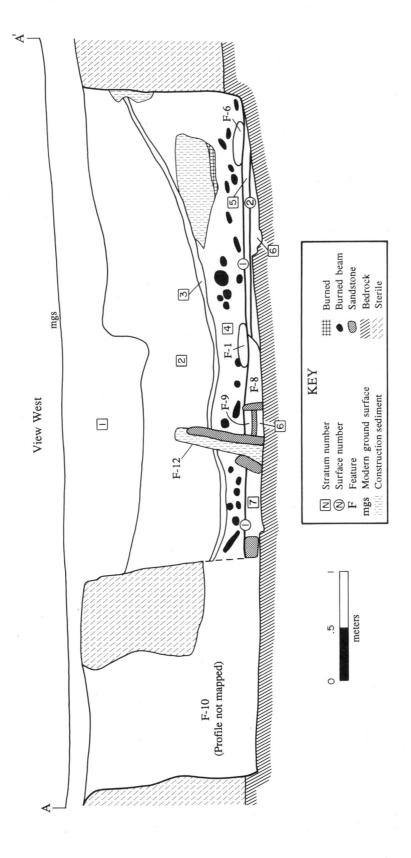

Figure 2.9. Pit Structure 1, stratigraphic profile. The location of the profile is shown in Figure 2.12.

Items found in postabandonment deposits above the roof fall layer (Strata 1 through 3) include sherds, chipped-stone debris, animal bone, and a human bone fragment. Formal tools include three manos, one axe, one hammerstone, one stone disk, numerous peckingstones, a projectile point, and several modified flakes. Many of the stone tools are fragments.

The distribution of artifacts in the burned roof fall layer (Stratum 4) is of interest because some of these materials may have been left on the roof during abandonment (Figure 2.10 and Table 2.7). Point-located items include 2 manos, 4 awls, and 14 jar sherds. Eight sherds are typed as Indeterminate Plain Gray, four as Moccasin Gray, and two as Chapin Black-on-white. A number of sherds that were not assigned PL numbers belong to one partial Bluff Black-on-red bowl and one partial Indeterminate White double bowl, the only double bowl recognized in the Duckfoot assemblage. An *Olivella* shell bead was recovered but was not mapped or assigned a PL number. Several fragments of human bone that probably belong to Burial 2 on the pit structure floor were found in the roof debris. Burned beam fragments—some point-located and collected as tree-ring samples, others not—show the pattern of roof collapse in Figure 2.10. Locations of tree-ring samples collected from the floor are also shown (marked with asterisks) to provide a more complete picture of the distribution of beams. The floor samples are listed on the floor PL table (Table 2.8).

Construction

As mentioned in the preceding section, Pit Structure 1 was excavated prehistorically through approximately 1.3 m of natural deposits to a layer of bedrock. Sterile sediments were then placed over portions of the bedrock substrate to level the floor, and at least some of the major floor features, such as the hearth, ash pit, and wing wall, were constructed. The wing wall is made of vertical sandstone slabs set on the bedrock subfloor. The slabs are supported by floor-construction sediments and plastered with a thick layer of mud. The hearth and ash pit were excavated into floor fill and partly lined with vertical slabs. The ventilation system consists of a vertical shaft extending to ground surface and a horizontal tunnel opening into the main chamber at floor level. The vent system was excavated through natural deposits at the same time as the main chamber. The roof of the pit structure was supported by four main posts. The northern posts were set into the floor approximately 70 to 90 cm in from the northern corners of the structure, and the southern roof-support posts were built into the northern face of the wing wall about 70 to 90 cm from the southern corners of the structure. The charred roof beams (Figure 2.10) preserved a remnant of the basic perpendicular beam pattern that is typical of Anasazi pit structures. The function of individual beams in the roof fall

Table 2.7. Point-Located Artifacts from Pit Structure 1, Roof Fall

No.	Description
Pottery Sherds	
5	white ware jar (2)
9	gray ware jar (12); not mapped
Metate	
6	missing
Two-Hand Mano	
8	incomplete, conglomerate
10	fragmentary, conglomerate; not mapped
Awl	
2	incomplete, *Odocoileus hemionus*
3	complete, canid (cf. *Vulpes macrotis?*)
4	incomplete, *Odocoileus hemionus*
7	complete, *Odocoileus hemionus*
Dating Samples	
11–128	tree-ring samples; the following point-locations are not mapped: 12–14, 27, 39, 51–56, 62, 63, 65, 66, 74, 81, 86, 88, 99, 103–112, 126–128
Miscellaneous	
1	human bone; from Burial 2

NOTES: Item counts appear in parentheses. Refer to Figure 2.10 for artifact locations.

stratum cannot be determined with certainty, but many are probably the remains of primary and secondary beams. The primaries would have spanned (east-west?) two main support posts, and their ends would have been cantilevered to the structure walls. The secondaries essentially were rafters that, supported by the primaries, would have spanned the entire structure and supported the roof-closing material. Closing material could have consisted of small, tightly spaced beams or split juniper planks or shakes. This material would have been covered with 20 to 30 cm of sediment that sealed and insulated the structure. It is possible that a cluster of small beams in the center of the chamber is the remains of the ladder used to enter the structure through a roof hatchway above the hearth.

Surface 1

Surface 1 is the final use surface of the structure. In the northeast corner, bedrock was left exposed as floor. In the central and eastern portions of the main chamber north of the wing wall, a constructional surface (Surface 2) was buried approximately 5 cm below Surface 1.

Artifacts and Samples. The collapse of the structure roof preserved many artifacts in use or abandonment contexts on Surface 1 (Figures 2.11 and 2.12 and Table 2.8). The floor contained a rich assemblage of stone and bone tools, reconstructible pottery vessels, and assorted debris related to everyday domestic activities (Figure 2.13).

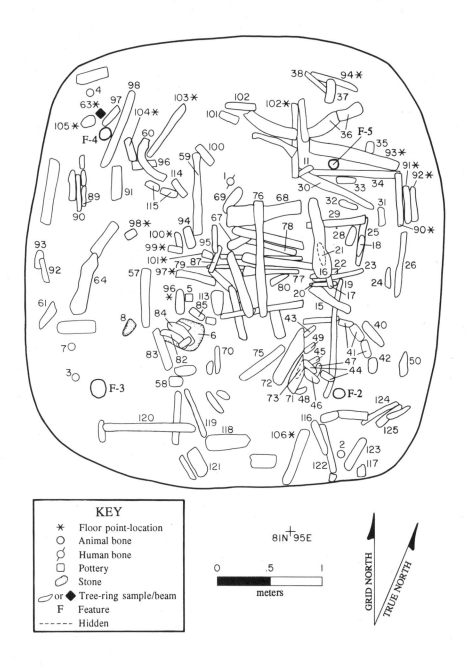

Figure 2.10. Pit Structure 1, roof fall. Point-located artifacts are described in Table 2.7. Tree-ring samples recovered from the floor are included to better illustrate the overall roofing pattern; these are listed in Table 2.8. Features 2, 3, 4, and 5 contained remnants of roof-support posts, which were collected as tree-ring samples. Beams not assigned PL numbers were not recovered.

Figure 2.11. Aerial view of Pit Structure 1 with the floor assemblage in place. The vessels were reconstructed from sherds and returned to their original locations for the photograph. Most of the floor features had not been excavated when the photograph was taken.

Point-located stone tools include 3 complete metates, 10 complete manos, an axe, 2 mauls, 2 hammerstones, 4 peckingstones, and 2 projectile points. In addition, chipped-stone debris, including several cores and edge-damaged flakes, was found in association with the floor. Four of six modified cobbles point-located on the floor were clustered in the west half of the structure, just north of the wing wall. Although this particular type of artifact was found in other structures, Pit Structure 1 had an unusually large number. Three bone awls and a complete Glycymeris shell pendant were also recovered. An unfired clay ball (PL 87) was found just north of the east half of the wing wall. One of the more unusual point-located items consists of a light gray, fine-textured material compressed into a large (diameter = 11.3 cm), dome-shaped "cake," very porous and light for its size (PL 65). Initially the object was identified as being made of clay or gypsum; subsequent laboratory analysis eliminated both of these possibilities but failed to establish an alternative. X-ray diffraction analysis might aid in identifying the substance but has not been undertaken.

Of the 1,170 sherds mapped and assigned PL numbers, 1,110 (94.9 percent) are gray ware, 35 (3.0 percent) are white ware, and 25 (2.1 percent) are red ware. The majority of the gray ware sherds are typed as Indeterminate Plain Gray, but 15 Chapin, 86 Moccasin, and 39 Mancos gray sherds are present in the assemblage as well. Of the decorated wares, 4 were typed as Piedra Black-on-white

and 10 as Bluff Black-on-red. Jar, bowl, and seed jar forms are represented. A whole pottery pipe (PL 81) was found in the northeast corner of the structure, and a modified sherd (PL 79) was recovered nearby. (A modified sherd is one that was modified on at least one but not all edges after the original vessel broke.)

The floor assemblage of Pit Structure 1 is remarkable for the number of whole and reconstructible vessels it contains. Counting several items that were not assigned point-location numbers, 22 whole, completely reconstructible, nearly complete, or partial vessels (jar, bowl, olla, seed jar, and pitcher) were recovered from the floor. Also present in the assemblage are 17 sherd containers, that is, large sherds that were recycled for continued use as containers after the original vessels broke. Eighteen of the 22 whole or reconstructible vessels are gray wares, identified to type as follows: 5 Chapin, 10 Moccasin, 2 Mancos, and 1 Indeterminate Plain Gray. Two vessels are Bluff Black-on-red, and two are local white wares, specific type indeterminate. Because sherd containers are vessel fragments, fewer of them could be identified to type: 9 of the 17 are Indeterminate Plain Gray, 4 are Moccasin Gray, 1 is Chapin Gray, 1 is Piedra Black-on-white, and 2 are Indeterminate White. As shown in Figure 2.12, many of the vessels were clustered along the east and west walls of the structure. This arrangement suggests storage in low-traffic peripheral areas of the chamber.

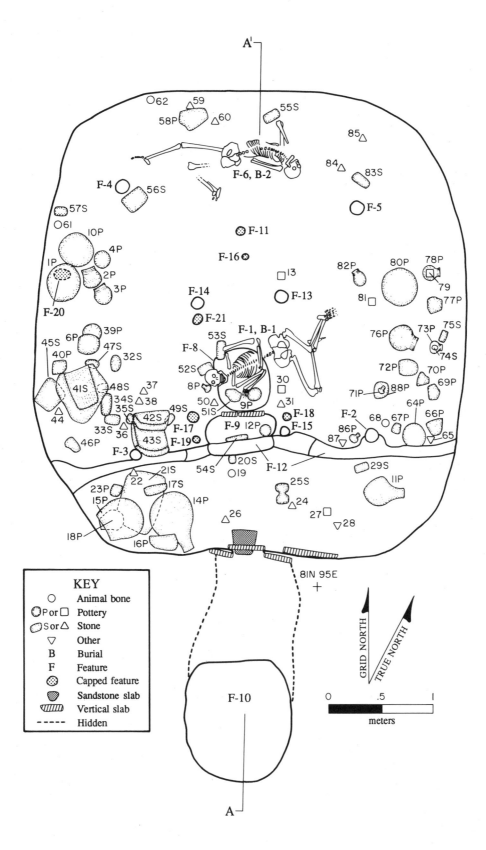

Figure 2.12. Pit Structure 1, Surface 1. Feature 16 is associated with Surface 2. Point-located artifacts are described in Table 2.8; features are described in Table 2.9. Profile A is shown in Figure 2.9.

Table 2.8. Point-Located Artifacts from Pit Structure 1, Surface 1

No.	Description	No.	Description
Pottery Sherds		42	complete, sandstone
1	gray ware jar (85); Vessel 12 (79)	43	complete, sandstone
2	gray ware jar (51); Vessel 13 (51)	51	fragmentary, sandstone
3	gray ware jar (61); Vessel 11 (58)	**Two-Hand Mano**	
4	gray ware jar (14); Vessel 15 (14)	17	complete, sandstone
	white ware jar (20); Vessel 15 (20)	29	complete, sandstone
	white ware other (7); Vessel 15 (7)	33	complete, igneous
6	gray ware jar (8); Vessel 57 (8)	34	complete, sandstone
8	gray ware jar (33); Vessel 4 (33)	52	complete, igneous
9	gray ware jar (29); Vessel 14 (29)	53	complete, conglomerate
10	gray ware jar (89); Vessel 29 (89)	54	complete, sandstone
11	gray ware jar (96); Vessel 17 (96)	55	complete, sandstone
	gray ware other (1); Vessel 17 (1)	75	complete, sandstone
12	gray ware jar (2); Vessel 3 (2)	83	complete, sandstone
	gray ware other (3); Vessel 3 (3)	**Abrader**	
13	gray ware jar (133); Vessel 27 (1), Vessel 28 (10)	56	complete, sandstone
	white ware bowl (6)	**Hammerstone**	
	red ware bowl (17); Vessel 32 (17)	22	fragmentary, quartz
14	gray ware jar (79); Vessel 25 (79)	35	complete, igneous
15	gray ware jar (63); Vessel 16 (63)	**Peckingstone**	
16	gray ware jar (6); Vessel 106 (6)	31	complete, Dakota orthoquartzite
18	gray ware jar (38); Vessel 19 (38)	38	complete, Morrison orthoquartzite
23	red ware bowl (1); Vessel 37 (1)	44	complete, Morrison orthoquartzite
27	red ware other (6); Vessel 20 (6)	84	complete, Morrison orthoquartzite
30	missing	**Maul**	
39	gray ware jar (4); Vessel 103 (4)	22	complete, Morrison orthoquartzite
40	gray ware jar (1); Vessel 104 (1)	25	complete, Dakota orthoquartzite
46	gray ware jar (5); Vessel 27 (5)	**Single-Bitted Axe**	
58	gray ware jar (2); Vessel 105 (2)	20	complete, igneous
64	gray ware jar (1); Vessel 6 (1)	**Projectile Point**	
66	gray ware jar (5); Vessel 99 (5)	24	complete, agate/chalcedony
67	gray ware jar (3); Vessel 101 (3)	85	complete, Morrison chert/siltstone
69	white ware bowl (1); Vessel 123 (1)	**Modified Cobble**	
70	red ware bowl (1); Vessel 114 (1)	32	complete, igneous
71	missing	47	complete, igneous
72	gray ware jar (23); Vessel 102 (23)	48	complete, unknown orthoquartzite
73	gray ware jar (29); Vessel 5 (29)	49	complete, igneous
76	gray ware jar (58); Vessel 9 (58)	57	fragmentary, igneous
77	gray ware jar (17); Vessel 98 (17)	74	complete, igneous
78	gray ware jar (7); Vessel 7 (7)	**Other Modified Stone/Mineral**	
80	gray ware jar (71); Vessel 10 (71)	45	fragmentary, sandstone
81	gray ware other, whole pipe (1)	87	condition unknown, clay
82	gray ware jar (46); Vessel 8 (46)	**Unmodified Stone/Mineral**	
86	gray ware jar (39); Vessel 28 (39)	65	unknown mineral
88	gray ware jar (7); Vessel 100 (7)	**Awl**	
Modified Sherd		19	complete, *Odocoileus hemionus*
79	complete, white ware	62	fragmentary, *Odocoileus hemionus*
Chipped-Stone Debris		68	fragmentary, indeterminate avian
22	cores (3)	**Pendant**	
	flakes (7)	28	complete, shell
26	edge-damaged flake (1)	**Dating Samples**	
36	flake (1)	63*	tree-ring sample; missing
37	flake (1)	90*–94*	tree-ring samples
50	edge-damaged flake (1)	96*–106*	tree-ring samples
59	flake (1)	107	tree-ring sample; not mapped
60	edge-damaged flake (1)	**Miscellaneous**	
Trough Metate		21	sandstone slab; not collected
41	complete, sandstone	61	animal bone

NOTES: Item counts appear in parentheses. For pottery, sherds belonging to vessels are also included in total item count. Refer to Figure 2.12 for artifact locations. Items marked with an asterisk (*) are shown on the roof fall map, Figure 2.10.

Table 2.9. Feature Summary, Pit Structure 1

No.	Type	Shape	Length (cm)	Width (cm)	Depth (Height) (cm)	Status
1	Burial 1	other	140	70	(10)	—
2	post hole	other	12	12	17	in use
3	post hole	other	10	10	17	in use
4	post hole	cylindrical	12	12	12	in use
5	post hole	nr	15	15	4	in use
6	Burial 2	other	160	75	(10)	—
8	hearth	basin	56	53	11	in use
9	ash pit	other	65	25	9	in use
10	vent	other	210	105	165	in use
11	sipapu	cylindrical	8	8	15	capped
12	wing wall	other	365	12	(77)	—
13	post hole	nr	13	13	~9	in use
14	post hole	cylindrical	11	11	12	unknown
15	post hole	cylindrical	11	10	15	in use
16[a]	pit	cylindrical	6	5	10	capped
17	pit	cylindrical	10	10	14	capped
18	pit	cylindrical	9	7	14	capped
19	pit	cylindrical	9	7	6	capped
20	pit	basin	17	10	7	capped
21	pit	cylindrical	10	10	7	capped

NOTE: Feature locations are shown in Figure 2.12.

nr = no record.

[a] Feature associated with Surface 2.

Figure 2.13. Pit Structure 1 work area consisting of trough metate (PL 41), two-hand manos (PLs 33 and 34), modified cobbles (PLs 32 and 47), and sherd containers (PLs 6, 39, and 40). See Figure 2.12.

Numerous flotation and pollen samples were collected from a variety of floor and floor-related contexts, including several features and many of the pottery vessels. Juniper, pine, sagebrush, and mountain mahogany remains recovered from flotation samples are probably derived from roof-construction materials, hearth fuel, and fuel carried into the structure to burn it at abandonment. Food remains recovered from flotation samples include cheno-am, prickly pear, groundcherry, corn, and a single bean cotyledon—the only evidence of domestic beans at Duckfoot. The one pollen sample that yielded interpretable results contained Cheno-am, prickly pear, cholla, and corn pollen.

Features. Twenty features were identified in Pit Structure 1 (Table 2.9 and Figure 2.12), including two human interments. Feature 1 (Burial 1) is the complete, articulated skeleton of a 30- to 39-year-old male that was found lying partly flexed across the hearth. The bones rested entirely on the floor and hearth fill and were covered by burned roof fall, which suggests that the individual was placed on the pit structure floor before the roof was burned. Only the bones of the left knee of this individual showed signs of burning. The skeleton was lying on its back with its head toward the southwest, facing up and slightly south. One mano (PL 52) was found under the head, and a second mano (PL 53) was under the left shoulder. A small Chapin Gray jar (PL 8) was immediately to the west of the cranium, and a Moccasin Gray sherd container (PL 9) was touched by the right hand.

Feature 6 is a second human skeleton (Burial 2), located near the north wall of the structure. The remains are those of a man 30 to 39 years of age. The bones were articulated but more fragmented and not as well preserved as those of Burial 1. The poorer preservation probably is due largely to the fact that the bones were more severely burned. The head was toward the east-northeast, the arms were flexed, the right leg was fully extended, and the left leg was flexed at the knee only. The skeleton rested partly on the floor, but several charred beams were under the torso. Possibly this body was on the roof when it burned, although if that were the case, one would expect more roof debris under the skeleton. Another possibility is that the structure was burned intentionally with the two bodies in it, and the beams under the Burial 2 torso were part of the fuel used to ignite the roof from inside the pit structure.

There were no grave goods, per se, associated with the burials in Pit Structure 1, but the utilitarian assemblage of tools and containers was abandoned along with the human remains. A study of the abandonment vessel assemblages for the site as a whole suggests that the Pit Structure 1 floor assemblage approximates the expected inventory of vessels that might have been used prehistorically in an architectural suite. Rather than the artifacts being interpreted as funerary goods added to the pit structure because of a mortuary ritual, it seems likely that the human bodies were added to the structure, which already contained a nearly complete inventory of utilitarian equipment. The one possible exception is the small jar category. Small jars are overrepresented in the assemblage, and several were found near the center of the structure, near Burial 1.

The central hearth (Feature 8) is a basin-shaped depression that was built into the pit structure floor. A ring of small sandstone slabs was set into the subfloor sediments and plastered over so that only the edges were exposed along the inside rim. The south edge of the hearth is defined by two slabs set on edge, and these separate the hearth from the adjoining ash pit (Feature 9). The hearth was filled with a homogeneous, dark gray to black, ashy fill with fine charcoal throughout. This matrix not only filled the hearth but extended onto the structure floor, covering an area about 20 cm beyond the rim, including the two vertical slabs and the ash pit to the south. Flotation samples from the hearth yielded both food and fuel remains; the results of analysis are reported in detail in Chapter 6.

Built between the south rim of the hearth and the north side of the wing wall–deflector, the ash pit (Feature 9) served as a receptacle for ashy debris periodically cleaned from the hearth. The large vertical slab that serves as the deflector also forms the south edge of the ash pit, and two smaller vertical slabs along the north edge of the ash pit separate it from the hearth. The east and west rims of the ash pit were sculpted from the floor-construction sediments, and there is no raised lip, or coping. The floor of the ash pit is a flat sandstone slab that was put in place before the floor-construction sediments were laid to form the east and west rims (Figure 2.9). The ash pit overflowed with the same type of ash and charcoal that filled the hearth. These materials were piled about 10 cm above the rim along the deflector and extended approximately 50 cm to the east and west on the floor, along the face of the wing wall. Results of flotation sample analysis are reported in Chapter 6.

Feature 12 is a wing wall–deflector complex that forms a continuous low wall that divides the main chamber into two separate areas. The wing wall and deflector apparently were built as separate units but later were joined. The feature was constructed of large sandstone slabs (roughly 60 × 50 × 5 cm) set on edge and covered with mud to form a wall approximately 10 to 12 cm thick and 77 cm high. The slabs were placed on bedrock and in some places were braced with floor-construction sediments, which indicates that the initial construction of the wing wall–deflector complex coincided with the construction of the floor. The 25- to 35-cm gaps between the deflector and the two halves of the wing wall appeared to have been filled later, as the sediment forming these "plugs" was darker than the wing

wall and deflector mud. These plugs were also above the floor, although the ash that overflowed the ash pit was piled against them. A few small sandstone slabs were set horizontally within these plugs, but mud was the primary construction material. Because the plugs were built only 30 to 35 cm above floor level, access between the two areas of the main chamber would have been retained even after the wing wall was remodeled.

Features 11 and 16 are small pits along the north-south center line of the structure and approximately 100 cm and 80 cm, respectively, north of the hearth. Feature 11 had been excavated through the floor-construction sediments to bedrock, filled with dark brown sandy loam, and capped. A well-made, corner-notched projectile point found roughly in the center of fill may have been placed in the pit as part of a ritual dedication. Feature 11 is interpreted to be a sipapu. Feature 16 was also excavated to bedrock and had a small sandstone slab set into the floor 2 cm north of its rim. Feature 16 was filled with coarse sand and apparently is associated with Surface 2, since it was effectively capped by the construction of Surface 1 above. Although this feature is described simply as a sand-filled pit, it may also have functioned as a sipapu. In the early ethnographic literature, the sipapu is described as a feature of special ritual significance, the "sacred cavity . . . through which comes the beneficent influence of the deities or powers invoked" (Mindeleff 1989:117). Archaeologically, the sipapu occurs throughout the Anasazi period and assumes several forms, ranging from a small, circular pit to a larger and more elaborate basin. The concavity frequently is sand-filled and sometimes is found in association with multiple prayer stick holes, or paho marks (Wilshusen 1988c:649–653, 1989). In addition, sipapus are consistently found north of the hearth, approximately along the north-south axis that divides the structure into east and west halves. The interpretation of Feature 11 and possibly Feature 16 as ritual features, then, is based on size, shape, sandy fill, and location. If both features indeed served a ritual function, they must have done so before the floor was replastered.

Feature 10 is the ventilation system. It consists of a vertical shaft, approximately 1 m south of the main chamber, and a horizontal tunnel that opens into the main chamber at approximately floor level. The shaft is a slightly irregular excavated cylinder that may have been partly lined with sandstone blocks, as these were common throughout fill (no rocks were found in place). An abundance of rubble in the upper fill may indicate that there was a raised rim around the top of the vent shaft to prevent runoff water from flowing into the pit structure through the vent system. The vent is large enough for a person to crawl into it from the shaft end, but the tunnel opening into the pit structure was constricted by a slab-and-mud aperture. A sandstone slab sealed the vent tunnel opening from inside the pit structure.

Seven post holes were identified on the floor of Pit Structure 1. Features 2, 3, 4, and 5 are believed to have held main roof-support posts. All four contained only rotted wood below floor level but yielded well-preserved, charred, juniper post fragments above. Features 2 and 3—the southeast and southwest members, respectively, of the four-post roof-support system—were partly socketed into the north face of the wing wall. The Feature 4 roof-support post in the northwest quadrant of the structure was held in place by the surrounding roof fall. Features 2, 3, and 4 had been excavated through constructional fill underlying the floor, and the posts were seated on bedrock. The last main roof support, Feature 5, is in the northeast quadrant of the structure, where the floor is solid bedrock. The shallow hole was pecked into the rock, and the post was supported by mud that had been packed around its base.

The functions of the remaining three post holes (Features 13, 14, and 15) are not clear. Features 13 and 14 are located approximately 40 cm north of the hearth and 40 cm east and west, respectively, of the north-south center line of Pit Structure 1. The placement of these two pits with respect to the structure's axis of symmetry suggests that their uses may have been associated. However, construction details and fill characteristics are quite different for the two features. Feature 13 was excavated down to, but not into, bedrock. A thin layer of black sediment was found at the bottom of the pit and was covered by a small sandstone slab (7 × 4 × 1 cm). The slab was covered by compact silt loam that filled about half the pit, and the upper half of the feature was filled with loose, dark brown, organic-rich sediment. The upper fill had been disturbed before excavation by a toad that used the loose sediment for a nest. Feature 14 was excavated through the floor-construction sediments and 5 cm into bedrock. The smoothness of the pit suggests that it had been ground as well as pecked. Feature fill was dark brown to gray, and small pieces of charcoal and a flake were found in the upper portion. Because of their locations with respect to one another, Features 13 and 14 might have served as ladder-pole sockets, but this interpretation does not explain the differences in construction and fill. An alternative possibility is that these pits contained auxiliary roof-support posts associated with the hatch/smokehole in the roof, but there is minimal support for this interpretation. A third alternative is that the pits may have supported some internal structure such as an altar (Wilshusen 1989) or a drying or cooking rack.

The last post hole, Feature 15, is located immediately north of the east wing wall–deflector passage, just east of the ash pit. Feature 15 contained dark brown, organic-rich fill but no charcoal. The organic matter may have been the remains of a rotted post.

Five small pits of unknown function were also found (Features 17, 18, 19, 20, and 21), and all were capped. Features 17 and 18 are cylindrical pits excavated to bedrock on either side of the ash pit. Feature 17 contained dark brown loam with a few lenses of sterile sediment (strong brown silt loam). Feature 18 had a dark gray, ashy-sand fill. Both pits were covered by the ash that overflowed the hearth and ash pit, which indicates that both pits were filled before abandonment. Features 17 and 18 could have paired with Features 13 and 14 to support some type of internal structure associated with the hearth. If such a structure existed, apparently it was not in place at the time of abandonment.

Feature 19, located just west of the ash pit, seems to be the logical counterpart of Feature 15, located just east of the ash pit. The two are similar in shape and fill characteristics. However, Feature 19 is so shallow it is unlikely that it could have supported a post; the pit does not extend to bedrock. Feature 20 is a shallow, trapezoidal basin excavated through floor-construction sediments to bedrock. The pit contained ashy loam with some tiny charcoal flecks and one chunk of yellow pigment, possibly carnotite, near the top. Feature 21 is a shallow pit filled with dark brown sandy loam, and it appears to have been filled before abandonment.

Surface 2

Surface 2 in Pit Structure 1 is the surface that lies approximately 5 cm below Surface 1 in the central and eastern portions of the main chamber. It consists of a layer of strong brown silt loam placed over bedrock, and it may have served as an earlier living or use surface. The only feature that clearly originated at Surface 2 is Feature 16, a small, sand-filled pit that may have been used as a sipapu (Feature 16 is described in more detail above, in the discussion of Surface 1 features). It is possible that the five other capped pits, also described above, were associated with this earlier surface, although conclusive evidence is lacking. No artifacts were point-located on Surface 2. The three sherds and two flakes recovered are believed to have been incidentally incorporated into construction sediments.

Dating

The 137 tree-ring samples collected from Pit Structure 1 yielded 74 dates, of which 62 are cutting dates (Appendix B). In spite of the substantial number of samples that yielded results, the interpretation of the date of construction of Pit Structure 1 is problematic. The date distribution contains two strong clusters, one consisting of 7 cutting and non-cutting dates in A.D. 853 and 854, and one consisting of 41 cutting dates in A.D. 865 and 866. Considered alone, this distribution suggests that the structure was built in A.D. 866 using some wood that was originally cut 12 to 13 years earlier. However, if one considers the construction of the architectural suite as a whole, a strong alternative argument can be made. Room Suite 1 has a cluster of five mid- to late A.D. 850s dates ending in A.D. 858 (Rooms 3 and 15).

If the construction of pit structures is assumed to be related to that of the associated rooms (that is, if the dirt removed during pit structure excavation is used in the construction of mud-and-stone or mud-and-post room walls [see Wilshusen 1984, 1987]), then the probable date of construction of the entire architectural suite, including Pit Structure 1, is A.D. 858. The strong terminal cluster of A.D. 865 and 866 dates in Pit Structure 1 could represent structure remodeling. The number of dates in this cluster suggests that remodeling was major and involved replacement of most of the roof as well as of at least three of the four main roof-support posts (the four posts yielded cutting dates of A.D. 863 and 866 and noncutting dates of A.D. 843 and 865). Single dates of A.D. 872 and 873 may relate to wood used in minor repairs or firewood that was inside the structure at the time of abandonment. These dates suggest that the occupation of Pit Structure 1 extended at least into the mid-A.D. 870s (see Lightfoot 1992b).

If the assumption that Pit Structure 1 and Room Suite 1 were built at the same time is correct, then dates from the entire architectural suite support the argument that Pit Structure 1 was constructed in the late A.D. 850s. This is the interpretation offered here. However, it is acknowledged that, taken in isolation, the dates from Pit Structure 1 strongly support an argument for construction in A.D. 866, or soon thereafter. If this argument is correct, it suggests that a suite of rooms was built almost 10 years before the associated pit structure was added.

Interpretations

The types of features and artifacts on the floor of Pit Structure 1 clearly reflect a wide range of domestic and ritual activities. Food preparation and cooking are indicated by the manos, metates, and cooking vessels in the floor assemblage and by the presence of a hearth and ash pit in the center of the main chamber. Distinct clusters of artifacts, including vessels, along the edges of the room suggest "active" storage of tools and other items within easy reach of the locations in which they were used. Artifact clusters south of the wing wall include items (projectile point, pendant, maul, and axe) used for activities other than food preparation and cooking and may reflect "passive" storage of items not used on a regular basis. Bone awls in the assemblage may have been used for sewing, weaving, or basket making.

The presence of two human bodies, which apparently were placed in the structure at abandonment, and at least one sipapu suggests that ritual activities took place in Pit Structure 1. Burning of the structure is interpreted as a deliberate act associated with the interring of the bodies and, therefore, may also have been part of an abandonment ritual. The traditional view of burned structures in the Southwest is that they represent accidental, if not catastrophic, events. However, a substantial amount of evidence contradicts this view, including two experimental burnings of aboriginal-style, earth-covered dwellings (Wilshusen 1986c:247). In Pit Structure 1, the strongest evidence that the fire was not an accident is that one body (Feature 1, Burial 1) was lying across the hearth, yet the only burning on the bones is on top of the left knee. It is unlikely that a fire in the hearth could have burned hot enough to accidentally ignite the earth-covered roof, while not burning the bones of the body. If Pit Structure 1 was intentionally destroyed, as appears to be the case, then all items, including the bodies within the pit structure, are part of the intentionally abandoned assemblage, and the burning itself may have been part of an abandonment ritual. Wilshusen (1986c) shows a correlation between the presence of certain ritual features and the intentional burning of pit structures between A.D. 860 and 900 in the Dolores River valley. He also shows a correlation between intentionally collapsed structures and the placement of human bodies on the pit structure floors. Thus, Pit Structure 1 abandonment is interpreted to have been part of a mortuary ritual. It is possible that Pit Structure 1 was abandoned and destroyed because of the death of the individuals left in it, which in turn suggests that one or both had relatively high status within the household (cf. Stodder 1987:357). The fact that the two individuals were older adult males is consistent with this interpretation. However, four of the other five skeletons found in pit structures at Duckfoot are those of two adult females and two children. Considering the age and sex mixture of all the individuals found on the floors of pit structures and the variety of floor assemblages associated with them, the case for status differentiation is weak.

Courtyard 1

Courtyard 1 is defined as the open area that surrounds Pit Structure 1 and extends between the pit structure and the west end of the roomblock (Figure 2.14). The eastern boundary of the courtyard is arbitrarily defined along the 97 east grid line. Where Feature 7 straddles this line, the courtyard boundary was expanded to include all of Feature 7 in Courtyard 1. Room 20, a small, poorly understood structure, is also located in Courtyard 1, and a description of the room is included in the following section.

Because stratigraphy was similar in all three courtyard areas, the same excavation strategy was employed throughout. Courtyard deposits consisted of a single stratum of organic-rich, A-horizon sediments (10 to 30 cm thick), which covered undisturbed sterile deposits. Stratum 1 was thickest near the roomblock, where wall collapse had contributed to the sediment that covered the courtyard, and became thinner with increasing distance from the roomblock. The prehistoric use surface was probably in the lower portion of the fill but was not discernible; most courtyard pit features were detectable only at the boundary with sterile sediments. The depressions created by the

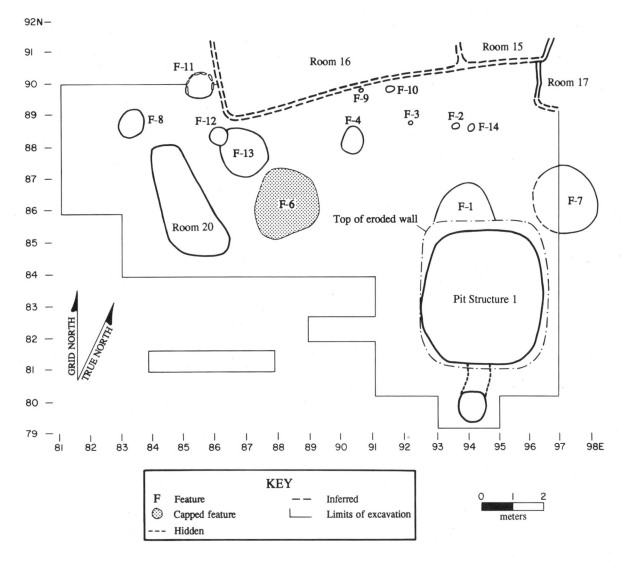

Figure 2.14. Courtyard 1, plan. The upper walls of Pit Structure 1 eroded after excavation of the pit structure was completed but before excavation of the courtyard began. The eroded area is indicated on the map. Features are described in Table 2.10. Feature 15 was not mapped.

collapse of the pit structures undoubtedly promoted erosion of the courtyard surface, and in the area immediately surrounding the pit structures, the original surface probably was completely missing. Excavation of the courtyard areas was conducted primarily in a series of 2-×-2-m grid squares. After loose sediment on modern ground surface was removed, Stratum 1 was excavated as a single unit. All sediments were screened through ¼-in mesh, except for those removed by backhoe. In Courtyard 1, two backhoe trenches were excavated west of Pit Structure 1 to confirm that no other large structures were present.

Artifacts

Although artifacts were recovered from courtyard sediments, and presumably at least some were associated with

the prehistoric use surface, the surface itself was too diffuse and irregular to define, and no materials were point-located. It is likely that erosion and soil-formation processes obliterated the prehistoric courtyard surface. What was found instead was an abrupt stratigraphic boundary between the dark brown loam A horizon (Stratum 1) and the strong brown silt loam sterile sediments below. This stratigraphic boundary was the level of detection for features, but artifacts were rarely found lying directly on it. The artifacts whose provenience is given as "courtyard" are those contained within the 10- to 30-cm-thick Stratum 1. Over 3,200 sherds and almost 350 pieces of chipped-stone debris, but relatively few formal tools, were recovered from courtyard deposits (excluding modern ground surface). Tools include one hammerstone, one polishing stone, one modified core, three peckingstones, two projectile points, a biface, and two modified flakes. A number of animal bones were also found.

Features

Fourteen features were defined in Courtyard 1 (Figure 2.14 and Table 2.10). One, Feature 11, is a hearth, and the remainder are pits of unknown function. The hearth was filled with cultural debris mixed with postabandonment deposits, and its presence suggests that some food preparation took place out of doors. Two nearby pit features, Features 8 and 12, may have been roasting pits. Both were relatively deep pits with distinct charcoal layers, and both showed signs of in situ burning (fire-reddened and -blackened earth and stones along the sides of the pit). Feature 8 was lined with sandstone slabs. During excavation of Feature 12, a patch of dark, organic-stained sediments was detected along the southeast edge of the feature and defined as a separate pit (Feature 13). It appears that the construction of Feature 12 truncated the existing Feature 13 pit. Because of time constraints, only the northwest corner of Feature 13 was excavated.

Features 1 and 7 are large, basin-shaped pits at or near the northwest and northeast corners of Pit Structure 1, respectively. Erosion damaged the edges of both features, especially the edges nearest Pit Structure 1. As a result, it is not known if or how the two features intersected with the pit structure, or whether they predate the structure or are contemporary with it. The fill of these two features revealed little about their use or function. Feature 1 contained silt loam with some charcoal, and Feature 7 contained refuse interpreted to have been intentionally dumped in the pit. Considering the height of the features above the Pit Structure 1 floor, it seems unlikely that they were alcoves or other roofed spaces integrated with the structure. Originally they may have been mud-mixing pits used for construction or maintenance of Pit Structure 1 and the west end of the pueblo. It is also possible that Features 1 and 7 were areas of disturbed soil resulting from foot traffic between the pit structure and the rooms. During wet periods it would have been especially muddy where the pit structure impeded drainage in a high-traffic area.

Feature 6 is a large, shallow pit that contained burned structural debris and numerous artifacts. The feature also had a distinct "floor," although there were few artifacts and no internal features. Feature 6 may be a small, isolated, mud-walled room that burned or a pit that was intentionally filled and capped with trash and burned structural debris.

Feature 4 is a medium-size, basin-shaped pit that was filled with clay loam mixed with caliche. Six small pits (Features 2, 3, 9, 10, 14, and 15) were found within 2 m of the south wall of the roomblock (the locations of all but Feature 15 are shown in Figure 2.14). All six may have been post holes, but none actually contained a post. Fea-

Table 2.10. Feature Summary, Courtyard 1

No.	Type	Shape	Length (cm)	Width (cm)	Depth (cm)	Status
1	pit	nr	190	135	~20	unknown
2	pit	cylindrical	23	19	19	unknown
3	pit	basin	11	11	10	unknown
4	pit	basin	76	55	7	unknown
6	pit	basin	218	195	16	capped
7	pit	basin	265	~200	20	in use
8	pit	basin	105	85	29	in use
9	pit	cylindrical	12	12	12	unknown
10	pit	other	21	17	15	in use
11	hearth	basin	80	80	24	unknown
12	pit	basin	50	50	15	in use
13[a]	pit	—	—	—	—	—
14[a]	pit	—	15	15	—	—
15[b]	pit	basin	28	27	16	unknown

NOTE: Feature locations are shown in Figure 2.14.
nr = no record.
[a] Not excavated.
[b] Not mapped.

tures 9, 10, and 15 are the best candidates for post holes because of their location near the south wall of the roomblock. Feature 15 was excavated but not mapped; Feature 14 was mapped but not excavated. Field notes indicate that Feature 15 was aligned with Features 9 and 10 along the south wall of Room 16.

Interpretations

Courtyard 1 probably served as a general use area for the occupants of Architectural Suite 1. Its northern and southern boundaries are well defined by the structures that make up the architectural suite; its eastern and western borders are defined along north-south grid lines. Although it is likely that many activities in this area (particularly those involving use of the features) were conducted by the occupants of Architectural Suite 1, it is also probable that at least some activities saw participation by other site inhabitants. As open, outdoor expanses, courtyards can probably be considered more fluid use areas than structures.

In all three courtyard areas at Duckfoot, the majority of features and artifacts were found between the roomblock and a line even with the north walls of the pit structures. South of this line, courtyard artifact and feature density diminished markedly. Within this general high-use area, the space immediately south of the roomblock appears to have been the area of most intense outdoor activity, and relatively clear spaces appear to have been maintained closer to the pit structures, perhaps to allow for unobstructed pedestrian traffic. However, this pattern is less clear in Courtyard 1 than in the other two courtyards.

Room 20

Length: 3.75 m
Width: 1.70 m
Area: 5.00 m²

Room 20 (Figure 2.15) is a shallow, amorphous, flat-bottomed pit in the west end of Courtyard 1 (Figure 1.5). It is interpreted as a structure on the basis of its level floor and vertical, excavated lower walls. Debris believed to be remnants of a collapsed roof filled the pit, and some artifacts were found on the floor. However, this "room" could also be nothing more than a large pit feature dug to supply mud for pueblo construction and repairs and then later filled with debris. The entire room was excavated as a single horizontal unit.

Stratigraphy

Room 20 fill, which consisted of dark brown, organic-rich sediments with abundant charcoal and burned mud inclu-

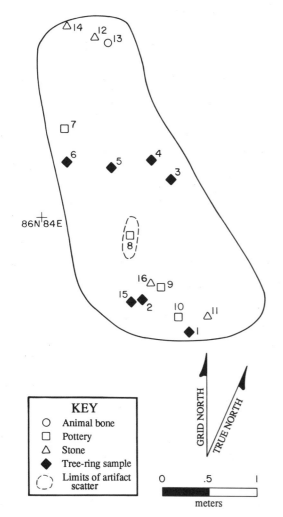

86N 84E

KEY
○ Animal bone
□ Pottery
△ Stone
◆ Tree-ring sample
⊂ ⊃ Limits of artifact scatter

GRID NORTH
TRUE NORTH

0   .5   1
meters

Figure 2.15. Room 20, surface. Point-located artifacts are described in Table 2.11.

sions, is interpreted to be collapsed roofing material. Items recovered from roof fall include sherds, chipped-stone debris, and a small piece of animal bone; formal tools include a metate, a peckingstone, and three modified sherds. Several tree-ring samples were collected as well. A sherd cluster, one flake, and four tree-ring samples are the only materials that were assigned point-location numbers.

Construction and Surface

Aside from the constructional mud and burned charcoal in fill, there are no clues as to how the walls and roof of Room 20 were built. Because of the absence of stone rubble and post holes, it is assumed that the walls and roof were made mostly of mud. The structure must have been very insubstantial, more of a "hut" than a formal room. There are no features in Room 20.

The floor was defined as the use-compacted surface at the contact between sterile deposits below and burned roof fall above. It was fairly level, although not perfectly flat, and a very slight circular depression was noted in the south end. Point-located items include sherds, one piece of chipped-stone debris, two peckingstones, a polishing/hammerstone, a bone awl, and several tree-ring samples (Figure 2.15 and Table 2.11). Additional items, including more sherds and chipped-stone debris and some burned corn, were recovered but not point-located. Of the 42 point-located sherds, 40 (95.2 percent) are gray ware and 2 (4.8 percent) are white ware. Nine sherds were identified to specific type categories: one Chapin Gray, six Moccasin Gray, one

Table 2.11. Point-Located Artifacts from Room 20, Surface

No.	Description
Pottery Sherds	
7	gray ware jar (1)
	white ware bowl (1)
8	gray ware jar (15)
	white ware bowl (1)
9	gray ware jar (3)
10	gray ware jar (21)
Chipped-Stone Debris	
14	flake (1)
Peckingstone	
11	complete, Morrison orthoquartzite
12	complete, Dakota orthoquartzite
Polishing/Hammerstone	
16	complete, unknown orthoquartzite
Awl	
13	fragmentary, medium mammal
Dating Samples	
1–6	tree-ring samples
15	tree-ring sample

NOTES: Item counts appear in parentheses. Refer to Figure 2.15 for artifact locations.

Mancos Gray, and one Piedra Black-on-white. Bowl and jar forms are represented. This assemblage, although small, is similar in composition to other floor sherd assemblages at Duckfoot.

Dating

Eleven tree-ring samples were collected from Room 20, four from roof fall and seven from the floor. Only one sample yielded a date, A.D. 874 (Appendix B), and although this is not a cutting date, it corresponds with a number of cutting dates yielded by samples collected from elsewhere on the site—for instance, Pit Structure 4. Pit Structure 4 probably was built in the early to mid-A.D. 870s, and several structures appear to have been remodeled or repaired during that same period. Perhaps Room 20 is a later addition to the site as well.

Architectural Suite 2

This is the central architectural suite at Duckfoot, and building styles and wall junctures indicate that it was the first architectural suite to be built. Tree-ring dates indicate construction in the late A.D. 850s, probably in A.D. 857. Rooms 4, 5, 6, 7, 11, 12, 13, and 14 probably were the original rooms belonging to this suite, and Pit Structure 2 probably was the original pit structure (Figure 2.1). Some-

time during the occupation, however, an additional room (Room 17) and a second pit structure (Pit Structure 4) were added to expand what was already the largest architectural suite at the site (Figure 1.5). It is not certain that Pit Structure 4 was built and used as a part of Architectural Suite 2, but it is being treated as such for the sake of this chapter's organization.

Room 11

Length: 3.15 m
Width: 3.10 m
Area: 9.99 m^2

Room 11, between Room 10 to the east and Room 12 to the west, is the easternmost front room in Room Suite 2, and it appears to be spatially associated with only one back room, Room 7 (Figure 2.1). It is rather small for a front room, and the floor area is further reduced by two large bins that occupy roughly 1.5 m^2 of space (Figures 2.16–2.18). The structure was burned at abandonment. The room was excavated as a single horizontal unit.

Stratigraphy

The fill of Room 11 consisted of two strata. The upper one, Stratum 1, was a 10- to 20-cm-thick layer of dark brown sandy loam that is an organic-rich A horizon. Beneath

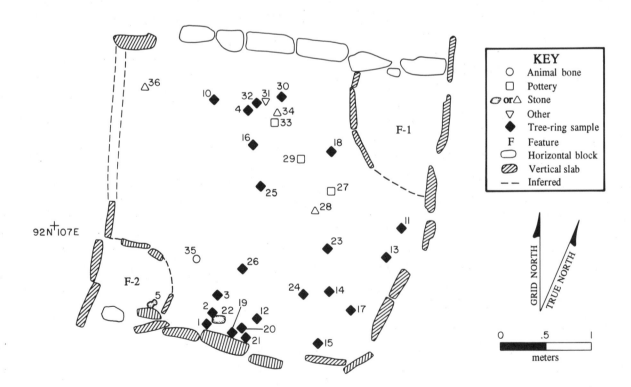

Figure 2.16. Room 11, roof fall. Point-located items are described in Table 2.12.

Stratum 1 were the remains of the burned roof, Stratum 2, which consisted of a layer of strong brown silt loam with concentrations of charred sediment and charred beams. The roof fall stratum lay directly on the floor of the room. Artifacts from the upper fill (Stratum 1) include sherds, chipped-stone debris, a small amount of animal bone, and a peckingstone.

Point-located items from the roof fall layer (Stratum 2) are mapped in Figure 2.16 and described in Table 2.12. These materials include sherds, animal bones, two peckingstones, an axe, a mano, a modified flake, and a small, highly polished fragment of turquoise that might be a piece of a pendant. Of the 78 point-located sherds, 1 is an Indeterminate White bowl sherd, and the remainder are Indeterminate Plain Gray jar sherds. Also shown in Figure 2.16 and listed in Table 2.12 are many of the tree-ring samples collected from roof fall, plus four from upper fill; these are described in more detail in the section on dating, later in this discussion.

Table 2.12. Point-Located Artifacts from Room 11, Roof Fall

No.	Description
Pottery Sherds	
27	gray ware jar (2)
29	gray ware jar (65)
	white ware bowl (1)
33	gray ware jar (10)
Chipped-Stone Debris	
9	missing; not mapped
Two-Hand Mano	
22	incomplete, sandstone
Peckingstone	
28	incomplete, Morrison orthoquartzite
34	condition unknown, Morrison orthoquartzite
Single-Bitted Axe	
5	complete, igneous
Modified Flake	
36	complete, Morrison orthoquartzite
Other Modified Stone/Mineral	
31	fragmentary, turquoise
Dating Samples	
1*–3*	tree-ring samples
4	tree-ring sample
6	tree-ring sample; not mapped
7*	tree-ring sample; not mapped
8	tree-ring sample; not mapped
10–21	tree-ring samples
23–26	tree-ring samples
30	tree-ring sample
32	tree-ring sample
Miscellaneous	
35	animal bone

NOTES: Item counts appear in parentheses. Refer to Figure 2.16 for artifact locations.
*Upper fill (Stratum 1) point-locations.

Construction

The north wall of Room 11 consists of three courses of horizontal-block masonry on top of a single course of double-row, slab-and-block masonry (Figure 2.17). The east wall, the south wall, and the southern end of the west wall are made of vertical sandstone slabs. However, most of the west wall is missing, and its location is inferred on the basis of the few rocks still in place and the location of the west wall of Room 7 to the north. The missing portion may have been built entirely of mud or of rock that later fell out of place when the room collapsed. All four walls are seated on earthen footings. It is likely that the upper walls for the entire room were made mostly of mud and stone. The roof must have been supported primarily by the structure walls, since no definite post holes were found during excavation. It is possible that two small pits (Features 14 and 15) held posts, but even if they did, it is not clear that the posts were used for roof support. Most likely, the floor in Room 11 was prehistorically excavated into sterile B-horizon sediments, but the surface was very irregular and difficult to define, which caused some problems in excavation and interpretation, as discussed below.

Surface

The irregularity of the floor in Room 11 may have been due in part to the number of pit features excavated into the floor and in part to the unevenness of the surface itself. Undulations of more than 5 cm in the floor surface made it necessary to evaluate the provenience of each item as it was found.

Artifacts and Samples. Floor-associated items that were point-located are shown in Figure 2.17 and listed in Table 2.13. The assemblage is composed almost entirely of gray ware jar sherds (total of 13), including 1 Moccasin Gray sherd. Several sherds, including the Moccasin piece, belong to a partial Moccasin Gray olla. A flotation sample from the floor of Room 11 yielded one groundcherry seed, possibly representing use of this plant as food, and several pieces of juniper and mountain mahogany wood, which may have been used in construction.

Features. Fifteen features were identified in Room 11, including 2 corner bins, a hearth, a large cist, and 11 miscellaneous pits, one of which probably served as a warming pit (Figure 2.17). The characteristics of these features are listed in Table 2.14. Feature 1, in the northeast corner of the room, is a slab-lined bin that was filled partly with burned roof fall and partly with naturally deposited sediments that accumulated on top of roof fall. The floor of the bin had been excavated prehistorically to 20 cm below the floor of the room, and vertical slabs were placed around the inside edges to form the bin wall. The west end of the

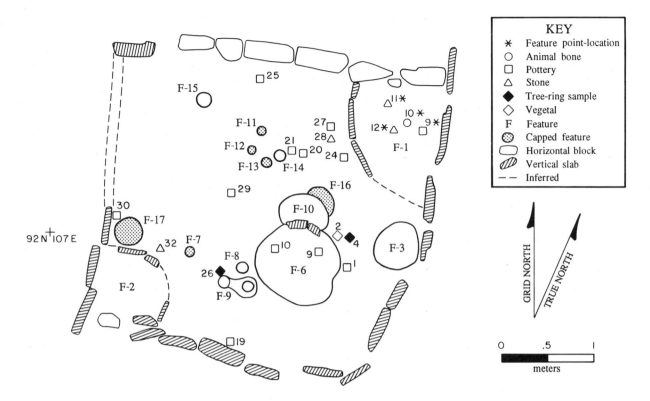

Figure 2.17. Room 11, surface. Point-located items are described in Table 2.13; features are described in Table 2.14.

feature had a sterile shelf approximately level with the room floor; in this area, the upright slabs had been set into a small trench. On the floor of the bin were three animal bones, a mano, a peckingstone, and three gray ware sherds that fit together to form a modified sherd (Figures 2.17 and 2.18). Flotation samples from the bin floor and fill yielded evidence of wood used in room construction (four-wing saltbush, juniper, pine, and mountain mahogany). Primarily on the basis of its size and shape, Feature 1 is interpreted as a storage bin, although it is not known whether it was used to store food or other items. It is possible that the artifacts found in direct contact with the floor were stored in the bin, but they may also have been discarded there after the original contents were removed.

Feature 2, the southwest corner bin, was filled with postabandonment deposits and did not contain any of the burned structural debris found in the rest of the room. The base of this bin was excavated approximately 5 cm above the floor of the room, and no artifacts were found in direct association with it. Again, the size and shape of this feature suggest a storage function, but the stored materials apparently were cleaned out before or during abandonment.

Feature 3 is a large, cylindrical cist near the east wall of the room. Two strata were recognized during excavation. The upper layer, Stratum 1, was approximately 10 cm thick and consisted of collapsed roofing material. Stratum 2 was a loose, organic-rich deposit with some charcoal and

numerous (more than 20) thin sandstone slabs, up to 25 cm in maximum dimension (not collected). Four peckingstones, a modified cobble, and 183 pieces of chipped-stone debris were among the items recovered from Stratum 2, and they are believed to have been stored, along with the sandstone slabs, in this feature. A flotation sample collected from Stratum 2 yielded several corn cupules, plus juniper, sagebrush, and four-wing saltbush charcoal, all indicative of fuel or roofing resources or both.

Feature 6 is a large hearth in the southeast quadrant of Room 11. The hearth is a deep, basin-shaped pit that was excavated into sterile subfloor sediments. Fill consisted of two strata, with the upper one, Stratum 1, being similar to the roof fall stratum of Feature 3. Stratum 2 consisted of dark, organic-rich fill with abundant ash and charcoal; it is interpreted as primary refuse associated with the use of the hearth. The feature had been badly disturbed by animal burrowing. Flotation samples were collected from Stratum 2 in areas that appeared less disturbed, and they yielded remains of charred food (cheno-am, prickly pear, groundcherry, and Indian rice grass) and fuel (juniper, pine, sagebrush, four-wing saltbush, and mountain mahogany). In addition, the only squaw apple/service berry wood identified in Duckfoot flotation samples was recovered from the hearth, where it is believed to have been used as fuel.

Feature 10 is a pit adjoining the north rim of the hearth. It was excavated prehistorically into the sterile subfloor and

Table 2.13. Point-Located Artifacts from Room 11, Surface

No.	Description
Pottery Sherds	
1	gray ware jar (1)
9	gray ware jar (1); Vessel 130 (1)
10	gray ware jar (1)
11	gray ware jar (1); Vessel 130 (1); not mapped
19	missing
20	gray ware jar (1)
21	gray ware jar (1)
24	gray ware jar (3); Vessel 130 (3)
25	gray ware jar (2)
27	gray ware jar (1)
29	missing
30	gray ware jar (1)
Chipped-Stone Debris	
28	flake (1)
32	missing
Modified Sherd	
9*	fragmentary, gray ware
Two-Hand Mano	
12*	complete, sandstone
Peckingstone	
11*	complete, Morrison orthoquartzite
Dating Samples	
4	tree-ring sample
26	tree-ring sample
Miscellaneous	
10*	animal bone
2	vegetal

NOTES: Item counts appear in parentheses. For pottery, sherds belonging to vessels are also included in total item count. Refer to Figure 2.17 for artifact locations.
*Feature 1 (bin) point-locations.

Table 2.14. Feature Summary, Room 11

No.	Type	Shape	Length (cm)	Width (cm)	Depth (Height) (cm)	Status
1	corner bin	other	130	120	20 (72)	in use
2	corner bin	other	90	85	(55)	unknown
3	cist	cylindrical	50	46	29	unknown
6	hearth	basin	93	90	20	in use
7	pit	basin	10	10	6	capped
8	pit	basin	13	12	9	in use
9	pit	other	42	24	21	in use
10	pit	other	56	36	14	in use
11	pit	basin	9	9	8	capped
12	pit	basin	8	8	6	capped
13	pit	basin	10	8	6	capped
14	pit	cylindrical	14	13	15	in use
15	pit	cylindrical	16	16	20	in use
16	pit	basin	36	12	~5	capped
17	pit	basin	30	30	12	capped

NOTE: Feature locations are shown in Figure 2.17.

5MT3868
ROOM 11
FEATURE 1
FLOOR
PD 452
22 OCT 86

Figure 2.18. View of artifacts on floor of bin, Feature 1, Room 11: two-hand mano (PL 12*), peckingstone (PL 11*), and modified sherd (PL 9*). See Figure 2.17.

is separated from the hearth by two sandstone slabs. The fill of Feature 10 was almost entirely a homogeneous, dark gray ash with fine charcoal inclusions. Two thin sandstone slabs lay near the surface of the ash. Feature 10 is interpreted to be a warming pit because of its contents and its association with the hearth. It is believed that hot coals were placed in the pit and covered with stones to retain heat and that food might have been warmed or cooked slowly on the stones. Indian rice grass remains, indicative of food preparation, were recovered from a flotation sample from the warming pit. Fuelwoods recovered from the same sample are similar to those found elsewhere in the room.

Features 14 and 15 are cylindrical pits that were partly filled with burned roof fall, which indicates that they were open at the time of abandonment. Feature 15 had a small sandstone slab set on edge on the south side of the pit. Slabs such as this often are seen alongside posts in post holes, where they are used as shims. In this case, there was no post, but if the building was intentionally destroyed, the post might have been deliberately removed from its seating pit in the floor. Feature 14, too, may have been a post hole, although it is rather shallow and, like Feature 15, did not contain a post.

Features 8 and 9 are two adjacent pits in the south half of the room. Like Features 14 and 15, they apparently were left open at abandonment. Feature 8 is a shallow basin.

Feature 9 consists of two distinct cylindrical pits joined by a shallow "trench." It is possible that Feature 9 was actually two separate features that extensive animal disturbance effectively joined into one.

Features 7, 11, 12, 13, 16, and 17 are basin-shaped pits that contained clean fill. All appear to have been intentionally capped. They might have served a variety of functions during the life of the structure but were no longer in use during the final stages of Room 13 occupation.

Dating

Thirty-four charred beam fragments were collected as tree-ring dating samples: 21 from roof fall, 5 from Stratum 1 deposits above roof fall, 2 from the floor, and the remainder from Features 1 and 3. Ten of these yielded dates, which ranged from A.D. 796 to 872 (Appendix B). The only two cutting dates, A.D. 839 and 844, are considered to be too early to represent the year in which the room was built and may instead indicate use of dead wood in construction. Given its construction style and wall-juncture characteristics, Room 11 is unlikely to predate Rooms 12 and 13. Room 13 yielded cutting dates in the mid- to late A.D. 850s. The four latest dates from Room 11 are noncutting dates of A.D. 853, 865, 871, and 872. These dates and others from Room Suite 2 and the site as a whole indicate that Room 11 was constructed in the late A.D. 850s. The three later dates may relate to wood used in later structure repairs or fuelwood left in the structure at abandonment.

Interpretations

Room 11 is interpreted as a domestic structure. The large hearth and associated warming pit indicate that food preparation and cooking took place in Room 11. The presence of two bins and a subfloor cist, plus the recovery of a variety of tools, particularly from features, lends further credence

to the interpretation of domestic use. Notably absent from the Room 11 floor assemblage, however, are metates, which are typically found in domestic structures. This absence may be explained by the presence of three metates (along with several manos, peckingstones, and modified cobbles) in the lower fill of adjacent back room 7. The Room 11 metates might have been tossed into Room 7 during the abandonment of these structures.

Room 12

Length: 3.60 m
Width: 3.30 m
Area: 11.75 m^2

Room 12 (Figures 2.19 and 2.20) is a medium-size front room that burned at abandonment. It is located between Room 11 to the east and Room 13 to the west and is associated with back room 6 (Figure 2.1). Initial excavations in Room 12 were part of a 2-m-wide exploratory trench, which eventually was widened to include the entire room. Artifacts on the floor were mapped, but point-location numbers were assigned retroactively after the implementation of the point-location proveniencing system. As a result, in most cases, specific artifacts cannot be linked with specific map locations.

Stratigraphy

Two strata were identified during excavation, both composed of a combination of wall fall and collapsed roofing material (Figure 2.19). However, the uppermost layer, Stratum 1, contained less organic material than Stratum 2 and probably consisted primarily of wall debris mixed with postabandonment deposits. Stratum 2, on the other hand, contained abundant ash, constructional mud, and burned wood and apparently consisted primarily of roof fall. On

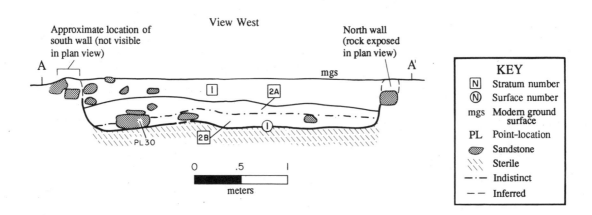

Figure 2.19. Room 12, stratigraphic profile. The location of the profile is shown in Figure 2.20. Feature 12 had not yet been excavated when the profile was drawn.

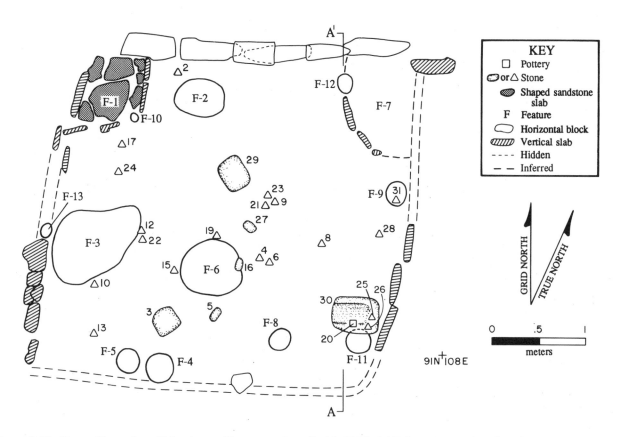

Figure 2.20. Room 12, surface. Point-located items are described in Table 2.15; features are described in Table 2.16. Profile A is shown in Figure 2.19.

Table 2.15. Point-Located Artifacts from Room 12, Surface

No.	Description	No.	Description
Pottery Sherds		6	complete, Morrison orthoquartzite
20	gray ware jar (10)	9	complete, Morrison orthoquartzite
Chipped-Stone Debris		10	complete, Morrison orthoquartzite
8	core (1)	12	fragmentary, Dakota orthoquartzite
14	edge-damaged flake (1); not mapped	13	complete, Morrison orthoquartzite
21	core (1)	15	complete, Morrison orthoquartzite
23	missing	19	complete, Morrison orthoquartzite
Trough Metate		28	complete, Morrison orthoquartzite
3	complete, igneous	31	complete, Morrison orthoquartzite
29	complete, sandstone	**Axe/Maul**	
30	complete, sandstone	7	fragmentary, igneous; not mapped
Two-Hand Mano		**Modified Cobble**	
2	fragmentary, conglomerate	26	complete, quartz
5	condition unknown, conglomerate	**Other Modified Stone/Mineral**	
16	incomplete, conglomerate	24	fragmentary, conglomerate
22	fragmentary, sandstone	**Gaming Piece**	
27	complete, sandstone	1	missing; not mapped
Abrader		**Miscellaneous**	
17	fragmentary, sandstone	25	pebble
Peckingstone			
4	complete, Morrison orthoquartzite		

NOTES: Item counts appear in parentheses. Refer to Figure 2.20 for artifact locations.

the basis of ash content and number of burned beam fragments, Stratum 2 was subdivided into two layers: Stratum 2A, which contained less ash and fewer pieces of charred wood, and Stratum 2B. Directly below Stratum 2B was the floor of the room.

Numerous artifacts were recovered from these deposits. The Stratum 1 assemblage includes a metate, a mano, a peckingstone, a hammerstone, a stone disk, a modified flake, and four human bone fragments. In addition, a nearly complete biface and a nearly complete drill, both made of materials not available locally, were recovered. Stratum 2 contained two metates, a mano, and several peckingstones, plus two pendants and a bone bead. Although it is likely that at least some of the Stratum 2 artifacts were deliberately stored or abandoned on the roof of the structure, none were point-located. Numerous animal bones, sherds, and pieces of chipped-stone debris were recovered from both strata. Flotation samples collected from two metates (one from Stratum 1, one from Stratum 2) yielded grass and corn remains that may have been ground for food or, alternatively, may have been incidentally incorporated into postabandonment sediments. Wood charcoal recovered from the metate-trough samples (cottonwood/willow, juniper, pinyon, sagebrush, and mountain mahogany) may represent construction wood.

Construction

The north and west walls of Room 12 are fairly well preserved and consist of horizontal blocks and vertical slabs, respectively (Figure 2.20). The north wall appears to have been seated on an earthen footing (Figure 2.19), and at most, two courses remain standing. The south and east walls are very poorly preserved. Three vertical slabs in alignment are all that remains of the east wall, and a single horizontal block on an earthen footing is all that remains of the south wall. The locations of the two walls are largely inferred. The upper walls were most likely of mud-and-stone construction. Five post holes were identified during excavation, and some or all of them probably held roof-support posts. Four of the post holes are large enough to suggest that the posts they contained carried the bulk of the weight of the roof.

Surface

The floor of Room 12, defined as the contact between sterile sediments below and collapsed structural debris above, had been prehistorically excavated into B-horizon sediments. Numerous artifacts and features were found in direct association with the floor.

Artifacts and Samples. Only some of the many artifacts found in direct association with the floor were actually point-located; locations of these items are shown in Figure

2.20 and listed in Table 2.15. Point-located items include 3 metates, 5 manos, 10 peckingstones, 1 axe or maul fragment, and several pieces of chipped-stone debris, including a number of cores and an edge-damaged flake. Only 10 pieces of pottery were assigned point-location numbers: 7 Moccasin Gray and 3 Indeterminate Plain Gray jar sherds, all of which fit together. Other sherds were found but not point-located. They too are gray wares, and one fits to a partial Moccasin Gray sherd container. A pollen sample collected from the floor just north of the hearth yielded Cheno-am, Umbelliferae, cholla, prickly pear, corn, and squash pollen.

Features. Thirteen features were excavated in Room 12: two corner bins, a hearth, five post holes, and five miscellaneous pits of unknown function (Figure 2.20 and Table 2.16). Features 1 and 7 are bins in the northwest and northeast corners of the room, respectively. The walls and floor of Feature 1 are lined with sandstone slabs. A post hole (Feature 10) in the southeast corner of the bin may have been associated with this feature but most likely held a roof-support post. The fact that the fill of Feature 1 consisted of roof fall and postabandonment deposits seems to indicate that the feature was left empty and open at abandonment. On the basis of its size and shape, it is interpreted as a storage bin. Feature 7 is not as well preserved as Feature 1; only three upright slabs remain in place to define the side facing the interior of the room. The base of Feature 7 is simply a shallow, unlined depression that probably was carved out through use and not deliberately constructed. Again, the pit appears to have been left empty and open at abandonment.

Feature 6 is a basin-shaped hearth that had been prehistorically excavated into sterile subfloor sediments. The

Table 2.16. Feature Summary, Room 12

No.	Type	Shape	Length (cm)	Width (cm)	Depth (Height) (cm)	Status
1	corner bin	other	80	77	(35)	in use
2	pit	basin	53	37	28	in use
3	pit	basin	113	74	15	in use
4	post hole	nr	30	30	nr	unknown
5	post hole	nr	27	22	nr	unknown
6	hearth	basin	65	57	15	in use
7	corner bin	other	96	75	nr	in use
8	post hole	nr	22	22	nr	unknown
9	post hole	nr	24	20	nr	unknown
10	post hole	nr	11	11	nr	in use
11	pit	nr	27	26	nr	unknown
12	pit	nr	21	17	nr	unknown
13	pit	nr	~16	~11	nr	unknown

NOTE: Feature locations are shown in Figure 2.20.
nr = no record.

lower fill consisted of gray ash and charcoal associated with the use of the hearth, and the upper fill consisted of mixed postabandonment deposits and cultural refuse. Flotation samples collected from the lower hearth fill yielded food (cheno-am, groundcherry, grass, corn) and fuel (juniper, pine, sagebrush, four-wing saltbush) remains.

Features 4, 5, 8, and 9 are fairly large post holes located near the edges of the room, and they are believed to have held roof-support posts. Features 4 and 5 contained rocks (including two peckingstones in Feature 4) that may have been used as shims. Feature 10 is a much smaller post hole incorporated into the side of the northwest corner bin, Feature 1; it contained the remains of a burned post. Feature 10 may be related to bin construction, or it might have held an auxiliary roof-support post.

The remaining features (Features 2, 3, 11, 12, and 13) are pits of unknown function. Features 2 and 3 are large basins that were filled with postabandonment deposits and collapsed structural debris, respectively. Features 11, 12, and 13 are smaller and may have functioned as post holes, but the physical evidence was not convincing enough to allow the inference of such a specific function.

Dating

Six tree-ring samples were collected from Stratum 1, and five were collected from Stratum 2. Ten produced dates, which range from A.D. 776 to 854 (Appendix B). All are noncutting dates. Given the dates from Room Suite 2 as a whole and the construction sequence inferred for this portion of the roomblock, the A.D. 854 date is the only one likely to come close to the actual date of construction. Room 13, to the west, yielded several samples with cutting dates in the mid- to late A.D. 850s, including an A.D. 857 cutting date. Because Room 12 appears to have been built no earlier than Room 13, a late A.D. 850s construction date seems plausible.

An archaeomagnetic dating sample was collected from the side of the hearth, Feature 6. Given architectural, pottery, and tree-ring sample evidence, the date estimates provided by this sample (pre-A.D. 700, A.D. 940 to 1010, and A.D. 1275 to post-1425) are not plausible.

Interpretations

With a hearth, two storage bins, cores, peckingstones, and numerous food-preparation tools (manos, metates) on its floor, Room 12 is interpreted to have been a domestic, or habitation, structure in which numerous everyday activities took place. The abundance and variety of materials found in the collapsed roof fall stratum suggest that items may have been stored on the roof of this structure as well. Like most of the other front rooms in Room Suite 2, Room 12 burned at abandonment.

Room 13

Length: 4.80 m
Width: 2.75 m
Area: 13.67 m²

Room 13 (Figures 2.21–2.23) is a large front room that burned at abandonment. It is located between Room 14 to the west and Room 12 to the east and is associated with Rooms 4 and 5 to the north (Figure 2.1). The room was excavated in two arbitrary horizontal subdivisions to allow pedestrian and wheelbarrow traffic across the roomblock. Fill from the eastern two-thirds of the room was removed first, then fill from the western third was excavated as a separate unit. The room was excavated as two strata down to 5 cm above the floor. Artifacts and samples collected within 5 cm of floor contact were provenienced as floor-associated, and floor-contact items were point-located.

Stratigraphy

The fill of Room 13 consisted of three strata (Figure 2.21). Stratum 1 was the dark brown soil A horizon. It graded into Stratum 2A, which consisted of a strong brown silt loam with clumps of caliche. Stratum 2B was similar to 2A, but it contained more charcoal and caliche inclusions.

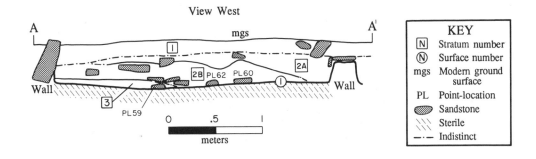

Figure 2.21. Room 13, stratigraphic profile. The location of the profile is shown in Figure 2.23.

Stratum 3, which did not extend over the entire room, had even more gray caliche inclusions. Sandstone rubble was abundant throughout Strata 2 and 3, which are interpreted as collapsed wall and roof materials mixed with postabandonment sediments. There was sufficient charcoal throughout fill to suggest that the structure burned, though not intensely.

Artifacts recovered from Stratum 1 include sherds, chipped-stone debris, animal bones, a peckingstone, and a modified flake. Artifacts recovered from Strata 2 and 3 include items that might have been on the roof when the structure burned. Among these items are four manos, a maul, and a peckingstone (Figure 2.22 and Table 2.17). Numerous burned beam fragments were mapped and collected as tree-ring samples; most, but not all, were assigned point-location numbers. Other materials found in Strata 2 and 3 but not believed to be directly associated with the roof surface include animal bones, sherds, chipped-stone debris, a biface, and a nearly complete projectile point made of nonlocal material.

Construction

The north wall of Room 13 consists of large, elongate blocks of horizontal masonry seated on sterile earthen footings; for the most part, only one course remains (Figure 2.23). The east wall has vertical sandstone slabs in its northern third, elongate sandstone blocks set on end (vertically) in the southern third, and a gap in the middle. The extant portion of the south wall includes blocks set on end, but sections of this wall are missing as well. The west wall has two vertical slabs at its north end, but the remainder is composed of small, coursed sandstone blocks, one course high for most of the length of the wall. The upper walls probably were of mud-and-stone construction. Two post holes located near the north wall of the room may have held roof-support posts, but since there are only two of them and both are rather small, the weight of the roof must have been carried primarily by the structure walls. The floor in Room 13 was excavated into sterile B-horizon sediments.

Surface

The living floor of Room 13 was defined as the stratigraphic break between underlying sterile sediments and overlying roof fall. The floor undulated, particularly in the area north and northwest of the hearth, but the many artifacts scattered across the room helped to define the floor level.

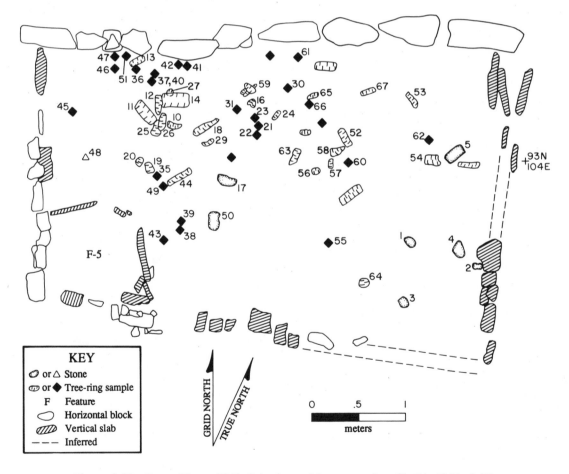

Figure 2.22. Room 13, roof fall. Point-located items are described in Table 2.17.

Artifacts and Samples. Diverse artifacts were scattered throughout the floor zone, but only those believed to be in direct association with the floor were point-located (Figure 2.23 and Table 2.18). Included in the point-located assemblage are sherds, chipped-stone debris, and a variety of formal tools: two metates, eight manos, and seven peckingstones. An edge-damaged flake and a core are among the pieces of chipped-stone debris. A pollen sample collected from a slab metate (PL 41) yielded Cheno-am, prickly pear, Umbelliferae, and corn pollen. Not point-located, mapped, or collected were numerous sandstone slabs, some shaped and others not, but all apparently in deliberate arrangements on the floor. A stack of thin (approximately 5 cm), shaped, rectangular slabs was found in the northwest corner. Unshaped slabs were found in two clusters, one surrounding two metates (PLs 41 and 42) and the other northwest of the hearth near four manos (PLs 56, 59, 60, and 62) and several sherd clusters (PLs 53, 54, 61, and 71). These slab piles did not appear to be structural collapse, because they lay directly on the floor and several rested on each other. Numerous artifacts found on top of these stones were provenienced as floor-associated items, even though they were more than 5 cm above the floor.

Of the 276 point-located sherds on the floor of Room 13, 273 (98.9 percent) are from gray ware jars and 3 (1.1 percent) are from white ware bowls. Included in the gray ware assemblage are 1 Chapin, 20 Moccasin, and 1 Mancos gray sherds; the remaining gray wares are indeterminate as to type. All three white ware sherds are Piedra Black-on-white. Although sherds were abundant and many were large, not a single reconstructible vessel is included in the floor assemblage. However, one Chapin Gray rim sherd that is probably from a dipper appears to have been recycled as a sherd container.

Features. Room 13 contains six features (Figure 2.23 and Table 2.19). Feature 1 is a basin-shaped, slab-lined hearth that contained ash and charcoal in lower fill (Stratum 2) and roof fall in upper fill (Stratum 1), which indicates that it was left open at abandonment. The ash and charcoal layer is interpreted as primary refuse related to the use of the hearth. A flotation sample from Stratum 2 yielded the usual variety of potential food items, such as cheno-am and groundcherry seeds, corn kernels, and Indian rice grass, and also contained tansymustard and purslane seeds, not as common in Duckfoot deposits. Charred remains of mountain mahogany, juniper, pine, and corn (cupules) suggest fuel use.

Feature 2 is a shallow, basin-shaped pit lined with small, irregular sandstone slabs. A complete mano was found slumped in the depression, as if it had been used to cover the pit. However, the feature was filled with roof fall, which suggests that it was empty before abandonment; falling roof debris may have dislodged the mano "lid." Mostly nonfood items were identified in a flotation sample collected from Feature 2: mountain mahogany/sagebrush, juniper, pine, and corn (cupule) remains are typical Duckfoot fuel resources. The one unusual item recovered from this sample was a sunflower-type fruit. This item may represent food, or it may have been fortuitously introduced into site deposits.

Features 3 and 4 are deep, cylindrical post holes near the north wall of the room. Both contained loose fill with abundant decayed organic matter, probably the remains of rotted posts. Toward the top of each post hole were charred fragments that may have been the remains of the burned portions of the posts at floor level. Both post holes appeared to have been in use at abandonment. Although the logical inference is that the holes held auxiliary roof-support posts, the two features are rather closely spaced (30 cm apart) to have been used simultaneously.

Feature 5 is a slab-lined bin in the southwest corner of Room 13. The floor of the feature was excavated prehistorically 30 to 32 cm into sterile deposits below the floor of the room; the sandstone slabs forming the feature wall rise 54 cm above the Room 13 floor. No artifacts were recovered from the bin floor. The fill within 10 cm of the floor was wind- and water-laid sediment that may have accumulated during the use of the feature. The upper portion of the feature contained clumps of silt loam, charcoal, and sandstone. Some of the slabs from the bin wall had collapsed into the fill as well. No pass-through

Table 2.17. Point-Located Artifacts from Room 13, Roof Fall

No.	Description
Mano	
50	missing
Two-Hand Mano	
3	fragmentary, conglomerate
4	fragmentary, sandstone
5	complete, sandstone
17	complete, sandstone
Peckingstone	
48	complete, Morrison orthoquartzite
Maul	
2	complete, igneous
Modified Cobble	
1	fragmentary, igneous
Dating Samples	
6–9	tree-ring samples; not mapped
10–14	tree-ring samples
15	tree-ring sample; not mapped
16	tree-ring sample
18–27	tree-ring samples
28	tree-ring sample; not mapped
29–31	tree-ring samples
32–34	tree-ring samples; not mapped
35–47	tree-ring samples
49	tree-ring sample
51–67	tree-ring samples

NOTE: Refer to Figure 2.22 for artifact locations.

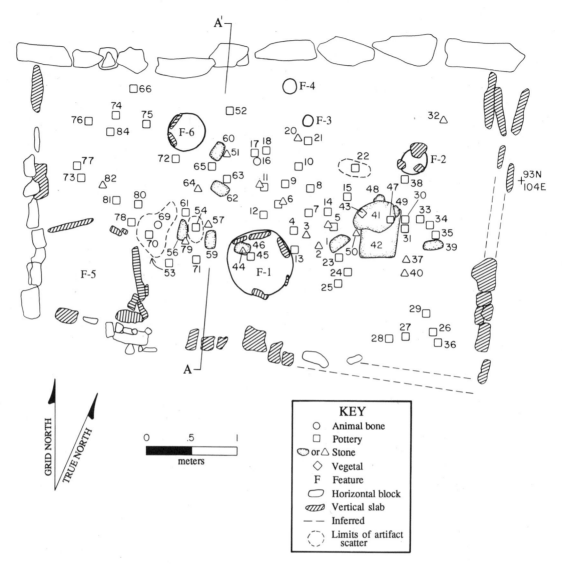

Figure 2.23. Room 13, surface. Point-located items are described in Table 2.18; features are described in Table 2.19. Profile A is shown in Figure 2.21.

was apparent in the sides of the bin, so the opening must have been through the top. The depth of the bin (approximately 85 cm from the top of the slabs to the bin floor) would have made it difficult to retrieve loose items from the bottom of the feature; therefore, if food was stored in the bin, it may have been in containers such as pots or baskets.

Feature 6 is a basin-shaped pit that was filled with roof fall and a few irregular sandstone slabs. Its function is unknown, and it is believed to have been left open when the room was abandoned.

Dating

Seventy tree-ring samples from Room 13 were submitted for dating, 67 from Strata 2 and 3, and 1 each from the floor, Feature 5, and Feature 3. The range for the 32 dated

samples is A.D. 774 to 866 (Appendix B). Sixteen cutting dates were obtained: one A.D. 847, one A.D. 849, one A.D. 850, one A.D. 852, one A.D. 854, four A.D. 855, four A.D. 856, two A.D. 857, and one A.D. 866. If these specimens are from structural components in Room 13, as is likely, then the cluster of dates between A.D. 855 and 857 supports the inference that Room 13 was built during or very soon after A.D. 857. The A.D. 866 date may be from wood used in minor remodeling or from fuel left in the structure at abandonment.

Interpretations

Room 13 is interpreted as a domestic structure associated with Rooms 4 and 5. The artifacts and features reflect a wide range of activities, including food preparation

Table 2.18. Point-Located Artifacts from Room 13, Surface

No.	Description	No.	Description
Pottery Sherds		75	gray ware jar (5)
4	gray ware jar (5)	76	white ware bowl (1)
5	gray ware jar (7)	77	gray ware jar (5)
6	gray ware jar (2)	78	gray ware jar (2)
7	gray ware jar (13)	80	gray ware jar (4)
8	gray ware jar (5)	81	gray ware jar (2)
9	gray ware jar (20)	83	gray ware jar (1); not mapped
10	gray ware jar (2)	84	gray ware jar (2)
	gray ware other (1); Vessel 55 (1)	**Chipped-Stone Debris**	
11	gray ware jar (1)	5	flake (1)
12	gray ware jar (2)	6	flake (1)
13	gray ware jar (1)	20	edge-damaged flake (1)
14	gray ware jar (1)	32	flake (1)
15	gray ware jar (1)	58	flake (1); not mapped
17	gray ware jar (1)	79	flake (1)
18	gray ware jar (1)	82	core (1)
21	gray ware other (1)	**Trough Metate**	
22	gray ware jar (3)	42	complete, sandstone
23	gray ware jar (2)	**Slab Metate**	
24	gray ware jar (1)	41	complete, sandstone
25	gray ware jar (4)	**Mano**	
26	gray ware jar (1)	46	missing
27	gray ware jar (1)	**Two-Hand Mano**	
28	gray ware jar (2)	1	complete, sandstone
29	gray ware jar (4)	39	complete, conglomerate
30	gray ware jar (2)	49	complete, sandstone
31	gray ware jar (1)	56	complete, sandstone
33	gray ware jar (1)	59	incomplete, sandstone
34	gray ware jar (1)	60	complete, sandstone
35	gray ware jar (1)	62	incomplete, sandstone
36	gray ware jar (1)	**Abrader**	
38	gray ware jar (2)	37	condition unknown, sandstone
45	gray ware jar (1)	**Peckingstone**	
47	gray ware jar (36)	3	complete, Dakota orthoquartzite
	white ware bowl (1)	11	complete, Morrison orthoquartzite
52	gray ware jar (5)	44	complete, Dakota orthoquartzite
53	gray ware jar (52)	50	complete, Dakota orthoquartzite
54	gray ware jar (23)	51	complete, Morrison orthoquartzite
55	gray ware jar (2); not mapped	57	complete, Morrison orthoquartzite
	white ware bowl (1); not mapped	67	complete, Morrison orthoquartzite; not mapped
61	gray ware jar (6)	**Modified Cobble**	
63	gray ware jar (1)	2	complete, sandstone
65	gray ware jar (1)	40	complete, igneous
66	gray ware jar (1)	48	complete, igneous
68	gray ware jar (5); not mapped	64	complete, unknown stone
70	gray ware jar (9)	**Miscellaneous**	
71	gray ware jar (4)	16	animal bone
72	gray ware jar (4)	69	animal bone
73	gray ware jar (1)	43	vegetal
74	gray ware jar (10)		

NOTES: Item counts appear in parentheses. For pottery, sherds belonging to vessels are also included in total item count. Refer to Figure 2.23 for artifact locations.

Table 2.19. Feature Summary, Room 13

No.	Type	Shape	Length (cm)	Width (cm)	Depth (Height) (cm)	Status
1	hearth	basin	70	70	35	in use
2	pit	other	31	25	9	in use
3	post hole	cylindrical	11	10	17	in use
4	post hole	cylindrical	13	12	25	in use
5	corner bin	other	110	90	31(54)	in use
6	pit	basin	47	40	15	in use

NOTE: Feature locations are shown in Figure 2.23.

(metates, manos), cooking (hearth), and storage (bin). Several features appear to have been emptied of their contents before the structure burned at abandonment, but many of the floor artifacts were left in place and sealed by the burning roof. Apparently some refuse was allowed to accumulate on the floor during the period of abandonment, as indicated by the abundance of broken pottery that did not refit to form vessels.

Room 14

Length: 2.60 m
Width: 1.50 m
Area: 3.85 m^2

Room 14, located between Room 13 to the east and Room 15 to the west, is a very small structure that was not burned at abandonment (Figure 2.24). It is regarded as a "front" room, although after Room 17 was added to the south side of the roomblock, Room 14 technically became a "middle" room, wedged between Room 17 to the south and Room 4 to the north (Figure 1.5). Room 14 was excavated as a single horizontal unit.

Stratigraphy

Two strata were defined in Room 14. Stratum 1 consisted of collapsed wall debris; Stratum 2, of silt loam that apparently was deposited on the floor after the room was abandoned but before the walls deteriorated. Materials recovered from Stratum 1 include sherds, chipped-stone debris, one hammerstone, a piece of worked gypsum, several animal bones, and several isolated human bone fragments. Stratum 2 contained sherds, chipped-stone debris, and a peckingstone, among other items.

Construction

All four walls are well preserved (Figure 2.24). The north and south walls are made entirely of horizontal sandstone blocks; only the basal course remains. The east wall consists of an alignment of horizontal blocks and vertical

slabs, with the former predominating. The construction of the west wall is similar, except that vertical slabs predominate. No post holes were present, and the roof probably was supported by the structure walls. The floor was excavated prehistorically into sterile B-horizon sediments.

Surface

The floor of Room 14 was defined as the contact between sterile sediments below and postabandonment fill above. Two Moccasin Gray jar sherds, 30 Indeterminate Plain Gray jar sherds, a peckingstone, a modified sherd, and a mano were recovered from the floor but were not point-located. There were no features in Room 14.

Dating

No tree-ring samples were collected from Room 14, but given its location with regard to other surface rooms and particularly to adjacent Room 13, it was probably constructed in the late A.D. 850s.

Interpretations

Despite its location in the front row of the roomblock, Room 14 is interpreted as a storage room. The lack of features and paucity of floor artifacts are typical of non-

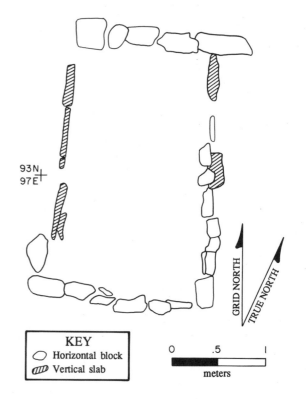

93N
97E

KEY
◯ Horizontal block
⬮ Vertical slab

GRID NORTH
TRUE NORTH

0   .5   1
meters

Figure 2.24. Room 14, surface. Artifacts were not mapped; refer to text for a description of floor-associated items.

habitation rooms in Pueblo I sites. Also, the room is too small to have served as a living room. Sometime during the occupation of the site, Room 17 was added to the south of Room 14, but there is no indication that this addition altered the function of the latter.

Room 17

Length: 2.75 m
Width: 2.00 m
Area: 5.22 m^2

Room 17 (Figure 2.25) is a small structure that protrudes from the south side of the Duckfoot roomblock, in front of Room 14 (Figure 1.5); it apparently was left empty at abandonment but was not burned. The structure appears to be a later addition to the roomblock, but no tree-ring samples were recovered, and it is not possible to date this construction precisely. Room 17 is at the boundary between Room Suites 1 and 2, and there is some question as to which side of Room 17 the line separating room suites should be drawn. The structure was excavated as a single horizontal unit.

Stratigraphy

The fill of Room 17 consisted of a single stratum of mixed natural and cultural postabandonment deposits. The organic-rich silt loam matrix contained flecks of charcoal and debris that might have derived from the collapsed walls and roof. Sherds, chipped-stone debris, animal bone, two metates, and two peckingstones were recovered from fill. The context of the artifacts, particularly the metates, was ambiguous, because the room had been built on top of a large pit feature (Feature 18) associated with Courtyard 2. The characteristics of pit fill made it difficult to determine the exact location of the room floor. The metates may have been abandoned on the structure roof, which later collapsed into the room. One of the metates is a massive piece of sandstone with a well-defined trough, but large, deep pits (5 cm in diameter by 2 to 3 cm deep) pecked into the surface had effectively destroyed the grinding surface. A flotation sample recovered from the second metate yielded sage-brush, juniper, pine, and corn remains. Because the metate contexts are ambiguous, these plant remains are difficult to interpret, but they probably are debris from postaban-donment fill rather than materials associated with the use of the tools.

Construction

Extant segments of the Room 17 walls consist of a combi-nation of vertical slabs and horizontal blocks (Figure 2.25). The north wall is made entirely of the latter, and only one horizontal course remains. The east and west walls incor-porate both vertical-slab and horizontal-block construction.

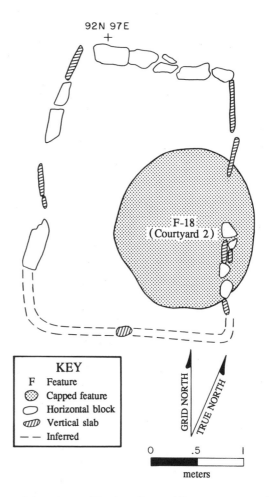

92N 97E

F-18
(Courtyard 2)

KEY
F   Feature
⊙   Capped feature
⬭   Horizontal block
▱   Vertical slab
— — Inferred

GRID NORTH

TRUE NORTH

0   .5   1

meters

Figure 2.25. Room 17, plan. Room 17 was constructed on top of Feature 18, which is associated with Courtyard 2.

The upper walls were probably made of mud and stone. Although the south wall is shown as an inferred line in Figure 2.25, it is possible that the room never had a south wall but, rather, was left open on this side. No post holes were present in Room 17, and the walls probably bore the weight of the roof without any auxiliary support. The floor, which had been excavated prehistorically into sterile B-horizon deposits, sloped downward to the southeast. The eastern portion of the room was constructed on top of a large pit feature associated with Courtyard 2. It is possible that this pit was deliberately filled to allow construction of Room 17.

Surface

A portion of the floor of Room 17 was a use-compacted surface defined by the stratigraphic break between sterile B-horizon sediments below and mixed postabandonment deposits above. Fewer than 100 sherds (gray, white, and red wares) and a small amount of chipped-stone debris were recovered from the floor zone, but no items were interpreted to be in direct association with the floor, and therefore none were point-located. No features were associated with the floor of Room 17. The large pit feature under the floor predates Room 17 and apparently was associated with the surface of Courtyard 2 (Feature 18 of Courtyard 2). That this feature was effectively sealed by the construction of Room 17 supports the inference that Room 17 was a late addition to the roomblock.

Dating

No tree-ring samples were collected from Room 17, but the construction of the room on top of an earlier Courtyard 2 feature indicates that the room was built sometime after initial site construction. Exactly how long after cannot be determined. The other main architectural addition to the Duckfoot site is Pit Structure 4, which, on the basis of tree-ring cutting dates, is believed to have been built in approximately A.D. 873. Perhaps Room 17 was built at the same time.

Interpretations

Room 17, like Room 14, is unusual for a front room because of its small size, paucity of artifacts, and lack of domestic features, such as a hearth or storage bin. It is interpreted as a storage room that was added to the roomblock sometime after initial site construction. This interpretation assumes that the room was completely walled. On the other hand, if the southern end of the room was left open, which is a possibility, then Room 17 would not have been suitable for storage, particularly food storage. In this case, the structure may have served as a partly enclosed ramada or work area.

Room 4

Length: 3.60 m
Width: 2.50 m
Area: 8.93 m^2

Room 4 (Figures 2.26 and 2.27) is the westernmost back room in Room Suite 2 (Figure 2.1). It is located between Room 3 to the west and Room 5 to the east and appears to be associated with Room 13 to the south. It is a typical back room only in terms of its small size. In terms of its feature and artifact assemblages, Room 4 is *not* typical, and it appears to have changed at some point from a storage room to a domestic room. The floor was littered with debris, and the roof was not burned. The lower 10 cm of fill was excavated as two arbitrary 5-cm levels. Materials from 5 to 10 cm above the floor were provenienced as Stratum 2-2, and materials from 0 to 5 cm, including those in contact with the floor, were provenienced as "surface" (see discussion of stratigraphy below).

Stratigraphy

The fill of Room 4 consisted of two strata differentiated more by soil-development processes than by differences in deposition (Figure 2.26). Stratum 1, the soil A horizon, was a 15-cm-thick layer of dark brown, organic-rich loam. Stratum 2 consisted of a strong brown silt loam with numerous sandstone blocks throughout. Both strata were derived from collapsed wall and roof materials that had been subjected to weathering and soil-development processes. Stratum 2 lay on the floor of Room 4. An ash lens associated with Feature 1 on the floor of the room was visible in profile at the contact between the floor and overlying Stratum 2.

Diverse materials were recovered from both fill strata. Sherds, chipped-stone debris, animal bone, a few human bone fragments, and a projectile point are among the items found in Stratum 1. Materials recovered from Stratum 2 include sherds, chipped-stone debris, and several isolated human bones. Formal tools found in Stratum 2 include a metate, a mano, several peckingstones, two projectile points (including one made of material not available locally), and a bone gaming piece. A flotation sample collected from the trough of the metate yielded grass and groundcherry remains, as well as juniper and mountain mahogany wood.

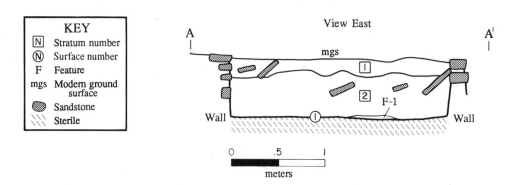

Figure 2.26. Room 4, stratigraphic profile. The location of the profile is shown in Figure 2.27.

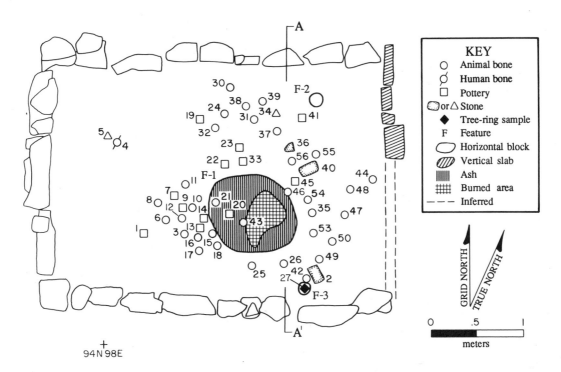

Figure 2.27. Room 4, surface. Point-located items are described in Table 2.20; features are described in Table 2.21. Profile A is shown in Figure 2.26.

Table 2.20. Point-Located Artifacts from Room 4, Surface

No.	Description	No.	Description
Pottery Sherds		Modified Cobble	
1	gray ware jar (11)	34	fragmentary, igneous
7	missing	Dating Samples	
9	gray ware jar (2)	27	tree-ring sample
13	gray ware jar (1)	Miscellaneous	
14	gray ware jar (1)	4	human bone
19	red ware bowl (1)	3	animal bone
20	gray ware jar (1)	6	animal bone
22	red ware bowl (1)	8	animal bone
23	gray ware jar (2)	10–12	animal bone
33	red ware bowl (1)	15–18	animal bone
41	gray ware jar (1)	21	animal bone
45	gray ware other (1)	24–26	animal bone
Two-Hand Mano		30–32	animal bone
2	complete, sandstone	35	animal bone
36	fragmentary, sandstone	37–39	animal bone
40	complete, conglomerate	42–44	animal bone
Abrader		46–50	animal bone
5	fragmentary, sandstone	53–56	animal bone

NOTES: Item counts appear in parentheses. Refer to Figure 2.27 for artifact locations.

Construction

The extant portions of the north, south, and west walls of Room 4 are made of massive, elongate sandstone blocks seated on sterile footings; in places, the walls stand two to three courses high (Figure 2.27). The north half of the east wall consists of huge vertical slabs and mud. The south half of the east wall apparently was built entirely of mud, as no stones were found in place. The upper walls around the room may have been constructed of a combination of coursed masonry and mud. Two small, circular pit features, one in the northeast corner and one in the southeast, are in the right locations for post holes. However, the characteristics of both were too ambiguous to infer specific functions. It is assumed that the roof was supported primarily, and perhaps entirely, by the structure walls. The floor of Room 4 was a use-compacted surface excavated into sterile B-horizon sediments.

Surface

The living floor of Room 4 was a use-compacted surface at the stratigraphic break between underlying sterile sediments and overlying cultural and postabandonment fill. It was littered with burned debris and ash related to the use of Feature 1, a burned spot just southeast of the center of the room.

Artifacts. Materials found directly on the floor were assigned point-location numbers, and these include three manos and relatively small amounts of pottery and chipped-stone debris (Figure 2.27 and Table 2.20). An isolated human rib was recovered from the northwest quadrant of the floor. Noteworthy is the quantity of small-mammal bones and bone fragments, many of them burned and crushed. These were scattered across the floor, particularly in the area around Feature 1. Much of the bone is identified simply as small mammal, but of the specimens that could be identified more precisely, many are hare or jackrabbit. A long-bone shaft that is ground and polished along its length and perforated at one end was found within 5 cm of the floor.

Of the 23 point-located and analyzed sherds, 20 (87.0 percent) are gray ware, and 3 (13.0 percent) are red ware. The gray wares consist primarily of Indeterminate Plain Gray jar sherds, but three Moccasin Gray jar sherds were recovered as well. All three red ware sherds are Bluff Black-on-red bowl pieces.

Features. Three features were identified in Room 4 (Figure 2.27 and Table 2.21). Feature 1, a burned spot on the floor, consisted of an 8-cm-high mound of oxidized, or fire-reddened, sediments covered by a 2-cm layer of ash. The ash was scattered on the floor around the feature as well. Burned small-mammal bones were found throughout

Table 2.21. Feature Summary, Room 4

No.	Type	Shape	Length (cm)	Width (cm)	Depth (Height) (cm)	Status
1	burned spot	other	102	84	(10)	in use
2	pit	basin	15	14	7	in use
3	pit	other	11	9	15	in use

NOTE: Feature locations are shown in Figure 2.27.

the ash scatter on the surrounding floor. Feature 1 is interpreted as an area where at least one and possibly several fires were built, evidently to cook small mammals.

Feature 2 is a shallow pit in the northeast quadrant of Room 4. It was completely filled with large chunks of charcoal, but there was no evidence of in situ burning. Feature 3 is a small pit in the southeast corner. Its size and location suggest that it might have been a post hole, but there was no remnant of a post.

Flotation samples were collected from Feature 1, Feature 2, and the floor near Feature 3. Between them, the three contexts yielded the usual variety of potential fuel resources (juniper, pine, mountain mahogany, and corncob). Feature 2 also contained charred squawbush wood. Potential food resources include cheno-am seeds and Indian rice grass remains.

Dating

Four tree-ring samples were collected from Room 4 (one from Stratum 2, one from the floor, and two from Feature 2). Two of these produced dates (Appendix B). The floor sample yielded a date of A.D. 828; one of the Feature 2 samples dated to A.D. 789. Both are noncutting dates, and on the basis of evidence gathered from the site as a whole, both appear to be too early to represent construction of Room 4. Given its location in the roomblock, Room 4 was probably built in the late A.D. 850s, along with the rest of Room Suite 2.

Interpretations

On the basis of its small size and location in the back row of the roomblock, Room 4 is believed to have originally been used for storage, but sometime, probably late in the occupation of the site, one or several fires were built in the room, presumably to cook food. The abundance of small-mammal bones, the frequency of crushed bones, and the presence of burned bones in the ash surrounding Feature 1 support the interpretation that small mammals were being cooked and consumed in Room 4, and room function is believed to have changed or at least diversified to include a domestic component. Apparently, little effort was made to clean up the resulting debris before abandonment.

Room 5

Length: 3.10 m
Width: 2.10 m
Area: 6.16 m^2

Room 5 (Figure 2.28) is located between Room 4 to the west and Room 6 to the east, in the back row of the roomblock (Figure 2.1). It is associated with Room 13 to the south. Room 5 did not burn at abandonment. Like Room 4, this room has feature and artifact assemblages that are unusual for a back room, and this evidence is used to argue for a functional shift similar to that proposed for Room 4. Room 5 was excavated without horizontal subdivisions.

Stratigraphy

Three strata were defined in the fill of Room 5. Stratum 1 consisted of postabandonment sediments deposited through natural forces on top of Stratum 2, a layer of collapsed wall debris. Stratum 3 was composed of naturally deposited sediments that washed in on top of the floor before major structural collapse, although a few wall rocks were mixed in with the sediments. Sherds and chipped-stone debris were recovered from throughout room fill. In addition, Stratum 1 yielded a small amount of animal bone and a mano; Stratum 2 contained a modified flake and numerous animal bones, including many identified as small mammals; and Stratum 3 contained a peckingstone and two manos.

Construction

The walls of Room 5 are fairly well preserved, particularly the north and south walls, both of which consist of large, elongate blocks laid horizontally on sterile footings (Figure 2.28). The north wall, which stands a maximum of three courses high, rests on a footing that is approximately 20 cm tall. The south wall footing is much higher, 30 to 40 cm, and is topped by only one course. The east wall is made of a combination of vertical slabs and horizontal blocks faced with a 10-cm layer of mud. The north half of the west wall is made of large vertical slabs; the south half of this same wall is missing and presumably was made mostly of mud. The upper walls around the room are likely to have been of mud-and-stone construction. No post holes were identified in Room 5, which suggests that the roof was supported by the structure walls. The floor was excavated prehistorically into sterile B-horizon sediments.

Surface

The floor of Room 5 was defined as the contact between sterile deposits below and postabandonment fill above. It was marked by a fairly even scatter of artifacts and the presence of several features.

Artifacts and Samples. Materials point-located on the floor of Room 5 include scattered sherds and chipped-stone debris, plus a small amount of animal bone and four very small fragments of human bone (Figure 2.28 and Table 2.22). Formal tools include five manos, two peckingstones, a hammerstone, a modified flake, and a bone awl. A projectile point fragment made of nonlocal stone was found within 5 cm of the floor but was not point-located. Several shaped sandstone slabs that may have been related to feature, wall, or doorway construction are shown in Figure 2.28; these were not collected.

Twenty-six sherds were point-located on the Room 5 floor: 23 Indeterminate Plain Gray jar sherds, 1 Chapin Gray bowl sherd, and 2 Indeterminate San Juan Red bowl sherds.

A pollen sample collected from under one of the shaped sandstone slabs on the floor yielded Cheno-am, prickly pear, and corn pollen. The slab was located in the northwest corner of the room, and it may have been associated with Feature 3, described below.

Features. Four features were excavated in Room 5 (Figure 2.28 and Table 2.23). Feature 1 is a fire pit slightly south of the center of the room. It is a shallow, basin-shaped depression with an irregular bottom. The bottom and sides of the pit were fire-reddened, and fill consisted almost entirely of ash. Given its small size, informal construction, and lack of associated food remains or food-preparation artifacts, Feature 1 may have been used primarily for heating or lighting rather than for cooking.

Feature 2, a bin in the northeast corner of the room, was filled with postabandonment deposits, including collapsed structural debris. The walls of the bin are lined with sandstone slabs except where the coursed-masonry room walls serve to define the bin on its north and east sides. Another slab lines the floor of the bin. Although the feature is interpreted as a storage bin, apparently it was empty at abandonment.

The remaining two features, Features 3 and 4, are pits of unknown function. Feature 3 is an amorphous pit in the northwest corner of the room, and it may be associated with a bin that is no longer recognizable as a feature. As shown in Figure 2.28, several shaped slabs were jumbled together in this corner of the room, and a single slab remained upright, roughly perpendicular to the structure wall. If a slab-lined corner bin once existed in this location, Feature 3 may have been a pit dug below floor level as part of its construction. The original contents had been removed from the pit, and it is not known whether the mixture of sand and silt loam it contained had been placed there to seal the feature or had washed in naturally after abandonment. Feature 4 is a basin-shaped pit near the east wall of the structure that apparently was sealed intentionally before abandonment.

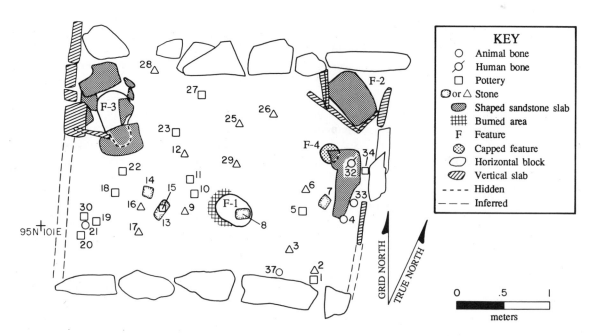

Figure 2.28. Room 5, surface. Point-located items are described in Table 2.22; features are described in Table 2.23.

Table 2.22. Point-Located Artifacts from Room 5, Surface

No.	Description	No.	Description
Pottery Sherds		Two-Hand Mano	
1	gray ware jar (2)	7	fragmentary, sandstone
5	gray ware jar (2)	8	fragmentary, sandstone
	gray ware bowl (1)	13	complete, sandstone
10	gray ware jar (1)	14	fragmentary, conglomerate
11	gray ware jar (1)	35	complete, sandstone; not mapped
15	gray ware jar (1)	Hammerstone	
18	gray ware jar (1)	9	complete, sandstone
19	gray ware jar (1)	Peckingstone	
20	gray ware other (1)	6	complete, Dakota orthoquartzite
22	red ware bowl (1)	25	complete, Dakota orthoquartzite
23	gray ware jar (1)	Modified Flake	
27	gray ware jar (2)	16	complete, Morrison orthoquartzite
	red ware bowl (1)	Modified Cobble	
30	gray ware jar (7)	12	fragmentary, igneous
34	gray ware jar (3)	Awl	
Chipped-Stone Debris		37	fragmentary, small mammal
2	flake (1)	Miscellaneous	
3	flake (1)	17	sandstone; missing
24	flake (1); not mapped	32	human bone
26	flake (1)	4	animal bone
28	missing	21	animal bone; missing
29	missing	31	animal bone; not mapped
36	missing; not mapped	33	animal bone

NOTES: Item counts appear in parentheses. Refer to Figure 2.28 for artifact locations.

Table 2.23. Feature Summary, Room 5

No.	Type	Shape	Length (cm)	Width (cm)	Depth (Height) (cm)	Status
1	fire pit	basin	37	24	4	in use
2	corner bin	other	67	56	(26)	in use
3	pit	basin	57	36	11	in use
4	pit	basin	21	17	12	capped

NOTE: Feature locations are shown in Figure 2.28.

Flotation samples were collected from all four features, and results of analysis are reported in Chapter 6. None contained any unusual materials; fuel resources found routinely in other Duckfoot contexts and a corn kernel that probably represents food were among the recovered plant remains.

Dating

No tree-ring samples were collected from Room 5. However, given its placement in the roomblock and its wall junctures with other rooms, Room 5 is clearly integrated architecturally with the other Room Suite 2 structures, which would place its construction in the late A.D. 850s.

Interpretations

Room 5 may well have started out as a typical Pueblo I storage room—its small size and location in the back row

of the roomblock suggest that this was the case—but if so, its function changed sometime during the occupation of the site to include a broader range of activities. The construction of a fire pit, at least one corner bin, and two other features, plus the variety of artifacts recovered, supports an interpretation of limited domestic activity (cooking, short-term storage) in the final stages of the room's use.

Room 6

Length: 3.20 m
Width: 2.40 m
Area: 7.53 m²

Room 6 (Figure 2.29) is a back room between Room 5 to the west and Room 7 to the east. It is associated with Room 12 to the south (Figure 2.1). Room 6 is small and contained few artifacts, typical for a back room. Although it does contain a feature, the feature probably is related to storage, which is consistent with the inferred function of the room. The room is not believed to have been intentionally destroyed. Burned vegetal material was noted in the fill of the room but is interpreted as originating from adjacent Room 12, which *was* burned at abandonment. Room 6 was excavated during the first field season at Duckfoot, and initial excavations in this room were in a north-south exploratory trench that crosscut Architectural Suite 2. These excavations eventually were expanded to include the entire room as a single horizontal unit. During these early excavations, artifacts occasionally were mapped, but the treatment was not systematic. Mapped artifacts were not

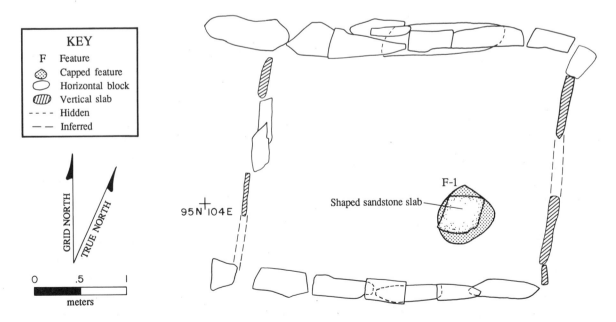

Figure 2.29. Room 6, surface. Artifacts were not mapped; refer to text for a description of floor-associated items and Feature 1.

assigned point-location numbers and therefore could not be tracked individually through laboratory cataloging and analysis.

Stratigraphy

The fill of Room 6 consisted mainly of collapsed structural debris. Wall rocks were abundant in upper fill (Stratum 1), and organic debris from the collapsed roof was more abundant in lower fill (Stratum 2). Stratum 1 contained numerous and diverse artifacts including two metates, several manos, a maul, two hammerstones, a polishing stone, several peckingstones, a projectile point, a modified flake, a bone pendant, a *Glycymeris* shell bracelet fragment, and a bone awl fragment. Sherds, chipped-stone debris, animal bones, and a single human bone were also recovered.

Fewer artifacts were recovered from Stratum 2, the roof fall zone. At least some of the items may have been used, stored, or abandoned on the roof. A polishing stone, a peckingstone, a nearly complete projectile point, several animal bones, chipped-stone debris (including an edge-damaged flake), and 129 pottery sherds were found in roof fall. One hundred twenty-four (96.1 percent) of the sherds are gray ware, four (3.1 percent) are red ware, and one (.8 percent) is white ware. Most are of indeterminate types, but 1 Chapin Gray, 13 Moccasin Gray, 1 Piedra Black-on-white, and 2 Bluff Black-on-red sherds are included in the assemblage. Both jar and bowl forms are present.

Construction

All four walls of Room 6 are very well preserved (Figure 2.29). The north and south walls are made of massive, elongate blocks laid horizontally on 25-cm-high sterile footings. The north wall has a maximum of three courses still standing, and the south wall has two. The east wall is marked by one horizontal block and three vertical sandstone slabs. The west wall consists of a combination of vertical slabs and horizontally coursed masonry. The upper walls probably were mostly of mud-and-stone construction. The roof would have been supported by the structure walls, since no post holes were identified. The floor was excavated prehistorically into sterile B-horizon deposits.

Surface

The floor of Room 6 was defined as the contact between sterile B-horizon sediments and postabandonment fill. Numerous artifacts are interpreted as being associated with the floor, but none were point-located. Floor artifacts include sherds, chipped-stone debris (including one core), six peckingstones, a maul, and two manos. Of the 45 sherds recovered, 42 (93.3 percent) are gray ware, 2 (4.4 percent)

are red ware, and 1 (2.2 percent) is white ware. The specific types identified are Chapin Gray (one), Moccasin Gray (two), Bluff Black-on-red (one), and Piedra Black-on-white (one). Both jar and bowl forms are represented.

The only feature in Room 6 is a large cist in the southeast quadrant of the floor. Feature 1 is 62 cm in diameter and 21 cm deep. It had been filled with charcoal-stained sediments and capped with a large sandstone slab. Although Feature 1 is thought to have been used as a storage pit, it was intentionally filled with sediment and capped before abandonment.

Dating

A tree-ring sample collected from the wall fall stratum yielded a noncutting date of A.D. 873 (Appendix B), which indicates that the tree from which it came could not have been cut any earlier than that year. This date is later than most other dates obtained for Room Suite 2 structures, and dates from the room suite as a whole suggest initial construction in the late A.D. 850s. If the Room 6 sample is from construction wood, it may indicate repair or remodeling of this room in the A.D. 870s. It is also possible that the sample was from wood used in repair or remodeling of adjacent Room 12, which burned at abandonment.

Interpretations

Room 6 is believed to be a storage room associated with living room 12 to the south. This interpretation is based on the size and location of the room, the small number of artifacts recovered, and the lack of a hearth or other thermal feature.

Room 7

Length: 3.50 m
Width: 2.60 m
Area: 8.92 m²

Room 7 (Figures 2.30–2.32) is the easternmost back room in Room Suite 2. It is located between Room 6 to the west and Room 8 to the east, and it is associated with Room 11 to the south (Figure 2.1). Like the other Room Suite 2 back rooms, Room 7 was not burned at abandonment. The fill of the room was excavated in two strata, without horizontal subdivisions. The bottom of Stratum 2 was arbitrarily divided into two 5-cm levels above the floor.

Stratigraphy

Two strata were recognized in the fill of Room 7. The upper one, Stratum 1, consisted of dark brown loam and rocks from the collapsed walls. The upper boundary of Stratum 2

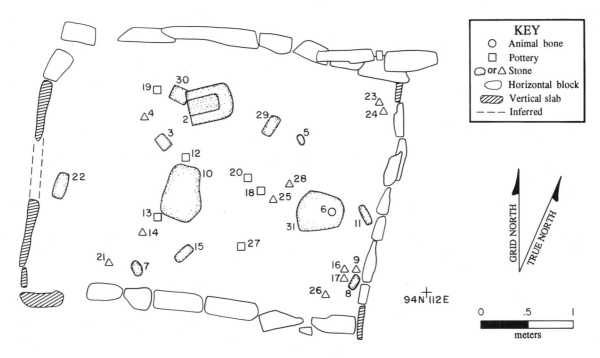

Figure 2.30. Room 7, roof fall. Point-located items are described in Table 2.24.

Table 2.24. Point-Located Artifacts from Room 7, Roof Fall

No.	Description	No.	Description
Pottery Sherds		29	complete, sandstone
12	gray ware jar (4)	**Peckingstone**	
13	red ware bowl (1); Vessel 113 (1)	17	complete, Morrison orthoquartzite
18	gray ware other (1)	24	complete, Dakota orthoquartzite
19	red ware bowl (2); Vessel 113 (2)	32	complete, Dakota orthoquartzite; not mapped
20	gray ware other (1)	**Modified Cobble**	
27	gray ware other (1)	5	fragmentary, sandstone
Chipped-Stone Debris		21	fragmentary, igneous
4	core (1)	25	fragmentary, Morrison orthoquartzite
9	core (1)	28	complete, igneous
26	core (1)	30	condition unknown, igneous
Trough Metate		**Other Modified Stone/Mineral**	
2	complete, sandstone	14	fragmentary, unknown stone
10	complete, sandstone	22	complete, conglomerate
11	fragmentary, sandstone	**Unmodified Cobble**	
31	complete, sandstone	23	complete, igneous
Two-Hand Mano		**Pendant**	
3	complete, sandstone	6	complete, medium carnivore
7	complete, igneous	**Miscellaneous**	
8	fragmentary, sandstone	16	pebble
15	complete, sandstone		

NOTES: Item counts appear in parentheses. For pottery, sherds belonging to vessels are also included in total item count. Refer to Figure 2.30 for artifact locations.

Figure 2.31. View of Room 7 after excavation. Note the construction of the north wall.

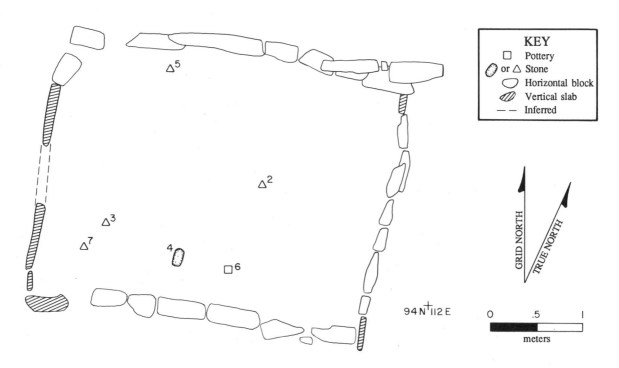

KEY
□ Pottery
◍ or △ Stone
◖ Horizontal block
▨ Vertical slab
– – Inferred

△5

△2

△3

△7

4 ◍

□6

GRID NORTH

TRUE NORTH

94N┼112E

0   .5   1

meters

Figure 2.32. Room 7, surface. Point-located items are described in Table 2.25.

was identified by a discontinuous layer of burned mud interpreted to be roof fall mixed with wall debris. The relatively small amount of burned debris (it was not found throughout Stratum 2) suggests that Room 7 itself was not burned at abandonment but that some wall and roofing materials from adjacent Room 11, which *was* burned, fell inside Room 7. Large wall rocks were abundant throughout Stratum 2.

Artifacts were relatively sparse in Stratum 1. Sherds, chipped-stone debris, three animal bone fragments, and one peckingstone are among the items recovered. Four metates, five manos, three peckingstones, three cores (chipped-stone debris), five modified cobbles, an animal tooth pendant, and a small number of sherds are among the point-located items from Stratum 2 (Figure 2.30 and Table 2.24). The most notable aspect of this assemblage is the number and diversity of stone tools. Most were recovered from lower fill, at approximately 10 cm above the floor. These items may have been stored on the roof and then deposited below when the roof deteriorated. Alternatively, the tools may have been dumped into Room 7 during or after abandonment. One potential source for these materials is Room 11, the associated front room, which was conspicuously lacking the grinding tools one would expect in a living room.

Of the 10 point-located sherds, 7 are Indeterminate Plain Gray jar sherds (including one handle piece). The remaining three belong to a partial Bluff Black-on-red bowl, whose other pieces also were recovered from Stratum 2 but were not point-located.

Construction

All four walls in Room 7 are well preserved. The north wall is unusual: it consists of three courses of large, elongate sandstone blocks placed horizontally on top of a single course of small vertical slabs (Figure 2.31). The line of vertical slabs is centered under the coursed-block masonry and is faced with 5 to 10 cm of mud. The east

wall is made mostly of mud-plastered vertical slabs and horizontal blocks, with the latter predominating. The stones are more exposed on the Room 8 side of this wall; the Room 7 side shows only mud. The south wall of Room 7 consists of three courses of horizontal-block masonry on top of a single course of double-row, slab-and-block masonry. The west wall is the least well preserved of the four, and it consists of one horizontal block and three vertical slabs plastered with mud. The upper walls around the room may have been a combination of horizontal masonry and mud. The roof must have been supported by the structure walls, since no post holes were located. The floor of the room had been excavated prehistorically into sterile B-horizon sediments and compacted through use.

Surface

The floor of Room 7 was a use-compacted surface defined at the contact between fill above and sterile sediments below. It had been disturbed by animal burrowing. Only a few artifacts were found in the floor zone, and only six of these were point-located (Figure 2.32 and Table 2.25). Point-located items include three stone tools (including two peckingstones) and four Indeterminate Plain Gray jar sherds found in a single point-located cluster near the south wall. No features were identified in Room 7, although it is possible that shallow features were destroyed by burrowing animals.

Dating

No tree-ring samples were collected from Room 7, but this structure is part of the core unit of the roomblock. On the basis of tree-ring dates from Room Suite 2 as a whole, Room 7 is inferred to have been constructed in the late A.D. 850s.

Interpretations

On the basis of size and placement in the back row of the roomblock, Room 7 is interpreted as a storage room associated with Room 11, the living room to the south. Although animal burrowing had disturbed the floor of the room, it is not likely that the items recovered from Stratum 2 (metates, manos, and cobbles) were displaced upward by this activity. It seems more likely that Room 7 was essentially empty when it began to deteriorate. The grinding tools and associated items in lower fill probably were dumped there during the period of abandonment or soon thereafter. Room 11 has features that are typical of a domestic, or habitation, room, but it lacks the food-processing tools (manos, metates) that generally accompany these features in a domestic room. The food-processing tools in the lower fill of Room 7 are probably those used in Room 11.

Table 2.25. Point-Located Artifacts from Room 7, Surface

No.	Description
Pottery Sherds	
6	gray ware jar (4)
Chipped-Stone Debris	
2	flake (1)
7	missing
Peckingstone	
3	condition unknown, Dakota orthoquartzite
5	fragmentary, Dakota orthoquartzite
Other Modified Stone/Mineral	
4	complete, sandstone

NOTES: Item counts appear in parentheses. Refer to Figure 2.32 for artifact locations.

Pit Structure 2

Length: 4.50 m
Width: 4.60 m
Area: 20.60 m^2

Pit Structure 2 (Figures 2.33–2.40) was the central pit structure at Duckfoot before the addition of Pit Structure 4 (Figures 1.5 and 2.1). It is the only pit structure that was not burned, although there is evidence to suggest that its roof was intentionally collapsed during abandonment. Two surfaces were defined in Pit Structure 2: the original living floor of the room (Surface 1) and a small area of hearth and ash pit overflow that effectively existed as a secondary surface in the immediate vicinity of these two features shortly before abandonment (Surface 2).

Because stratigraphic breaks were not discernible during excavation, the pit structure fill was excavated in arbitrary levels. Most of the fill levels were 30 cm thick down to 10 cm above the floor. The upper fill was excavated entirely with trowels and shovels without screening; within 10 cm of floor, all sediments were screened through ¼-in mesh. Initial excavations in Pit Structure 2 were part of an exploratory trench across Architectural Suite 2. The east wall of the pit structure was discovered in this trench, which was excavated to about 10 cm above the structure floor. Next, the west half of the structure, exclusive of a 50-cm-wide balk along the north-south center line, was excavated in levels comparable to those in the original trench. A stratigraphic profile was drawn, after which the balk was removed as a single unit from modern ground surface to 10 cm above floor. Once the fill of the structure had been removed to 10 cm above floor, a 1-×-1-m grid was established inside the structure. The 5- to 10-cm-above-floor level was excavated and screened in a series of 1-×-1-m squares, but the artifacts were grouped across the entire structure. Fill within 5 cm of floor was excavated and provenienced within the same series of 1-×-1-m squares, and items in contact with the floor were point-located.

Stratigraphy

Figure 2.33 presents a stratigraphic profile of Pit Structure 2. Four major abandonment and postabandonment events are represented in the five strata identified in profile. The layer immediately overlying the floor of Pit Structure 2, designated Stratum 5, consisted of a grayish brown, gravelly deposit interlaced with mottled, orangish brown fill. The latter contained clumps of strong brown and dark brown sediment and small chunks of charcoal. Stratum 5 was a massive deposit that, because it was distinctly mounded in the center of the structure, is interpreted as roof fall resulting from intentional, prehistoric human destruction of the structure roof and not as a natural deposit that slumped in from the sides of the pit. Immediately overlying

the collapsed roof sediments were two strata (Strata 3 and 4) that apparently were the result of natural slumping from the sides of the excavated pit. Stratum 3 was a strong brown silt loam resembling the sterile B-horizon deposits outside the pit structure. It had subangular blocky peds and fibrous networks of caliche. Stratum 4 was similar to Stratum 3, except it (Stratum 4) consisted of gray C-horizon deposits from outside the structure. Overlying these slump deposits were wind- and water-deposited sediments that completely filled the pit structure depression. Stratum 2 was a strong brown silt loam with abundant caliche and some laminations. Stratum 1 was a dark brown silt loam identified as the soil A horizon. Some thin, sorted laminations are visible toward the bottom of the deposit.

Artifacts found in postabandonment fill above the roof and wall fall layers include sherds, chipped-stone debris, a few animal bones, one mano, and two modified flakes. Because these deposits were not screened, this assemblage is not comparable to assemblages from screened deposits. Artifacts were more numerous and diverse in the roof and wall fall layer, which was screened, but few items were mapped or assigned point-location numbers. Materials recovered include sherds, chipped-stone debris (including two edge-damaged flakes and a core), four metates, two manos, three peckingstones, two modified flakes, a calcite pendant, and several pieces of animal bone. Gray, white, and red wares are present in the sherd assemblage; specific types include Chapin, Moccasin, and Mancos gray; Chapin and Piedra black-on-white; and Abajo Red-on-orange and Bluff Black-on-red. Both jar and bowl forms are present in the sherd assemblage. Several sherds classified as "jar" during sherd analysis were found to make up a partial beaker during the vessel-reconstruction project. Several other sherds belong to three partial Chapin, Moccasin, and Mancos gray jars, one partial Moccasin Gray olla, a partial red ware pitcher, and a Moccasin Gray sherd container.

Construction

Pit Structure 2 was excavated prehistorically through approximately 1.35 m of natural sediment and 15 cm into bedrock. Sand, gravel, and sterile silt/silt loam sediments were placed in low spots to even out irregularities in the bedrock surface (Figures 2.34 and 2.35). Next, layers of caliche, strong brown silt loam, and sand were added to form the living floor of the structure. Many of the floor features were constructed at this time as well. The hearth was constructed to take advantage of one of the many natural bedrock ledges, and several pit features were constructed with subfloor "collars" of sterile sediment (Figures 2.35 and 2.36).

For the most part, Pueblo I pit structure walls were simply excavated to form smooth, vertical faces that exposed the natural soil profile. Pit Structure 2 at Duckfoot is unusual in that the lower portions of the north and east

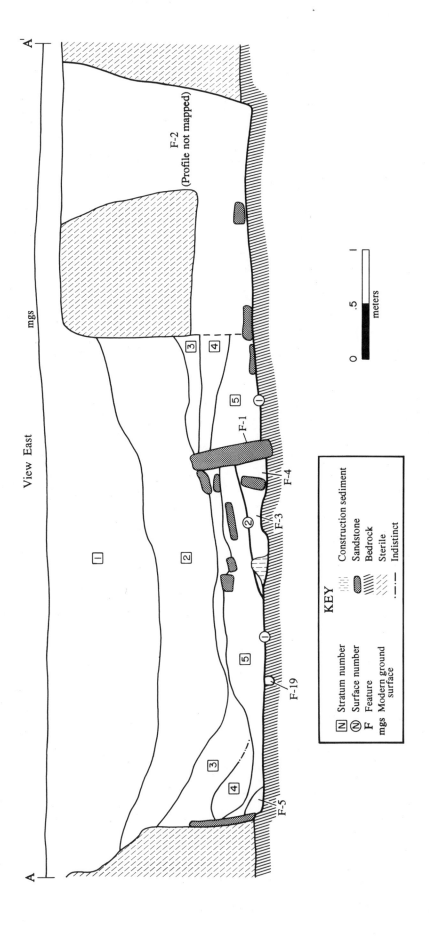

Figure 2.33. Pit Structure 2, stratigraphic profile. The location of the profile is shown in Figure 2.39.

walls are faced with tall vertical slabs. All the slabs are still in place along the north wall, but only three remain along the east wall. The "missing" stones from the east wall were found on the floor along the eastern edge of the structure. The construction of the north wall of Pit Structure 2 merits its own discussion. The lower portion of this wall is faced with a single row of shaped sandstone slabs that stand 60 to 75 cm high (Figure 2.37). Slightly over a meter from the east end of this wall a metate was used instead of a sandstone slab. Behind the wall metate and the adjacent slabs to the east is a coursed-masonry wall that reaches a height of 1.10 m above the floor. When the coursed masonry was examined in more detail, four peckingstones, six manos, and one hammerstone were found to have been used in wall construction (Table 2.26). All but one of the items are usable tools, including the complete metate with a trough that is worn only a few millimeters deep. It is not clear why complete (or nearly complete) and usable tools would have been used as construction materials, nor is it certain that they were part of the original construction of Pit Structure 2. The coursed-masonry construction was

contemporary with or earlier than the vertical-slab construction. Perhaps the tools were incorporated as a dedicatory cache.

Several post holes in the pit structure floor probably held beams that supported the roof. Feature 28, in the northeast corner, and Feature 31, in the southeast corner north of the wing wall, are two such post holes (Figure 2.39). It is not clear what supported the roof on the west side of the structure, because no definite post holes were identified; however, two pit features (Features 10 and 8) are located in approximately the right places for main roof supports. The shallowness of the post holes and possible post holes suggests that they could not have anchored support posts estimated to stand approximately 1.5 m high unless the superstructure was tied securely to the ground. However, the excavated walls were high enough to allow the construction of a nearly flat roof. Cantilevered primaries that extended from wall to wall across the pit would have provided the lateral stability or rigidity necessitated by the shallow post holes (cf. Wilshusen 1984, 1988a). Because the Pit Structure 2 roof did not burn, there is little additional information about roof construction.

Figure 2.34. View of Pit Structure 2, looking north, after floor sediments were removed and underlying bedrock was exposed. Note the natural irregularities in the bedrock surface, the depressions indicating where features were pecked into bedrock, and the vertical-slab lining of the north wall of the structure.

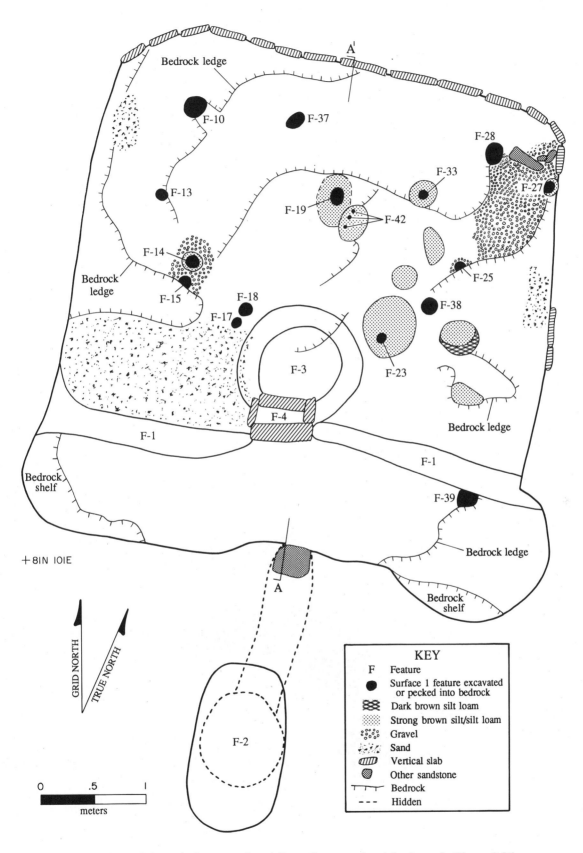

Figure 2.35. Pit Structure 2, subfloor. Cross section A is shown in Figure 2.36.

Bedrock ledge

F-10

F-37

F-28

F-33

F-13

F-19

F-42

F-27

F-14

Bedrock
ledge

F-25

F-15

F-18

F-38

F-17

F-23

F-3

Bedrock ledge

F-4

F-1

F-1

Bedrock
shelf

F-39

Bedrock ledge

+8IN IOIE

Bedrock
shelf

A'

A

GRID NORTH

TRUE NORTH

F-2

KEY

F   Feature

⬤   Surface 1 feature excavated or pecked into bedrock

▨   Dark brown silt loam

░   Strong brown silt/silt loam

∘∘∘   Gravel

∴   Sand

▥   Vertical slab

▦   Other sandstone

⊥⊤   Bedrock

---   Hidden

0   .5   1

meters

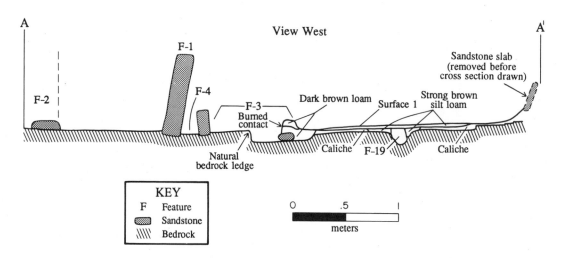

Figure 2.36. Pit Structure 2, architectural cross section. The location of the cross section is shown in Figure 2.35.

Surface 1

Surface 1, the original living floor of Pit Structure 2, was made by applying thin layers of caliche, silt loam, and sand to the excavated bedrock surface to smooth and level it. The thin layers were not continuous across the structure, and they may have represented several episodes of floor construction and remodeling. In several places, subfloor sediments were prepared for the construction of pit features. Many features were pecked and ground into bedrock beneath the floor. Of particular interest is the care with which subfloor sediments were applied in some areas where features were later excavated. In some cases it appears that the features were preplanned. For example, an area of bedrock would be cleared and a patch of silt loam applied so that the rim and upper half of the pit were lined with silt loam rather than with bedrock.

Artifacts and Samples. The floor was littered with broken pottery and other artifacts, many of which were point-located. Point-located items are mapped in Figure 2.38 and listed in Table 2.27. The assemblage includes over 900 sherds (including 7 modified sherds), a smaller amount of chipped-stone debris (including cores and edge-damaged flakes), numerous animal bones or bone fragments, 3 metates, 4 manos, 26 peckingstones, 2 modified flakes, 1 hammerstone, 2 axes, and 5 awls. An incomplete and fragmented human skeleton was on the floor in the northwest quadrant of the main chamber. It is likely that several isolated human bones in the west half of the structure also belong to this skeleton. In general, the artifacts and other debris were clustered along the east and west walls, and a relatively clear area extended from the hearth to the north wall.

Of the 912 point-located sherds from Surface 1, 802 (87.9 percent) are gray ware, 44 (4.8 percent) are white ware, and 64 (7.0 percent) are red ware. The majority are indeterminate types, but 8 Chapin Gray, 63 Moccasin Gray, 44 Mancos Gray, 1 Chapin Black-on-white, 17 Piedra Black-on-white, 4 Cortez Black-on-white, and 11 Bluff Black-on-red sherds are part of the assemblage as well. Many of the sherds were found in clusters, but during vessel reconstruction, it became clear that most sherds either did not refit with other sherds in the structure or refit with only a few sherds. The only vessel that was classified as "nearly complete" was a miniature bird-effigy jar. Also in the Pit Structure 2 floor assemblage are nine partial vessels (three bowls, one pitcher, one olla, and four jars) and six sherd containers. One sherd container (PL 138), made of approximately half of a longitudinally split Mancos Gray jar, contained a deep red mineral pigment.

Because Pit Structure 2 did not burn, few charred plant remains were recovered from a flotation sample collected from floor deposits. Juniper, corn, and pine remains are the only items recovered, and these probably are incidental introductions. A pollen sample from a metate (PL 108) yielded Cheno-am, cholla, prickly pear, corn, and Umbelliferae pollen, which might suggest food processing. A sample collected from floor sediments beneath a slab (not point-located) near the north wall of the pit structure yielded Cheno-am and beeweed pollen. Ethnographically, beeweed is known to have been used as both food and paint (cf. Robbins et al. 1916:58–59).

Features. Forty-one features were identified on the floor of Pit Structure 2 (Figure 2.39 and Table 2.28). A partial human skeleton in the northwest quadrant of the pit structure was designated Feature 41, Burial 3. The remains are those of a female, age 30 to 39 years. Only the head and torso elements are present, and many of the bones are broken and show signs of carnivore chewing. The body was oriented face down with the head to the southeast. The

View North

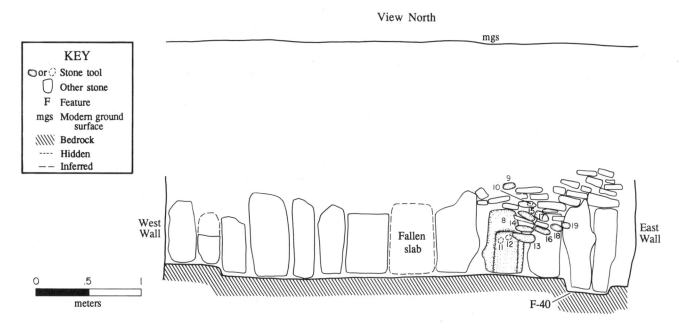

Figure 2.37. Pit Structure 2, detail of north wall construction. Point-located items are described in Table 2.26.

Table 2.26. Point-Located Artifacts from Pit Structure 2, North Wall

No.	Description
Trough Metate	
8	complete, sandstone
Two-Hand Mano	
13	complete, sandstone
14	complete, sandstone
15	complete, sandstone
16	complete, sandstone
17	incomplete, conglomerate
18	fragmentary, sandstone
Hammerstone	
9	complete, igneous
Peckingstone	
10	complete, Morrison orthoquartzite
11	complete, Morrison orthoquartzite
12	complete, Morrison orthoquartzite
19	complete, unknown chert/siltstone

NOTE: Refer to Figure 2.37 for artifact locations.

remains were not in a burial pit, but numerous sandstone rocks were clustered over and around the body. Figure 2.39 shows the placement and orientation of the skeleton; Figure 2.38 shows the skeleton in relation to the sandstone rocks associated with it. The absence of limb bones and the careful arrangement of the remaining skeletal elements indicate that Feature 41 is a secondary interment. The arms and legs evidently had already been detached (possibly by the same animal or animals that left the gnaw marks on the

recovered bones) before the body was deliberately and carefully placed on the pit structure floor and covered with rocks. Numerous artifacts were found in this quadrant of the pit structure, but it was not clear whether any were directly associated with the burial.

Feature 1 is the wing wall–deflector complex that forms a continuous east-west wall across the southern end of the structure. The wall stands roughly 90 cm high and is made of a combination of vertical slabs, horizontal blocks, and mud. The base consists of a double row of massive sandstone slabs held in place with mud. The top 20 to 30 cm of the wing wall is primarily mud with numerous sandstone blocks laid horizontally within it, though not in a uniform masonry pattern. In the middle of the wall, the deflector consists of a single thick slab and a stack of horizontal blocks that form a wedge beside it. As in Pit Structure 1, there are two low (45-cm-high) passages between the deflector and the two halves of the wing wall. In order to facilitate easier movement across the wing wall–deflector passages, sandstone block steps were set into the floor on each side of the wall next to both passages. When moving from one side of the wing wall to the other, excavators found these steps to be perfectly placed and very stable. It appears that the entire wing wall–deflector complex, including the passages, was built at the same time.

Feature 2 is a ventilation system consisting of a vertical shaft approximately 1 m south of the pit structure and a horizontal tunnel connecting the bottom of the shaft to the main chamber at floor level. A cluster of gray ware jar sherds and three stone tools were point-located on the floor of the vent shaft (Figure 2.38 and Table 2.27). Almost all the sherds, plus additional sherds that were not

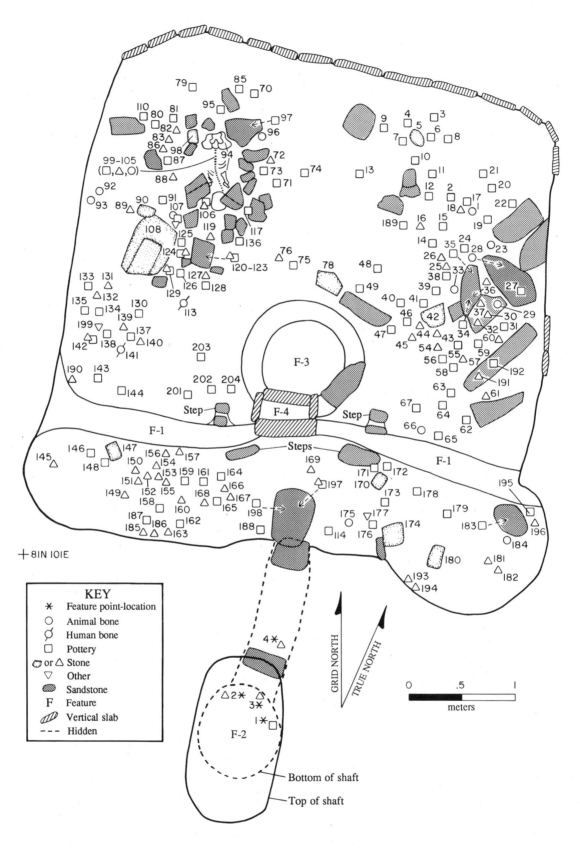

Figure 2.38. Pit Structure 2, Surface 1 artifacts. Refer to Table 2.27 for descriptions of point-located items, including "grouped" point-locations (for example, PLs 99–105).

Table 2.27. Point-Located Artifacts from Pit Structure 2, Surface 1

No.	Description	No.	Description
Pottery Sherds			white ware bowl (3)
1*	gray ware jar (34); Vessel 90 (32)	74	gray ware jar (2)
2	gray ware jar (6)	75	gray ware jar (4)
3	gray ware jar (14)	79	gray ware jar (4)
4	red ware bowl (1)	80	gray ware jar (1)
6	gray ware jar (10); Vessel 87 (1)		red ware jar (1); Vessel 118 (1)
7	gray ware jar (5); Vessel 87 (2)	81	gray ware jar (9); Vessel 87 (6)
8	gray ware jar (7)	85	gray ware jar (1)
9	gray ware jar (5)	87	gray ware jar (2)
10	gray ware jar (1)	91	gray ware other (11); Vessel 58 (11)
11	missing	95	gray ware jar (8); Vessel 87 (7)
12	gray ware jar (29); Vessel 88 (20)	97	gray ware jar (4)
13	missing	102	white ware bowl (1)
14	white ware bowl (6); Vessel 124 (6)	103	white ware bowl (1); Vessel 125 (1)
15	gray ware jar (29); Vessel 93 (5)	106	gray ware jar (6)
17	gray ware jar (54); Vessel 87 (2), Vessel 93 (5)	107	gray ware jar (16); Vessel 88 (6)
19	gray ware jar (2)	110	red ware jar (2); Vessel 118 (2)
	red ware jar (3); Vessel 118 (3)	114	white ware bowl (3)
20	gray ware jar (11)	117	gray ware jar (3)
21	red ware jar (1); Vessel 118 (1)	118	gray ware jar (4); Vessel 87 (1); not mapped
22	gray ware jar (13); Vessel 130 (3)		white ware jar (1); not mapped
	red ware jar (27); Vessel 118 (27)	120	gray ware jar (3)
24	gray ware jar (10); Vessel 93 (2)		gray ware other (1)
27	gray ware jar (48); Vessel 93 (5), Vessel 95 (41)	121	gray ware jar (6)
31	gray ware jar (12); Vessel 88 (7)	122	gray ware jar (5)
34	gray ware jar (6)	123	gray ware jar (1)
	white ware bowl (1)	124	gray ware jar (1)
35	gray ware jar (29)	125	gray ware jar (6); Vessel 88 (1)
38	gray ware jar (2); Vessel 93 (1)	126	gray ware jar (3)
39	gray ware jar (55); Vessel 87 (12),	128	gray ware jar (1)
	Vessel 130 (1)	129	gray ware jar (3)
40	gray ware jar (17)	130	gray ware jar (2)
41	red ware jar (1)	133	gray ware jar (4)
	red ware bowl (5)	134	gray ware jar (2)
46	red ware bowl (4); Vessel 112 (4)	135	red ware jar (1); Vessel 118 (1)
47	gray ware jar (2)	136	gray ware jar (19)
48	gray ware jar (1)	137	gray ware jar (12); Vessel 94 (9)
	red ware bowl (11); Vessel 110 (10)		white ware bowl (1); Vessel 125 (1)
49	gray ware jar (25); Vessel 130 (3)	138	gray ware jar (13); Vessel 92 (13)
	white ware jar (1)	142	gray ware jar (8)
55	gray ware jar (12); Vessel 130 (1)		red ware jar (3); Vessel 118 (1)
56	gray ware jar (2); Vessel 130 (1)	143	gray ware jar (11); Vessel 94 (4)
58	red ware bowl (2)	144	gray ware jar (3)
59	gray ware jar (11); Vessel 87 (1), vessel 91 (7)	146	gray ware jar (1)
	white ware bowl (1)	148	gray ware jar (1)
62	gray ware jar (1); Vessel 130 (1)	158	gray ware jar (4)
	white ware bowl (12); Vessel 125 (10)	159	gray ware jar (3)
63	gray ware jar (13); Vessel 130 (2)	161	gray ware jar (7); Vessel 89 (6)
	white ware bowl (4); Vessel 125 (3)	162	gray ware jar (35); Vessel 89 (2)
64	gray ware jar (1)	164	gray ware jar (17); Vessel 89 (9)
65	white ware jar (2)		white ware jar (2)
67	gray ware jar (4)	165	gray ware jar (13); Vessel 89 (6)
70	gray ware jar (1)	171	red ware jar (1); Vessel 118 (1)
	white ware bowl (1)	172	gray ware jar (1)
71	gray ware jar (4); Vessel 130 (3)	173	gray ware jar (2)
73	gray ware jar (3)	176	gray ware jar (6)
	white ware jar (1)	178	gray ware jar (4)

Table 2.27. Point-Located Artifacts from Pit Structure 2, Surface 1 (continued)

No.	Description	No.	Description
179	gray ware jar (19); Vessel 87 (11)	Two-Hand Mano	
183	gray ware jar (21); Vessel 87 (21)	98	fragmentary, sandstone
187	gray ware jar (5)	147	complete, sandstone
188	gray ware jar (1); Vessel 89 (1)	170	incomplete, igneous
189	gray ware jar (1)	180	complete, conglomerate
192	white ware jar (1)		
195	gray ware jar (2); Vessel 87 (2)	Abrader	
197	gray ware jar (10)	145	fragmentary, sandstone
198	gray ware jar (1)	151	fragmentary, sandstone
201	gray ware jar (17)	153	fragmentary, sandstone
	white ware jar (1)	Hammerstone	
	red ware bowl (1)	166	complete, igneous
202	gray ware jar (1)	Peckingstone	
203	gray ware jar (1)	26	complete, Morrison orthoquartzite
204	gray ware jar (1)	43	complete, Morrison orthoquartzite
Modified Sherd		61	complete, Dakota orthoquartzite
15	fragmentary, gray ware	76	complete, Dakota orthoquartzite
15	fragmentary, gray ware	83	complete, Morrison orthoquartzite
15	fragmentary, gray ware	86	complete, Morrison orthoquartzite
39	fragmentary, gray ware	88	complete, Morrison orthoquartzite
39	fragmentary, gray ware	89	complete, Dakota orthoquartzite
189	fragmentary, gray ware	90	complete, Morrison orthoquartzite
197	fragmentary, white ware	100	complete, Morrison orthoquartzite
Chipped-Stone Debris		104	complete, Dakota orthoquartzite
16	flake (1)	119	complete, Dakota orthoquartzite
25	flake (1)	152	complete, Morrison orthoquartzite
26	flake (1)	154	complete, Dakota orthoquartzite
32	flake (1)	155	complete, Morrison orthoquartzite
36	flake (1)	156	complete, Dakota orthoquartzite
37	flake (1)	157	complete, unknown chert/siltstone
44	flake (1)	168	complete, Morrison orthoquartzite
45	flake (1)	169	complete, Dakota orthoquartzite
54	flake (1)	181	complete, Morrison orthoquartzite
57	flake (1)	182	complete, Morrison orthoquartzite
82	flake (1)	186	complete, Morrison orthoquartzite
99	edge-damaged flake (1)	190	complete, Dakota orthoquartzite
105	flake (1)	191	complete, Dakota orthoquartzite
106	flake (1)	193	complete, Dakota orthoquartzite
116	flakes (2); not mapped	196	complete, Morrison orthoquartzite
122	flake (1)	Single-Bitted Axe	
124	flake (1)	163	complete, igneous
127	core (1)	185	complete, Morrison orthoquartzite
129	flake (1)	Modified Flake	
131	edge-damaged flake (1)	3*	complete, Morrison orthoquartzite
132	flake (1)	60	complete, Morrison orthoquartzite
139	core (1)	72	complete, Morrison orthoquartzite
140	other (1)	Other Chipped-Stone Tool	
142	flake (1)	4*	complete, Morrison orthoquartzite
149	missing	30	complete, obsidian
150	core (1)	Modified Cobble	
160	flake (1)	2*	complete, igneous
197	flake (1)	5	incomplete, igneous
Trough Metate		18	complete, unknown stone
42	fragmentary, sandstone	174	complete, unknown orthoquartzite
78	fragmentary, sandstone	Other Modified Stone/Mineral	
108	complete, sandstone	167	complete, sandstone

Table 2.27. Point-Located Artifacts from Pit Structure 2, Surface 1 *(continued)*

No.	Description	No.	Description
194	complete, sandstone	Miscellaneous	
Unmodified Stone/Mineral		94	human bone; Burial 3
107	azurite	113	human bone
177	caliche	141	human bone
199	iron oxide	1	animal bone
Awl		23	animal bone
28	incomplete, *Odocoileus hemionus*	66	animal bone
29	complete, *Odocoileus hemionus*	93	animal bone
33	complete, cf. *Odocoileus*	107	animal bone
96	complete, *Odocoileus hemionus*	116	animal bone; not mapped
101	incomplete, *Odocoileus hemionus*	175	animal bone
Other Modified Bone		184	animal bone
92	fragmentary, indeterminate mammal		

NOTES: Item counts appear in parentheses. For pottery, sherds belonging to vessels are also included in total item count. Refer to Figure 2.38 for artifact locations.
*Feature 2 (ventilation system) point-locations.

point-located, belong to a partial Moccasin Gray jar. Feature fill appeared to be natural, and it contained only a few artifacts. The cover of the vent tunnel was found on the floor of the pit structure immediately in front of the opening. The shaft was not lined, faced, or plastered, but the 15 to 20 sandstone rocks in the fill may have formed some type of raised rim or stone lining around the top.

Feature 3 is the hearth in the center of Pit Structure 2. The hearth is relatively large and well made and has a raised rim of sandstone and mud. The base of the pit is on very uneven bedrock, the north rim is built of mud, the southeast and southwest rims are sandstone blocks, and the southern rim is formed by a rectangular sandstone slab that is more a part of the ash pit (Feature 4) than of the hearth proper. A thin layer of clean sand covered the bottom of the hearth, and this, in turn, was covered by layers of charcoal and ash. The uppermost layer of ashy fill completely overflowed the hearth, spilled out onto the floor, and covered the top of the ash pit to the south. The top of this ash layer would have been the final use surface in this part of the pit structure and was designated Surface 2. Surface 2 is described in the next section. Fuel (juniper, pine, sagebrush, four-wing saltbush, mountain mahogany, and corncob) and food (cheno-am and corn) remains were recovered from flotation samples collected from hearth fill.

Feature 4 is the ash pit associated with the hearth. In this case, it might best be called an ash box, because its sides are constructed of vertical slabs. The south edge of Feature 4 is formed by the massive sandstone deflector, and the other three sides are formed by well-fitted vertical slabs, approximately 23 cm high and 8 cm thick. Fill consisted of a single stratum of fine, gray ash that was piled at least 10 cm above the sides of the feature and onto the north face of the deflector. It appears that, like the hearth, the ash pit was allowed to overflow late in the occupation

of the site, when abandonment was imminent. Little was recovered from flotation samples collected from the ash pit, but what was recovered was consistent with remains found in the hearth samples.

Features 5, 6, and 7 were piles of sand on the floor. Feature 5 was a large pile of clean, tan, coarse sand located against the north wall. Feature 6 was a circular pile of sand in the northeast corner of the structure. It covered the shallow post hole (Feature 28) that would have held the main roof support in this part of the structure. This post hole had been deliberately filled with sand and capped with mud. Feature 7 was another circular pile of clean sand in the northwest corner of the pit structure. This pile covered a shallow, deliberately filled pit, Feature 10, that may be another main support post hole. Feature 10 was excavated 8 cm into the bedrock floor and was filled with a layer of clean sand. The fact that Features 6 and 7 covered the post holes or possible post holes (Features 28 and 10) supports the argument that the posts had been pulled from their holes and the structure deliberately collapsed at abandonment. Feature 5 may be sand that was stockpiled for feature construction or some other (unknown) purpose. In all cases, the sand in these features had to have been brought to the site from a stream or other water-sorted depositional context. Pit structures in the Dolores River valley often had small piles of sand near the northern posts (Breternitz and Morris 1988:364–365; Varien 1988:144, 166, 243; Lightfoot et al. 1988:644). The sand was thought to have been stockpiled for temporary filling of pits as needed. Sipapus and more complex ceremonial features are often filled with sand (Wilshusen 1988c), and the sand itself may have a ritual purpose or value.

Feature 19 is a pit that is interpreted to be a sipapu on the basis of its small size, circular shape, sand fill, and location north of the hearth. Three small, conical, sand-

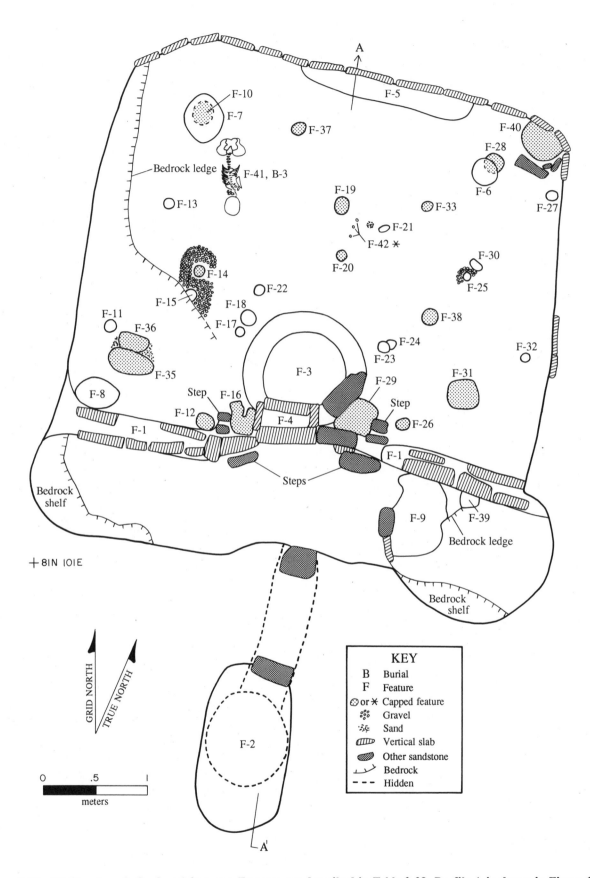

Figure 2.39. Pit Structure 2, Surface 1 features. Features are described in Table 2.28. Profile A is shown in Figure 2.33.

Table 2.28. Feature Summary, Pit Structure 2

No.	Type	Shape	Length (cm)	Width (cm)	Depth (Height) (cm)	Status
1	wing wall	other	440	30	(90)	—
2	vent	other	265	85	207	in use
3	hearth	basin	115	95	20	in use
4	ash pit	other	62	26	6 (17)	in use
5	sand pile	other	150	25	(23)	in use
6	sand pile	other	26	24	(8)	in use
7	sand pile	other	40	38	(10)	in use
8*	pit	basin	36	27	15	in use
9	pit	basin	80	50	9	in use
10*	pit	basin	20	20	8	capped
11	pit	basin	10	10	4	in use
12	pit	basin	19	16	5	capped
13	pit	basin	10	9	7	in use
14	pit	cylindrical	12	12	7	capped
15	pit	cylindrical	14	10	7	unknown
16	pit	cylindrical	30	24	16	capped
17	pit	cylindrical	9	8	10	in use
18	pit	cylindrical	12	10	12	in use
19	sipapu	other	17	14	17	capped
20	pit	cylindrical	8	8	2	capped
21	pit	basin	11	7	2	in use
22	pit	basin	10	10	2	in use
23	pit	cylindrical	12	12	12	unknown
24	pit	other	11	9	4	in use
25	pit	cylindrical	8	7	12	in use
26	pit	basin	12	8	3	capped
27	pit	cylindrical	9	9	10	in use
28	post hole	basin	20	18	7	capped
29	pit	other	32	30	10	capped
30	pit	other	15	8	8	in use
31	post hole	basin	30	27	9	capped
32	pit	cylindrical	8	8	9	in use
33	pit	cylindrical	10	8	9	capped
35	pit	basin	44	22	5	capped
36	pit	basin	30	20	3	capped
37	pit	basin	16	11	7	capped
38	pit	basin	17	16	7	capped
39	post hole	cylindrical	16	15	17	in use
40	pit	basin	43	36	15	capped
41	Burial 3	other	73	34	(10)	—
42	paho marks	other	20	5	nr	capped

NOTE: Feature locations are shown in Figure 2.39.

nr = no record.

* Possible posthole.

filled pits clustered immediately to the southeast of the sipapu are believed to be paho marks, which may indicate where pahos, or prayer sticks, were placed around the sipapu. Similar small holes were associated with sipapus and vault features in pit structures in the Dolores River valley (Wilshusen 1986c:250; Brisbin et al. 1988; Varien 1988). Wilshusen (1988c) uses ethnographic data to argue that these holes may have been produced by thrusting prayer

sticks, or pahos, into the floor around the sipapu. Sipapus and paho marks are often filled with sand. The paho mark cluster in Pit Structure 2 is designated Feature 42, and its presence further strengthens the interpretation that Feature 19 functioned as a sipapu. Like many other Pit Structure 2 floor features, Feature 42 seems to have been constructed in a carefully prepared area of subfloor: the tiny holes were dug into a patch of strong brown silt loam subfloor sediments (Figure 2.35). As with several other features, the paho marks apparently were deliberately filled with sand.

Features 28, 31, and 39 are post holes. Feature 28, as discussed above, is inferred to have held the northeast main roof-support post. Feature 31 is interpreted as the main support post hole in the southeast corner of the structure north of the wing wall. It too had been intentionally filled, in this case with brown loam, and capped with a layer of strong brown silt loam before or during abandonment. Feature 39 is a small post hole between the southern edge of the wing wall and a natural bedrock ledge in the southeastern corner of the structure. It is not clear how it functioned, but it may have held an auxiliary roof-support post.

The remaining twenty-eight features are pits of unknown function. Thirteen of them appear to have been open at abandonment, and another 13 were deliberately capped. The characteristics of the remaining two were too ambiguous to determine status at abandonment. The difficulty in inferring the status of these two may be partly the result of the excavated structure standing open (though covered with heavy plastic) for a winter before the floor-construction sediments were removed. The intentionally capped features had been filled with sand, loam, or (occasionally) mixed sediments, including some "dirty" fill. Four pits (Features 14, 15, 22, and 27) are described as having "complex" fill because they were carefully constructed with several types of sediment, including gray gravel, strong brown silt loam, and tan sand. Sand-filled pits are ubiquitous in pit structures from Basketmaker to Pueblo I (Brew 1946; Brisbin et al. 1986:564–565; Brisbin 1986:666; Bullard 1962:172; Lancaster et al. 1954:50; Lightfoot et al. 1988; Varien 1988; Morris 1988b, 1988c). Wilshusen (1988c) argues that many may have been used for erecting altars, an interpretation implying that the sand was merely a convenient way to fill a pit that was used regularly, allowing it to be easily reexcavated when needed. Other pits may have been associated with temporary constructions such as looms (Bullard 1962:172) or cooking racks. Lancaster et al. (1954:50) suggest that these pits could have been used as pot rests, that is, basins into which round-bottomed pots were placed and held upright. Brew (1946:157) notes that sand filling in pits does not preclude the use of such features as containers; sand could have been used to fill the pits when they were not in use. The type of sand used to fill the Pit Structure 2 pits is not available immediately around Duckfoot, although the sand could have been transported from nearby Alkali Canyon.

Several of the pit features merit individual discussion because of their possible functions. Feature 10 has already been mentioned as a candidate for a main roof-support post hole. Feature 8 is another basin-shaped pit that, given its location in the corner formed by the juncture of the wing wall with the west wall of the structure, may have served a similar function. The broad, shallow form of Feature 8 would be unusual for a main roof support in most pit structures, but it is not unlike Features 10 and 28. The pit was excavated into floor-construction sediments, and bedrock formed the base of the pit. It appears to have been open at abandonment, as it contained the same mottled roof fall sediments that covered the floor. If Feature 8 indeed held a main support post, then, like the other roof-support features (Features 28, 31, and possibly 10), the post probably was removed as part of the intentional dismantling of the structure at abandonment. Unlike the other post holes or possible post holes, however, Feature 8 was not deliberately filled after the post was removed. Portions of a Moccasin Gray sherd container, a nearly complete Chapin Gray pitcher, and a partial red ware pitcher were recovered from pit fill.

Feature 9 is a broad, shallow, D-shaped pit south of the east wing wall. The pit was carved out of decomposing platy bedrock. A mud rim was added along the west side, and one small sandstone slab was set into this rim at the southwest corner of the pit. The pit contained roof fall sediments as well as sherds, two peckingstones, and a small fragment of animal bone. The size and shape of the pit resemble those of a mealing bin, but no metate was found within it. However, several pieces of ground sandstone (abraders) were found on the floor behind the west wing wall. It is possible that these items are pieces of broken slab metates that were too fragmented to be identified as such. Mealing bins are generally associated with slab metates rather than with trough metates.

Surface 2

Surface 2 was defined as the top of the layer of ash that overflowed the hearth and ash pit onto the surrounding floor (Surface 1). It appears that, during the time when the occupants anticipated abandoning the structure or the site, ash was allowed to accumulate, effectively creating a new use surface in the area immediately adjacent to these two features. Artifacts were deposited or discarded on this surface, and all were point-located during excavation (Figure 2.40 and Table 2.29). Among the materials recovered are a metate and numerous sherds, some of which belong to four of the same partial or nearly complete vessels (a Moccasin Gray olla, a Moccasin Gray jar, a Chapin Gray pitcher, and a Piedra Black-on-white bowl) already discussed in the Surface 1 and Feature 8 descriptions. The

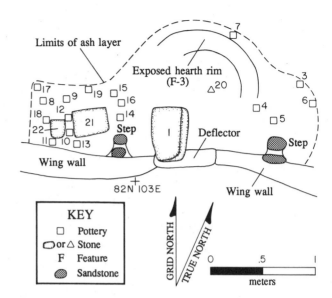

Figure 2.40. Pit Structure 2, Surface 2. Point-located items are described in Table 2.29.

Surface 2 sherd assemblage consists of 61 (89.7 percent) gray ware and 7 (10.3 percent) white ware sherds. Specific types identified are one Chapin Gray, four Moccasin Gray, one Mancos Gray, and two Piedra Black-on-white sherds. The remaining sherds are local indeterminate types, even though many belong to reconstructed vessels that are typed more specifically. Jar and bowl vessel forms are recognized in the sherd assemblage. A pollen sample collected from the metate yielded Cheno-am, cholla, prickly pear, and corn pollen, which may indicate processing of these plants for food.

Dating

Of the four tree-ring samples collected from Pit Structure 2, only one yielded a date, A.D. 822 (Appendix B). This noncutting date is much earlier than the construction date inferred on the basis of other Architectural Suite 2 dates. Dates for the site as a whole and cutting dates from Room 13 indicate that Pit Structure 2 was built in the late A.D. 850s.

Interpretations

On the basis of the variety of artifacts and features discovered, Pit Structure 2 is interpreted to be primarily a habitation structure. The presence of manos, metates, a hearth, and an ash pit is typical of pit structures at Duckfoot and supports an inference of domestic use. However, much of the artifact assemblage is likely to have been dumped into the structure or allowed to accumulate as refuse during the period of abandonment. No complete or nearly complete cooking or storage vessels are present in the assem-

Table 2.29. Point-Located Artifacts from Pit Structure 2,
Surface 2

No.	Description
Pottery Sherds	
3	gray ware jar (5); Vessel 130 (3)
4	gray ware jar (13); Vessel 130 (1)
5	gray ware jar (18)
6	gray ware jar (3)
7	gray ware jar (6)
8	gray ware jar (3)
9	gray ware jar (2); Vessel 87 (1)
10	gray ware jar (5); Vessel 60 (5)
11	gray ware other (1); Vessel 60 (1)
12	gray ware jar (1)
13	gray ware jar (1)
14	white ware jar (1)
	white ware bowl (1); Vessel 125 (1)
15	gray ware jar (1)
16	gray ware jar (1)
17	white ware jar (5)
18	gray ware jar (1)
19	missing
Chipped-Stone Debris	
20	flake (1)
Trough Metate	
1	complete, sandstone
Abrader	
21	complete, sandstone
Unmodified Stone/Mineral	
22	condition unknown, conglomerate

NOTES: Item counts appear in parentheses. For pottery, sherds belonging to vessels are also included in total item count. Refer to Figure 2.40 for artifact locations.

blage, and even the sherd containers were reconstructed from sherds scattered across the floor. Only about 25 percent of the sherds from floor contexts (including fill within 5 cm of floor) are accounted for by sherd containers or partially reconstructible vessels. The remaining 75 percent are sherds that probably represent secondary refuse (dumped into the structure from elsewhere) or abandonment refuse (allowed to accumulate during the period of abandonment). The presence of a sipapu, paho marks, sand-filled pits, and a partial human skeleton on the floor suggests that ritual activities were conducted in the structure in both the preabandonment and abandonment phases of its use.

Abandonment of Pit Structure 2 involved deliberate dismantling of the structure as well as deliberate interment of human remains, and these activities may have been conducted as a part of an abandonment ritual. The lower fill of Pit Structure 2 clearly consisted of materials that were piled into the structure depression and not washed in through natural means. In addition, the main support post holes and possible post holes in the floor had been deliberately filled and capped, demonstrating that they no longer

had posts in them when the structure began to fill. These observations support the argument that the roof of Pit Structure 2 was intentionally dismantled and collapsed onto the floor at abandonment. Wilshusen (1988c, 1986c) notes that pit structures with certain ritual features and human bodies on the floor often are intentionally destroyed without burning. If the structure had been intentionally collapsed, the main posts might have been pulled out and the post holes filled as a part of an abandonment ritual.

Pit Structure 4

Length: 3.45 m
Width: 3.05 m
Area: 8.90 m²

Pit Structure 4 (Figures 2.41–2.45) was discovered two weeks before the end of the final field season during excavation of what originally was believed to be an open courtyard between Pit Structures 2 and 3 (Figure 1.5). Pit Structure 4 is the smallest and shallowest pit structure at Duckfoot, characteristics dictated at least in part by constraints imposed by existing architecture. Pit Structure 4 is believed to be a late addition to the site, and there was little space between existing Pit Structures 2 and 3 to accommodate yet another building. Like Pit Structures 1 and 3 and several surface rooms, Pit Structure 4 was burned at abandonment.

Excavation of Pit Structure 4 was accelerated because of the limited time available before the end of the final field season. Once the pit structure was identified in a 2-×-2-m excavation square, a 30-cm-wide trench was extended along the central north-south axis to the top of burned roof fall. Fill in this trench was screened as a single layer; obvious stratification was not apparent in the fill above roof fall. The remainder of the upper fill was removed with a backhoe and was not screened. Burned roof fall was excavated with hand tools as a single stratum down to 5 cm above the floor. Tree-ring samples, flotation samples, pollen samples, and selected artifacts (tools and large or diagnostic sherds) in the roof fall layer were individually point-located, and all roof fall sediments were screened. Fill within 5 cm of the floor was excavated and provenienced as floor, and all artifacts in contact with the floor were mapped and point-located. Floor features were mapped and excavated, and finally, the floor-construction sediments were removed to expose capped features and to document construction and remodeling.

Stratigraphy

Figure 2.41 is a stratigraphic profile of Pit Structure 4, and it reflects construction, occupation, abandonment, and postabandonment events. As usual, the pit structure was excavated prehistorically through natural deposits to bed-

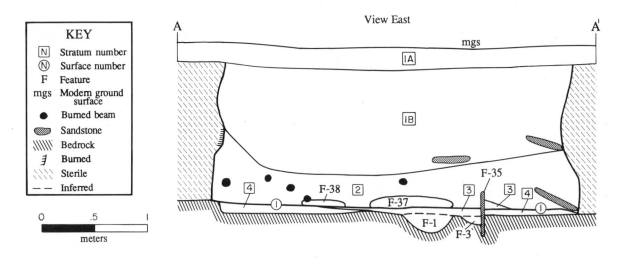

View East

Figure 2.41. Pit Structure 4, stratigraphic profile. The location of the profile is shown in Figure 2.44. Refer to text for a discussion of the Stratum 3–Surface 1 interface.

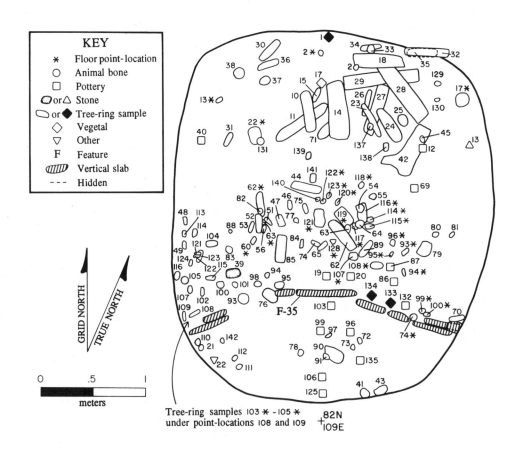

Figure 2.42. Pit Structure 4, roof fall. Point-located items are described in Table 2.30. Tree-ring samples recovered from the floor are included to better illustrate the overall roofing pattern; these are listed in Table 2.31.

rock. A layer of strong brown silt loam (Stratum 4) was placed over bedrock to form the floor. Certain areas of the floor were overlain by a layer of ash, designated Stratum 3. This ash lens covered the hearth, the ash pit, and approximately half the pit structure floor north of the wing wall; ash was also found in two piles in the northwest corner of the structure. Apparently, once abandonment was imminent, the occupants stopped cleaning the hearth and ash pit as often as they might have otherwise, and ash was allowed to overflow these features and accumulate on top of the living floor. (Artifacts found in Stratum 3 are inferred to be associated with the use of the floor and therefore are included in the discussion of floor artifacts below.) Directly on top of the floor and Stratum 3 were the remains of the burned and collapsed roof (Stratum 2). Roof fall consisted of burned beams, smaller burned vegetal materials, mud with beam impressions, unburned sediments, and artifacts. The boundary between underlying and overlying deposits was very distinct. After the roof was destroyed, sediments began to accumulate naturally in the pit structure depression (Strata 1A and 1B). Stratum 1B consisted of a strong

brown to dark brown silt loam with faint laminations toward the base, which indicate deposition primarily by wind and water. Stratum 1A was the soil A horizon, a 20-cm-thick stratum of dark brown loam and silt loam with few inclusions other than charcoal flecks and roots. These sediments are slightly darker in color and coarser in texture than the underlying deposit, and the boundary between the two is diffuse. Stratum 1A was excavated, before the discovery of Pit Structure 4, in a series of 2-×-2-m squares that included portions of Courtyards 2 and 3. As a result, Stratum 1A artifacts were assigned to the two courtyard areas rather than to the pit structure.

Artifacts recovered from Stratum 1B include numerous sherds, a few pieces of chipped-stone debris, a few pieces of animal bone, a modified flake, and a bone awl. Materials recovered from the roof fall stratum (Stratum 2) are of greater interest because they are more likely to relate to the use and abandonment of the structure than are the artifacts in the washed-in fill above roof fall. Point-located roof fall items include sherds, a peckingstone, a mano, and two bone awls (Figure 2.42 and Table 2.30). Of the 242

Table 2.30. Point-Located Artifacts from Pit Structure 4, Roof Fall

No.	Description	No.	Description
Pottery Sherds		**Other Modified Bone**	
12	gray ware jar (13)	126	fragmentary, *Odocoileus hemionus*; not mapped
19	gray ware jar (6)	**Dating Samples**	
20	gray ware jar (7)	1–2	tree-ring samples
40	gray ware jar (1)	3–9	tree-ring samples; not mapped
69	gray ware jar (7); Vessel 86 (7)	10–11	tree-ring samples
86	gray ware jar (1)	14–15	tree-ring samples
87	gray ware jar (10); Vessel 86 (3)	18	tree-ring sample
	white ware bowl (1)	21	tree-ring sample
96	missing	23–38	tree-ring samples
99	gray ware jar (42); Vessel 75 (3), Vessel 77 (3), Vessel 80 (3), Vessel 86 (30)	41–49	tree-ring samples
		51–56	tree-ring samples
103	gray ware jar (7); Vessel 85 (1), Vessel 128 (3)	57–61	tree-ring samples; not mapped
106	gray ware jar (107); Vessel 75 (18), Vessel 77 (20), Vessel 80 (35), Vessel 81 (11), Vessel 86 (1)	62–65	tree-ring samples
		66–68	tree-ring samples; not mapped
	white ware bowl (1)	70–85	tree-ring samples
125	gray ware jar (8); Vessel 85 (1), Vessel 86 (5)	88–91	tree-ring samples
132	gray ware jar (1)	93–95	tree-ring samples
135	gray ware jar (27); Vessel 75 (8), Vessel 78 (3), Vessel 80 (3), Vessel 86 (1)	97–98	tree-ring samples
		100–102	tree-ring samples
Two-Hand Mano		104	tree-ring sample
39	complete, sandstone	107–116	tree-ring samples
Peckingstone		121–124	tree-ring samples
13	complete, Morrison orthoquartzite	129–130	tree-ring samples
Awl		133–134	tree-ring samples
105	fragmentary, cf. *Odocoileus*	137–142	tree-ring samples
131	incomplete, cf. *Odocoileus*	**Miscellaneous**	
		17	vegetal
		22	adobe (construction sediment)

NOTES: Item counts appear in parentheses. For pottery, sherds belonging to vessels are also included in total item count. Refer to Figure 2.42 for artifact locations.

point-located sherds, 240 (99.2 percent) are gray ware and 2 (.8 percent) are white ware. Most are indeterminate types, but 12 Chapin Gray, 32 Moccasin Gray, and 9 Mancos Gray sherds are present as well. Both jar and bowl forms are recognized in the point-located assemblage, and numerous sherds belong to four partial vessels (one Chapin Gray olla, one Chapin Gray jar, two Mancos Gray jars) and three sherd containers. Only one, the partial Chapin Gray olla, is made of pieces recovered exclusively from roof fall. Six items are made of sherds recovered from both roof fall and floor contexts, which suggests that there was some mixing of deposits as the roof collapsed onto the floor. Also shown in Figure 2.42 are the locations of beams collected as tree-ring samples, including samples collected from the floor (flagged with asterisks). The latter are listed in the floor PL table (Table 2.31).

A pollen sample collected from roof fall yielded, among other things, a very high percentage of sagebrush pollen. Sagebrush remains were recovered from flotation samples as well. The evidence suggests that sagebrush may have been used in roof construction or may have been used as fuel to deliberately burn the structure at abandonment.

Construction

Pit Structure 4 was excavated prehistorically through approximately 1.25 m of natural sediments to the top of bedrock; in the north half of the structure, the bottom of the pit was pecked about 10 cm into bedrock (Figure 2.43). The floor was constructed by applying a layer of silt loam over the exposed rock surface. However, in some areas bedrock showed through (Figure 2.44) either because it had never been plastered over or because the plaster had worn off through use. Several of the main floor features, such as the hearth, ash pit, and wing wall–deflector complex, were constructed at the same time as the floor. Structure walls consist of native sediments, and the lower 5 cm of each wall is slightly undercut.

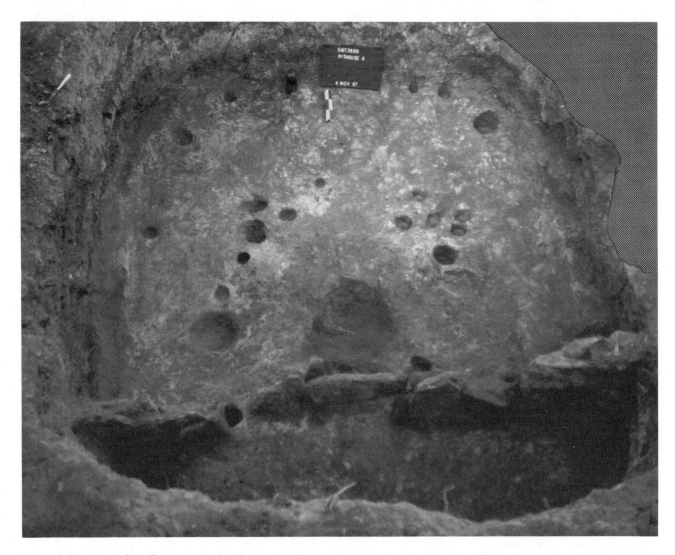

Figure 2.43. View of Pit Structure 4 after floor sediments were removed and underlying bedrock was exposed. The stippled area in the upper right corner is where the film was overexposed.

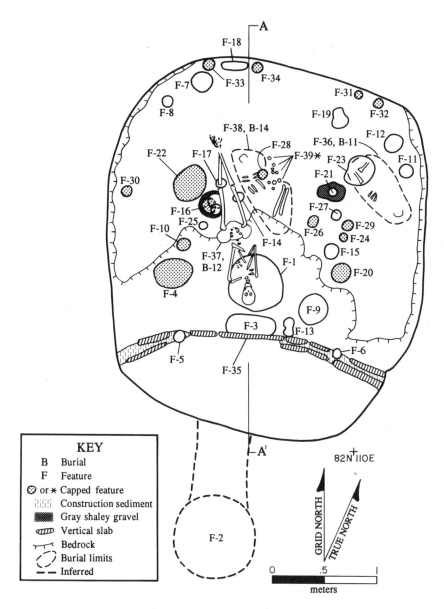

Figure 2.44. Pit Structure 4, surface features and subfloor. Bedrock was exposed from the bedrock line south to the wing wall. Features are described in Table 2.32. Profile A is shown in Figure 2.41.

Four post holes contained rotted wood fragments, the remains of roof-support posts. Many tree-ring samples from both the roof fall stratum and the pit structure floor were assigned point-location numbers, and many are shown in Figure 2.42. Beam diameters range from 2 to 18 cm. The orientation of beams does not provide much evidence of roof construction. However, there were many roof casts (slabs of burned construction mud) with impressions of closely spaced beams larger than 5 cm in diameter. This indicates that mud was applied directly to the beams without the intervening layer of "closing material" that often is assumed to have been used in the construction of pit structure roofs.

Surface

The Pit Structure 4 living floor was prepared largely by applying a layer of strong brown silt loam over the exposed bedrock surface. The silt loam floor sediments were 16 cm thick along the north wall and thinner toward the south. In the center of the structure, either the floor was never plastered or the plaster wore off through use, and the top of the bedrock served as floor. North of the wing wall–deflector, the central portion of the floor was covered by a layer of ash that was up to 4 cm thick, thinner toward the edges. In the stratigraphic description, this is defined as Stratum 3, but during excavation it was treated as part of the floor. The ash is assumed to be primary refuse that accumulated on the floor over time, and items in and on the ash layer were point-located.

Artifacts and Samples. Figure 2.45 shows the point-located artifacts on the floor and in the ash layer (Stratum 3) above the floor. Table 2.31 provides brief artifact descriptions. The floor assemblage is rich in terms of the number and diversity of artifacts present. Point-located materials include sherds (including six modified sherds and several pieces of unfired pottery), chipped-stone debris (including two cores and several edge-damaged flakes), and animal

bone. Three human burials were found as well. Formal tools in the assemblage include 5 metates or metate fragments, 14 manos, 3 axes, 3 hammerstones, 7 modified cobbles, 8 peckingstones, 2 modified flakes, 8 bone awls, 1 calcite pendant, 1 gaming piece, and 1 small piece of cordage made of plant fibers. Of the 678 sherds recovered and assigned point-location numbers, 636 (93.8 percent) are gray ware, 34 (5.0 percent) are red, and 8 (1.2 percent) are white. Twelve Chapin Gray, 47 Moccasin Gray, 13 Mancos Gray, 20 Bluff Black-on-red, and 4 Piedra Black-on-white sherds are present in the assemblage; the remaining sherds are of indeterminate types. Jar and bowl vessel

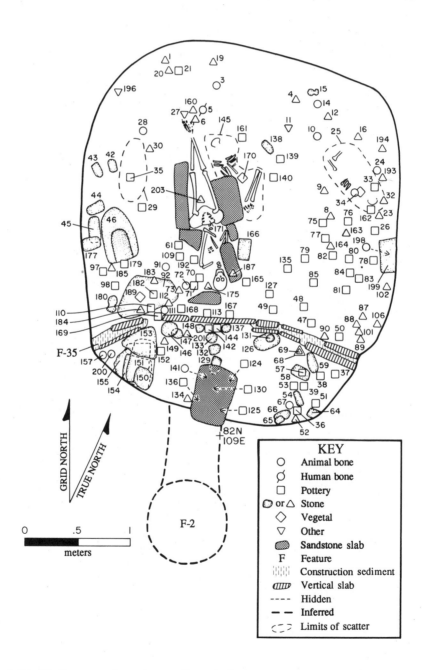

Figure 2.45. Pit Structure 4, surface artifacts. Point-located items are described in Table 2.31.

Table 2.31. Point-Located Artifacts from Pit Structure 4, Surface

No.	Description	No.	Description
Pottery Sherds		182	gray ware jar (5); Vessel 78 (1)
21	gray ware jar (1)		red ware bowl (4)
26	gray ware jar (4); Vessel 76 (4)	192	gray ware jar (17); Vessel 75 (1)
29	gray ware jar (1)		red ware bowl (1)
	red ware bowl (16); Vessel 59 (12)	**Unfired Pottery Sherds**	
35	gray ware jar (41); Vessel 81 (33)	33	unfired sherds
36	gray ware jar (19); Vessel 82 (17)	84	unfired sherds
37	gray ware jar (1)	142	unfired sherds
38	gray ware jar (1)	**Modified Sherd**	
39	gray ware jar (9); Vessel 75 (1), Vessel 82 (7)	47	condition unknown, gray ware
47	gray ware jar (45); Vessel 79 (16), Vessel 83 (9)	57	condition unknown, white ware
48	gray ware jar (7); Vessel 83 (4)	80	condition unknown, white ware
49	gray ware jar (2)	83	condition unknown, red ware
50	red ware bowl (5); Vessel 59 (5)	127	complete, white ware
51	gray ware jar (1); Vessel 82 (1)	127	condition unknown, gray ware
53	gray ware jar (2); Vessel 82 (1)	**Chipped-Stone Debris**	
61	gray ware jar (1); Vessel 75 (1)	1	core (1)
70	gray ware jar (1); Vessel 75 (1)	6	flakes (2)
71	gray ware jar (1)	12	flake (1)
75	gray ware jar (1)	16	edge-damaged flake (1)
76	gray ware jar (1)	20	flakes (2)
77	gray ware jar (1)	23	flake (1)
78	gray ware jar (1)	32	edge-damaged flake (1)
79	white ware bowl (1)	73	flake (1)
80	gray ware jar (5)	87	flake (1)
81	gray ware jar (28); Vessel 83 (9)	89	flake (1)
	red ware bowl (2); Vessel 109 (2)	90	flake (1)
82	white ware bowl (2); Vessel 126 (2)	101	flake (1)
85	gray ware jar (49); Vessel 74 (3), Vessel 83 (11)	102	flake (1)
97	gray ware jar (1)	106	flake (1)
98	gray ware jar (2)	158	flake (1); not mapped
	gray ware other (3)	160	flake (1)
109	gray ware jar (10)	163	flake (1)
110	gray ware jar (115); Vessel 75 (1)	164	flake (1)
113	gray ware jar (9)	175	flake (1)
124	gray ware jar (7); Vessel 75 (1), Vessel 78 (4)	183	flake (1)
125	gray ware jar (15); Vessel 75 (6), Vessel 78 (3), Vessel 80 (3)	184	core (1)
		193	flake (1)
127	gray ware jar (156); Vessel 74 (119), Vessel 83 (8), Vessel 84 (9), Vessel 85 (3)	194	flake (1)
		202	edge-damaged flake (1); not mapped
130	gray ware jar (8); Vessel 75 (1)	203	flake (1)
135	gray ware jar (20); Vessel 74 (2), Vessel 84 (6), Vessel 85 (6)	**Metate**	
		199	complete, conglomerate
136	gray ware jar (9); Vessel 75 (2)	**Trough Metate**	
139	red ware bowl (1)	46	incomplete, conglomerate
140	red ware bowl (1)	65	fragmentary, conglomerate
144	gray ware jar (2)	92	complete, sandstone
152	gray ware jar (17); Vessel 75 (3), Vessel 80 (7)	153	fragmentary, sandstone
	white ware jar (1)	**One-Hand Mano**	
158	gray ware jar (11); Vessel 75 (1); not mapped	42	complete, igneous
161	gray ware jar (1)	**Two-Hand Mano**	
162	gray ware jar (1)	44	complete, sandstone
165	gray ware jar (1)	45	complete, sandstone
167	white ware bowl (1)	52	complete, sandstone
168	red ware bowl (1)	54	complete, conglomerate
169	gray ware jar (1)	58	complete, igneous
179	red ware bowl (2)	59	complete, conglomerate

Table 2.31. Point-Located Artifacts from Pit Structure 4, Surface *(continued)*

No.	Description	No.	Description
64	complete, sandstone	196	siltstone with hematite
66	complete, igneous	**Awl**	
126	complete, sandstone	14	fragmentary, artiodactyl
129	complete, conglomerate	24	fragmentary, *Odocoileus hemionus*
138	complete, sandstone	24	fragmentary, *Odocoileus hemionus*
166	complete, unknown orthoquartzite	24	fragmentary, *Odocoileus hemionus*
177	complete, sandstone	28	complete, cf. *Odocoileus*
Abrader		157	fragmentary, artiodactyl
4	fragmentary, sandstone	198	fragmentary, artiodactyl
Hammerstone		200	fragmentary, artiodactyl
112	complete, igneous	**Gaming Piece**	
131	complete, Dakota orthoquartzite	3	complete, indeterminate mammal
149	incomplete, quartz	**Pendant**	
Peckingstone		27	complete, calcite
68	complete, Dakota orthoquartzite	**Modified Vegetal**	
69	complete, Morrison orthoquartzite	170	cordage, fragmentary
88	complete, Morrison orthoquartzite	**Dating Samples**	
106	condition unknown, Morrison orthoquartzite	2*	tree-ring sample
133	complete, Morrison orthoquartzite	13*	tree-ring sample
134	complete, Dakota orthoquartzite	17*	tree-ring sample
185	complete, Dakota orthoquartzite	22*	tree-ring sample
187	complete, Dakota orthoquartzite	60*	tree-ring sample
Single-Bitted Axe		62*–63*	tree-ring samples
15	complete, igneous	74*	tree-ring sample
180	incomplete, Morrison orthoquartzite	93*–96*	tree-ring samples
201	complete, Morrison orthoquartzite	99*–100*	tree-ring samples
Modified Flake		103*–105*	tree-ring samples
19	complete, Morrison orthoquartzite	107*–108*	tree-ring samples
30	complete, Morrison orthoquartzite	114*–123*	tree-ring samples
Modified Cobble		128*	tree-ring sample
8	condition unknown, unknown orthoquartzite	**Miscellaneous**	
43	fragmentary, igneous	5	human bone
132	fragmentary, sandstone	25	human bone; Burial 11
137	incomplete, igneous	145	human bone; Burial 14
146	complete, igneous	171	human bone; Burial 12
147	complete, igneous	10	animal bone
148	fragmentary, igneous	34	animal bone
Polishing/Hammerstone		67	animal bone
9	complete, unknown orthoquartzite	72	animal bone
Other Modified Stone/Mineral		91	animal bone
150	incomplete, sandstone	111	animal bone
151	other modified stone; not collected	141	animal bone
154	other modified stone; not collected	190	animal bone; not mapped
155	other modified stone; not collected	34	vegetal
Unmodified Stone/Mineral		189	vegetal
11	unknown mineral		

NOTES: Item counts appear in parentheses. For pottery, sherds belonging to vessels are also included in total item count. Refer to Figure 2.45 for artifact locations. Items marked with an asterisk (*) are shown on the roof fall map, Figure 2.42.

forms are recognized in the sherd assemblage. Many of the point-located sherds belong to eight sherd containers and six partial or nearly complete jars and bowls (one Chapin Gray jar, two Moccasin Gray jars, one Mancos Gray jar, and two Bluff Black-on-red bowls). Eight of the vessels are made of pieces found exclusively on the Pit Structure 4

floor, and five include sherds from both floor and roof fall proveniences.

Pollen samples collected from the floor yielded evidence of prehistoric use of cholla, prickly pear, corn, and squash. Most of the samples were collected from beneath stone tools, contexts least likely to have been contaminated by postabandonment pollen rain. Flotation samples from a variety of floor contexts yielded plant remains (cheno-am, groundcherry, cactus, Indian rice grass, juniper, squaw apple, and corn) that could represent use as food or accumulation of debris during abandonment.

Features. Thirty-nine features, 15 of them capped, were found in Pit Structure 4 (Figure 2.44 and Table 2.32). Three of the features are human skeletons. Burial 11 (Feature 36) consists of the incomplete, fragmented, and burned remains of a three- to four-year-old child. The bone fragments were found in direct contact with the floor and were covered with burned roof fall. Because the remains were not articulated, determining position and orientation was difficult; however, cranial fragments were found at the southeast end of the scatter.

Burial 12 (Feature 37) is the skeleton of an adult female, at least 50 years of age, found on top of the Stratum 3 ash deposit, which was the final use surface in this part of the structure. The remains are better preserved than those of the other two individuals, and most of the bones are articulated. The skeleton was face up, fully extended, with the head and shoulders lying across the hearth and the arms flexed over the chest. The head was oriented to the southeast. Surrounding the body were several large, unmodified pieces of sandstone that appear to have been deliberately placed as part of burial preparation. (Figure 2.45 shows Burial 12 in relation to these rocks and in relation to other floor artifacts; Figure 2.44 illustrates the skeletal elements only). The left ribs and right shoulder were partly covered by two of the sandstone pieces, and the right leg was slightly on top of a slab that partly covered adjacent Burial 14. The arrangement of the skeleton and the absence of unmodified sandstone in other parts of the structure indicate that the body and the stones were placed deliberately. It is significant that some bones of Burial 12 are only partly burned and that these are consistently burned on the top rather than on the bottom. The bones of the right shoulder are burned on top, except where they were protected by the sandstone slab. The cranium was socketed slightly into the ash that covers the hearth, yet the bone is unburned. Clearly, the fire that burned the bones and the roof was entirely on top of the skeletons. Even though the body was on top of the hearth, the underside of the skeleton is not burned, indicating that the hearth did not cause the fire that destroyed the roof.

Burial 14 (Feature 38), like Burial 11, is severely burned, fragmented, and disarticulated. The remains are those of a child, five to six years of age, that had been placed alongside the remains of the adult, Burial 12. The bones were in such poor condition that little could be determined with regard to placement and orientation, although the head appeared to be to the northwest.

Feature 1 is the central hearth, which was constructed by pecking a basin-shaped pit into bedrock. The sides of

Table 2.32. Feature Summary, Pit Structure 4

No.	Type	Shape	Length (cm)	Width (cm)	Depth (Height) (cm)	Status
1	hearth	basin	50	50	13	in use
2[a]	vent	—	—	—	—	—
3	ash pit	other	44	21	8	in use
4	pit	basin	35	26	10	capped
5	post hole	cylindrical	13	12	26 (25)	in use
6	post hole	cylindrical	11	9	22 (10)	in use
7	pit	basin	20	22	7	in use
8	post hole	other	12	10	17	in use
9	pit	basin	28	28	2	in use
10	pit	cylindrical	13	11	5	capped
11	pit	basin	13	13	4	in use
12	post hole	cylindrical	16	16	25	in use
13	pit	other	20	13	10	in use
14	pit	cylindrical	12	11	12	in use
15	pit	basin	15	14	13	in use
16	pit	other	21	20	14	capped
17	pit	other	9	7	13	in use
18	pit	other	27	6	5	in use
19	pit	other	17	16	10	unknown
20	pit	basin	21	19	7	capped
21	pit	other	8	7	10	in use
22	pit	basin	35	30	4	capped
23	pit	basin	29	27	6	unknown
24	pit	cylindrical	11	10	11	capped
25	pit	cylindrical	8	8	12	in use
26	pit	basin	13	13	12	capped
27	pit	other	9	8	8	in use
28	sipapu	cylindrical	9	9	10	capped
29	pit	basin	10	10	5	capped
30	pit	basin	10	10	6	capped
31	pit	basin	7	7	8	capped
32	pit	cylindrical	9	9	8	capped
33	pit	basin	9	9	8	capped
34	pit	nr	9	8	9	capped
35	wing wall	other	267	15	(90)	—
36	Burial 11	other	65	30	(3)	—
37	Burial 12	other	155	35	(10)	—
38	Burial 14	other	75	40	(4)	—
39	paho marks	other	30	10	7	capped

NOTE: Feature locations are shown in Figure 2.44.
nr = no record.
[a] Not excavated.

the hearth are predominantly bedrock, but a portion of the southern wall is lined with silt loam. The base and the rim of the hearth are fire-reddened. Two strata were recognized during excavation. The lower stratum (Stratum 2) was an 8-cm-thick layer of silt loam mottled with ash, and it is interpreted to be construction or remodeling fill that reduced the original size of the hearth. The top of this layer was heavily oxidized. Stratum 1, which overlay Stratum 2, was a 7-cm-thick deposit of compressed, white ash deposited during the latter stages of hearth use.

The ash pit, Feature 3, is a shallow pit between the hearth and the deflector. Like the hearth, the ash pit was pecked into bedrock below the floor. The bottom of the pit is fairly flat, except for a cylindrical depression (5 cm in diameter and 3 cm deep) in the northeast corner that may have been used with adjacent Feature 13 as a ladder socket. Fill was similar to the Stratum 3 ash that covered the ash pit, hearth, and central floor of the structure.

Feature 28 is interpreted as a sipapu on the basis of its location north of the hearth and the associated paho marks, Feature 39. The 1- to 2-cm-diameter holes that make up Feature 39 are clustered around the rim and scattered to the east of the sipapu. Cross sections of eight of the nine small holes reveal that they are 4 to 7 cm deep and taper to points. Both the sipapu and the small holes contained clean sand and were capped with a 1-cm-thick layer of silt loam, which suggests that they were not in use at the time of abandonment.

The wing wall–deflector complex, Feature 35, encases the two southern roof-support posts, Features 5 and 6. The wall was constructed of a double row of upright slabs set in mud; a single upright slab in the center functioned as a deflector. The wing wall is 90 cm high where it abuts the pit structure walls. During excavation of the upper fill of the structure, the backhoe removed a portion of the center of the wing wall–deflector, and the original height of the feature in this area is unknown.

The four main roof-support posts are defined as such by their locations, the presence of charred post stubs above floor level, and rotted wood in the fill. The southeast and southwest post holes, imbedded in the wing wall, have just been described. Two additional post holes inferred to have held main roof-support posts are in the northeast and northwest corners of the structure (Features 12 and 8, respectively).

Feature 2 is the ventilation system, but because it was not excavated, it is known only from the tunnel opening identified in the south wall of the main chamber. Presumably it consisted of a horizontal tunnel and a vertical shaft, similar to the ventilation systems used in the other three Duckfoot pit structures. The top of the vertical shaft was noted, but not mapped, in a shallow test trench extending south from Pit Structure 4.

The remainder of the Pit Structure 4 floor features are pits of unknown function. Ten were filled with sand. Three basin-shaped pits (Features 4, 9, and 20) close to the hearth may have been used as pot rests.

A rough symmetry in the distribution of floor features is apparent when comparing the east and west halves of Pit Structure 4; that is, features in one half of the structure tend to have counterparts in the other half of the structure. For example, Features 9, 20, 15, 24, 29, 26, 27, 21, and 23 in the east half are similar in terms of feature type, fill, and distribution to Features 4, 10, 25, 16, 14, 17, and 22 in the west half. Similarly, Features 11, 32, and 31 in the

east half are similar to Features 7, 33, and 34 in the west half. This symmetrical feature distribution may be the result of pits having been used in pairs for temporary structures that spanned the north-south axis, or it may be the result of similar activities having taken place in the two halves of the structure.

Dating

Pit Structures 1, 2, and 3 are evenly spaced in relation to one another, and with the roomblock they form a well-organized site plan. Pit Structure 4 is squeezed between Pit Structures 2 and 3, disrupting the symmetry of the site layout. The size and location of Pit Structure 4 indicate that it may have been constructed relatively late in the occupation of the site, after the other structures had been built.

Of the 142 tree-ring samples collected from roof fall and floor proveniences, 81 yielded dates (Appendix B). The dates range from A.D. 689 to 875; cutting dates range from A.D. 864 to 875. A distinct cluster of A.D. 871 to 873 cutting dates strongly argues for construction of Pit Structure 4 in the early A.D. 870s, probably in A.D. 873. The structure was probably occupied for only a few years. The one A.D. 875 cutting date may represent roof repair or maintenance.

Interpretations

Pit Structure 4 is the smallest Duckfoot pit structure, and from its location and tree-ring sample evidence, it is interpreted to be a late addition to the site. A similar pattern was noted in the Dolores area, where small, late structures were sporadically located among strongly patterned pre-existing architectural units. At Grass Mesa Village, late (A.D. 880–910), small pit structures were built between existing structures, into the fill of abandoned pit structures, and in abandoned areas of the roomblocks (Lipe et al. 1988). This same situation was seen on a smaller scale at Rio Vista Village in the Dolores River valley (Wilshusen 1986b) and at Morris Site 33 near Johnson Canyon, approximately 38 km southeast of Duckfoot (Reed et al. 1985).

Most artifacts on the Pit Structure 4 floor are interpreted as primary or de facto refuse reflecting a wide range of domestic and specialized activities. The presence of a hearth and the recovery of a wide variety of food-processing tools and equipment (metates, manos, and storage and cooking vessels) support the argument that the structure was used for domestic activities. However, the presence of a sipapu and related paho marks, plus the deliberate destruction of the structure involving the interment of human remains, strongly suggests ritual use and abandonment. This leaves open the possibility that the floor assemblage was altered by these ritual activities, an issue that is addressed in Volume 2.

Courtyard 2

Courtyard 2 is the area around Pit Structure 2 and south of Room Suite 2 (Figure 2.46). Its eastern boundary is the 97 east grid line, and its western boundary is "stepped" between the 107 and 111 east lines. The latter boundary bisects Pit Structure 4, which was not discovered until after the courtyards were defined and excavation was nearly complete. Stratigraphy and excavation strategy are identical to those described in the Courtyard 1 discussion.

Artifacts

Artifacts recovered from Courtyard 2 fill, exclusive of modern ground surface, include over 3,800 sherds, approximately 500 pieces of chipped-stone debris, and numerous animal bones. Formal tools include 3 manos, 10 pecking-stones, 3 modified flakes, and 1 *Glycymeris* shell bracelet fragment.

Features

Seventeen features were found in Courtyard 2 (Figure 2.46 and Table 2.33), the majority of them (11) located in the northeast corner. A smaller cluster of five features is in the northwestern corner of the courtyard, and one feature is located in the center. One feature is classified as a bin and one as a fire pit; the remaining features are pits of unknown function.

Feature 1 is a small, shallow, basin-shaped pit in the center of Courtyard 2. It appears to have been intentionally filled and capped with sediments that contained large amounts of caliche, and its function is unknown.

Features 2, 4, 5, 6, and 18 make up the cluster of features in the northwest corner of the courtyard. Feature 2 is an unusual bin with several vertical slabs on the north and east sides and another slab placed horizontally to form the bin floor. This bottom slab is imbedded approximately 3 cm into sterile sediments. Originally, additional slabs may have been used to delineate the other sides of the bin, but if so, they had been removed or displaced. No signs of burning or grinding were evident on the slabs, and the fill appears to have been deposited naturally. The function and associations of Feature 2 are unknown, but it is possible that this feature was built into postabandonment wall fall deposits.

Features 4, 5, and 6 are small, cylindrical or basin-shaped pits filled with loam and few cultural inclusions. Fill in all three cases is interpreted as postabandonment in origin. Feature 18 is a large, deep pit that originally was thought to be associated with Room 17, under which it is located. However, after the Courtyard 2 surface outside the room was exposed, it became apparent that the wall of Room 17 was built across the fill of the pit. Thus, the construction, use, and filling of Feature 18

predate the construction of Room 17. Feature 18 is funnel-shaped, with the wide circular opening tapering to a deep cylindrical pit at the bottom. Fill consisted of two strata. Stratum 1 was a strong brown loam mottled with charcoal, constructional mud, and small chunks of sandstone; Stratum 2 was a dark brown loam that filled the lower cylindrical portion of the pit. Stratum 2 lacked the pockets of charcoal and mud that characterized Stratum 1. Feature 18 may have been filled intentionally to prepare for the construction of Room 17. Neither fill nor artifacts found in fill are interpreted to be associated with the use of Feature 18.

A larger cluster of features is found in the northeast corner of Courtyard 2. Feature 3 is a large, basin-shaped pit comparable to Feature 7 in Courtyard 1, except that it is slightly farther from the rim of its associated pit structure. The fill consisted of gray, ashy sediments with scattered charcoal, but there were no signs of in situ burning. Possibly the pit served as a borrow pit or mixing basin for mud used in building construction or maintenance. The gray, ashy fill may have been dumped intentionally to eliminate a water-trapping depression in the Courtyard 2 surface.

Feature 7 is a shallow, basin-shaped pit that contained gray, ashy fill with little charcoal or other cultural materials. The pit did not appear to have been burned but apparently had been filled intentionally with ash or ash-rich sediments.

Features 8, 9, and 10 are broad, shallow pits that intersect one another. Feature 8 had dark, organic fill with pockets of charcoal at the base. The sides of the pit were oxidized, indicating in situ burning, and the feature is interpreted as a fire pit. A flotation sample collected from Feature 8 yielded large quantities of corncobs and cupules, which suggests that corncobs were used as fuel. Feature 9 appears to truncate the west edge of Feature 8. Features 9 and 10 also intersect, but it is not possible to tell which, if either, was earlier, and the two may have been used simultaneously. Both contained a homogeneous, gray-brown loam with some small chunks of charcoal and tiny clumps of sterile sediment. Neither Feature 9 nor Feature 10 appeared to have been burned, yet they were sharply defined around the rims. The function of these pits is unknown, and the fill is not interpreted as being associated with their use.

Features 11, 12, 13, and 14 are small, circular pits clustered approximately 1 m south of the south wall of Room 11. Features 11, 12, and 13 were filled with a brown loam that, on the basis of its similarity to courtyard fill, was inferred to be postabandonment in origin. Feature 14 fill contained ash and had been capped before abandonment. In all four cases, feature function is unknown.

Feature 15 is a deep pit with nearly vertical walls and a flat bottom. The base and the walls are lined with large, thin, shaped sandstone slabs, and several other large slabs

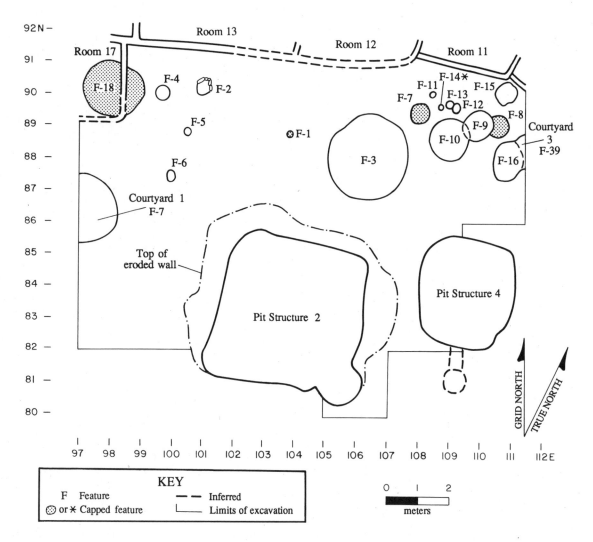

Figure 2.46. Courtyard 2, plan. The upper walls of Pit Structure 2 eroded after excavation of the pit structure was completed but before excavation of the courtyard began. The eroded area is indicated on the map. Features are described in Table 2.33.

are in the fill. Fill consisted of two strata. Stratum 1 was a gray, ashy loam mottled with clumps of sterile sediment, caliche, and charcoal. The sterile sediments and caliche are interpreted as structural debris, probably from Rooms 10 and 11. Stratum 2 was a gray-brown loam with fewer inclusions than Stratum 1 and finer bits of charcoal and caliche. Sandstone slabs were scattered throughout Stratum 2—perhaps the remains of a cover that once protected the walls and contents of Feature 15. The southeast quadrant of the feature contained a burned, black, organic-rich loam that lay directly on the basal sandstone slab, although sediment color did not appear to be the result of in situ burning. There were many large sherds at the base of the pit, and some appeared to belong to the same two, or maybe three, vessels, including one heavily sooted jar. A variety of potential food items (cheno-am, Indian rice grass,

groundcherry, and corn) were recovered from two flotation samples collected from Stratum 2. Two functional interpretations may be offered for Feature 15. Its size, shape, and construction suggest that it may have been used for storage. The sherds may represent vessels that contained the stored goods. On the other hand, the burned lower fill and the sooted pot suggest that the slab-lined pit may have been used as a warming oven, although intense fires do not appear to have been built in Feature 15.

Feature 16 is a broad, shallow, basin-shaped pit with an irregular bottom. The fill consisted of charcoal, burned mud, sherds, and chipped-stone debris in a loam matrix. The contents of the pit were burned, but the sides and bottom were not. This feature may have been a small borrow pit that was backfilled with trash.

Table 2.33. Feature Summary, Courtyard 2

No.	Type	Shape	Length (cm)	Width (cm)	Depth (Height) (cm)	Status
1	pit	basin	15	15	5	capped
2	bin	other	50	48	3 (31)	unknown
3	pit	basin	250	250	30	unknown
4	pit	basin	44	44	26	unknown
5	pit	cylindrical	18	17	11	in use
6	pit	basin	35	28	5	unknown
7	pit	basin	60	60	11	capped
8	fire pit	basin	75	75	10	capped
9	pit	basin	87	60	16	unknown
10	pit	basin	124	120	23	unknown
11	pit	other	15	15	6	unknown
12	pit	basin	28	25	12	unknown
13	pit	cylindrical	19	18	16	unknown
14	pit	cylindrical	12	10	13	capped
15	pit	other	70	60	56	in use
16	pit	basin	130	90	28	in use
18	pit	other	195	167	74	capped

NOTE: Feature locations are shown in Figure 2.46.

Interpretations

Like Courtyard 1, Courtyard 2 was probably a focal point for a variety of everyday activities ranging from storage to food preparation, and possibly many others not identifiable from the available feature evidence. Much of this outdoor activity seems to have been concentrated within a 2-m-wide strip immediately south of the roomblock, and particularly in front of Room 11. Feature density decreases to the south, and a relatively clear space seems to have been maintained closer to the pit structure, perhaps to allow pedestrian traffic. Although the eastern and western boundaries of Courtyard 2 were established along grid lines, it is inferred that these limits approximately define space used primarily by the inhabitants of Architectural Suite 2. However, the fact that the boundaries crosscut structures and features underscores the extent to which courtyard boundaries are arbitrary.

Architectural Suite 3

The third architectural suite at Duckfoot is at the east end of the pueblo. As shown in Figure 2.1, Architectural Suite 3 consists of two front rooms (Rooms 10 and 19), three back rooms (Rooms 8, 9, and 18), and an associated pit structure (Pit Structure 3) and courtyard (Courtyard 3). The rooms of Architectural Suite 3 appear to have been added on to those of the core suite (Architectural Suite 2) in the late A.D. 850s, possibly in A.D. 859. Whether Suite 3 was added at the same time as Architectural Suite 1 at the west end of the pueblo is not known, but it is unlikely that the two construction events were widely separated in time.

Room 10

Length: 4.40 m
Width: 2.90 m
Area: 11.98 m²

Room 10, a relatively small front room, is unusual in that it is longer along its north-south axis than along its east-west axis (Figures 2.47 and 2.48). The room is between Room 11 to the west and Room 19 to the east and is associated architecturally with only one back room, Room 8 (Figure 2.1). On the basis of changes in its feature assemblage, Room 10 is inferred to have served as a domestic structure early in its use but as a storage facility by the end of site occupation. The room was burned at abandonment. The eastern third of Room 10, and particularly the northeast corner, had been excavated by pot hunters before Crow Canyon's work on the site. Fill and floor deposits were disturbed, and this damage is reflected in plan maps of the room, which show few artifacts near the east wall of the structure. The dirt from the looter's pit was removed from modern ground surface around the pit and screened. The items in this dirt were recorded as having been found in a recently disturbed area (i.e., no stratigraphic context) within Room 10. The remainder of the room was excavated as a single horizontal unit in two strata (described below). At approximately 10 cm above the floor, a 1-×-1-m grid was established, and the remaining fill was excavated within individual squares. Artifacts in the zone 5 to 10 cm above the floor were grouped horizontally with other artifacts from the same level throughout the room. Artifacts within 5 cm of the floor were provenienced with the grid square in which they were found, and all items found in contact with the surface were point-located.

Stratigraphy

The fill of Room 10 consisted of two strata. The upper one, Stratum 1, was a 15- to 20-cm-thick layer of dark brown to strong brown silt loam with sandstone, charcoal, and relatively few artifacts. This stratum, which had undergone some soil development, is interpreted to be the upper portion of the collapsed structure, primarily wall fall. Stratum 2 was 30 to 40 cm thick and was predominantly a strong brown to dark brown silt loam with charred beams and clumps of burned sediment. Stratum 2 is interpreted as burned roof fall, but the absence of burned beams on the floor, or within 5 cm of the floor, suggests that the room may have been abandoned for a brief time before burning. Artifacts recovered from Stratum 1 include sherds, chipped-stone debris, and a few pieces of animal bone. Artifacts found in the roof fall stratum might have been on the roof when the structure was abandoned, or they could have been discarded into the recently destroyed room.

Point-located items are mapped in Figure 2.47 and described in Table 2.34, and they include 11 sherds, 2 manos, a maul, 3 peckingstones, and 2 animal bone fragments. In addition, many tree-ring samples were collected from roof fall, and results of analysis are presented in the discussion on dating, below. The point-located sherd assemblage consists of five Moccasin Gray, two Chapin Gray, and four Indeterminate Plain Gray sherds; both jar and bowl sherds are present. Several sherds found in a cluster in the west half of the room (PLs 13 and 31) belong to a partial Moccasin Gray jar whose other pieces were found in Stratum 1 and on the Room 10 floor.

Construction

The preserved portions of all four walls of Room 10 are constructed of vertical sandstone slabs and mud (Figure 2.48). The north and west wall bases are largely intact, and enough remains of the east and south walls to infer their locations. The upper walls were probably of mud-and-stone construction. Two post holes (one in the southwest corner of the room, one in the southeast) and a shallow pit that may have served as a post hole (in the northwest corner)

probably held roof-support posts. The floor was excavated prehistorically into sterile B-horizon sediments and compacted through use.

Surface

The floor of Room 10 was a use-compacted surface at the contact of undisturbed sterile sediments below and post-abandonment fill above.

Artifacts. Point-located artifacts found within 5 cm of the floor are described in Table 2.35, but only the few in direct contact with the floor are mapped in Figure 2.48. The point-located assemblage as a whole consists mostly of sherds, including one modified sherd, but one piece of chipped-stone debris and a peckingstone were recovered as well. Of the 19 sherds assigned point-location numbers, 18 (94.7 percent) are gray ware and 1 (5.3 percent) is red ware. Two Moccasin Gray sherds are the only gray wares identified to a specific type; the red ware is a Bluff Black-on-red bowl sherd. With the exception of one piece of a small bowl or ladle, the gray ware assemblage consists entirely of jar fragments. Two sherds belong to a partial Moccasin Gray

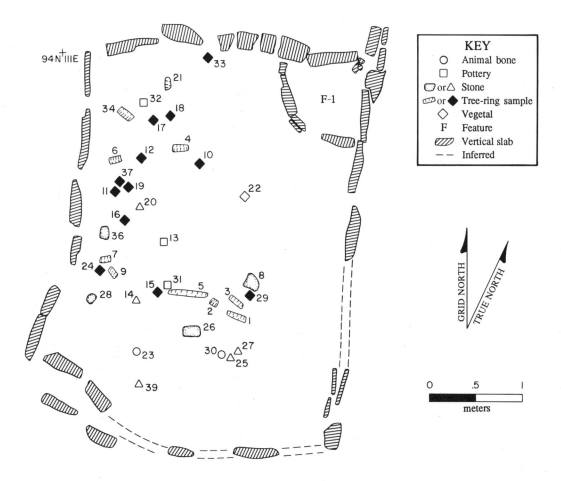

Figure 2.47. Room 10, roof fall. Point-located items are described in Table 2.34.

jar whose other pieces were scattered throughout room fill. It is possible that none of the sherds on the floor relate to the use of the room. Sherds could have accumulated as primary refuse while the room was in use, or they could have accumulated after the room was abandoned.

Features. Sixteen features were identified in Room 10: three bins, an abandoned hearth, two cists, two post holes, and eight pit features whose functions are unknown (Figure 2.48 and Table 2.36). The hearth was capped, which indicates that a functional shift took place sometime during the use of the room.

The northeast corner of Room 10 is one of only two areas on the site where recent disturbance by looters is apparent. The slabs of Feature 1, an above-floor corner bin, were visible on modern ground surface and apparently

Table 2.34. Point-Located Artifacts from Room 10, Roof Fall

No.	Description
Pottery Sherds	
13	gray ware jar (8); Vessel 73 (7)
31	gray ware jar (1); Vessel 73 (1)
32	gray ware jar (1)
	gray ware bowl (1)
Mano	
36	fragmentary, sandstone
One-Hand Mano	
28	incomplete, Dakota orthoquartzite
Peckingstone	
14	complete, Morrison orthoquartzite
27	complete, Morrison orthoquartzite
39	complete, Morrison orthoquartzite
Maul	
25	complete, unknown stone
Other Modified Stone/Mineral	
8	missing
26	fragmentary, sandstone
Unmodified Stone/Mineral	
20	complete, sandstone
Dating Samples	
1–7	tree-ring samples
9–12	tree-ring samples
15–19	tree-ring samples
21	tree-ring sample
24	tree-ring sample
29	tree-ring sample
33–34	tree-ring samples
37	tree-ring sample
38	tree-ring sample; not mapped
Miscellaneous	
23	animal bone
30	animal bone
22	vegetal

NOTES: Item counts appear in parentheses. For pottery, sherds belonging to vessels are also included in total item count. Refer to Figure 2.47 for artifact locations.

attracted a looter who probed into the fill, and possibly the floor, of the bin. Feature 1 is made of vertical slabs and mud, and the existing walls stand 58 cm high. The floor of the bin is approximately 6 cm lower than the floor of the room. Bin fill had been completely excavated by looters, and although much of the floor appeared to be intact, an oval pit (Feature 15) excavated into the floor may have been the result of the looter's digging. No artifacts were found on the Feature 1 floor or in Feature 15 fill.

Feature 2 is a large bin against the center of the east wall. The walls of the bin are delineated by large vertical slabs that still stand up to 60 cm high. The floor of the bin is 9 cm lower than the floor of the room. Some of the fill of Feature 2 apparently was removed by the looter; again, the floor of the bin seemed to be mostly intact, but an oval pit (Feature 7) dug into the floor may have been recent damage. Feature 7, like its counterpart in the northeast corner bin, contained no artifacts.

Feature 3 is a bin in the southeast corner of Room 10. Only a few vertical slabs (maximum height, 58 cm) remain as evidence of walls, but the floor of the bin was excavated prehistorically 12 cm lower than the floor of Room 10, which helped to delineate feature boundaries. No artifacts were found on the floor of the bin. Feature 3 contained natural postabandonment deposits rather than the burned roof materials found in the main portion of the room. This could mean that the feature was not maintained during the final use of the structure and began to fill before abandonment. Alternatively, the bin could have been roofed or otherwise sealed, which would have prevented burned roof fall from collapsing into the feature at abandonment.

Pollen samples collected from Features 1, 2, and 3 yielded many of the same pollen types identified from other contexts at the site, but an interesting difference in the specific distributions may shed light on what was stored in the three bins. Prickly pear and Cheno-am pollen predominate in the Feature 1 sample, and beeweed and corn predominate in the Feature 2 sample. Flotation samples collected from Features 1, 3, and 5 yielded a number of possible fuel and construction resources similar to those documented elsewhere at the site.

Feature 4 is a large, basin-shaped hearth that was abandoned and capped sometime during the occupation of the room. The basin was filled with sandy loam and small bits of charcoal. Fill was not the ashy sediment typical of primary hearth refuse, but the rim of the pit was fire-reddened. Apparently, after the pit was used as a hearth, it was cleaned out, filled with dirt, and capped. This event is the basis for proposing a functional shift in Room 10 from a domestic structure earlier in the occupation to a storage facility later. Sometime after Feature 4 was filled in, it was truncated by Feature 17.

Feature 17 is a large, cylindrical cist excavated below the floor of Room 10. The cist was filled with roof fall

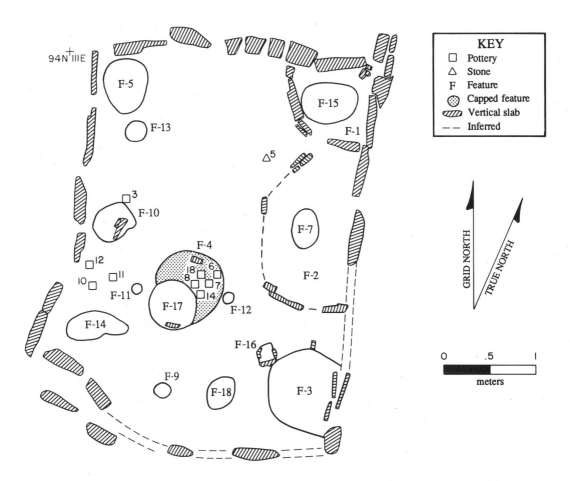

Figure 2.48. Room 10, surface. Point-located items are described in Table 2.35; features are described in Table 2.36.

deposits, indicating that the pit was open at abandonment. Fill contained, among other items, a projectile point made of nonlocal stone. A mano was found on the cist floor. It is because of its size and shape that Feature 17 is interpreted as a storage cist. Feature 17 truncates the hearth (Feature 4), a remodeling event that signals a functional change in Room 10.

Feature 5 is a cist in the northwest corner of Room 10. The cist contained burned roof fall and is believed to have been left open at abandonment. The size, shape, and location are the only bases for interpreting Feature 5 as a storage cist.

Features 14 and 16 are post holes. Feature 16 is nearly square in plan and is lined with small sandstone shims. Feature 14 is unusual in that it is bilobed, consisting of two overlapping pits. The east half is about 39 cm deep and has small, sandstone-slab shims set on edge around the sides of the pit. The west half of the feature is only 20 cm deep and has two sandstone slabs at the base. The relationship between the two halves is not clear, but it is possible that a post was removed from its original hole and reset in an overlapping hole as part of roof remodeling or repair.

The remaining features are pits whose functions are not known. One, Feature 13, is a cylindrical pit in the northwest corner of the room that might be a post hole. However, it is rather shallow to have supported a post, particularly a main roof-support post. Features 9, 10, 11, 12, and 18 are pits in the south half of the room; all apparently were open at abandonment and were filled with natural deposits or a combination of roof fall and natural deposits.

Dating

Twenty-three tree-ring samples were collected from the roof fall stratum in Room 10. Only five yielded dates—A.D. 825, 841, 850, 861, and 873—and of these only one, A.D. 861, is a cutting date (Appendix B). Adjacent Room 19 yielded a sample with a cutting date of A.D. 859, and given the layout of the roomblock, Room 10 could not have been built after Room 19. Also, tree-ring samples collected from Pit Structure 3, the pit structure associated with the eastern end of the roomblock, yielded a cluster of late A.D. 850s cutting dates, including many A.D. 859 dates. If the construction of the pit structures is related to that of the rooms,

Table 2.35. Point-Located Artifacts from Room 10, Surface

No.	Description
Pottery Sherds	
1	gray ware jar (1)
2	gray ware jar (1)
3*	gray ware jar (1)
6*	gray ware jar (1)
7*	gray ware jar (1)
8*	gray ware jar (1)
11*	gray ware jar (3)
12*	gray ware jar (1)
14*	gray ware other (1)
15	gray ware jar (1)
16	gray ware jar (1)
17	gray ware jar (2); Vessel 73 (2)
18*	gray ware jar (1)
35	gray ware jar (1)
	red ware bowl (1)
Chipped-Stone Debris	
5*	other (1)
Modified Sherd	
10*	incomplete, gray ware
Peckingstone	
9	complete, Morrison orthoquartzite

NOTES: Item counts appear in parentheses. For pottery, sherds belonging to vessels are also included in total item count.

*Items found in direct contact with the surface; locations are shown in Figure 2.48. Other point-locations were not mapped.

Table 2.36. Feature Summary, Room 10

No.	Type	Shape	Length (cm)	Width (cm)	Depth (Height) (cm)	Status
1	corner bin	other	78	76	6 (58)	in use
2	bin	other	173	95	9 (60)	in use
3	corner bin	other	84	67	12 (58)	in use
4	hearth	basin	77	75	20	capped
5	cist	cylindrical	56	46	22	in use
7	pit	basin	43	29	50	unknown
9	pit	basin	20	16	4	in use
10	pit	other	44	42	12	in use
11	pit	basin	11	10	4	in use
12	pit	cylindrical	12	12	7	in use
13	pit	cylindrical	20	20	12	in use
14	post hole	other	68	29	39	in use
15	pit	other	58	42	28	unknown
16	post hole	cylindrical	24	20	30	in use
17	cist	cylindrical	60	60	65	in use
18	pit	basin	40	30	12	in use

NOTE: Feature locations are shown in Figure 2.48.

a late A.D. 850s building date is indicated for Room 10. The A.D. 861 and 873 dates yielded by Room 10 samples might date structural repair or remodeling; the latter date suggests that the room continued to be used until at least the early A.D. 870s.

Interpretations

Because Room 10 is located in the front row of the roomblock and because it initially had a hearth, it is believed to have served originally as a domestic room. However, sometime during the occupation of the structure, the hearth was cleaned out, filled, and capped, effectively removing it from the suite of available features. A storage cist was built in its stead, partly truncating the earlier hearth. When this remodeling occurred is not known, but it is believed that, by the end of the occupation, Room 10 had ceased to serve as a domestic room and was used instead as a storage structure.

Room 19

Length: 4.85 m
Width: 2.55 m
Area: 12.61 m^2

Room 19 (Figures 2.49–2.52) is the southeasternmost structure in the roomblock, and it adjoins Rooms 9 and 18 to the north (Figure 2.1). Like many other rooms in the front row of the roomblock, Room 19 burned at abandonment. Initial excavations in Room 19 were in a 35-cm-wide trench that extended north-south across the room along the 117.5 east grid line. This trench was trowel-excavated to the floor for the purpose of defining stratigraphy and to locate stratigraphic evidence of the south room wall. The upper fill of the room was excavated in strata down to 10 cm above the floor. At that point, a 1-×-1-m grid was established across the room, and the remaining fill was excavated in grid squares. The 5- to-10-cm-above-floor level was provenienced as a single horizontal unit, even though it was excavated in 1-×-1-m squares. All artifacts found in place within 5 cm of the floor were point-located, and those found in the screen were provenienced to the appropriate excavation square.

Stratigraphy

Figure 2.49 is a north-south profile that shows the three strata recognized in Room 19. Stratum 1 was a brown to dark brown sandy loam interpreted to be the soil A horizon that developed in the upper fill above wall and roof fall deposits. Artifacts recovered from this stratum include numerous sherds, chipped-stone debris, animal bone, two peckingstones, and two modified flakes. Stratum 2 was dark brown loam mottled with charcoal, burned sediment, and caliche. Stratum 3 was a discontinuous layer of strong brown silt loam similar to Stratum 2, but it (Stratum 3) had even more caliche inclusions. Strata 2 and 3 are interpreted as collapsed structural debris (wall and roof fall) and were differentiated in profile in spite of there being only subtle differences in their appearance and composition. Because

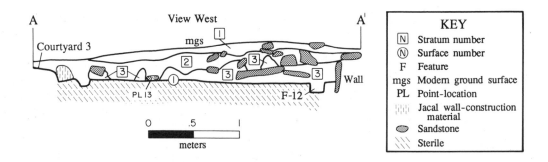

Figure 2.49. Room 19, stratigraphic profile. The location of the profile is shown in Figure 2.51.

it was not possible to consistently separate Stratum 2 from Stratum 3 during excavation, they were treated more like arbitrary levels, with Stratum 3 being 10 to 5 cm above the floor. Figure 2.50 shows point-located items from Strata 2 and 3 combined, and the items are listed in Table 2.37. The jacal wall sediment shown in Figure 2.49 and interpreted to be an in situ wall remnant was visible only in profile and does not appear on the corresponding plan map. Point-located items in roof/wall fall include sherds, two metates, six manos, two projectile points, and one modified flake. The modified

flake is made of obsidian, and one of the projectile points is of agate or chalcedony. In addition, numerous tree-ring samples were collected from roof fall, and the results of analysis are discussed in the section on dating, below.

Of the 34 Strata 2 and 3 sherds analyzed, 32 (94.1 percent) are gray ware and 2 (5.9 percent) are red ware. The only sherds identified to specific type are one Bluff Black-on-red and four Moccasin Gray sherds. All 34 pieces are classified as jar fragments, and two belong to a gray ware sherd container.

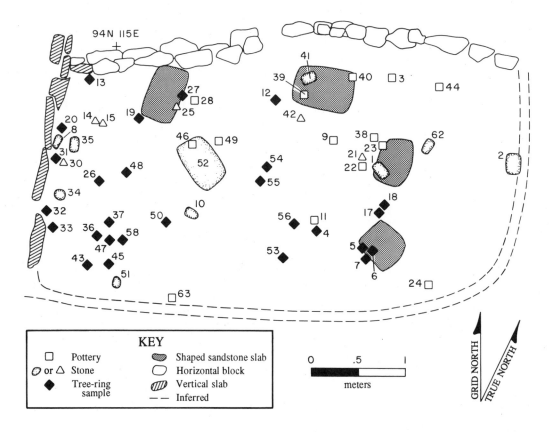

Figure 2.50. Room 19, wall and roof fall. Point-located items are described in Table 2.37.

Construction

Three different construction styles are evident in the walls of Room 19 (Figure 2.51). The north wall, made of coursed-block masonry, still stands two to three courses high. The west wall is constructed of large, vertical sandstone slabs supported by mud. The south and east walls were of wattle-and-daub construction, as evidenced by an alignment of small post holes around the edges of the room. The upper portions of the north and west walls were probably of wattle-and-daub construction also. The roof would have been supported by four posts (represented by Features 9, 10, 11, and 14) and by the structure walls. The floor was excavated prehistorically into sterile B-horizon sediments and compacted through use.

Surface

The living floor of Room 19 was recognized as an abrupt stratigraphic break between collapsed structural debris above and sterile sediment below. The floor contained among the richest and most varied artifact and feature assemblages in front rooms at Duckfoot.

Artifacts and Samples. Point-located materials are mapped in Figure 2.51 and described in Table 2.38; they include sherds, chipped-stone debris, an animal bone fragment, seven manos, a peckingstone, an axe, a projectile point, an awl, and four *Glycymeris* shell bracelet fragments. An additional *Glycymeris* bracelet fragment was recovered but not point-located. Pottery was abundant on the floor, and the assemblage includes numerous large sherds and clusters of sherds. Of the 328 point-located sherds recovered, 305 (93.0 percent) are gray ware, 14 (4.3 percent) are red ware, and 9 (2.7 percent) are white ware. Most sherds are of indeterminate types, but 3 Chapin Gray, 34 Moccasin Gray, 2 Mancos Gray, 4 Chapin Black-on-white, and 10 Bluff Black-on-red sherds are present as well. Jar sherds predominate, but bowl and ladle forms also are recognized. Several sherds belong to three partial or nearly complete Moccasin gray jars, a partial Chapin Gray ladle, a partial white ware jar, and a gray ware sherd container. Also included in the vessel assemblage is a complete, unbroken Moccasin gray gourd jar.

Pollen samples were collected from the floor beneath stone tools found either directly on the floor or in lower roof fall. Results are reported in Chapter 7.

Table 2.37. Point-Located Artifacts from Room 19, Wall and Roof Fall

No.	Description	No.	Description
Pottery Sherds		35	complete, conglomerate
3	gray ware jar (7)	62	complete, igneous
9	gray ware jar (1)	Peckingstone	
11	gray ware jar (1)	42	missing
22	gray ware jar (1)	Projectile Point	
23	gray ware jar (1)	14	incomplete, agate/chalcedony
24	gray ware jar (1)	21	complete, Dakota orthoquartzite
28	gray ware jar (2)	Modified Flake	
38	gray ware jar (2); Vessel 64 (2)	30	complete, obsidian
39	gray ware jar (3)	Modified Cobble	
	red ware jar (2)	34	complete, igneous
40	gray ware jar (1)	51	fragmentary, igneous
44	gray ware jar (4)	Dating Samples	
46	gray ware jar (1)	4–7	tree-ring samples
49	missing	12–13	tree-ring samples
63	gray ware jar (7)	16	tree-ring sample; not mapped
Chipped-Stone Debris		17–20	tree-ring samples
15	missing	26–27	tree-ring samples
Trough Metate		31–33	tree-ring samples
25	fragmentary, sandstone	36–37	tree-ring samples
52	complete, conglomerate	43	tree-ring sample
Mano		45	tree-ring sample
41	missing	47–48	tree-ring samples
Two-Hand Mano		50	tree-ring sample
1	complete, sandstone	53–56	tree-ring samples
2	complete, conglomerate	57	tree-ring sample; not mapped
8	fragmentary, conglomerate	58	tree-ring sample
10	fragmentary, sandstone		

NOTES: Item counts appear in parentheses. For pottery, sherds belonging to vessels are also included in total item count. Refer to Figure 2.50 for artifact locations.

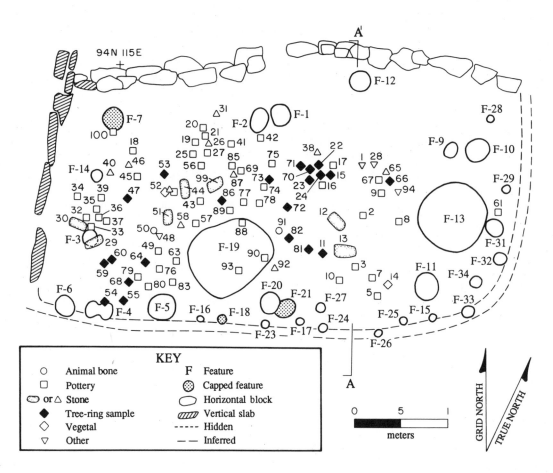

Figure 2.51. Room 19, surface. Point-located items are described in Table 2.38; features are described in Table 2.39. Profile A is shown in Figure 2.49.

Features. Thirty-one features were found on the floor of Room 19 (Figures 2.51 and 2.52 and Table 2.39). Most are post holes or pits of unknown function, but a hearth and two cists were excavated as well.

Feature 19 is a basin-shaped hearth just south and west of the center of the room. It appears as an irregular oval in plan, and the sides and bottom of the pit were fire-reddened. The base of the pit was lined with sandy loam, and upper fill consisted of ash with small charcoal inclusions. The ashy deposits extended beyond the rim of the hearth approximately 10 to 20 cm onto the surrounding floor. A flotation sample from Feature 19 yielded remains of juniper, pine, mountain mahogany/sagebrush, corn (cupule), and cottonwood/willow, probably indicative of fuel use. Cottonwood/willow was rare in Duckfoot flotation samples.

Features 12 and 13 are bell-shaped cists near the north and east walls of the room, respectively. Feature 12 is the smaller of the two, and it contained charcoal, sandstone slabs, and artifacts in a silt loam matrix. The fill appeared to be burned roof fall, and therefore the feature is inferred to have been left open at abandonment. Feature 13 is much

larger and it contained unusual fill. The lower 25 cm consisted of black, ashy charcoal, and several charred wood fragments large enough to be tree-ring samples were recovered. This stratum was covered by 45 cm of silt loam mixed with charcoal and sandstone interpreted to be collapsed roofing material. Although large bell-shaped cists usually are thought to have been used for storage, the charred, ashy deposit at the bottom of Feature 13 was similar to primary refuse fill in hearths and fire pits. It is possible that Feature 13 originally functioned as a storage cist but later was reused as a roasting pit or oven. If so, the fires built inside could not have attained very high temperatures, since the walls of the pit were not fire-reddened. Because it is not known whether the potential food and fuel resources represented in flotation samples were related to feature use or were incidentally incorporated into fill when the roof burned and collapsed, the results of sample analysis provided little information to clarify feature function. Pollen identified in samples collected from Feature 13 fill and floor include corn, squash, beeweed, Cheno-am, cholla, prickly pear, and buckwheat.

Table 2.38. Point-Located Artifacts from Room 19, Surface

No.	Description	No.	Description
Pottery Sherds		93	gray ware jar (7)
2	gray ware jar (3)		gray ware other (2); Vessel 61 (1)
3	gray ware jar (1)	100	gray ware other (1); Vessel 22 (1)
5	gray ware jar (6); Vessel 69 (2)	**Chipped-Stone Debris**	
7	gray ware jar (6)	38	missing
8	gray ware jar (1)	40	flake (1)
9	gray ware jar (1)	52	flakes (2)
	red ware bowl (2)	65	flake (1)
10	gray ware jar (1)	**One-Hand Mano**	
16	gray ware jar (2); Vessel 68 (2)	105	complete, sandstone; not mapped
17	gray ware jar (1)	**Two-Hand Mano**	
18	gray ware jar (1)	12	complete, sandstone
19	gray ware jar (1)	29	complete, sandstone
20	gray ware jar (1); Vessel 70 (1)	30	complete, sandstone
21	gray ware jar (1)	44	complete, conglomerate
25	gray ware jar (6)	51	complete, sandstone
27	gray ware jar (30); Vessel 70 (13)	99	complete, sandstone
32	gray ware jar (1)	**Abrader**	
33	gray ware jar (1)	13	complete, sandstone
34	gray ware jar (1)	**Peckingstone**	
35	missing	58	complete, Morrison orthoquartzite
36	gray ware jar (1)	**Single-Bitted Axe**	
37	gray ware jar (3)	26	complete, Morrison orthoquartzite
39	gray ware jar (7)	**Projectile Point**	
41	gray ware jar (22); Vessel 70 (10)	92	missing
42	white ware bowl (1)	**Modified Cobble**	
43	gray ware jar (12); Vessel 53 (8)	31	complete, unknown orthoquartzite
	white ware jar (3); Vessel 53 (2)	87	complete, igneous
45	gray ware jar (12)	**Other Modified Stone/Mineral**	
	red ware bowl (1)	46	complete, sandstone
49	gray ware jar (1)	**Awl**	
52	gray ware jar (75); Vessel 53 (4), Vessel 70 (12)	91	incomplete, cf. *Odocoileus*
56	gray ware jar (3)	**Bracelet**	
57	gray ware jar (10); Vessel 53 (8)	1	fragmentary, shell
	white ware jar (1)	28	fragmentary, shell
61	gray ware jar (2)	48	fragmentary, shell
63	gray ware jar (1); Vessel 66 (1)	94	fragmentary, shell
67	gray ware jar (32)	**Dating Samples**	
	red ware jar (9)	11	tree-ring sample
69	gray ware jar (8); Vessel 70 (7)	15	tree-ring sample
74	gray ware jar (1)	22–24	tree-ring samples
75	gray ware jar (1)	47	tree-ring sample
76	gray ware jar (1); Vessel 66 (1)	53–55	tree-ring samples
77	gray ware jar (2)	59–60	tree-ring samples
78	missing	64	tree-ring sample
79	gray ware jar (1)	66	tree-ring sample
80	red ware bowl (1)	68	tree-ring sample
83	gray ware jar (20); Vessel 66 (2)	70–73	tree-ring samples
84	red ware jar (1); not mapped	81–82	tree-ring samples; missing
85	gray ware jar (4); Vessel 70 (2)	86	tree-ring sample
88	gray ware jar (9); Vessel 53 (2)	**Miscellaneous**	
	white ware jar (4); Vessel 53 (4)	50	animal bone
89	missing	14	vegetal
90	gray ware jar (2)	52	vegetal

NOTES: Item counts appear in parentheses. For pottery, sherds belonging to vessels are also included in total item count. Refer to Figure 2.51 for artifact locations.

Figure 2.52. View of Room 19, showing excavated features.

On the basis of pit size, shape, location, and, in some cases, contents, 13 post holes were identified in Room 19. Nine post holes (Features 4, 15, 16, 17, 18, 23, 25, 26, and 33) are interpreted to be part of the southern wall of Room 19, which is believed to have been of wattle-and-daub construction. The characteristics of the post holes vary, but most of the pits are cylindrical, some had rotted wood in fill, and at least one had small, sandstone-slab shims that, wedged into the pit, would have helped support a wall post. The general alignment of these features is the main basis for interpreting them as part of wall construction. Four other post holes

(Features 9, 10, 11, and 14) are within the boundaries of the room and may have held roof-support posts. Some post holes may represent both wall construction and roof support, and at least one was capped before abandonment (Feature 18).

The remaining features are pits of unknown function. Several located near or under the structure walls may have functioned as post holes. Many probably were open at abandonment, although two had been capped, and the status of several is indeterminate. Two features, Features 20 and 21, intersect, with the former truncating the fill of the latter.

Table 2.39. Feature Summary, Room 19

No.	Type	Shape	Length (cm)	Width (cm)	Depth (cm)	Status
1	pit	basin	23	22	11	in use
2	pit	basin	26	18	10	in use
3	pit	basin	20	18	12	unknown
4	post hole	other	30	16	19	in use
5	pit	other	30	26	10	in use
6	pit	basin	20	21	12	unknown
7	pit	other	26	26	16	capped
9	post hole	cylindrical	17	13	14	in use
10	post hole	other	24	23	23	unknown
11	post hole	other	33	26	26	in use
12	cist	other	23	21	23	in use
13	cist	other	73	73	76	in use
14	post hole	cylindrical	13	10	13	in use
15	post hole	cylindrical	8	8	12	in use
16	post hole	cylindrical	7	7	17	in use
17	post hole	cylindrical	9	9	17	in use
18	post hole	cylindrical	9	9	16	capped
19	hearth	basin	90	73	17	in use
20	pit	cylindrical	20	20	21	in use
21	pit	cylindrical	21	21	30	capped
23	post hole	cylindrical	10	7	7	in use
24	pit	other	11	9	10	unknown
25	post hole	cylindrical	6	6	12	unknown
26	post hole	other	8	8	14	unknown
27	pit	other	9	9	9	unknown
28	pit	basin	8	8	6	unknown
29	pit	cylindrical	8	8	9	unknown
31	pit	basin	20	18	5	unknown
32	pit	cylindrical	9	9	8	unknown
33	post hole	other	12	14	13	in use
34	pit	basin	12	12	10	in use

NOTE: Feature locations are shown in Figure 2.51.

Dating

Fifty-five tree-ring samples were collected from Room 19 (30 from the mixed wall and roof fall stratum, 22 from collapsed roof materials on the floor, and 3 from Feature 13). Only nine yielded dates, and these range from A.D. 800 to 868 (Appendix B). An A.D. 859 cutting date corresponds very well to cutting dates obtained for Pit Structure 3 samples and is used to date the construction of Room 19 in that year. A cutting date of A.D. 849 may reflect use of old or salvaged wood. The latest cutting date in the group is A.D. 868, and it suggests that the structure was repaired or remodeled almost 10 years after its initial construction.

Interpretations

Room 19 is a large front room interpreted to be the habitation room associated with back storage rooms 9

and 18. Its size and location, as well as feature and artifact assemblages, support an interpretation of domestic use. The burning of the structure at abandonment is believed to have been deliberate and related to the destruction of the four pit structures and numerous other rooms at the site.

Room 8

Length: 3.00 m
Width: 2.05 m
Area: 5.71 m^2

Room 8, a small back room between Room 7 to the west and Room 9 to the east (Figures 2.1 and 2.53), is the only back room spatially associated with Room 10 in the front row of the roomblock. It is a typical storage room, with few features or artifacts, and it was not burned at abandonment. The entire room was trowel-excavated as a single horizontal and vertical unit. The change from fill to sterile was first detected at the west end of the room, but the floor was so much higher than the floors of adjacent rooms that the typical strategy of excavating an arbitrary 5-cm level above the floor could not be used. Instead, in Room 8, the floor provenience applies only to items in direct contact with the floor.

Stratigraphy

The fill of Room 8 consisted of a single stratum of strong brown silt loam with numerous sandstone rocks, chunks of constructional mud, and charcoal inclusions. This stratum, which was about 30 cm thick, is interpreted to be the remains of the collapsed structure walls. Materials recovered include sherds, chipped-stone debris, several animal bones, a polishing stone, a modified core, a projectile point (made of nonlocal material), an *Olivella* shell bead, and two awls. Many of the items are fragmentary. In addition, 12 small, unmodified igneous cobbles or cobble fragments were recovered from fill. It is possible that the "crushed igneous" pottery temper common in the Mesa Verde region during Pueblo I times (Blinman 1988b; Hegmon 1991) was made from crushed weathered cobbles like these. This type of rock is not available in the immediate vicinity of the site, however. It is possible that some of the items found in fill were stored on the structure roof and were incorporated into postabandonment deposits as the structure gradually deteriorated, but a distinct roof fall layer was not detected during excavation.

Construction

Room 8 has vertical-slab walls on the north, east, and south sides (Figure 2.53). The west wall is predominantly horizontal-block masonry. No post holes were found, so the roof is inferred to have been supported by the structure walls. The use-compacted floor was defined at the boundary between sterile B-horizon sediments and postabandonment fill. This floor is 15 to 20 cm higher than the floor of adjacent Rooms 7 and 9, and it is possible that clean B-horizon sediments were deliberately placed in Room 8 to create a higher surface.

Surface

The floor of Room 8 was defined as the contact between sterile sediments below and postabandonment fill above. Only six items were found in contact with the floor: two manos, a peckingstone, a sherd, and two pieces of unmodified igneous rock, similar to those recovered from fill (Table 2.40 and Figure 2.53). The sherd was lost before analysis, but it was identified in the field as a local plain

Table 2.40. Point-Located Artifacts from Room 8, Surface

No.	Description
Pottery Sherds	
6	missing
Two-Hand Mano	
3	incomplete, sandstone
4	complete, conglomerate
Peckingstone	
2	incomplete, Morrison orthoquartzite
Unmodified Cobble	
1	fragmentary, igneous
Unmodified Stone/Mineral	
5	condition unknown, igneous

NOTE: Refer to Figure 2.53 for artifact locations.

gray ware. All the artifacts were recovered from the east half of the room.

The only feature in Room 8 is a small, irregular pit excavated into the floor near the east wall. The pit is 10 cm in diameter at floor level, but it expands to 18 cm long by 15 cm wide beneath the floor. The pit is only 7 cm deep. Fill characteristics suggest that this feature may actually be a rodent burrow.

Dating

No tree-ring samples were collected from Room 8, so the only basis for estimating the construction date is association with other Architectural Suite 3 structures. Tree-ring samples collected from Room 19 and Pit Structure 3 yielded cutting dates or date clusters in the A.D. 850s, and a good case can be made for a construction date of A.D. 859. Given its position in the roomblock, Room 8 could not have been built any later than Room 19, and so the late A.D. 850s is suggested as the latest possible construction date for this room.

Interpretations

Room 8 is interpreted as a storage room because of its location in the back row of the roomblock and its limited feature and artifact assemblages. As with many of the back-row rooms at Duckfoot, there were more artifacts in the fill than on the floor, which suggests that the room interior may have been cleaned of its contents before abandonment and that items stored or abandoned on the roof fell into the structure as the walls and roof gradually deteriorated.

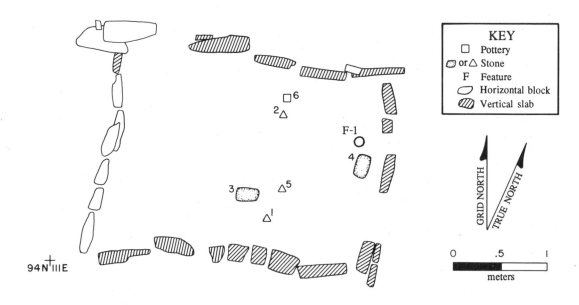

Figure 2.53. Room 8, surface. Point-located items are described in Table 2.40; Feature 1 is described in text.

Room 9

Length: 2.40 m
Width: 1.75 m
Area: 4.06 m²

Room 9 (Figure 2.54) is located between Room 8 to the west and Room 18 to the east (Figure 2.1). Rooms 9 and 18 as a pair appear to be spatially associated with front habitation room 19. Room 9 is a small back room by Duckfoot standards. Its floor area is about one-third less than the average in the Dolores River valley during the same time period (A.D. 850–900) but is equal to the average back-room floor area for Dolores pueblos that date to A.D. 760–850 (Wilshusen 1988a). The room was not burned at abandonment. Because the wall separating Rooms 9 and 18 was not apparent when excavation began, artifacts recovered from modern ground surface and Stratum 1 in the two rooms were combined and provenienced as Room 9. When, at the top of Stratum 2, it became apparent that there were two rooms, artifacts from the two structures were separated for the remainder of the excavation.

Stratigraphy

Two strata were recognized in the fill of Room 9. Stratum 1 was a layer of collapsed wall rubble mixed with postabandonment sediments; Stratum 2 was the underlying layer of mixed wall and roof fall materials. Artifacts recovered from Stratum 1 include sherds, chipped-stone debris, animal bone, a metate, a mano, a peckingstone, and a biface (made of nonlocal material). The metate was found upside down, and the heavily used mano was recovered from its inverted trough. Items recovered from Stratum 2 that may have been stored or abandoned on the roof include a metate, three manos, and a hammerstone. The locations of these artifacts are shown in Figure 2.54, and each item is described in Table 2.41.

Construction

The walls of Room 9 reflect a variety of building techniques (Figure 2.54). The west half of the north wall is a double-wall construction of vertical slabs and horizontal coursed masonry. The east half of this same wall may have been constructed predominantly of mud, as all that remains is a burned earthen footing. The south wall still stands two to three courses high and is composed entirely of horizontal coursed masonry. The west wall is made of vertical slabs and mud, and the east wall is almost entirely mud, incorporating only one large vertical slab and one horizontal block. The upper walls probably were at least partly of mud-and-stone construction. The north, east, and south walls were heavily burned, but because there

was no burned roof fall debris in the fill of the room, they may have been deliberately burned during construction to harden the mud. The structure roof probably was supported by four posts, represented by four post holes, one in each corner of the room. The floor was excavated prehistorically into sterile B-horizon sediments and then compacted through use.

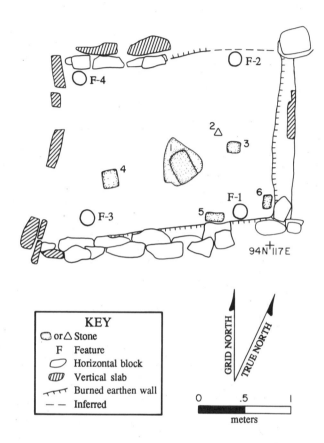

Figure 2.54. Room 9, roof fall artifacts and surface features. Point-located items are described in Table 2.41; Features 1–4 are described in text.

Table 2.41. Point-Located Artifacts from Room 9, Roof Fall

No.	Description
Trough Metate	
1	incomplete, sandstone
Two-Hand Mano	
3	fragmentary, sandstone
4	fragmentary, sandstone
5	complete, sandstone
Abrader	
6	incomplete, sandstone
Hammerstone	
2	complete, unknown stone

NOTE: Refer to Figure 2.54 for artifact locations.

Surface

The floor of Room 9 was a use-compacted surface at the contact between sterile sediments below and roof fall above. A handful of artifacts, including sherds, chipped-stone debris, and two animal bones, was collected from the floor zone, but no items were in direct contact with the floor and therefore none were point-located. The only features associated with the floor are the four post holes mentioned above. They range in diameter from 13 to 17 cm and in depth from 7 to 14 cm. All are basin-shaped; one contained the remains of a burned post, and the remainder contained postabandonment fill.

Dating

One tree-ring sample was collected from Stratum 1, but it did not yield a date. However, wall junctures and placement in the roomblock indicate that Room 9 was built at the same time as the other Architectural Suite 3 rooms, particularly adjacent Rooms 18 and 19. A tree-ring sample collected from Room 19, which burned at abandonment, yielded a cutting date of A.D. 859. In the absence of other evidence, this date is given as a reasonable estimate for the construction of Room 9.

Interpretations

Room 9 appears to be a typical Pueblo I storage room. It is small, located in the back row of the roomblock, has few artifacts (none associated with the floor), and no features except for post holes used to hold the main roof-support posts. The room appears to have been cleaned of its contents before abandonment, but like many of the other surface rooms, it had numerous artifacts in fill that could have been stored on its roof.

Room 18

Length: 2.30 m
Width: 1.65 m
Area: 3.69 m^2

Room 18 (Figures 2.55 and 2.56) is a small back room at the northeast corner of the pueblo. It is associated with Room 19, the adjacent large habitation room in the front row of the roomblock (Figure 2.1). Room 18 did not burn at abandonment. When excavation began in this area of the roomblock, Rooms 9 and 18 were thought to be a single large back room, designated Room 9. Because the wall dividing the two was not discovered until fill had been removed to the bottom of Stratum 1, artifacts from the two rooms were combined and assigned to Room 9 (modern ground surface and Stratum 1 artifacts only). The room was excavated as a single horizontal unit.

Stratigraphy

Three strata were identified in the fill of Room 18 (Figure 2.55). Stratum 1 was the soil A horizon that developed in postabandonment deposits on top of wall fall, and it consisted of dark brown sediments with few artifacts or other cultural inclusions. Stratum 2, which consists of numerous sandstone rocks in an orangish red matrix, is interpreted to be a layer of collapsed wall debris. Below Stratum 2 was a layer of gray-brown silt loam with charcoal and caliche inclusions, interpreted to be the remains of the roof, which gradually deteriorated over time and collapsed onto the floor of the structure. The roof fall layer was designated Stratum 3. Very few artifacts were recovered from room fill. Stratum 2 contained sherds, chipped-stone debris, a mano, and two animal bones; Stratum 3 contained sherds and chipped-stone debris. None of the roof fall items were point-located.

Construction

The north, south, and east walls are made of horizontal-block masonry on top of 15- to 35-cm-high sterile footings; at most, two courses remain standing (Figure 2.56). The west wall is almost entirely mud except for a single vertical slab and one horizontal block. On the basis of the amount of sandstone rubble in fill, it is inferred that the upper walls were a combination of horizontal masonry and mud. Any of five small, circular pit features located along the north and south walls might have functioned as post holes that held roof-support posts, but clearly the roof must have been supported primarily by the structure walls. The floor of Room 18 was excavated prehistorically into B-horizon sediments.

Surface

The floor of Room 18 was a use-compacted surface at the stratigraphic break between underlying sterile deposits and overlying roof fall.

Artifacts. Only a few items were found in direct contact with the floor: a cluster of sherds (PL 1), two animal bone fragments (PL 2), and a mano (PL 3). The sherd cluster consists of 11 Indeterminate Plain Gray jar sherds, 1 Moccasin Gray jar sherd, 2 Indeterminate San Juan Red bowl sherds, and 1 Indeterminate White bowl sherd. Several of the gray ware pieces fit together to form a sherd container. Point-located items are mapped in Figure 2.56.

Features. Five small pit features of unknown or ambiguous function were identified in Room 18 (Figure 2.56). The pits range in diameter from 9 to 13 cm and in depth from 2 to 7 cm. Features 1, 2, and 5 are very shallow, basin-shaped pits; Features 3 and 4 are slightly deeper and

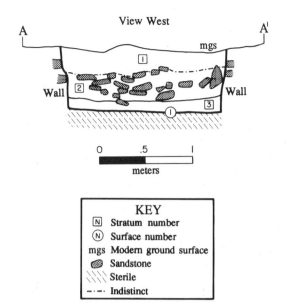

Figure 2.55. Room 18, stratigraphic profile. The location of the profile is shown in Figure 2.56.

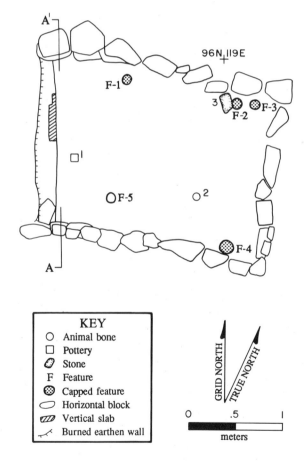

Figure 2.56. Room 18, surface. Point-located items and Features 1–5 are described in text. Profile A is shown in Figure 2.55.

cylindrical. Features 1, 2, 3, and 4 had been capped. Feature 4 had a small sandstone slab covering the base of the pit and a small vertical slab along its east edge, which suggests that it may have functioned as post hole, albeit a shallow (7 cm) one.

Dating

Only one tree-ring sample was collected from Room 18, and it yielded a noncutting date of A.D. 853 (Appendix B). This date is slightly earlier than the late A.D. 850s cutting dates yielded by samples from Room 19 and Pit Structure 3, also in Architectural Suite 3. It is possible that the non-cutting date of A.D. 853 is slightly too early to be a construction date, but it would not contradict a late A.D. 850s date based on association with other structures in the same architectural suite.

Interpretations

Room 18 is interpreted to be a storage room associated with Room 19 to the south. As in the case of Room 9 and several other Duckfoot back rooms, the paucity of artifacts and interpretable features requires that function be inferred primarily on the basis of negative evidence. Few materials were left on the floor or on the roof at abandonment. Like most of the other back rooms at Duckfoot, Room 18 was not burned at abandonment but, rather, deteriorated gradually over time.

Pit Structure 3

Length: 5.10 m
Width: 5.10 m
Area: 23.95 m²

Pit Structure 3 (Figures 2.57–2.62) is the easternmost pit structure at Duckfoot (Figure 1.5). It is also the largest, and like Pit Structures 1 and 4, it burned at abandonment. Many artifacts on the floor are interpreted as having been left in their use or active-storage contexts, and a human skeleton was found in the center of the main chamber floor (Surface 1). An intrusive pit feature in postabandonment fill is evidence for later, nonintensive use of the structure after the roof collapsed and the structure partly filled. The use surface implicitly associated with the construction and use of this intrusive pit was too indistinct to be identified in plan or profile, but it has been designated "Surface 2" as a convenient proveniencing device.

Several excavation methods were used to remove Pit Structure 3 fill above roof fall. The northwest quadrant of the pit structure was excavated in arbitrary 20-cm levels, and fill was screened through ¼-in mesh. Arbitrary levels were used because the boundaries between naturally deposited strata were too diffuse to be recognized during excavation. The southwest quadrant was also excavated in 20-cm levels, and the east half of the structure was excavated by backhoe; however, sediments from these two excavations were not screened.

Burned roof fall, a stratum approximately 50 cm thick, was screened. To maintain vertical control within roof fall, the stratum was subdivided into two 20-cm levels and a third level consisting of deposits that were approximately 5 to 10 cm above the floor. Tree-ring samples, whole tools, large or diagnostic sherds, cores, and large flakes were mapped and assigned point-location numbers. All sediments from 0 to 5 cm above the floor were screened, and items found in floor contact were point-located.

Stratigraphy

Several construction, occupation, and postabandonment events are represented in Figure 2.57, a stratigraphic profile of Pit Structure 3 along the central north-south axis. The pit was excavated prehistorically through approximately 1.2 m of natural deposits and an additional 40 cm into bedrock. The hearth and ash pit depressions were pecked into bedrock, and north of the deflector the floor was prepared by the application of a layer of strong brown silt loam, shown as Stratum 6 in profile. Various items

accumulated on the floor during the occupation of the structure. At abandonment, a human body (Feature 65, Burial 4) was placed on the floor, and the roof of the structure was burned. As the roof collapsed onto the floor, the structure contents were sealed in place with a thick layer of debris. This roof fall layer, Stratum 5, consisted of burned beams, smaller burned twigs and brush, hard-baked sediments with beam impressions, and unburned sediments. Unburned sediments included reddish brown silt loam, similar to the undisturbed B horizon, and a gray, shaley gravel, similar to the undisturbed C horizon. Unburned sediments were roof components that were not exposed to heat intense enough to harden or discolor the dirt. Feature 2, a slab-lined pit, was excavated prehistorically into the roof fall deposit, and its presence is evidence that the structure was reused, if only briefly, after the main occupation ended.

Stratum 4 was a mottled deposit piled against the pit structure walls but not present in the center of the structure. Stratum 4 consisted of strong brown silt loam, shaley-gravel, chunks of caliche, and charcoal. Wall slump, natural wind and water deposition, and the collapse of the last remnants of the roof combined to form Stratum 4.

A clear boundary separated Stratum 4 from overlying Stratum 3. The latter was a brown to dark brown laminated

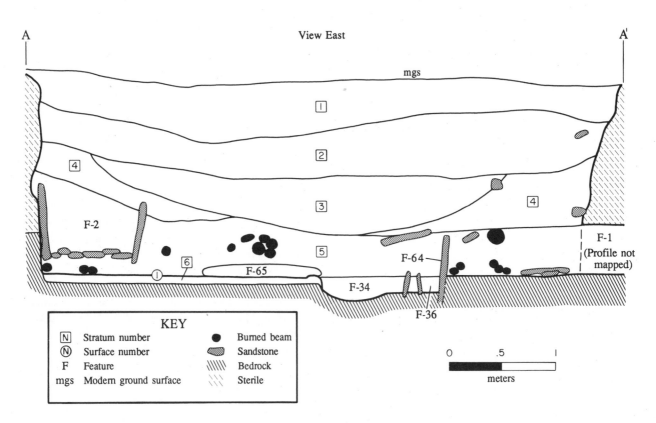

Figure 2.57. Pit Structure 3, stratigraphic profile. The location of the profile is shown in Figure 2.61; Feature 2 is shown in plan in Figure 2.58.

deposit characterized by separate charcoal and caliche lenses ranging in thickness from 5 mm to 3 cm. The laminae formed as the result of natural depositional processes, especially alluviation. A gradual boundary separated Strata 2 and 3.

Stratum 2 was a layer of naturally deposited sediments that ranged from silt loam to clay loam in texture and from brown to dark brown in color. Stratum 1, the soil A horizon that developed in naturally deposited sediments, graded from a brown sandy loam to a dark brown silt loam. Eolian and alluvial processes were responsible for the deposition of Strata 1 and 2 sediments. The slight color and texture differences between the two resulted from pedogenic processes (primarily the breakdown of organic material on the surface and translocation of clay particles) that altered the profile after deposition of the sediments.

Materials recovered from postabandonment fill above roof fall (Strata 1 through 4) include sherds, chipped-stone debris, animal bone, a peckingstone, a biface, a pottery pipe, and two modified flakes. The assemblage of roof fall artifacts is far richer. The point-located assemblage includes relatively few sherds and pieces of chipped-stone debris (including two cores and an edge-damaged flake) but does include a wide variety of tools: 11 metates, 5 manos, 1 peckingstone, 1 modified flake, 1 modified sherd, 1 shaped sherd, and 5 bone awls. A polishing stone and an axe/maul are among the items recovered from roof fall but not mapped. Other point-located items in the roof fall stratum include numerous animal bones and a single human tooth, probably associated with Burial 4 on the floor of the pit structure. Many of the tools, including all of the metates, four of the manos, and four of the awls, are fragments. Of the 56 point-located sherds recovered from roof fall, 49 (87.5 percent) are gray ware, 5 (8.9 percent) are red ware, and 2 (3.6 percent) are white ware. Specific types identified are Chapin Gray (eight sherds), Moccasin Gray (seven sherds), Chapin Black-on-white (one sherd), and Piedra Black-on-white (one sherd). Both jar and bowl sherds are present in the assemblage; several sherds belong to two sherd containers. Point-located items from roof fall are mapped in Figure 2.58 and described in Table 2.42.

Several depositional processes could account for the presence of artifacts in roof fall. Artifacts might have been (1) used on top of the roof, (2) stored on the roof, (3) suspended from roof beams, (4) incorporated into roof fall sediments during construction, (5) deliberately placed or dumped as part of an abandonment activity, (6) dumped into the pit structure depression as postabandonment trash, and/or (7) deposited in the pit structure depression during reuse of the site. Metate fragments were found with clusters of stones in the northwest and northeast quadrants and in the center of the pit structure. Stones in these clusters appear to have been burned, and tree-ring specimens lie above and below them. It appears that these items were deposited when the roof burned and collapsed. The metates, together with the manos found in roof fall, may represent a rooftop mealing area that fell with the roof. Some of the stone concentrations, especially those in the northwest and northeast quadrants, may be related to the postabandonment reuse of the site indicated by Feature 2 (refer to the description of this feature in the discussion of Surface 2 below).

Smaller tools in the roof fall assemblage—for example, the bone awls—may be items that were stored in the pit structure rafters. Several roof fall sherds fit with several floor sherds to form two reconstructible sherd containers (one Moccasin Gray, one Chapin Gray) in the northeast quadrant of the structure. Falling beams may have crushed these containers and displaced some of the sherds upward into the roof fall stratum.

Construction

Pit Structure 3 was excavated through 1.2 m of natural deposits and 40 cm into bedrock. Structure walls, therefore, are composed primarily of native sediments, although a yellow-brown sandy plaster partly covers the lower bedrock portion of the wall. The hearth, ash pit, and deflector were constructed at the same time that the floor north of the wing wall was prepared by the application of a layer of strong brown silt loam over bedrock. The ventilation system, consisting of a vertical shaft and horizontal tunnel, probably was excavated at this time also.

Figure 2.58 shows the pattern of burned and collapsed roof beams. Charred beams from both roof fall and floor contexts are included in this map to better illustrate the overall distribution, although the point-located items are listed in their separate, respective tables (Table 2.42 and Table 2.43). In addition to tree-ring samples, other vegetal materials were point-located as "vegetal samples." Some vegetal samples were pieces of wood that were too fragmentary to be submitted as tree-ring specimens; others were smaller twigs or pieces that may have been used as closing material or as tinder to burn the roof.

Floor features and the pattern of beams in the roof fall stratum permit inferences about roof design and construction. Such inferences assume that a relationship exists between the roof fall pattern and the arrangement of beams in the original roof. Figure 2.58 shows that most roof beams were oriented parallel to the nearest wall. Beams closest to the east and west walls were aligned north-south, and beams closest to the north and south walls were aligned east-west. Orientation of individual specimens indicates that leaner beams (timbers that lean in from the ground around the perimeter of the structure and rest on either primary or secondary beams [cf. Roberts 1929; Glennie 1983]) were not a part of the Pit Structure 3 roof design. Given the absence of leaner beams, the roof design probably incorporated cantilevered primary and secondary

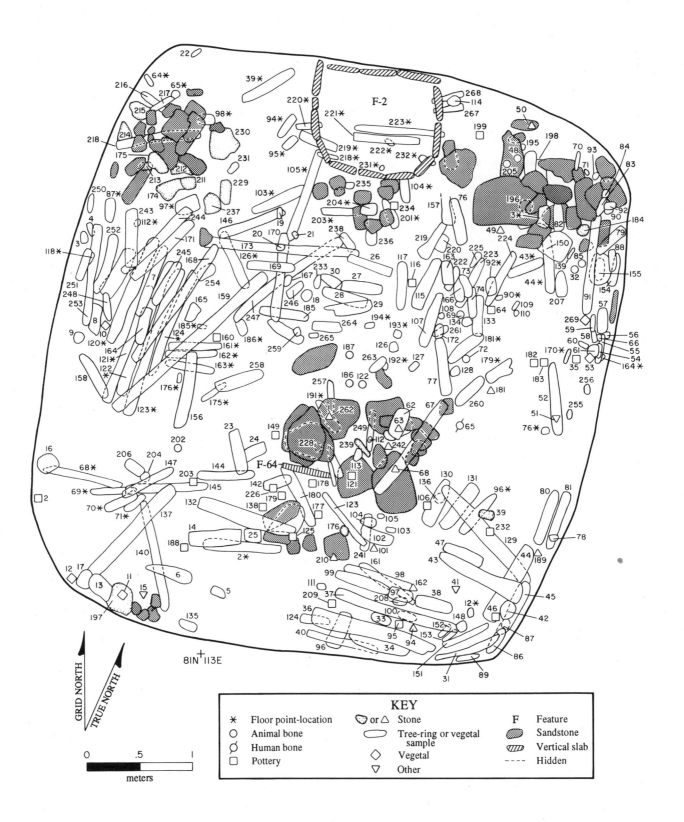

Figure 2.58. Pit Structure 3, roof fall. Point-located items are described in Table 2.42. Tree-ring samples recovered from the floor are included to better illustrate the overall roofing pattern; these are listed in Table 2.43. Feature 2 is associated with Surface 2 and postdates the main occupation of the site.

Table 2.42. Point-Located Artifacts from Pit Structure 3, Roof Fall

No.	Description	No.	Description
Pottery Sherds		**Modified Flake**	
2	missing	63	complete, Dakota orthoquartzite
46	gray ware jar (2)	**Modified Cobble**	
64	gray ware jar (2)	162	fragmentary, igneous
95	gray ware jar (1)	176	incomplete, igneous
106	gray ware jar (6)	**Other Modified Stone/Mineral**	
116	white ware jar (1)	68	incomplete, sandstone
121	gray ware jar (1)	215	fragmentary, sandstone
125	gray ware jar (1)	**Unmodified Stone/Mineral**	
138	gray ware jar (2)	175	fragmentary, sandstone
	gray ware other (2)	**Awl**	
149	gray ware jar (1)	32	fragmentary, cf. *Odocoileus*
160	gray ware other (1)	48	complete, *Odocoileus*
177	gray ware jar (10)	69	fragmentary, medium mammal
179	gray ware jar (2)	72	fragmentary, medium mammal
182	white ware bowl (1)	205	fragmentary, cf. *Odocoileus*
183	red ware jar (1)	**Other Modified Bone**	
	red ware bowl (2)	85	fragmentary, cf. *Odocoileus*
188	gray ware jar (1)	202	fragmentary, medium mammal
199	gray ware jar (7); Vessel 71 (7)	**Dating Samples**	
203	red ware bowl (1)	1	tree-ring sample; not mapped
209	gray ware jar (5)	3–9	tree-ring samples
226	gray ware bowl (1)	13–14	tree-ring samples
232	gray ware jar (1)	16–31	tree-ring samples
234	gray ware jar (2); Vessel 52 (2)	34	tree-ring sample
Modified Sherd		36–38	tree-ring samples
178	condition unknown, gray ware	40	tree-ring sample
Shaped Sherd		43–45	tree-ring samples
35	complete, red ware	47	tree-ring sample
Chipped-Stone Debris		52	tree-ring sample
50	core (1)	54	tree-ring sample
	flakes (3)	57	tree-ring sample; missing
101	other (1)	58–60	tree-ring samples
189	core (1)	62	tree-ring sample
210	edge-damaged flake (1)	66–67	tree-ring samples
	other (3)	70–71	tree-ring samples
Metate		74	tree-ring sample
211	fragmentary, sandstone	76–83	tree-ring samples
237	missing	86	tree-ring sample
Trough Metate		88	tree-ring sample
174	fragmentary, sandstone	89–93	tree-ring samples
184	fragmentary, conglomerate	96–100	tree-ring samples
197	fragmentary, sandstone	102–105	tree-ring samples
212	fragmentary, sandstone	107–115	tree-ring samples
213	fragmentary, sandstone	117	tree-ring sample
214	fragmentary, sandstone	123–124	tree-ring samples
229	fragmentary, sandstone	126–137	tree-ring samples
230	fragmentary, conglomerate	139	tree-ring sample; missing
239	fragmentary, sandstone	140	tree-ring sample
Two-Hand Mano		142	tree-ring sample
39	complete, sandstone	143	tree-ring sample; not mapped
49	fragmentary, conglomerate	144–148	tree-ring samples
181	fragmentary, conglomerate	150–159	tree-ring samples
242	fragmentary, conglomerate	161	tree-ring sample
262	fragmentary, sandstone	163–173	tree-ring samples
Peckingstone		180	tree-ring sample
94	complete, Morrison orthoquartzite	185	tree-ring sample

Table 2.42. Point-Located Artifacts from Pit Structure 3, Roof Fall *(continued)*

No.	Description	No.	Description
190	tree-ring sample; not mapped	Miscellaneous	
195–196	tree-ring samples	65	human bone
198	tree-ring sample	122	animal bone
204	tree-ring sample	186–187	animal bone
206–208	tree-ring samples	270	animal bone; not mapped
216–220	tree-ring samples	10–12	vegetal
222–225	tree-ring samples	33	vegetal
227	tree-ring sample; not mapped	42	vegetal
228	tree-ring sample	53	vegetal
231	tree-ring sample	55–56	vegetal
233	tree-ring sample	61	vegetal
235–236	tree-ring samples	73	vegetal
238	tree-ring sample	84	vegetal
241	tree-ring sample	87	vegetal
243–261	tree-ring samples	269	vegetal
263–265	tree-ring samples	15	adobe (construction sediment)
267–268	tree-ring samples	41	adobe (construction sediment)
271	tree-ring sample; not mapped	51	adobe (construction sediment)

NOTES: Item counts appear in parentheses. For pottery, sherds belonging to vessels are also included in total item count. Refer to Figure 2.58 for artifact locations.

beams that extended the essentially flat roof out to the surrounding ground surface (cf. Wilshusen 1984, 1988a).

Along the west wall there were two distinct layers of beams, both of which were oriented north-south. A set of five closely spaced parallel beams lay on the floor (PLs 120*, 121*, 122*, 123*, and 124*), and another set of beams of similar size was just above them (PLs 7, 164, 168, 171, 243, 244, and 245). As stated above, the orientation argues against their being leaner beams, and their size and spacing argue against their being primary beams (cf. Roberts 1929; Glennie 1983); therefore, these specimens probably are secondary beams. The absence of leaners and the concentration of secondaries along the walls is best explained by Wilshusen's (1984, 1988a) model of a roughly ground-level roof with primaries that are cantilevered beyond the support posts to the walls of the structure. The spacing of the secondaries, which was observed not only in the placement of the beams on the floor but also in casts of burned roof sediments, supports the inference that they were close enough together that mud could have been applied directly to them. This mud-on-beams roof is more similar to roofs seen on some Pueblo II/III pit structures (cf. Lekson 1986:32–34; Cattanach 1980:78–81) than it is to the mud-on-closing-material construction frequently seen in Basketmaker structures with high, domed roofs (e.g., Morris 1980). The beams that preserved may be only a small subset of the original total. Wood may be totally consumed by fire if ample oxygen and heat are maintained, or it will decay if it is not sufficiently charred to prevent bacteria, microbe, and insect activity. Glennie (1983) built a replica of a Dolores-area pit structure, and a year later

he burned it down. In this experiment, the large primary beams did not burn thoroughly enough to collapse. After most of the earth-covered roof was destroyed, the two large primary beams remained intact atop the four main posts. If this occurred prehistorically, it would explain the absence of clearly identifiable primary beams in the roof debris.

Four post holes, three containing remains of posts, provide evidence for roof support. Features 3, 4, 5, and 6 are post holes in the southwest, southeast, northwest, and northeast corners of the structure, respectively. The post in Feature 4 measured 13 cm in diameter at a height of 30 cm above the floor and 11 to 12 cm in diameter at floor level. Below floor level, the post had rotted, but the post hole, which is 20 cm deep, tapers to 10 cm in diameter at its base. This roof-support post appears to have been placed so that the base of the tree was toward the roof. Each of the other post holes also tapers toward the base, which suggests that inverting the roof-support post was standard. This strategy would place the widest portion of the post at the top where it was needed to support a primary beam.

Remodeling of the Pit Structure 3 roof is evidenced by Features 7, 8, and 49. These post holes are located adjacent to the southeast, southwest, and northeast roof-support post holes, respectively, and they may indicate where earlier roof-support posts had been. All three post holes (Features 7, 8, and 49) contained sandstone shims, and all had been intentionally capped. In the northwest corner of the structure, another post hole (Feature 48) and a cylindrical pit (Feature 59) may also have held earlier roof-support posts.

In fact, they appear to be in better locations for roof-support post holes than does Feature 5. The location of Feature 5 is asymmetrical in relation to the locations of the other post holes with in situ posts, which supports the interpretation that the roof was remodeled. Features 48 and 59 contain mixed fill and were intentionally capped.

Surface 1

Surface 1 in Pit Structure 3 was constructed by covering bedrock with a layer of strong brown silt loam similar in color and texture to the sterile B horizon. These floor sediments were approximately 5 cm thick in the northern two-thirds of the pit structure, but they thinned to the south and were absent within a meter of the south wall. The floor was smooth but not level, since it followed the contours of the underlying bedrock.

Artifacts and Samples. Figure 2.59 shows the distribution of point-located items on the floor of Pit Structure 3, and Table 2.43 lists and describes each item. Tree-ring and vegetal samples collected from the floor are included in the floor table but are mapped on the roof fall map (Figure 2.58), where they complete the overall roof-beam pattern.

The floor assemblage of Pit Structure 3 contains many complete tools and reconstructible pottery vessels. Sherds (including one shaped and three modified sherds), chipped-stone debris (including one core and one edge-damaged flake), animal bone, and numerous formal tools and implements reflect a variety of everyday activities. Point-located tools include 8 metates, 5 metate rests (rocks used to prop up a metate at an angle suitable for grinding), 13 manos, 2 hammerstones, 7 modified cobbles, and 1 each of the following items: stone disk, modified core, pecking-stone, projectile point, drill, axe, pottery pipe (in pieces), and awl. In addition, a small basketry fragment (PL 160) was recovered from near the east wall of the structure, and a nearly complete pendant was found in the southwest corner. A small piece of yellow ochre (possibly carnotite) could have been used as pigment. Burial 4 (PL 177, Feature 65) is located directly on the floor in the center of the pit structure and partly covers the north rim of the hearth.

The point-located pottery assemblage consists of 540 sherds: 526 gray wares (97.4 percent), 13 white wares (2.4 percent), and 1 red ware (.2 percent). Most (463) are Indeterminate Plain Gray, but 3 Chapin Gray, 45 Moccasin Gray, 15 Mancos Gray, 1 Piedra Black-on-white, and 2 Chapin Black-on-white sherds were recovered as well. The remaining sherds are indeterminate white or red types. Jar sherds predominate, but a few bowl pieces are also present. Two complete, unbroken Chapin Gray vessels (one miniature pitcher, one miniature gourd jar) were found in the northeast corner of the structure. In addition, 4 nearly complete jars, 2 partial vessels (a gourd jar and a pitcher), and 11 sherd containers were reconstructed during vessel analysis. With the exception of one partial white ware gourd jar, all vessels are gray wares—Chapin, Moccasin, and Mancos gray.

Flotation samples collected from various locations on the floor yielded materials that may have been used in roof construction and/or as fuel to deliberately ignite the structure at abandonment (remains of juniper, pine, sagebrush, and four-wing saltbush). Pollen samples collected from beneath several tools on the floor yielded a wide variety of pollen types that may represent food plant use. Notable among these is one of only two occurrences of squash pollen in the four Duckfoot pit structures.

Features. Sixty-five features were recognized in Pit Structure 3, including an intrusive, postabandonment pit (Feature 2), which is described in the Surface 2 discussion. Features associated with Surface 1 are shown in Figures 2.60 and 2.61, and the location of the intrusive pit is shown in Figure 2.58. Characteristics of all 65 features are listed in Table 2.44.

Feature 65 is a human burial (Burial 4) that was found on the floor in the center of the main chamber. The remains are those of an adult male, 30 to 39 years of age. The body was oriented north-south and was on its back, with the head to the north and facing up. The knees and left arm were flexed, and the legs overlapped the north rim of the hearth. The skeleton was mostly complete and articulated, although the cranium was crushed, apparently by the impact of the collapsing roof. Portions of the body burned as the result of coming into contact with burning roof debris. As with other burials in burned pit structures at Duckfoot, the Burial 4 body appears to have been deliberately placed on the floor before the intentional destruction of the roof. Again, the hearth clearly was not the source of the fire that destroyed the structure, because the charred portions of bones were not those in contact with the hearth.

Feature 34, the central hearth, is a basin-shaped pit lined with silt loam similar to that used in floor construction. A sandstone slab forms the south wall of the pit, and a raised mud rim encircles the remainder of the pit. The presence of a thin lens of ash on a use-compacted surface beneath the rim suggests that the hearth was remodeled by the addition of the rim sometime after initial construction. Fill consisted of two distinct ash strata. Dark gray ash with numerous charcoal inclusions lined the sides of the pit, and light gray ash with fewer charcoal inclusions filled the center. A sandstone slab separates the hearth from a slab-constructed ash pit, Feature 36. The slab that forms the south wall of the ash pit is also the deflector, Feature 64. The ash pit contained a homogeneous deposit of dark gray ash with charcoal inclusions and numerous sherds. The ash in the hearth is interpreted to be primary refuse associated with its use. The adjacent ash pit may have

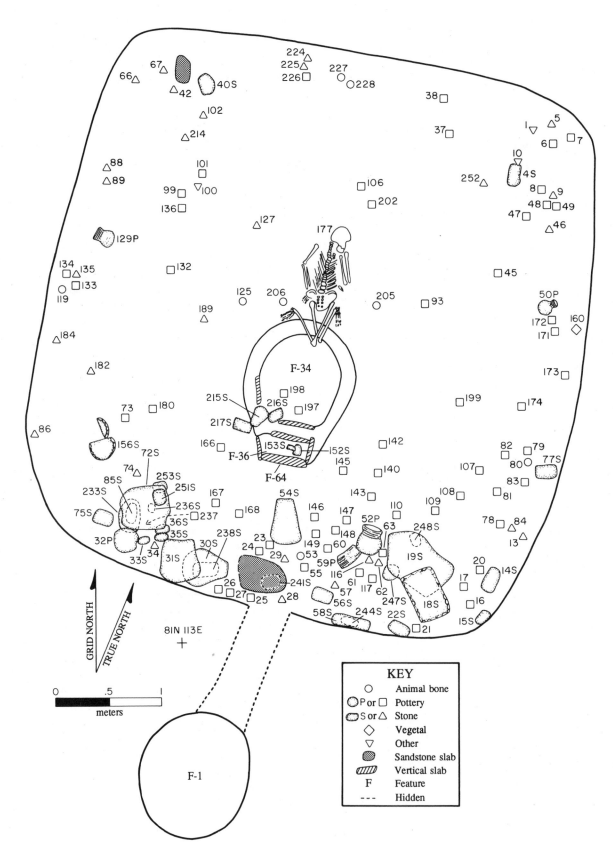

Figure 2.59. Pit Structure 3, Surface 1 artifacts. Point-located items are described in Table 2.43.

Table 2.43. Point-Located Artifacts from Pit Structure 3, Surface 1

No.	Description	No.	Description
Pottery Sherds		167	gray ware jar (1)
6	gray ware jar (5); Vessel 63 (3)	168	gray ware jar (4); Vessel 44 (3)
7	gray ware jar (1)	171	gray ware jar (3); Vessel 48 (3)
8	gray ware jar (1); Vessel 63 (1)	172	gray ware jar (6); Vessel 48 (4)
16	gray ware jar (3)		white ware jar (1)
17	gray ware jar (2)	174	gray ware jar (8); Vessel 48 (1)
20	gray ware jar (1)	180	white ware jar (7); Vessel 97 (7)
21	gray ware jar (1)	197	gray ware jar (2)
23	gray ware jar (4); Vessel 41 (3)	198	gray ware jar (3)
24	gray ware jar (2); Vessel 41 (2)	199	gray ware jar (18); Vessel 43 (15)
25	gray ware jar (24); Vessel 41 (21)	202	gray ware jar (26); Vessel 52 (24)
26	gray ware jar (1); Vessel 41 (1)	226	gray ware other, pipe fragments (2)
27	gray ware jar (4); Vessel 41 (4)	237	gray ware jar (1)
32	gray ware jar (7); Vessel 45 (7)	**Modified Sherd**	
37	gray ware jar (1); Vessel 71 (1)	81	condition unknown, gray ware
38	gray ware jar (3); Vessel 71 (3)	132	fragmentary, gray ware
45	gray ware jar (2)	173	complete, gray ware
47	white ware jar (2); Vessel 97 (2)	**Shaped Sherd**	
	white ware other (2); Vessel 97 (2)	99	complete, gray ware
48	gray ware other (1); Vessel 24 (1)	**Chipped-Stone Debris**	
49	gray ware other (1); Vessel 23 (1)	5	flakes (3)
50	gray ware jar (46); Vessel 26 (46)	9	flake (1)
52	gray ware jar (85); Vessel 40 (84), Vessel 43 (1)	13	edge-damaged flake (1)
55	red ware bowl (1)	28	other (1)
59	gray ware jar (4); Vessel 43 (4)	29	flake (1)
60	gray ware jar (32); Vessel 41 (3), Vessel 43 (18)	42	flake (1)
61	gray ware jar (2)	66	missing
63	gray ware jar (1)	67	missing
73	gray ware jar (18)	84	flake (1)
78	gray ware other (1)	88	flake (1)
79	gray ware jar (7)	89	flake (1)
81	gray ware jar (1)	116	flake (1)
82	gray ware jar (6)	184	core (1)
83	gray ware jar (6)	214	flake (1)
93	gray ware jar (1)	252	flake (1)
101	gray ware other (1); Vessel 50 (1)	**Trough Metate**	
106	gray ware jar (12); Vessel 52 (12)	19	complete, conglomerate
107	gray ware jar (22); Vessel 41 (1), Vessel 44 (13)	30	complete, sandstone
108	gray ware jar (6); Vessel 44 (4)	31	complete, sandstone
109	gray ware jar (10); Vessel 44 (7)	40	fragmentary, sandstone
110	gray ware jar (3); Vessel 44 (3)	58	complete, sandstone
117	gray ware jar (1)	72	complete, sandstone
129	gray ware jar (19); Vessel 42 (17)	215	fragmentary, sandstone
133	gray ware jar (1); Vessel 49 (1)	**Slab Metate**	
134	gray ware jar (3); Vessel 46 (3)	54	complete, sandstone
136	gray ware jar (7); Vessel 47 (7)	**Metate Rest**	
137	gray ware jar (1); not mapped	35	missing
140	gray ware jar (8); Vessel 44 (3), Vessel 51 (3)	36	complete, sandstone
	white ware jar (1)	233	condition unknown, sandstone
142	gray ware jar (14); Vessel 44 (14)	247	condition unknown, sandstone
143	gray ware jar (12); Vessel 44 (3)	248	condition unknown, sandstone
145	gray ware jar (8); Vessel 43 (2)	253	complete, sandstone
146	gray ware jar (30); Vessel 43 (29)	**One-Hand Mano**	
147	gray ware jar (3); Vessel 43 (3)	33	complete, unknown orthoquartzite
148	gray ware jar (8); Vessel 43 (6)	**Two-Hand Mano**	
149	gray ware jar (5); Vessel 43 (2)	4	complete, conglomerate
166	gray ware jar (3)	14	complete, sandstone

Table 2.43. Point-Located Artifacts from Pit Structure 3, Surface 1 *(continued)*

No.	Description	No.	Description
15	complete, sandstone	228	fragmentary, medium mammal
22	complete, conglomerate	Pendant	
56	complete, conglomerate	86	incomplete, unknown chert/siltstone
75	complete, sandstone	Basketry	
77	complete, sandstone	160	fragmentary
85	complete, sandstone	Dating Samples	
216	complete, sandstone	3*	tree-ring sample
238	complete, sandstone	12*	tree-ring sample
244	incomplete, sandstone	39*	tree-ring sample
251	complete, sandstone	43*–44*	tree-ring samples
Abrader		64*–65*	tree-ring samples
153	condition unknown, sandstone	68*–71*	tree-ring samples
217	condition unknown, sandstone	76*	tree-ring sample
Hammerstone		87*	tree-ring sample
57	complete, unknown orthoquartzite	90*	tree-ring sample
182	complete, igneous	94*–96*	tree-ring samples
Peckingstone		97*	tree-ring sample; missing
74	complete, Morrison orthoquartzite	98*	tree-ring sample
Single-Bitted Axe		103*–105*	tree-ring samples
236	complete, igneous	112*	tree-ring sample
Projectile Point		118*	tree-ring sample
224	incomplete, Dakota orthoquartzite	120*–124*	tree-ring samples
Drill		126*	tree-ring sample
225	complete, Dakota orthoquartzite	161*–164*	tree-ring samples
Modified Core		170*	tree-ring sample
46	complete, Morrison orthoquartzite	175*	tree-ring sample; missing
Modified Cobble		176*	tree-ring sample
34	complete, igneous	179*	tree-ring sample
62	complete, igneous	181*	tree-ring sample
102	complete, igneous	185*–186*	tree-ring samples
127	complete, igneous	191*–194*	tree-ring samples
135	complete, unknown orthoquartzite	201*	tree-ring sample
189	complete, igneous	203*–204*	tree-ring samples
241	complete, unknown stone	218*–223*	tree-ring samples
Stone Disk		231*–232*	tree-ring samples
156	complete, sandstone	Miscellaneous	
Other Modified Stone/Mineral		177	human bone; Burial 4
18	complete, sandstone	53	animal bone
152	complete, Dakota orthoquartzite	80	animal bone
Unmodified Stone/Mineral		125	animal bone
1	unknown mineral	205–206	animal bone
10	carnotite?	2*	vegetal
Awl		92*	vegetal
119	fragmentary, medium mammal	100	adobe (construction sediment)
Other Modified Bone			
227	fragmentary, medium mammal		

NOTES: Item counts appear in parentheses. For pottery, sherds belonging to vessels are also included in total item count. Refer to Figure 2.59 for artifact locations. Items marked with an asterisk (*) are shown on the roof fall map, Figure 2.58.

served as a warming pit (using hot coals from the hearth), a storage pit (for ash being saved for use in recipes), or a temporary receptacle for ash removed from the hearth during cleaning. Possible fuel resources identified in flotation samples from the hearth and ash pit include juniper, pine, four-wing saltbush, and corn (cobs). Possible food items include corn kernels, cheno-am seeds, and Indian rice grass seeds.

Feature 1 is the structure's ventilation system. It consists of a vertical shaft that opened at ground surface about a meter south of the main chamber and a horizontal tunnel that opened into the main chamber at floor level. There is

Figure 2.60. View of Pit Structure 3, Surface 1, showing excavated features visible at floor level (additional features were discovered after the floor sediments were removed). View is looking south.

no wing wall, but the sandstone slab deflector (Feature 64) would have diverted air entering the structure from the vent. An east-west alignment of pits on either side of the deflector weakly suggests that a post-and-mud wing wall may have existed at one time; however, the characteristics of these features were ambiguous, and the alignment itself is vague. Most of the pits are basin-shaped or cylindrical in cross section, and all were capped. If a wing wall ever existed in Pit Structure 3, it was dismantled before abandonment.

Two adjacent sand-filled pits, Features 27 and 33, are believed to be sipapus. Feature 27 apparently was still in use at abandonment; Feature 33 had been capped. Interpretation of these features as sipapus, or ceremonial features, is based on their size and shape, location north of the hearth, and association with a cluster of small holes, or paho marks. Each small paho mark (1 to 2 cm in diameter and 2 to 5 cm deep) characteristically tapered to a point, as if a sharp object had been jabbed into the floor.

These marks were not mapped, and a separate feature number was not assigned to them.

Four main support post holes (Features 3, 4, 5, and 6) and five pits that may have seated the original roof-support posts (Features 7, 8, 48, 49, and 59) have already been described in the discussion of construction. The remaining 48 features on the Pit Structure 3 floor are pits of unknown function; all but the shallowest penetrate bedrock below the floor. Not all of these pits were in use in the pit structure at the same time. Seventeen features were visible when the floor was initially exposed during excavation, and these are interpreted as having been in use, or available for use, immediately before the structure was abandoned. Twenty-eight features could not be seen at that stage of excavation, because they had been capped prehistorically with sediments identical to the floor-construction sediments. When the floor sediments were peeled away to the level of bedrock, the capped pits were exposed. These features are

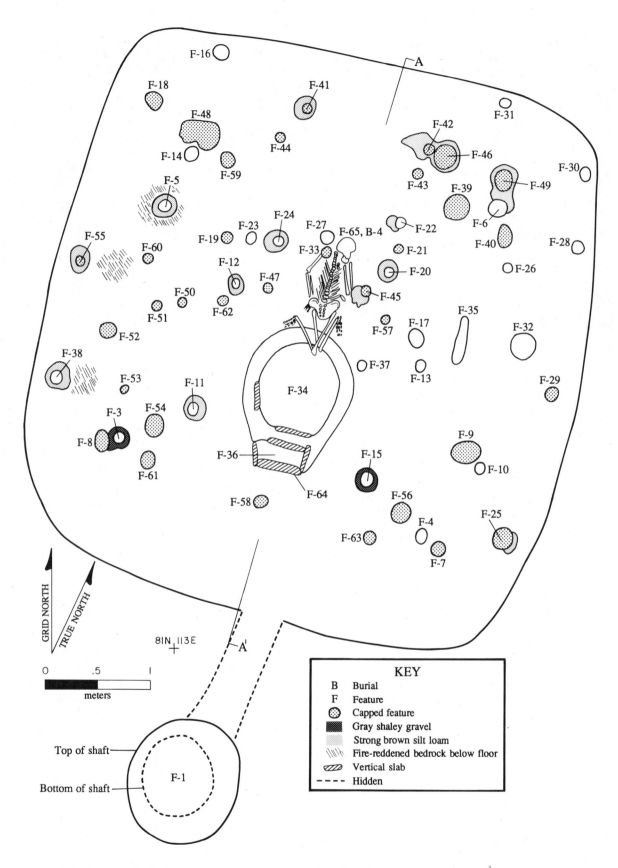

Figure 2.61. Pit Structure 3, Surface 1 features. Features are described in Table 2.44. Profile A is shown in Figure 2.57.

Table 2.44. Feature Summary, Pit Structure 3

No.	Type	Shape	Length (cm)	Width (cm)	Depth (Height) (cm)	Status	No.	Type	Shape	Length (cm)	Width (cm)	Depth (Height) (cm)	Status
1	vent	other	215	105	155	in use	34	hearth	basin	108	101	20	in use
2[a]	pit	other	126	94	44	—	35	pit	basin	45	12	7	in use
3	post hole	cylindrical	12	12	18	in use	36	ash pit	other	45	17	17	in use
4	post hole	cylindrical	12	11	20	in use	37	pit	other	9	8	10	in use
5	post hole	cylindrical	11	10	15	in use	38	pit	cylindrical	11	10	17	in use
6	post hole	cylindrical	14	14	32	in use	39	pit	basin	26	25	21	capped
7	post hole	cylindrical	12	12	30	capped	40	pit	basin	24	13	12	capped
8	post hole	cylindrical	18	16	15	capped	41	pit	cylindrical	8	8	14	capped
9	pit	basin	23	19	10	capped	42	pit	cylindrical	9	9	13	capped
10	pit	cylindrical	11	10	6	in use	43	pit	other	9	9	9	capped
11	pit	cylindrical	9	9	16	in use	44	pit	basin	8	8	4	capped
12	pit	cylindrical	9	9	14	in use	45	pit	other	11	10	10	capped
13	pit	cylindrical	11	10	26	in use	46	pit	cylindrical	23	23	8	capped
14	pit	basin	15	15	4	in use	47	pit	cylindrical	9	8	14	capped
15	pit	cylindrical	12	12	26	in use	48	post hole	other	43	26	25	capped
16	pit	basin	13	13	11	in use	49	post hole	cylindrical	20	17	31	capped
17	pit	basin	15	14	5	in use	50	pit	basin	9	8	4	capped
18	pit	other	17	17	8	capped	51	pit	cylindrical	8	8	10	capped
19	pit	cylindrical	9	8	6	capped	52	pit	basin	16	14	11	capped
20	pit	cylindrical	9	9	17	unknown	53	pit	basin	7	6	3	capped
21	pit	cylindrical	8	8	11	capped	54	pit	basin	20	17	9	capped
22	pit	cylindrical	8	8	11	in use	55	pit	other	9	9	16	capped
23	pit	basin	10	8	2	in use	56	pit	cylindrical	21	20	19	capped
24	pit	cylindrical	8	8	20	in use	57	pit	cylindrical	8	8	13	capped
25	pit	cylindrical	18	18	20	capped	58	pit	other	13	12	5	capped
26	pit	other	7	5	4	unknown	59	pit	cylindrical	16	16	17	capped
27	sipapu	cylindrical	13	11	16	in use	60	pit	other	8	8	6	capped
28	pit	cylindrical	12	12	12	in use	61	pit	basin	15	12	10	capped
29	pit	cylindrical	17	16	24	capped	62	pit	cylindrical	10	8	9	capped
30	pit	cylindrical	14	10	15	unknown	63	pit	cylindrical	10	9	12	capped
31	pit	cylindrical	10	10	14	in use	64	deflector	other	85	48	18 (67)	—
32	pit	basin	24	24	5	in use	65	Burial 4	other	110	70	(10)	—
33	sipapu	cylindrical	10	9	17	capped							

NOTE: Feature locations are shown in Figures 2.58 and 2.61.
[a] Feature associated with Surface 2.

interpreted as having been used earlier in the occupation of the structure and then capped when they were no longer needed or when the structure was remodeled. The status (in use versus capped) of the remaining three pit features was not determined. Nineteen of the 48 pit features had been intentionally filled with sand. Clean sand is not readily available in the natural sediments surrounding the Duckfoot site, so the sand must have been imported from a stream bed such as Alkali Canyon.

Many features on the Pit Structure 3 floor appear to be arranged in pairs or sets, and there is a certain degree of bilateral symmetry in their distribution. Many features in the east half of the pit structure have a counterpart in the west half. The following pairs are clear examples of this type of bilateral symmetry: Features 25 and 38, 29 and 55, 10 and 53, 15 and 11, and 31 and 16. In other cases, clusters of features, rather than individual features, are symmetrically patterned. For example, the cluster consisting of Features 54, 11, and 61 in the southwest corner appears to have a counterpart in the cluster formed by Features 56, 15, and 63 in the southeast corner. Three pits (Features 20, 21, and 22) aligned north-south on the east side of the sipapu (Feature 27) correspond in location to three pits (Features 19, 23, and 24) aligned east-west on the west side of the sipapu. Similarly, the locations of Features 45, 57, 17, and 13 in the east half of Pit Structure 3 mirror the locations of Features 12, 62, 50, and 51 in the west half. These patterned sets of features may support Wilshusen's (1988c) argument that some sand-filled pits were used for erecting temporary altars, whereas others may have been used for temporary structures for nonritual purposes, such as cooking, drying, or smoking foods.

Surface 2

Surface 2 was a postabandonment surface in the fill of Pit Structure 3. The reuse of the pit structure depression occurred after the roof burned and collapsed but before much additional sediment accumulated in the basin. Surface 2 was not observed during excavation or in profile, but its existence is inferred on the basis of a slab-lined pit (Feature 2) dug into roof fall (Stratum 5) against the north wall of the structure (Figures 2.57, 2.58, and 2.62; Table 2.44). This feature was dug from somewhere in Stratum 4, and its base lies approximately 20 cm above the pit structure floor. Upright slabs form the feature walls, and a closely fit pavement of slabs forms the base. The walls of the feature were sooted, and the north wall of the pit structure was oxidized above the feature rim, which suggests that Feature 2 was used as a hearth, roasting pit, or oven. A 2-cm-thick layer of ash and charcoal covered the slab base,

and this material is interpreted as primary refuse related to the use of the feature (not shown in Figure 2.57). An 18-cm-thick stratum of charcoal in a matrix of reddish brown clay loam overlay the ash. The charcoal in this layer differed from that in the roof fall stratum of the pit structure. The reddish brown clay loam in this stratum may be postabandonment fill that mixed with the fuel or may be sediments used to cover the pit during use. No artifacts are associated with Surface 2.

Dating

Of the 286 tree-ring samples collected from Pit Structure 3 contexts, 264 were from floor, roof fall, or floor features, and 22 were from the postabandonment bin, Feature 2. Of the samples related to the main occupation, 113 yielded dates (Appendix B). The dates range from A.D. 766 to 876, and cutting dates range from A.D. 787 to 876. Although the

Figure 2.62. View of Feature 2, Pit Structure 3. The feature is associated with Surface 2.

cutting dates indicate the deaths of trees, those in the A.D. 700s through early A.D. 800s are clearly too early to be the construction dates for Pit Structure 3. In particular there is a cluster of 18 dates (including 13 cutting dates) between A.D. 830 and 838. The samples for which this cluster of dates was obtained may be from reused beams that were scavenged from earlier sites, or they may be from older dead wood that was collected on the prehistoric ground surface. A very strong cutting date cluster in the mid- to late A.D. 850s, and particularly a cluster of 16 cutting dates in A.D. 859, is interpreted as representing the construction of Pit Structure 3. This date is complemented by an A.D. 859 cutting date among the few dated samples from Room 19. Pit Structure 3 samples yielded an almost continuous string of 41 dates from A.D. 860 to 876, and there are particularly strong clusters in A.D. 862, 863, and 866. These dates suggest frequent maintenance, repair, and remodeling of the structure after its initial construction. The terminal cutting dates of A.D. 876 following such a continuous string of dates suggest that the structure was abandoned within a few years after A.D. 876.

Eight of the samples from Feature 2 yielded dates, all noncutting dates ranging from A.D. 802 to 861. The feature clearly postdates the use of the structure, yet the dates appear to be contemporaneous with or to predate the structure. The dates probably reflect reuse of charred pit structure roof beams as fuel in the postabandonment feature. Although the exact date of bin construction is unknown, it is clear that the pit structure had not entirely filled with sediment when the bin was built.

Interpretations

When the roof of Pit Structure 3 burned and collapsed, it sealed the floor artifacts in place and reduced the probability of postabandonment contamination and mixing of materials. Artifacts on the floor probably remained in their final use or abandonment locations. For the most part, the locations of artifacts on the floor of Pit Structure 3 might be best explained by domestic use of the structure. A wide variety of food-preparation and storage implements (e.g., manos, metates, and storage and cooking vessels) indicates everyday domestic activities. However, the presence of two sipapus in the structure and the apparently deliberate destruction of the roof on top of a human body on the floor strongly argue for a ritual component to the use and abandonment of the structure (cf. Wilshusen 1986c; Stodder 1987).

Several observations indicate that Pit Structure 3 was remodeled, possibly more than once. A thin layer of ash under the raised rim of the hearth suggests that the original floor extended to the edges of a rimless hearth. Deliberately filled and capped features are the clearest examples of remodeling in Pit Structure 3. Several capped features are interpreted as the original seating pits

for the four main roof-support posts. If this interpretation is correct, it provides evidence for reconstruction of the Pit Structure 3 roof, a scenario also supported by tree-ring sample evidence.

Postabandonment use of Pit Structure 3 is indicated by the presence of a slab-lined pit in the roof fall stratum. This pit lies 20 cm above the pit structure floor, and its construction and use did not alter or disturb the floor assemblage of Pit Structure 3.

Courtyard 3

Courtyard 3 is the area around Pit Structure 3 and south of Room Suite 3 (Figure 2.63). The western boundary is "stepped" between the 109 and 111 east grid lines; the eastern boundary is also stepped and extends from the 115 to 123 east lines. Courtyard limits were defined and excavations almost completed before Pit Structure 4 was discovered, at which time it became apparent that the western boundary of the courtyard bisected the structure. Thus, although courtyard limits were defined in part by room suite boundaries and pit structure locations, the limits of the courtyard are clearly arbitrary. Courtyard 3 stratigraphy and excavation strategy are the same as those described for Courtyard 1, and the reader is referred to that discussion for details.

Artifacts

Artifacts recovered from the fill of Courtyard 3, exclusive of modern ground surface, include over 5,100 sherds, almost 900 pieces of chipped-stone debris, and numerous animal bones. Formal tools are few but include a hammerstone, a modified core, a peckingstone, 15 modified flakes, and an awl.

Features

Thirty-two features were identified in Courtyard 3 (Figure 2.63 and Table 2.45). The features may be grouped into the following broad categories: large, basin-shaped pits (Features 5, 18, 39, and 40); cists (Feature 19); post holes (Features 12, 13, 14, and 15); and other pits (all remaining features).

Feature 5 is a large, basin-shaped pit with a flat bottom and gently sloping sides. The floor and walls of the pit were fire-reddened, and lower fill contained abundant charcoal, ash, and burned daub with beam impressions. The burned structural debris suggests that Feature 5 may have been roofed, a supposition that is further supported by the presence of several post holes at the perimeter of the feature. Two post holes, Features 14 and 15, were incorporated into the rim of the pit. Two other post holes, Features 12 and 13, were located several centimeters back from the rim, and a shallow, basin-shaped pit that might be

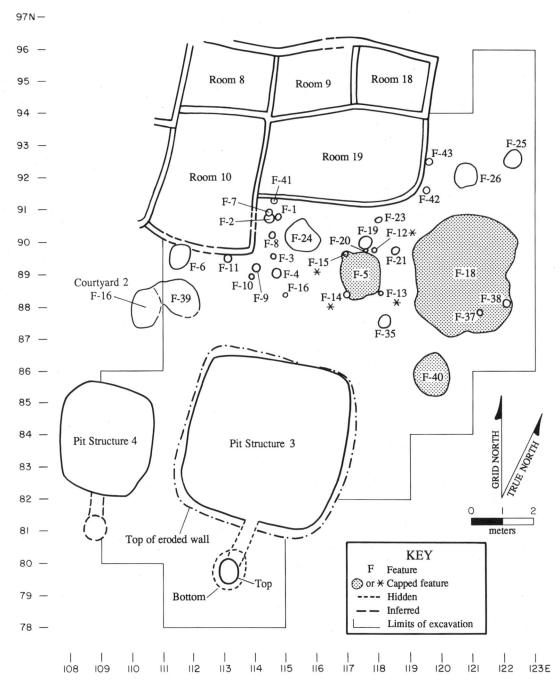

Figure 2.63. Courtyard 3, plan. The upper walls of Pit Structure 3 eroded after excavation of the structure was completed but before excavation of the courtyard began. The eroded area is indicated on the map. Features are described in Table 2.45.

a post hole, Feature 20, was excavated near the center of the north rim. All five of these post holes or possible post holes contained a similar gray-brown, ashy loam fill with flecks of charcoal. Above the burned roof fall layer in Feature 5 was a layer of mixed fill, predominantly silt loam. The mixed sediments were covered by a layer of ash containing burned corncobs, burned dirt, and burned sherds. Several of the sherds make up a complete Moccasin

Gray sherd container. Two different scenarios may explain the fill sequence of Feature 5, and both assume that the pit was roofed. In the first scenario, the roof burned and collapsed, and sometime later a smaller basin was dug into fill and used as a fire pit. The ash and burned corn would then be interpreted as being associated with the reuse of the pit. In the second scenario, charred material in lower fill is inferred to be the remains of fuel used to ignite the

Table 2.45. Feature Summary, Courtyard 3

No.	Type	Shape	Length (cm)	Width (cm)	Depth (cm)	Status
1	pit	basin	18	16	17	in use
2	pit	cylindrical	26	26	20	unknown
3	pit	cylindrical	15	15	20	unknown
4	pit	basin	33	27	20	unknown
5	pit	basin	130	123	21	capped
6	pit	basin	80	65	13	in use
7	pit	basin	23	16	20	unknown
8	pit	basin	17	18	11	in use
9	pit	cylindrical	25	25	23	in use
10	pit	basin	10	10	8	in use
11	pit	basin	25	25	22	in use
12	post hole	cylindrical	12	12	17	capped
13	post hole	cylindrical	9	9	14	capped
14	post hole	cylindrical	17	~17	19	capped
15	post hole	cylindrical	15	~15	30	capped
16	pit	basin	9	9	8	unknown
18	pit	basin	340	325	50	capped
19	cist	other	40	41	26	in use
20	pit	basin	8	8	9	unknown
21	pit	cylindrical	20	20	18	unknown
23	pit	cylindrical	13	16	15	in use
24	pit	other	110	92	25	in use
25	pit	other	57	57	29	unknown
26	pit	other	80	70	23	in use
35	pit	basin	40	34	25	unknown
37	pit	basin	14	13	9	in use
38	pit	basin	28	27	18	in use
39	pit	basin	175	100	50	unknown
40	pit	basin	135	115	29	capped
41	pit	basin	20	15	6	unknown
42	pit	basin	18	17	14	unknown
43	pit	other	28	26	15	in use

NOTE: Feature locations are shown in Figure 2.63.

fire that burned the roof, the mixed fill layer is the roof itself, and the ash is simply roof materials that burned more completely. The function of Feature 5 is unknown, but it is possible that it was a storage facility whose roof served as a protective cover to prevent flooding and contamination. Seven tree-ring samples were collected from Feature 5, one yielding a noncutting date of A.D. 862 (Appendix B).

Feature 18 is the largest pit feature at the Duckfoot site. Although other large pit features may actually have been roofed "structures," Feature 18 clearly was not. The base of the pit is irregular and does not appear to have been used as a floor or living surface. The fill consisted of a variety of discontinuous lenses. Four major stratigraphic units were recognized. The lowest layer, Stratum 4, was a compact silt loam deposit in the southeast quadrant of the pit. Stratum 4 was almost identical to the natural soil B horizon except that it contained tiny flecks of charcoal. Stratum 3 was a mottled layer composed of many small lenses that perhaps represented basket loads of fill. Stratum 3 is interpreted as backfill intentionally dumped into the pit. Stratum 2 was an ash layer that yielded numerous artifacts, including the second pottery "duck foot," the mate to the first found several years earlier in Arbitrary Unit 2. Stratum 1, the uppermost layer, was a mottled stratum just below the soil A horizon. Feature 18 is interpreted to be a borrow pit from which the desired B-horizon sediments were mined for structural repairs, and feature fill is believed to be debris that was deliberately dumped into the pit as backfill. Two small, basin-shaped pits (Features 37 and 38) were detected at the base of Feature 18. It is not clear whether these two pits are associated with Feature 18 or with the Courtyard 2 surface. Two tree-ring samples collected from Feature 18 yielded noncutting dates of A.D. 860 and 864 (Appendix B).

Feature 39 is a large, irregular basin filled with dark brown, organic-rich loam. It is located in an area of root disturbance and apparently was truncated along its western edge by Feature 16 in Courtyard 2. Because of the late discovery of Pit Structure 4, which of necessity became a higher priority at the end of site excavations, Feature 39 was not completely investigated. Feature 40 is a large basin that was filled with caliche, decomposing sandstone, and gray, shaley sediments. These sediments were similar to the deeper, natural deposits into which all four pit structures were excavated. Feature 40 may have been a borrow pit or a mud-mixing pit that was backfilled with natural sediments that were unsuitable for construction.

Feature 19, just north of Feature 5, is the only cist identified in Courtyard 3. It has a shallow shelf along the north edge, which gives way to a deeper pit in the southern two-thirds of the feature. The southern edge of the pit is partly lined with three sandstone slabs. Feature fill consisted of postabandonment sediments and debris; the pit is believed to have been left open at abandonment. Its function is inferred on the basis of feature size and shape.

The remaining Courtyard 3 features are pits of unknown or ambiguous function. Three, Features 11, 25, and 35, may have functioned as cists, but this interpretation cannot be proposed with confidence on the basis of the available evidence. Feature 11, just outside the south wall of Room 10, is the smallest of the three. Its basin was filled with postabandonment sediments mixed with charcoal, artifacts, small chunks of sandstone, and two larger blocks of sandstone. This fill is believed to be collapsed wall material from Room 10.

Feature 25 consists of a deep basin to the northeast and a 10-cm-deep shelf to the southwest. Two small, parallel sandstone slabs rest on the shelf against the southwestern rim of the feature. The upper fill of the pit was a very compact silt loam with tiny flecks of charcoal. A burned corncob (with kernels) and a clump of gray sediment that resembles unfired, tempered pottery clay were recovered from the bottom of the pit. There was no evidence of in situ burning. The clay and corn may have been left in the bottom of the pit before it filled naturally.

Feature 35 is similar in size to Feature 19 but is a simple basin-shaped pit. Fill consisted of three layers. The fill at the bottom of the pit was gray, ashy sediment with no artifacts. The middle layer was dark brown silt loam with numerous caliche inclusions. The uppermost fill layer was similar to the middle layer but had more charcoal, more burned dirt, and more artifacts. The contents of Feature 35 were probably postabandonment in origin, with the possible exception of the ashy sediment at the bottom, which may have accumulated during use of the feature.

Two Courtyard 3 pits, Features 24 and 26, may have served as fire pits. Feature 24 is an irregular depression with fire-reddened sediments at its base and burned sandstone in its fill. Lower fill consisted of a thin layer of loam covered by a 6-cm layer of ash and fine charcoal. The remainder of the pit was filled with sediments similar to the Stratum 1 deposit that covers the courtyard. Feature 26 is an irregular pit with burned sediment at its base. The fill was predominantly ash with small pieces of charcoal and burned sandstone. The rim of the pit did not appear to be burned, but a burned corncob with kernels was found in upper fill.

The remaining features are a collection of small- to medium-size pits, some cylindrical, some irregular in profile, but most consisting of fairly shallow basins. Features 2, 3, 9, and 23 are cylindrical, and they are located near the walls of Rooms 10 and 19. Some or all may be post holes, but there is nothing about their fill characteristics that strongly supports such an inference. Given the meager evidence, no interpretations of the other pits are warranted.

Interpretations

Although most of the features in Courtyard 3 are of indeterminate function, there are enough of sufficiently distinct character and distribution to allow several general statements with regard to courtyard activities. Several everyday domestic functions, including storage (Feature 19 and possibly Features 11, 25, and 35), heating or cooking (Features 24 and 26), and construction or maintenance of structures (Features 18 and 40), are indicated by the feature assemblage. In addition, at least one feature (Feature 5) may have served as a roofed storage facility in the central part of the courtyard. Most of the features appear to have been left open at abandonment; only seven—Features 5, 12, 13, 14, 15, 18, and 40—had been capped.

Most features in Courtyard 3 are located within 2 to 3 m of the roomblock, and a relatively clear space appears to have been maintained closer to the pit structure; Courtyard 3 is very similar to Courtyard 2 in this respect. When the feature distribution in Courtyard 3 is considered in conjunction with the feature distributions in Courtyards 1 and 2, the overall pattern reflects the basic tripartite division of the site: three clusters, one at each end of the site and a third in front of Room 11, correspond to the three architectural suites.

Midden

Midden deposits begin roughly 10 m south of the south walls of the pit structures, approximately along the 71 north grid line; however, selected squares as far north as the 78 north line were included in midden excavations (Figure 2.64). Caliche, decomposing sandstone, and shale were abundant in the higher deposits around the pit structure vent shafts, in the northern reaches of the midden, and in a backhoe test trench that connected the courtyard excavations to the midden excavations (Figure 1.5). This rocky layer was up to 30 cm thick. In undisturbed soil profiles, these types of sediment occur much deeper, in the B and C horizons. Their occurrence near modern ground surface in the area between the pit structures and the midden suggests that they were backdirt from prehistoric excavation of the pit structures. The silt loam B horizon removed during prehistoric excavation of the pit structure was probably used in the construction of the mud walls of the pueblo. However, the lower rocky sediments removed during prehistoric excavation might have been unsuitable for room construction and therefore left where they were originally dumped on prehistoric ground surface.

The midden itself is a 20- to 30-cm-thick deposit of dark brown humic soil (anthropic epipedon) with a very high artifact density. Midden excavation encompassed a 430-m^2 area from the 50 to 78 north grid lines and from the 91 to 115 east lines. Most of the work involved excavation in contiguous squares, although a few isolated units were excavated to the east of the main block. A single 2-×-2-m square was excavated in two arbitrary 7-cm levels to sterile deposits. When no cultural or natural strata were observed, and when no differences were noted in pottery type frequencies from one level to the next, it was decided to excavate the remaining squares in single cultural strata.

Artifacts

The Duckfoot midden deposit yielded a rich and varied artifact assemblage. Among the materials recovered are 67,300 sherds (including 24 shaped and 111 modified sherds), over 17,000 pieces of chipped-stone debris (including 238 cores and 608 edge-damaged flakes), and almost 700 pieces of animal bone. Formal tools and ornaments include 30 manos, 250 peckingstones, 11 hammerstones, 13 polishing stones, 7 stone disks, 21 modified cobbles, 22 bifaces, 42 projectile points, 7 drills, 275 modified flakes, 1 double-bitted axe, 1 maul, 2 modified cores, 4 beads, 2 pendants, 1 Glycymeris shell bracelet,

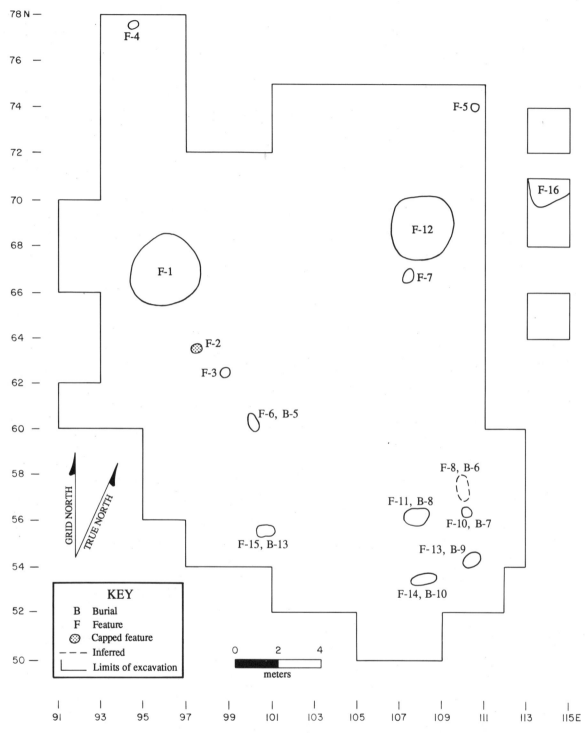

Figure 2.64. Excavated midden area. Features are described in Table 2.46.

4 awls, 5 gaming pieces, and 1 partly drilled piece of ironstone. Of the 929 formal tools in the assemblage, 52.8 percent are complete, 5.5 percent are incomplete (nearly complete), 31.5 percent are fragmentary, and 10.1 percent are of unknown condition. These figures include materials recovered from modern ground surface, midden fill, and features.

Gray wares account for 88.4 percent of the midden sherd assemblage (N = 59,475), white wares 5.7 percent (N = 3,806), and red wares 6.0 percent (N = 4,013). The majority are Indeterminate Plain Gray sherds (55,656), but the following more specific types are present as well: Chapin Gray (658), Moccasin Gray (2,537), Mancos Gray (555), Mesa Verde Corrugated (2), Chapin Black-on-white

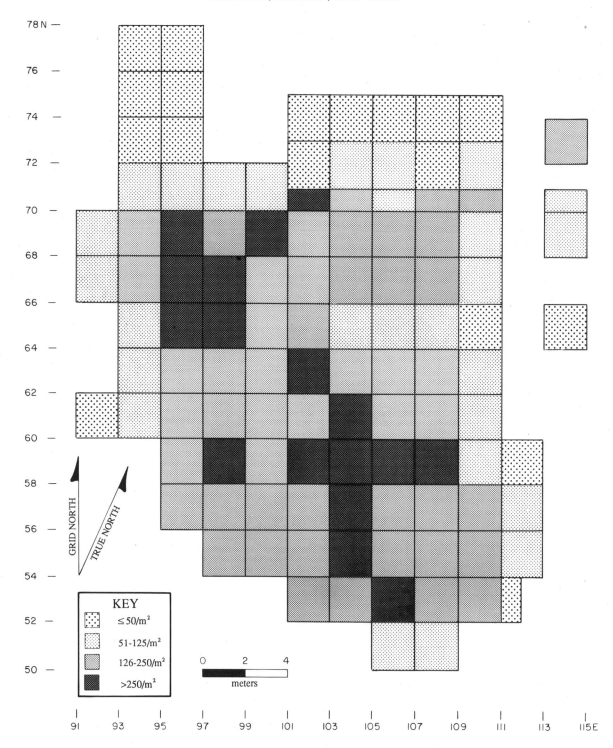

Figure 2.65. Midden sherd density, by individual excavation unit.

(214), Piedra Black-on-white (842), Cortez Black-on-white (18), Mancos Black-on-white (21), Abajo Red-on-orange (26), Bluff Black-on-red (1,330), and Deadmans Black-on-red (20).

In Figure 2.65, sherd frequencies are used as a proxy for overall artifact density and distribution in the midden

area. As shown in this figure, sherd density is generally higher in the central portion of the gridded area and gradually thins toward the periphery. Within the area of higher density, two concentrations are evident, one in the northwest and the other in the southeast. Many of the peripheral squares show generally lower density, but it is also apparent that the midden deposit continues beyond the gridded area, especially to the southwest of the excavated block. It is estimated that approximately 16 m^2 of midden deposits were not excavated.

Features

Fifteen features were discovered in the trash area, and seven of them are shallow burial pits (Figure 2.64 and Table 2.46). The human burials (Features 6, 8, 10, 11, 13, 14, and 15) were concentrated in the southern and southeastern portions of the gridded midden area. Feature 6 is a pit that was excavated into sterile sediments below the midden deposits, although its surface of origin may have been somewhere in the midden stratum. The pit contained the partial remains of a child (Burial 5), five to six years of age. Only the cranium and one humerus remained, and these were incomplete and fragmented. Their placement in the pit suggests that the body was buried intact and that the missing bones decomposed or were destroyed or removed by animals. The fill of the pit was dark, organic-rich loam; the few artifacts recovered are interpreted as debris from the midden rather than as grave goods.

Feature 8 is an inferred pit in the midden deposit. Its upper limits were not discernible, and the lower limits only slightly penetrated the underlying sterile sediments. The presence of the pit is inferred on the basis of a nearly complete skeleton of an adult male, 40 to 49 years of age (Burial 6). The skeleton was in a supine position with its head toward the south-southeast. The arms were straight alongside the torso, and the legs were tightly flexed with both knees to the west (left side of skeleton). The remains

Table 2.46. Feature Summary, Midden

No.	Type	Shape	Length (cm)	Width (cm)	Depth (cm)	Status
1	pit	other	330	290	26	in use
2	pit	cylindrical	32	32	47	capped
3	pit	cylindrical	30	24	40	unknown
4	ash pit	basin	52	33	4	in use
5	pit	nr	26	32	16	in use
6	Burial 5	other	90	34	18	—
7	pit	other	72	63	nr	unknown
8	Burial 6	other	115	55	15	—
10	Burial 7	other	60	38	20	—
11	Burial 8	other	140	79	15	—
12	pit	basin	300	300	50	in use
13	Burial 9	other	100	87	15	—
14	Burial 10	other	138	45	17	—
15	Burial 13	other	95	52	37	—
16[a]	pit	—	~102	~75	—	—

NOTE: Feature locations are shown in Figure 2.64.

nr = no record.

[a] Not excavated.

were discovered very near modern ground surface, and the cranium was covered by only a few centimeters of sediment. No artifacts are interpreted as having been intentionally interred with the burial.

Feature 10 is a pit that was excavated into the sterile sediments beneath the midden. The pit contained the fully flexed skeleton of a six- to seven-year-old child (Burial 7) lying on its left side, with its head to the east. The arms, the middle of the vertebral column, and the feet were in disarray, probably the result of rodent activity. The fill was dark, organic-rich loam, and none of the artifacts in the immediate vicinity are believed to be associated with the burial.

Feature 11 is a shallow pit that originated somewhere in the midden deposit and extended 15 cm into underlying sterile sediments. The pit contained the partly flexed skeleton of an adult male, at least 50 years of age (Burial 8). The head was to the east-northeast, and the knees were to the south. The torso was slightly rotated or twisted: the legs and pelvis were on their left side, the shoulders were nearly flat, the left humerus was under the chest, and the head was face down and turned slightly to the right (facing southeast). The legs and feet were essentially intact except for the right femur, which was missing all but its distal end. The hand bones were also missing. The missing bones possibly were removed or destroyed by animals. No grave goods are associated with Burial 8.

Feature 13 is a shallow pit that was excavated through the midden deposit and 15 cm into underlying sterile sediments. The pit contained the fully flexed skeleton of a child, 10 to 11 years of age (Burial 9). The relatively complete upper torso was in a supine position, with the head to the southwest and the hands folded onto the upper abdomen and lower chest. All the leg bones were missing except the right tibia and a fragment of the left fibula, which were in anatomically plausible positions if the legs were flexed. Some bones of both feet were present as well. The fill of the pit was a dark, organic-rich loam with a few miscellaneous sherds and flakes not believed to be associated with the burial. However, a partial Piedra Black-on-white bowl lay just above the chest cavity and may have been a funerary offering.

Feature 14 is a pit in the midden deposit that cut 17 cm into underlying sterile sediments. The pit was covered by unshaped sandstone slabs and contained the skeleton of an adult female, 18 to 20 years old (Burial 10). The supine body was partly flexed, with the head toward the northeast. The arms were straight beside the torso, and the legs were slightly flexed, with knees to the right (north). The skeleton was nearly complete. No grave goods were associated with this burial, but it is the only burial in the trash area covered by slabs.

Feature 15 is a deep, oval pit excavated into sterile sediments, with the deepest portion of the base on bedrock. The pit contained the skeleton of an adolescent, possibly

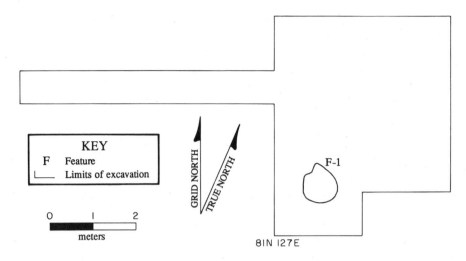

Figure 2.66. Arbitrary Unit 4. Feature 1 is described in text.

female, 12 to 15 years of age (Burial 13). The body was tightly flexed in the small pit, and it lay on its right side, with the head pointing to the northeast and facing slightly north. The skeleton was incomplete and in poor condition; ground moisture and root disturbance contributed to the poor state of preservation. No grave goods were associated with the remains.

The remaining midden features consist of one ash pit and seven miscellaneous pits. The ash pit, Feature 4, is located in one of the two northernmost squares excavated as midden, but this part of the site is more appropriately considered part of the open area surrounding Pit Structure 1. Because there were no signs of in situ burning in Feature 4, this shallow pit is assumed to have served as a receptacle for materials removed from a hearth or fire pit elsewhere on the site.

Features 1 and 12 are two very large pits that originated in the midden stratum and extended into sterile deposits. Feature 1 had been excavated prehistorically into sterile sediments and then filled with trash. The trash included abundant charred wood and burned mud, materials that appeared to be burned structural debris. Tree-ring samples from Feature 1 produced two A.D. 858 cutting dates, which indicate contemporaneity with several structures in the pueblo and eliminate the possibility that Feature 1 contained the remains of an older structure. There were no features or post holes to suggest that Feature 1 was a structure, and the origin of the debris is unknown. Feature 12 was also filled with trash, but in this case the trash was identical to that in the surrounding general midden deposit.

The five remaining pits in the trash offer little evidence of function. Features 2, 3, 5, and 7 were completely excavated; Feature 16 was identified in the final days of the investigation, and there was not enough time to completely excavate and document it. A human mandible was found near the edge of Feature 16, and excavation into the pit proceeded far enough to determine that the pit did not contain a burial. Most of Feature 16 remains unexcavated.

Other Excavations

"Arbitrary Unit" is a provenience category that applies to (1) exploratory excavations where the cultural study unit is unknown, (2) excavation units that crosscut cultural study units, and (3) postabandonment deposits that overlie definable cultural study units. Examples include grid squares or exploratory trenches north of the roomblock, grid squares excavated to search for the tops of pit structure walls, trenches across the roomblock to define the tops of room walls, and backhoe trenches excavated to search for additional pit structures around the perimeter of the site.

Arbitrary Units 1, 2, and 3

Arbitrary Units 1, 2, and 3 are general areas of the site that correspond to the areas defined as Architectural Suites 1, 2, and 3, respectively. For the most part, these provenience designations were applied to fill above the walls of rooms and pit structures. Once wall lines were visible in the roomblock and pit structure limits were roughly defined in the courtyard area, the arbitrary unit designations were dropped in favor of structure or courtyard designations. As a result, only those materials recovered from the uppermost fill were assigned to Arbitrary Units 1 through 3. These units were excavated with hand tools, and the sediments were sieved through ¼-in mesh. Materials recovered from Arbitrary Units 1, 2, and 3 include 3,574 sherds, 477 pieces of chipped-stone debris, and a small amount of animal bone. Formal tools include a metate, 9 peckingstones, 10 modified flakes, 2 stone disks, a drill, a biface, 2 projectile points, and an awl. The pottery "duck foot" after which the site was named was recovered from Arbitrary Unit 2, near Pit Structure 2. The artifact totals given here are for all three arbitrary units combined and include items from modern ground surface as well as from subsurface deposits.

Arbitrary Unit 4

Arbitrary Unit 4 is a small excavation block due east of the row of pit structures. It is connected to the main excavation block by a 6-m-long backhoe trench, also considered part of the arbitrary unit (Figures 1.5 and 2.66). The main part of Arbitrary Unit 4 was chosen for excavation because it contained an artifact scatter on modern ground surface. The four 2-×-2-m squares and one 1-×-2-m square that make up this portion of Arbitrary Unit 4 were excavated using trowels, and all sediments were screened. Fill consisted of a single 15-cm-thick stratum that was the soil A horizon. Artifact density was light, and most materials were recovered from modern ground surface. Among the items recovered are 565 sherds (408 from modern ground surface) and 117 pieces of chipped-stone debris (97 from modern ground surface). One peckingstone and one modified flake are among the formal tools recovered from modern ground surface, and two modified flakes are among items found in subsurface deposits.

Feature 1, a fire pit, is the only feature in Arbitrary Unit 4. It is a shallow basin, 90 cm long, 70 cm wide, and 20 cm deep. A burned slab at the bottom of the pit was covered by a layer of burned sediment and charcoal. The east side of the rim was burned. Because the burning throughout the feature appeared to be in situ, the feature is interpreted as a fire pit. Potential fuel (sagebrush, juniper, pine, four-wing saltbush, and corn cupule) and food (cheno-am and corn) items were identified in a flotation sample collected from Feature 1.

After the block excavation was complete, a backhoe trench was excavated between it and the block excavation in Courtyard 3. The purpose of the backhoe trench was to verify that no other structures were in the intervening area. The trench was excavated to a depth of only 25 cm, deep enough to penetrate the sterile soil B horizon and to confirm the absence of structures and features. The sediments from the backhoe trench were not screened, artifacts were not collected, and the profiles were not drawn.

Arbitrary Unit 5

Arbitrary Unit 5 consists of three exploratory trenches that abut the north wall of the roomblock (Figure 1.5). These trenches, which were excavated early in the investigation for the purpose of defining the north wall of the roomblock and testing for outdoor occupation surfaces, were dug with hand tools, and sediments were screened through ¼-in mesh. Wall fall was found on top of the prehistoric ground surface, but artifact densities were low, and no features were present. The prehistoric ground surface showed little evidence of use as a cultural surface. Among the materials recovered from modern ground surface and subsurface deposits are 904 sherds, 331 pieces of chipped-stone debris, and numerous pieces of animal bone. Formal tools include six modified flakes and a projectile point.

3

Pottery

Mary C. Etzkorn, Lisa K. Shifrin, and Michelle Hegmon

Pottery constitutes the single largest category in the Duckfoot artifact assemblage: 112,197 sherds were recovered from proveniences across the site. In this chapter, the results of sherd, vessel, and pottery tool analysis provide basic descriptive information about the Duckfoot assemblage. In addition, pottery production, distribution, and style, which were the subjects of three special studies involving comparisons of red and white wares from several sites in the Four Corners area, are discussed. Finally, the results of these special studies are used to address the topic of social relationships and interaction at local, regional, and interregional levels.

Sherds

Mary C. Etzkorn

Standard sherd analysis was designed to provide basic information about ware, type, and form for the Duckfoot pottery assemblage and was conducted on the following types of artifacts: sherds, sherds that fit together to form reconstructed vessels, vessels recovered whole and unbroken from the field, and pottery tools (for example, sherds that were modified for use as scrapers). The only pottery artifacts excluded from analysis were objects that did not appear to be parts of vessels or tools (for example, pottery figurines) and unfired sherds, which, because they represent an incomplete stage of manufacture, do not lend themselves to the same type of evaluation used for classifying fired pottery. Only 38 unfired sherds were recovered from the site.

Sherds were submitted from the field in lots according to the specific proveniences from which they were recovered. A single lot may consist of one sherd or many

hundreds of sherds, depending on the number found in a given context. After the materials were cleaned and assigned lot catalog numbers, they were sorted by ware, type, vessel form, and vessel part. Counts were recorded for each analytic category, and all sherds, regardless of size, were included in the count. When recognized, sherds broken during excavation or laboratory processing were counted as single items.

In a sherd-based analysis, all sherds are evaluated individually, and whole, unbroken vessels are counted as single sherds of no greater analytic import than any other sherd. The advantage of this approach is that it provides comparable data on sherd assemblages that include reconstructed and/or whole vessels and those that do not. The disadvantage of a sherd-based analysis is that it results in apparent inconsistencies when sherd data are compared with reconstructed-vessel data for the same assemblage. For instance, a sherd classified as Indeterminate Plain Gray during sherd analysis may be incorporated into a vessel typed as Moccasin Gray during vessel reconstruction, and a vessel classified as Piedra Black-on-white may be reconstructed of sherds typed as Piedra Black-on-white, Cortez Black-on-white, and Indeterminate White during sherd analysis. Also, the number of vessel forms recognized in vessel analysis exceeds that identified in sherd analysis, and it is not unusual for a sherd typed as "jar" to be incorporated into a pitcher, seed jar, or other form during vessel reconstruction.

Analytic Criteria

Local Wares and Types

Local Mesa Verde–region pottery constitutes the majority (94.8 percent) of the Duckfoot assemblage, and typological assignments were made largely on the basis of criteria and definitions outlined by Breternitz et al. (1974). Assignment to a specific named type was contingent on a particular sherd's having a definitive characteristic or suite of characteristics. For gray wares, the necessary characteristics for type identification are found only on neck and rim pieces. For decorated (white and red) wares, painted designs are requisite for type identification. Because most sherds in the assemblage lack the characteristics necessary for identification to a specific type, several grouped types were defined to allow at least general statements about assemblage composition. The wares and types recognized in the Duckfoot assemblage are described below. Individual descriptions are provided to document criteria and procedures specific to the Duckfoot analysis. Date ranges for traditional Mesa Verde types (and the San Juan red ware types described in the next section) follow Wilson and Blinman (1991a).

Mudware. Mudware is a general term usually used to describe the earliest pottery manufactured in the region ("Basketmaker Mudware"), recognized on the basis of vegetal fiber temper and (sometimes) basket impressions on the exterior surface. However, the same name can be applied to any crude, poorly fired ware dating to any time period, and it is in this latter sense that the term is used in this report. Mudware ranges from reddish brown to gray-brown in color, and although it is never painted, polished, or slipped, it is so distinct from the gray wares described below that it warrants a separate classification.

Mesa Verde Gray Wares. Gray ware sherds are not painted, polished, or slipped. The color of the paste is generally gray, but depending on the type of clay used and the atmosphere (oxidizing, reducing, or neutral) in which a vessel is fired, the color may range from gray or black to white, buff, or pink. The texture of the paste is often coarser than that of the decorated wares, and the temper most often consists of crushed igneous rock. Two general categories of gray ware are recognized: plain (made before A.D. 900/975) and corrugated (made after A.D. 900/975). The coils used to build plain gray ware bowls were completely smoothed over; the coils used to build plain gray ware jars were smoothed over from the base to at least the vessel shoulder. Plain gray ware jars may or may not have neckbands, that is, coils left exposed around the vessel neck and rim. On corrugated vessels, which are always jars, the coils were left exposed over the entire exterior of the vessel and were manipulated by pinching or tooling, which produced a rippled surface pattern. The specific plain gray ware types in the Duckfoot assemblage are Chapin Gray (A.D. 575–950), Moccasin Gray (A.D. 775–950), and Mancos Gray (A.D. 850–975); the grouped type is Indeterminate Plain Gray. The specific corrugated gray ware types in the assemblage are Mancos Corrugated (A.D. 930–1100)

and Mesa Verde Corrugated (A.D. 1100–1300); the grouped type is Indeterminate Corrugated.

To be classified as Chapin Gray, sherds must be from the rim of the vessel (jar or bowl) and must be scraped smooth. In the case of a jar, there must be no evidence of neckbands, and a rim sherd must be large enough to allow the analyst to confidently infer that it is not simply the top fillet from a neckbanded vessel. Jar neck sherds that do not include the rim proper are occasionally classified as Chapin Gray if enough of the neck is present to indicate that there is no banding. Moccasin Gray sherds must be from the neck or rim of a jar, and they are identified on the basis of their relatively wide, flat, nonoverlapping neckbands. For the degree of overlap to be evaluated, at least two bands must be present. To be classified as Mancos Gray, sherds must also be from the neck or rim of a jar, but they must have relatively narrow and distinctly rounded coils or wider and flatter, but overlapping, bands (similar to clapboard siding on a building). A single rounded coil from the neck may be classified as Mancos Gray, but a single flat band may not, because the degree of overlap between it and the adjacent band cannot be observed. Gray ware sherds that are not corrugated and are not identifiable as one of the specific types listed above are classified as Indeterminate Plain Gray. This category includes (1) body sherds that lack the neck or rim portions, (2) rim sherds that are too small to show the presence or absence of neckbanding, (3) neckbanded jar sherds whose characteristics are intermediate between Moccasin and Mancos gray, and (4) sherds consisting of only a single flat neckband.

The two specific corrugated types are distinguished solely on the basis of rim eversion. The rim of a Mancos Corrugated jar is relatively vertical or only slightly everted; the rim of a Mesa Verde Corrugated jar is markedly flared. Sherds assigned to Mancos or Mesa Verde corrugated must be large enough to allow a confident assessment of the degree of rim eversion. Corrugated sherds that are not identifiable as either Mancos or Mesa Verde corrugated are assigned to the Indeterminate Corrugated category. This category is composed primarily of jar body sherds that lack the rim portions needed to assess eversion. However, rim sherds that are too small to show degree of eversion or that are intermediate between the two specific types are included as well.

Mesa Verde White Wares. White wares are painted, polished, slipped, or any combination of the three. Early (pre–A.D. 900) white wares often have a relatively coarse paste, and they are usually tempered with crushed igneous rock. After A.D. 900, white ware paste became finer and crushed sherd temper was introduced, eventually becoming the predominant white ware temper. Like gray wares, white wares vary in color depending on the clay used and the amount of oxygen present during firing, but even a misfired "pink" white ware is usually distinguishable in color and texture from a true red ware. The majority of white ware vessels have painted designs somewhere on their surfaces, but individual sherds from painted vessels frequently do not. Slip usually is found only on vessels that have been polished, although a polished vessel will not always have slip. White wares made before A.D. 900 are less likely to be as well polished and slipped as those made after that date, and some early white wares are not polished or slipped at all. This results in an inherent and unavoidable bias when working with pre–A.D. 900 assemblages such as that recovered from Duckfoot: by definition, unpainted sherds from unpolished, unslipped white ware vessels are classified as gray wares. The specific white ware types identified in the Duckfoot assemblage are Chapin Black-on-white (A.D. 575–900, but rare after A.D. 800), Piedra Black-on-white (A.D. 775–900), Cortez Black-on-white (A.D. 900–1000+), Mancos Black-on-white (A.D. 1000–1150), McElmo Black-on-white (A.D. 1075–1300), and Mesa Verde Black-on-white (A.D. 1180–1300).

Chapin Black-on-white sherds usually are not polished or slipped. Painted designs tend to be crudely executed in mineral paint and incorporate simple elements such as dots, z's, triangles, and tick marks. On bowls, designs often radiate upward and outward from the center of the vessel; on jars, designs may be wrapped around the vessel body. Piedra Black-on-white sherds may or may not be lightly polished and/or slipped, and designs are painted with predominantly mineral paint. A distinguishing feature of Piedra Black-on-white is the use of sets of parallel lines, relatively close together and embellished with flags, ticks, and triangles. The layout tends to be more regular and the overall design execution more refined than on Chapin Black-on-white vessels.

Sherds identified as Cortez Black-on-white frequently are distinguished by a very white, "crackled" slip on a polished surface, in addition to characteristic designs painted (usually) with mineral pigments. Sets of parallel lines, thinner and more widely spaced than those seen on Piedra sherds, serve as the basis for designs that also include dots, triangles, ticks, flags, ric-rac, scrolls, interlocking scrolls, stepped triangles, squiggles, and squiggle hatching. The design layout is often based on a division of the vessel surface into sections (for example, thirds or quarters).

Mancos Black-on-white is perhaps the most variable of all the decorated Mesa Verde types. Sherds are usually polished and slipped, but the quality of both is variable. Although the surface treatment often appears relatively crude, very finely finished examples occasionally are noted. Common design elements include straight-line hatching and crosshatching, checkerboards, dots, triangles, stepped frets, scrolls, and "drip lines" extending down from the rim. The layout is variable, ranging from broad panels to "all-over" designs, and the paint is almost always mineral.

McElmo and Mesa Verde black-on-white are distinguished from the earlier types not only by their painted designs but by their overall physical characteristics as well. They tend to be thicker than the types that preceded them and usually have squarer rims. The rims sometimes are decorated with simple tick marks or, in the case of Mesa Verde Black-on-white, occasionally with more elaborate designs. Both types usually are well polished and slipped, but Mesa Verde Black-on-white is characterized by a particularly fine, "pearly" finish, which helps to distinguish it from McElmo. McElmo designs are arranged in bands, frequently without framing lines (detached lines that parallel the banded design). Common design elements, although not exclusively diagnostic of McElmo Black-on-white, include triangles, checkerboards, stepped frets, and straight-line hatching between sets of parallel lines. Mesa Verde designs are usually arranged in bands as well, but most often they are bordered by single or multiple framing lines. Occasionally, unbanded "all-over" designs are also seen. Specific design elements include many of the same ones found on McElmo sherds; however, one notable distinction is the use of straight-line hatching as a background filler (Mesa Verde) rather than as a part of the main design (McElmo). Paint type for both McElmo and Mesa Verde black-on-white is usually carbon, but mixed carbon and mineral is also seen.

Although white ware type assignments are based partly on surface treatment (polish and slip), the emphasis is on painted designs, and assignment of a sherd to a specific type requires that enough of a design be present to reveal individual elements and overall layout. As a result, relatively few painted white ware sherds are assigned to a specific type. Sherds that cannot be identified to specific type, either because no painted design is present or because the painted design is too small or ambiguous, are classified as Indeterminate White. No distinction is made between painted and unpainted sherds assigned to this category.

Nonlocal Wares and Types

San Juan Red Wares. Most red wares recovered from Duckfoot are thought to have been manufactured in the area of southeastern Utah that encompasses the San Juan River drainage, hence the appellation "San Juan," which is used to describe this group of red wares (see Hegmon's discussion of red wares later in this chapter). San Juan red wares are polished, they may or may not be slipped, and, although whole vessels are virtually always painted, individual sherds may or may not include painted designs. The color of the paste is generally a deep brick-orange or red. The texture is usually finer than that noted for Mesa Verde white wares dating to the same time period, and the temper is often sand or a combination of sand and crushed rock. Specific types present in the Duckfoot assemblage are

Abajo Red-on-orange (A.D. 700–850), Bluff Black-on-red (A.D. 780–940), and Deadmans Black-on-red (A.D. 880–1100). Red ware sherds that cannot be assigned to a specific type are classified as Indeterminate San Juan Red. Typological assignments are made primarily on the basis of paint color and, in some instances, the presence or absence of slip; less emphasis is placed on specific design elements and layout, and in this respect, red ware analysis differs markedly from white ware analysis. Because the assignment of a painted red ware sherd to a specific type is contingent primarily on the ability to identify paint color, regardless of design, almost all painted red ware sherds are assigned to a specific type.

In the Duckfoot assemblage, red ware sherds that have a red or reddish brown paint on an unslipped, orange background are classified as Abajo Red-on-orange. Bluff Black-on-red sherds have black or brownish black paint on an orange background. Although the type names suggest a difference in background color between Abajo and Bluff, the two are very similar in this respect, and background color was not used as a distinguishing feature in this analysis. Duckfoot sherds typed as Bluff Black-on-red are not slipped. Deadmans Black-on-red, on the other hand, is recognized on the basis of black paint on a deep red slip; individual sherds must have both features to be assigned to this type. Red ware sherds that are unpainted or that have "muddy" paint colors, neither red nor black, are classified simply as Indeterminate San Juan Red.

Other Nonlocal Wares. Any sherd whose physical characteristics (texture, color, temper, paint type, painted design), singly or in combination, do not allow assignment to any of the local Mesa Verde or nonlocal San Juan types is categorized as "other nonlocal" gray, white, or red, depending on its ware. When possible, a more specific designation (e.g., "Tsegi Orange Ware") is provided in the comments section of the analysis form.

Vessel Form and Part

Three vessel-form categories are recognized during sherd analysis: jar, bowl, and "other." Jars are vessels with more-or-less spherical bodies and cylindrical necks. Bowls are neckless vessels that consist of a portion of a sphere but not more than half of a sphere. "Other" refers to vessels that are neither jars nor bowls. Because generally it is not possible to evaluate complete vessel attributes during sherd analysis, assignment to a form category depends on the characteristics observed on a given sherd. Although the criteria used to assign sherds to a particular vessel-form category differ for the different wares, assignment is generally based on curvature and, when applicable, the location (interior versus exterior surface) of paint, polish, and/or slip.

The default form category for gray ware sherds is jar. That is, all gray ware sherds, plain and corrugated, are recorded as jars unless a particular sherd is from the rim of the vessel and is curved in such a way that it could be only from a bowl or some other vessel form. Thus, the count of gray ware jar sherds is inflated by the number of body sherds from bowls and other vessels, with the consequence that these forms are underrepresented in the assemblage.

Assignment of white and red ware sherds to vessel-form category is based on the location of paint, polish, and slip. Jars typically are decorated on their exterior (convex) surfaces; bowls, on their interior (concave) or both interior and exterior surfaces. Therefore, a white or red ware sherd that is painted, polished, or slipped on its exterior surface is typed as a jar; one that is painted, polished, or slipped on its interior, or on both its interior and exterior, is typed as a bowl. The only exceptions are red and white ware sherds, usually from vessel rims, whose curvatures clearly indicate that they are from neither jars nor bowls. These sherds are assigned to the "other" form category. Because generally there are fewer rim sherds than body sherds per vessel, the jar and bowl form categories in the red and white ware assemblages are overrepresented, and the "other" category is underrepresented.

Any sherd, regardless of ware, is typed as "other" if, usually on the basis of curvature, the analyst is certain that it is from neither a jar nor a bowl. This category includes sherds from seed jars, ladles, dippers, and effigy vessels, when those sherds are from the portion of the vessel that allows recognition of the form represented. For example, a sherd from the body of a seed jar would be classified as a jar, but a sherd from the rim of the same vessel showing the distinctive constricted opening characteristic of seed jars would be classified as "other." Similarly, a ladle body sherd that does not have the handle or handle attachment would be identified as a bowl, but a sherd that includes the handle or attachment would be classified as "other." Handles that are detached from their vessels sometimes are recorded as "other" as well, particularly when the type of vessel to which they were originally attached cannot be identified. However, there was some inconsistency in how these items were treated in the assemblage, with some jar handles being recorded as jar sherds and the remainder as "other" sherds. The specific type of vessel represented by a sherd classified as "other" was not recorded systematically but, when recognized, was usually included as a comment on the analysis form.

Vessel part refers to the portion of the vessel (rim or body) from which a particular sherd came. A sherd that retains the original finished edge, or lip, of the vessel is classified as a rim sherd, and a piece that does not is classified as a body sherd. Handles are generally classified as body sherds, unless one happens to be attached to the vessel at the rim and actually includes a portion of the rim.

The Sherd Assemblage

Table 3.1 summarizes sherd frequencies by ware, type, and form for general provenience groupings at the Duckfoot site; the same information is given for individual structures and other excavation units in Appendix A. Mesa Verde gray wares predominate, making up 89.8 percent of the total assemblage, followed by San Juan red (5.2 percent) and Mesa Verde white (5.0 percent) wares. Other nonlocal wares make up less than .1 percent of the total assemblage. In all cases, sherds that could not be identified to specific type predominate: 92.9 percent of the gray wares, 71.2 percent of the white wares (Mesa Verde), and 63.6 percent of the red wares (San Juan) were classified as indeterminate types. When considering the identifiable assemblage only—that is, the sherds for which traditional type assignments could be made—Pueblo I types clearly, and not unexpectedly, dominate the assemblage. Of the 7,204 typeable gray wares, 17.0 percent are Chapin, 67.8 percent are Moccasin, and 15.1 percent are Mancos (see Figure 3.1). The two specific corrugated types combined constitute less than 1 percent of the identifiable gray ware assemblage. In addition to being the predominant identifiable gray ware type overall, Moccasin also consistently outnumbers all other traditional named types in all contexts at the site.

Of the 1,617 Mesa Verde White Ware sherds identified to specific type, 72.7 percent are Piedra Black-on-white, which is the standard Pueblo I style throughout the Mesa Verde region. Much smaller quantities of both earlier and later types were recovered: 21.5 percent of the identifiable assemblage is Chapin Black-on-white, 2.3 percent is Cortez Black-on-white, and 3.2 percent is Mancos Black-on-white. McElmo and Mesa Verde black-on-white combined make up less than 1 percent of the identifiable assemblage. The recovery of Basketmaker pottery (Chapin Black-on-white) in the quantities seen here can be explained most easily by the presence of numerous Basketmaker sites in the immediate vicinity of Duckfoot: prehistoric scavenging and inadvertent mixing of deposits from nearby or adjacent sites are likely. Selected white ware sherds are shown in Figure 3.2.

The typeable San Juan Red Ware collection is composed overwhelmingly of Bluff Black-on-red sherds (2,023 of 2,107 sherds, or 96.0 percent). Abajo Red-on-orange and Deadmans Black-on-red make up only 2.1 and 1.9 percent of the identifiable red ware assemblage, respectively.

Only 10 sherds, less than .1 percent of the total Duckfoot assemblage, were identified as "other nonlocal." One red ware sherd has mica inclusions, which probably indicates nonlocal manufacture. The remaining pieces (four red ware, five white ware) do not conform exactly to any of the foregoing type definitions but could be from Mesa Verde or San Juan vessels that happened to have nonstandard characteristics.

The composition of the Duckfoot pottery assemblage is generally similar to that of Dolores-area collections dating to roughly the same time period (A.D. 860–880) (Blinman 1986a:70–75, Table 2.5), with a few exceptions that probably reflect differences in analytic criteria more than actual differences in the assemblages. For both Duckfoot and Dolores, indeterminate types make up the majority of the collections, and within the identifiable assemblages, Moccasin Gray, Piedra Black-on-white, and Bluff Black-on-red predominate. Percentages calculated on the basis of the combined typed and indeterminate assemblages show almost identical values for Piedra Black-on-white (1.0 percent for Duckfoot, .9 percent for Dolores) and similar figures for Bluff Black-on-red (1.8 percent for Duckfoot, 1.0 percent for Dolores). The greatest differences between the two assemblages are in the proportions of the two neckbanded types, Moccasin and Mancos gray. Moccasin Gray constitutes 14.3 percent of the total A.D. 860–880 sample from Dolores but accounts for only 4.4 percent of the total Duckfoot collection. Similarly, 3.1 percent of the Dolores collection was typed as Mancos Gray, as opposed to only 1.0 percent at Duckfoot. However, these differences are at least partly, and perhaps wholly, related to differences in analytic criteria: in the Dolores Archaeological Program (DAP) analytic system, all neckbanded sherds were assigned to either Moccasin or Mancos gray categories (Blinman 1984:129), whereas in the Crow Canyon system, neckbanded sherds with characteristics intermediate between the two types were placed in the Indeterminate Plain Gray category. This analytic bias is also reflected in the percentages of Indeterminate Plain Gray sherds (called "Early Pueblo Gray" at Dolores): 83.2 percent for Duckfoot, as opposed to 72.4 percent for Dolores.

The Duckfoot sherd assemblage is composed largely of jar fragments (91.4 percent), with bowls and "other" forms constituting much smaller percentages (7.6 and .9 percent, respectively). Included in the "other" form category are seed jar, dipper, double dipper, and double-bowl sherds; handles and appliqués detached from their vessels; pipes and pipe fragments; and the two pottery "feet" after which the site was named.

Vessels

Lisa K. Shifrin and Mary C. Etzkorn

After the completion of basic sherd analysis, a vessel-reconstruction, or "refitting," study was undertaken to provide more detailed information on assemblage composition. Included in the Duckfoot vessel study were containers recovered whole and unbroken from individual archaeological contexts and containers reconstructed in the laboratory from sherds recovered from a variety of proveniences across the site. All sherds from floor and roof fall

Table 3.1. Sherd Data Summary

WARE AND TYPE	Room Surfaces N	%	Room Roofs N	%	Pit Structure Surfaces N	%	Pit Structure Roofs N	%	Midden N	%	Modern Ground Surface N	%	Other N	%	Site Total N	%
Mesa Verde																
Mudware													3	.0	3	.0
SUBTOTAL													3	.0	3	.0
Gray Ware																
Chapin Gray	29	.9	49	1.2	49	1.1	38	2.4	658	1.0	45	.9	357	1.3	1,225	1.1
Moccasin Gray	207	6.7	230	5.6	340	7.6	112	7.2	2,537	3.8	192	4.0	1,265	4.7	4,883	4.4
Mancos Gray	50	1.6	44	1.1	134	3.0	26	1.7	555	.8	24	.5	253	.9	1,086	1.0
Indeterminate Plain Gray	2,651	85.2	3,423	83.0	3,563	80.1	1,228	78.7	55,656	82.7	4,195	87.3	22,620	84.3	93,336	83.2
Mancos Corrugated									2	.0			6	.0	6	.0
Mesa Verde Corrugated											2	.0			4	.0
Indeterminate Corrugated			5	.1	2	.0			67	.1	43	.9	111	.4	228	.2
SUBTOTAL	2,937	94.4	3,751	91.0	4,088	91.9	1,404	90.0	59,475	88.4	4,501	93.6	24,612	91.7	100,768	89.8
White Ware																
Chapin Black-on-white	6	.2	17	.4	6	.1	13	.8	214	.3	15	.3	77	.3	348	.3
Piedra Black-on-white	14	.4	51	1.2	50	1.1	12	.8	842	1.3	23	.5	183	.7	1,175	1.0
Cortez Black-on-white			1	.0	8	.2			18	.0	3	.1	7	.0	37	.0
Mancos Black-on-white									21	.0	3	.1	23	.1	51	.0
McElmo Black-on-white			4	.1							1	.0	3	.0	4	.0
Mesa Verde Black-on-white											1	.0	1	.0	2	.0
Indeterminate White	42	1.3	153	3.7	131	2.9	64	4.1	2,711	4.0	121	2.5	782	2.9	4,004	3.6
SUBTOTAL	62	2.0	226	5.5	195	4.4	89	5.7	3,806	5.7	167	3.5	1,076	4.0	5,621	5.0
Nonlocal																
San Juan Red Ware																
Abajo Red-on-orange			3	.1			2	.1	26	.0	1	.0	13	.0	45	.0
Bluff Black-on-red	62	2.0	67	1.6	53	1.2	28	1.8	1,330	2.0	34	.7	449	1.7	2,023	1.8
Deadmans Black-on-red									20	.0			19	.1	39	.0
Indeterminate San Juan Red	51	1.6	77	1.9	112	2.5	37	2.4	2,637	3.9	104	2.2	670	2.5	3,688	3.3
SUBTOTAL	113	3.6	147	3.6	165	3.7	67	4.3	4,013	6.0	139	2.9	1,151	4.3	5,795	5.2
Other Nonlocal Red									3	.0			2	.0	5	.0
Other Nonlocal White									3	.0	1	.0	1	.0	5	.0
SUBTOTAL									6	.0	1	.0	3	.0	10	.0
FORM																
Jar	2,938	94.4	3,887	94.3	4,200	94.4	1,441	92.4	61,265	91.0	4,519	94.0	24,352	90.7	102,602	91.4
Bowl	151	4.9	209	5.1	203	4.6	108	6.9	5,713	8.5	253	5.3	1,937	7.2	8,574	7.6
Other	23	.7	28	.7	45	1.0	11	.7	322	.5	36	.7	556	2.1	1,021	.9
TOTAL	3,112	2.8	4,124	3.7	4,448	4.0	1,560	1.4	67,300	60.0	4,808	4.3	26,845	23.9	112,197	100.0

NOTE: Structure surfaces include surfaces and surface-associated features containing cultural or constructional fill. Roofs include roof fall and mixed roof and wall fall deposits. Midden includes modern ground surface, fill, and all midden features. Modern ground surface includes all modern surface proveniences except for the midden. "Other" includes all other site deposits (e.g., structure and feature fills not included elsewhere, courtyard deposits, arbitrary units).

Figure 3.1. Gray ware sherds (*top to bottom*): *first row,* Chapin Gray; *second row,* Moccasin Gray; *third row,* Mancos Gray; *fourth row,* Indeterminate Corrugated.

contexts in pit structures and front-row rooms were included in the refitting study. Materials from back rooms, courtyards, and the midden were not examined systematically, although some red and white ware sherds from these contexts were recognized in a cursory scan as belonging to partially reconstructed vessels from other locations. The resulting assemblage of 126 vessels represents only a small portion of the total that would have been used during the occupation of the site but a portion that, because of its context, is potentially of great interpretive value.

Analytic Criteria

The term *vessel* refers to any pottery container, and in the Duckfoot study two broad categories were recognized: (1) whole or reconstructed original containers that were at least 30 percent present and (2) sherd containers, or sherds that, regardless of how much of the original vessel they represented, were believed to have been recycled for use as platters or other shallow receptacles. The inference that a sherd served as a container was based on the physical attributes of the sherd (size, shape, modification, location of stains and sooting; see Figure 3.21) and on the context in which it was found: sherds from proveniences (for example, surfaces) in which it was demonstrable that the sherd, rather than the original vessel, was being used were more likely to be identified as sherd containers than those recovered from more ambiguous contexts (for example, postabandonment fill). When a sherd container was identified, it, rather than the original vessel from which it derived, became the vessel of primary interest, although data were collected on both when possible. Observations recorded during vessel analysis include the following: pottery ware and type, vessel form, modified form (condition), use wear, and other signs of modification. Numerous measurements were recorded as well, including rim diame-

Figure 3.2. White ware sherds (*top to bottom*): *first row,* Chapin Black-on-white; *second and third rows,* Piedra Black-on-white; *fourth row,* Cortez Black-on-white (*first and second from left*) and Mancos Black-on-white.

ter and height, inflection-point diameter and height, maximum diameter and height, thickness, and volume. A detailed explanation of analytic procedures, observations, and measurements used or recorded in the vessel study is on file at the Anasazi Heritage Center.

The Vessel Assemblage

Pottery Ware/Type

Traditional ware and type designations were assigned to the vessel as a whole, regardless of the type name or names assigned to the constituent sherds during sherd analysis. When a vessel could not be identified as one of the

traditional types, it was assigned to one of the grouped types described in the section on sherd analysis. Many vessels in the assemblage are composed of sherds of more than one type, a natural result of fragment's of the same vessel having very different characteristics (for example, some sherds from a given vessel are painted, but others from the same vessel are not).

All vessels in the Duckfoot assemblage are Mesa Verde–region pottery types. Of the 126 vessels recorded, 96 (76.2 percent) are gray wares, 15 (11.9 percent) are red wares, and 15 (11.9 percent) are white wares. In marked contrast to the pattern seen during sherd analysis, in which most of the individual gray ware sherds were assigned to the grouped indeterminate type, almost three-quarters of the gray ware vessels were identified to specific named types, as follows: 21 Chapin Gray, 40 Moccasin Gray, and 8 Mancos Gray. The remaining 27 gray ware vessels, primarily sherd containers made of broken jars, were assigned to the Indeterminate Plain Gray category. A similar pattern is apparent in the red ware vessel assemblage, with 14 of the 15 vessels typed as Bluff Black-on-red and only 1 assigned to the Indeterminate San Juan Red category. The opposite is true of white ware vessels: five are Piedra Black-on-white and one is Mancos Black-on-white, but the remaining nine vessels were classified simply as Indeterminate White. Pottery type distributions are illustrated in Figure 3.3.

Vessel Form

Thirteen vessel forms were recognized during vessel analysis: wide-mouth jar, olla (narrow-mouth jar), seed jar, gourd jar, jar (not further specified), pitcher, beaker, bowl, double bowl, ladle, dipper, effigy, and "other." The criteria used to assign individual vessels are based on those used by Blinman (1988a:453–468, 1988b:127–143) and Shepard (1976:224–248), although adjustments were made to tailor the system to the Duckfoot assemblage specifically. The different vessel forms are illustrated in Figure 3.4, and selected whole and reconstructed vessels are shown in Figures 3.5 through 3.21. For purposes of this discussion, "sherd container," which during analysis was recorded as a modified form rather than as the vessel form proper,

is treated as the equivalent of each of the 13 vessel forms just named, to bring the total number of forms recognized to 14.

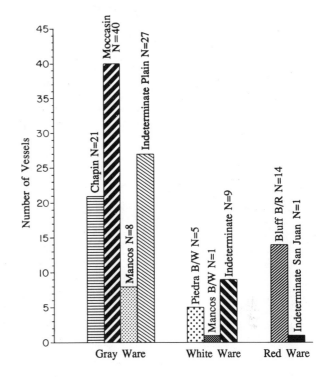

Figure 3.3. Duckfoot vessels by pottery type.

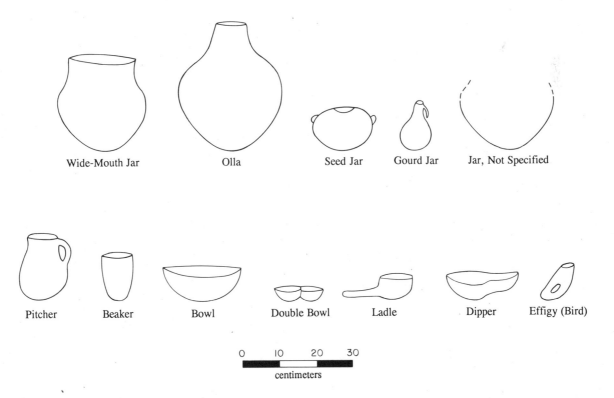

Figure 3.4. Vessel forms in the Duckfoot assemblage.

Figure 3.5. Chapin Gray olla (Vessel 17). Height = 41.0 cm.

Figure 3.6. Modified Chapin Gray olla (Vessel 25). The neck is pecked and scored, and the rim is lightly ground. Height = 45.1 cm.

Table 3.2 summarizes vessel-form frequencies by ware and type. Sherd containers and wide-mouth jars dominate the assemblage, and most vessel-form classes are composed primarily of gray wares. Moccasin Gray is the predominant specific type, particularly in the wide-mouth jar category (Figures 3.9–3.11). Indeterminate Plain Gray makes up roughly half of the sherd container assemblage, which is to be expected, since, by definition, this vessel-form class is made up of broken vessels. The only form classes not dominated by gray wares are seed jar, dipper, double bowl, and bowl. However, three of these form classes—seed jar, dipper, and double bowl—are represented by only one or two vessels, which makes it difficult to generalize about assemblage composition as a whole. Bowls, on the other hand, are represented by 18 vessels. Twelve of these are red wares, five are white wares, and only one is a gray ware, a distribution that, given the relatively large sample size, strongly suggests that this vessel class is composed predominantly of decorated (white and red) wares.

Functional inferences for specific vessel types are based on several factors in addition to form: size; use wear; contents; stability; and ease in placing, securing, and retrieving vessel contents are among the considerations

relevant to any analysis of function (cf. Blinman 1988a:450–453). Although a detailed evaluation is not within the purview of the present discussion, it is clear from even a cursory examination of the Duckfoot vessel assemblage that a range of uses is represented. Traditionally, jars and ollas are believed to have been used for cooking and/or short- and long-term storage; bowls, for serving and short-term storage. Functions of less commonplace vessels, such as pitchers and beakers, are not as clearly or consistently defined but may include all three functions when considering each form class as a whole or when considering the different purposes for which an individual vessel may have been used during its use life.

A relatively unambiguous observation that may be brought to bear on the most basic functional distinction—that between cooking and noncooking (storage or serving)—is the presence or absence of sooting. Table 3.3 compares the relative occurrence of sooting among several Duckfoot form classes for which sample sizes are large enough to show patterns. Only sooting that, because of location, coverage, and patterning, was inferred to be the result of use in or over a cooking fire was recorded; sooting as the result of building destruction or other abandonment or postabandonment processes was excluded from considera-

Table 3.2. Vessel Form by Pottery Ware and Type

Vessel Form	Gray Ware								White Ware						Red Ware				Total	
	Chapin		Moccasin		Mancos		Ind Plain		Piedra B/W		Mancos B/W		Indeterminate		Bluff B/R		Ind San Juan			
	N	%	N	%	N	%	N	%	N	%	N	%	N	%	N	%	N	%	N	%
Sherd container	4	8.2	13	26.5	1	2.0	25	51.0	1	2.0	1	2.0	4	8.2					49	38.9
Wide-mouth jar	3	9.1	22	66.7	7	21.2							1	3.0					33	26.2
Olla	4	57.1	2	28.6			1	14.3											7	5.6
Seed jar													1	50.0	1	50.0			2	1.6
Gourd jar	1	33.3	1	33.3									1	33.3					3	2.4
Jar, not specified			1	100.0															1	.8
Pitcher	3	50.0	1	16.7			1	16.7									1	16.7	6	4.8
Beaker	1	100.0																	1	.8
Bowl	1	5.5							4	22.2			1	5.5	12	66.7			18	14.3
Double bowl													1	100.0					1	.8
Ladle	1	100.0																	1	.8
Dipper	1	50.0													1	50.0			2	1.6
Effigy	1	100.0																	1	.8
Other	1	100.0																	1	.8
TOTAL	21	16.7	40	31.7	8	6.3	27	21.4	5	4.0	1	.8	9	7.1	14	11.1	1	.8	126	100.0

Ind = Indeterminate; B/W = Black-on-white; B/R = Black-on-red.

Figure 3.7. Small Chapin Gray vessels: *left,* pitcher (Vessel 3); *right,* wide-mouth jar (Vessel 4). Pitcher height = 15.5 cm; jar height = 16.3 cm.

tion. The results of this comparison support traditional functional inferences for most of the vessel forms examined and allow a fairly clear separation of cooking and noncooking containers. Sooting was observed on 72.7 percent of the wide-mouth jars, a finding that strongly supports the inference that this vessel form was used primarily for cooking. Ollas, on the other hand, apparently were used for storage or some other noncooking purpose, an interpretation supported not only by the absence of sooting but also by the neck and rim configurations of these oversize jars (narrow necks and mouth openings make ollas particularly suitable for the secure storage and transport of a variety of materials, particularly liquids) (Figures 3.5 and 3.6). The evidence for pitchers is mixed, with a third of the assemblage showing signs of having been used in or

over fires. It is possible that, as a class, pitchers were multipurpose, used as circumstances dictated for a variety of functions. With only 11.1 percent of the bowls in the assemblage showing evidence of sooting, this type of vessel appears to have been used primarily for serving or other noncooking functions. Sooting was recorded for sherd containers only when it was apparent that the discoloration was acquired after the sherd became the primary vessel

Figure 3.8. Chapin Gray bowl with handle (Vessel 117). Diameter = 30.5 cm.

Table 3.3. Sooting on Selected Vessel Forms

Vessel Form	Sooting	
	N	%
Sherd container (N = 49)	37	75.5
Wide-mouth jar (N = 33)	24	72.7
Olla (N = 7)	0	0.0
Pitcher (N = 6)	2	33.3
Bowl (N =18)	2	11.1

Figure 3.9. Large Moccasin Gray wide-mouth jar (Vessel 29). Height = 26.5 cm.

Figure 3.11. Small Moccasin Gray wide-mouth jar with coil handle (*left*) and pinch handle (*right*) (Vessel 13). Height = 20.2 cm.

Figure 3.10. Small Moccasin Gray wide-mouth jar (Vessel 26). Height = 17.0 cm.

(Figure 3.21). The percentage of sherd containers with sooting (75.5 percent) is similar to the percentage for wide-mouth jars. Given the wide range of sizes and shapes of sherd containers—some broad and shallow, others smaller and deeper—it seems likely that some of these vessels served as platters and dishes but that the majority were used for cooking.

Modification

Modification refers to any alteration of the vessel not interpreted to be the direct result of use. Four types of modification were recognized during analysis of the Duck-foot assemblage: edge grinding, pecking (to create a more regular edge or to create a controlled area of weakness, as in scoring), drilling, and "other." A fifth category, flaking (chipping), was excluded from consideration because of the difficulty in recognizing this type of modification. In an experiment to determine whether or not it would be possible to recognize flaking, four Duckfoot sherds were flaked (by percussion and pressure) in the same manner that some stone artifacts are flaked. The resulting sherd edges and debris were examined for diagnostic features, but none were observed, and it was decided not to include this type of modification as an analytic option, although it probably did occur.

Figure 3.12. Large Mancos Gray wide-mouth jar with narrow, rounded coils (Vessel 10). Height = 28.5 cm.

Figure 3.13. Medium Mancos Gray wide-mouth jar with incised neckbands (Vessel 116). Height = 24.5 cm.

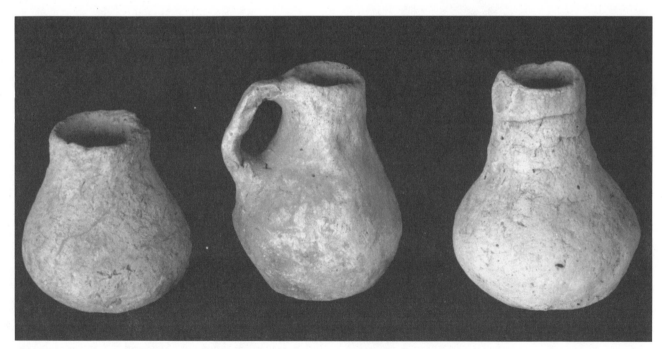

Figure 3.14. Miniature gray ware vessels: *left to right,* Chapin Gray gourd jar (Vessel 24, height = 5.9 cm); Chapin Gray pitcher (Vessel 23, height = 7.5 cm); Moccasin Gray gourd jar (Vessel 22, height = 7.3 cm).

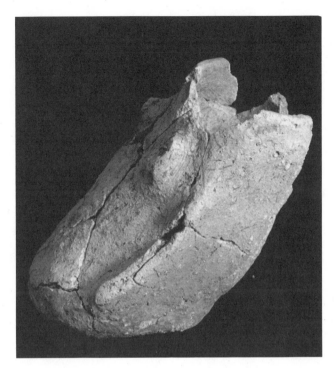

Figure 3.15. Chapin Gray effigy vessel (Vessel 58). Although this container is believed to represent a duck, it rests securely on a flat base and is not associated with the paired duck feet shown in Figure 1.3. Height = 8.5 cm.

Figure 3.16. Indeterminate White double bowl fragment (*left*, Vessel 2); Chapin Gray dipper fragment (*right*, Vessel 67). Bowl diameter (intact portion) = 7.9 cm.

Figure 3.17. Piedra Black-on-white bowls: *left,* Vessel 33 (diameter = 19.5 cm); *right,* Vessel 125 (diameter = 19.2 cm).

Figure 3.18. Indeterminate White seed jar (Vessel 15). Height = 10.5 cm.

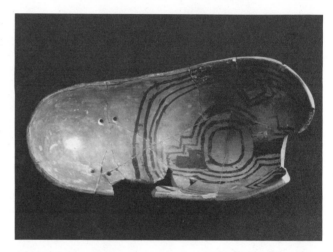

Figure 3.19. Bluff Black-on-red dipper with repair holes (Vessel 31). Length = 19.7 cm.

Figure 3.20. Bluff Black-on-red bowls: *left,* Vessel 59 (diameter = 20.9 cm); *right,* Vessel 35 (diameter = 19.0 cm). Note repair holes in Vessel 59.

Figure 3.21. Sherd containers. The location of the sooting suggests that the stains were acquired after the original vessels broke and the sherds were recycled for use as cooking containers. Vessel 46, on the left, was found on the floor of Pit Structure 3; Vessel 52, on the right, was recovered from floor and roof fall deposits in the same structure. Maximum dimension, both pieces = 16.0 cm.

Eleven vessels are modified: three ollas, two pitchers, one bowl, one dipper, and four sherd containers. The assemblage is composed of eight gray wares, two red wares, and one white ware. Signs of grinding and/or pecking are evident on 6 of the 11 pieces in the assemblage and are most common on gray ware ollas and sherd containers. In two cases, ollas whose necks were broken had been modified by pecking and/or grinding the jagged edges smooth to form a new rim, which may have afforded greater ease in placing and removing vessel contents (see Figure 3.6 for one example). A similar strategy evidently was employed in the case of two gray ware pitchers whose broken necks were "evened out" by deliberately breaking off additional pieces to produce a more regular opening. The newly formed rims were also smoothed, but possibly from use, as the smoothing was not even or regular enough to suggest deliberate modification. Drilled holes representing repair are present on only two pieces in the assemblage, both red wares. Attempts to repair a dipper and a bowl that had cracked are evident in the form of paired holes drilled from both the interior and the exterior surfaces on both sides of the cracks (Figures 3.19 and 3.20). It is inferred that fibers laced through the holes and pulled taut would have held the vessel segments together and prevented further cracking (cf. Morris 1986:410–412). On both vessels, two sets of

two holes were used, for a total of four holes per vessel. Given the relatively small sample size, generalizations are difficult, but it is interesting that the only examples of this relatively complex type of repair in the Duckfoot vessel collection are on nonlocal decorated wares that might be argued to have had greater economic or intrinsic value than the local plain gray wares.

Other Pottery Tools

Mary C. Etzkorn

The term *other pottery tool* refers to pottery artifacts that were not used as containers. This assemblage consists of (1) pottery artifacts that are not, and never were, vessels or parts of vessels and (2) sherds that were recycled for use as tools after the original vessels broke. Two types of sherd tools are recognized: modified sherds and shaped sherds. A modified sherd is one that has been modified on at least one but not all edges or on one or both surfaces. A shaped sherd has been modified on all edges, and the modification is continuous or nearly continuous around the circumference of the sherd. In both cases, modification may be incidental (as the result of use), deliberate (as part of the manufacture of the tool), or a combination of the two. However, modification on modified sherds is more likely to be the result of use, and that on shaped sherds the result of manufacture or a combination of manufacture and use.

Analytic Criteria

Information about ware, type, vessel form, and vessel part was recorded for all pottery tools (except "miscellaneous objects," discussed separately) during basic sherd analysis, and this information is imbedded in the summary sherd data presented in the section entitled "The Sherd Assemblage." During basic analysis of these same items as tools, additional descriptive information was recorded, including condition and type of modification. Condition was recorded as *complete, incomplete* (enough present to infer the final-use size and shape of the tool), *fragmentary* (not enough present to infer tool size and shape), and *unknown*. Because most of the pottery tools in the assemblage are modified sherds, which, by definition, are pieces of broken vessels, it was not always clear whether the tool itself was whole or broken, and "unknown" was recorded in those cases. Modification, the result of manufacture and/or use, can take several forms and was recorded as follows: ground/polished (no distinction made), flaked (chipped), and perforated (drilled). Even though, on the basis of subsequent experimentation during vessel analysis it was decided not to include "flaked" as a modification option for vessels, it *was* recorded for tools; flaking in this context should be regarded as any breaking of an edge to deliberately shape the piece. Modification may occur anywhere on the tool, and more than one type of modification may appear on a single tool.

The Tool Assemblage

Of the 232 pottery tools analyzed, 190 are modified sherds, 38 are shaped sherds, and 4 are pipes. One shaped and eight modified sherds incorporated into reconstructed vessels are excluded from consideration here. Although not analyzed as sherds, a number of unusual pottery objects that cannot be considered tools in the strict sense are also described in this section.

Modified Sherds

Although gray wares predominate (72.6 percent) in the modified sherd assemblage, the decorated wares are more strongly represented than in the total sherd population (Table 3.4). Red and white wares make up roughly 5 percent (each) of the total sherd assemblage but account for 12.1 and 15.3 percent, respectively, of the modified sherd collection. Modified sherds were recovered in roughly equal numbers from rooms and pit structures (combined floor and roof fall contexts), but when floor deposits alone

are considered, it is seen that twice as many modified sherds were recovered from pit structure floors than from room floors.

By far the most common form of modification is grinding and/or polishing of the sherd edge, but sherd surfaces are occasionally ground as well (Table 3.4). The most obvious and plausible functional inference for edge-ground items is that they were used as scraping tools, possibly to smooth clay coils during pottery manufacture. The question of function is examined in more detail in the following discussion of shaped sherds.

Shaped Sherds

Like the modified sherd collection, the shaped sherd assemblage is composed primarily of gray wares (Table 3.5). Relative proportions of decorated wares, however, differ markedly in the two artifact types. Whereas red and white wares are represented in roughly similar proportions in the modified sherd assemblage (12.1 and 15.3 percent, respectively), the former make up a much higher proportion of the shaped sherd assemblage (31.6 percent red, as opposed to only 7.9 percent white).

Twenty-five of the 38 shaped sherds in the Duckfoot assemblage are round or approximately round. The smallest measures only 2.1 cm in diameter, the largest 7.0 cm. Other shapes include oblong/oval (six), rectangular (three), tear-drop (one), and irregular (two). One unusual piece is a gray ware jar handle that had been flaked around the edges and ground on both ends. Selected shaped sherds are shown in Figure 3.22.

As seen in Table 3.5, grinding/polishing is the most common form of modification, and it occurs frequently in conjunction with flaking. In those cases in which both types of modification occur, flaking often appears to have been used to initially shape the piece, whereas grinding seems to have been used to produce a smooth, regular finished edge. The functions of the various specimens are unknown, although at least some of the smaller items may be gaming pieces or "counters" (cf. Rohn 1971:246) or unfinished or broken pendants (cf. Kidder 1932:150–151; Hayes and Lancaster 1975:142; Judd 1981:104). Other small- to medium-size pieces might have been used as scrapers in pottery manufacture, and larger disks could have served as jar lids or small "plates."

Preliminary results of microwear analysis of the combined modified and shaped sherd assemblages from Duckfoot (Shifrin 1991) indicate that these two types of artifact probably served a variety of purposes. Analysis involved the microscopic examination of worn sherd edges for "drumlins," that is, areas of paste shielded by projecting temper particles during unidirectional abrasion (Waterworth and Blinman 1986:4). Depending on the location of drumlins in relation to the beveled edge of the sherd, use wear can be identified as having resulted from

Table 3.4. Modified Sherd Summary

| | Room | | | | Pit Structure | | | | Midden | | Modern Ground Surface | | Other | | Site Total | |
| | Surfaces | | Roofs | | Surfaces | | Roofs | | | | | | | | | |
WARE AND TYPE	N	%	N	%	N	%	N	%	N	%	N	%	N	%	N	%
Mesa Verde																
Gray Ware																
Chapin Gray									1	.9					1	.5
Moccasin Gray			1	8.3	1	7.1	3	37.5	2	1.8					7	3.7
Indeterminate Plain Gray	7	100.0	5	41.7	9	64.3	4	50.0	82	73.9	1	100.0	22	59.5	130	68.4
SUBTOTAL	7	100.0	6	50.0	10	71.4	7	87.5	85	76.6	1	100.0	22	59.4	138	72.6
White Ware																
Chapin Black-on-white					2	14.3			2	1.8			2	5.4	4	2.1
Piedra Black-on-white			1	8.3					5	4.5			1	2.7	9	4.7
Indeterminate White			4	33.3	2	14.3	1	12.5	4	3.6			5	13.5	16	8.4
SUBTOTAL			5	41.7	4	28.6	1	12.5	11	9.9			8	21.6	29	15.3
Nonlocal																
San Juan Red Ware																
Bluff Black-on-red			1	8.3					7	6.3			5	13.5	13	6.8
Indeterminate San Juan Red									8	7.2			2	5.4	10	5.3
SUBTOTAL			1	8.3					15	13.5			7	18.9	23	12.1
MODIFICATION																
Ground/polished	4	57.1	10	83.3	12	85.7	8	100.0	95	85.6	1	100.0	31	83.8	161	84.7
Flaked	1	14.3							2	1.8					3	1.6
Multiple[a]	2	28.6	2	16.7	2	14.3			14	12.6			6	16.2	26	13.7
TOTAL	7	3.7	12	6.3	14	7.4	8	4.2	111	58.4	1	.5	37	19.5	190	100.0

NOTE: Structure surfaces include surfaces and surface-associated features containing cultural or constructional fill. Roofs include roof fall and mixed roof and wall fall deposits. Midden includes modern ground surface, fill, and all midden features. Modern ground surface includes all modern surface proveniences except for the midden. "Other" includes all other site deposits (e.g., structure and feature fills not included elsewhere, courtyard deposits, arbitrary units).

[a] Items for which more than one type of modification was observed. In all cases, items were both ground/polished and flaked.

Table 3.5. Shaped Sherd Summary

	Room Surfaces		Room Roofs		Pit Structure Surfaces		Pit Structure Roofs		Midden		Modern Ground Surface		Other		Site Total	
	N	%	N	%	N	%	N	%	N	%	N	%	N	%	N	%
WARE AND TYPE																
Mesa Verde																
Gray Ware																
Indeterminate Plain Gray	1	100.0	1	50.0					14	58.3	1	100.0	6	75.0	23	60.5
SUBTOTAL	1	100.0	1	50.0					14	58.3	1	100.0	6	75.0	23	60.5
White Ware																
Piedra Black-on-white			1	50.0											1	2.6
Indeterminate White							1	50.0	1	4.2					2	5.3
SUBTOTAL			1	50.0			1	50.0	1	4.2					3	7.9
Nonlocal																
San Juan Red Ware																
Bluff Black-on-red									2	8.3			1	12.5	3	7.9
Indeterminate San Juan Red							1	50.0	7	29.2			1	12.5	9	23.7
SUBTOTAL							1	50.0	9	37.5			2	25.0	12	31.6
MODIFICATION																
Ground/polished	1	100.0					2	100.0	12	50.0	1	100.0	4	50.0	20	52.6
Flaked			1	50.0					1	4.2					2	5.3
Multiple[a]			1	50.0					11	45.8			4	50.0	16	42.1
TOTAL	1	2.6	2	5.3			2	5.3	24	63.2	1	2.6	8	21.1	38	100.0

NOTE: Structure surfaces include surfaces and surface-associated features containing cultural or constructional fill. Roofs include roof fall and mixed roof and wall fall deposits. Midden includes modern ground surface, fill, and all midden features. Modern ground surface includes all modern surface proveniences except for the midden. "Other" includes all other site deposits (e.g., structure and feature fills not included elsewhere, courtyard deposits, arbitrary units).

[a] Items for which more than one type of modification was observed. In 15 cases, items were both ground/polished and flaked; in 1 case, the item was ground/polished, flaked, and perforated (drilled).

Figure 3.22. Shaped sherds.

either scooping or scraping motions. In the Duckfoot modified and shaped sherd assemblages, 3 items appear to have been used as scoops, and 14 appear to have been used as scrapers. Use of sherds in pottery manufacture to scrape the coils smooth has been inferred for various prehistoric collections (Hayes and Lancaster 1975:142; Judd 1981:184–185; Waterworth and Blinman 1986:4–6).

Pipes

Pipes are conical pottery objects small enough to be held in one hand, with a small hole at the proximal end and a flared opening at the distal end. Four such objects, all gray wares, were recovered from Duckfoot: two from pit structure floors (Pit Structures 1 and 3), one from postabandonment deposits in Pit Structure 3, and one from the fill of Room 2. Two complete specimens (from the floor of Pit Structure 1 and postabandonment fill in Pit Structure 3) are very similar in size, measuring roughly 6 cm in length and between 3 and 4 cm in diameter at their flared ends. The remaining two pipes are fragmentary. Figure 3.23 illustrates the complete specimen from Pit Structure 3. Judd (1981:299) describes the use of conical pipes, or "cloudblowers," in historic Pueblo ritual, and numerous pipes similar to those recovered from Duckfoot have been recorded for other sites in the region (e.g., Martin 1936:62, Figure 15; Hayes and Lancaster 1975:141–142; Blinman 1988a:467).

Miscellaneous Pottery Objects

Two pottery effigies, both fragmentary, were recovered from Feature 1 in the midden. One is clearly a four-legged

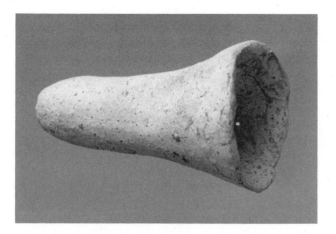

Figure 3.23. Pipe. Length = 6.2 cm.

animal, but because the head is missing, more precise identification is difficult (Figure 3.24). The length of the existing piece is 3.3 cm. The second effigy is so fragmented that it is difficult to identify the specific form represented, although overall it is similar to the other effigy. No measurements were taken of this piece.

Two fired pottery balls, one 1.2 cm in diameter and the other 2.3 cm, were recovered from midden deposits, as were three unusual objects of unknown function: a poorly

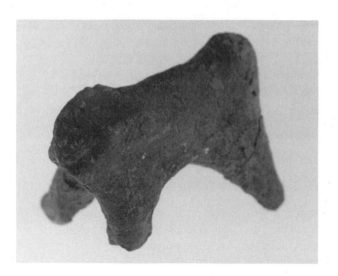

Figure 3.24. Pottery effigy. Length = 3.3 cm.

fired and oddly shaped piece resembling a fish tail; a small, molded, cone-shaped object; and a fragment of a thin, flattened "patty" with an incised crosshatch pattern on its surface. None of the miscellaneous pottery objects were polished, slipped, or painted, but because these items are not vessels or parts of vessels, traditional ware and type assignments were not made.

Analyses of Production and Stylistic Information

Michelle Hegmon

Because the Duckfoot site represents a single-component occupation that is securely dated with hundreds of tree-ring samples, its pottery assemblage provides a superb database for intensive analyses of style and technology. This section describes how Duckfoot pottery was used in studies of production, distribution, and stylistic communication at local, regional, and interregional levels.[1]

During the late Basketmaker III and Pueblo I periods in the Mesa Verde region, pottery became much more elaborate and its distribution more complex. In addition to the plain gray wares of earlier times, red and white wares (usually polished, sometimes slipped, and often painted) entered the repertoire. Most researchers assume or argue that the gray and white wares were made locally (by most communities if not most households), whereas the red wares were made in southeastern Utah and distributed widely (e.g., Blinman 1988b:76–95; Hegmon et al. 1991; Hurst 1983; Lucius and Breternitz 1981; Wilson and Blinman 1991b). These conclusions are based primarily on three lines of evidence: distributional data (i.e., the abundance of red wares in southeastern Utah); the distribution of pottery-making tools and occasional raw materials, which are found on most sites; and temper analyses using low magnification. Data from Duckfoot and several other sites in the Four Corners area were used to examine these theories of pottery production and distribution.

Three studies that include material from the Duckfoot site are described here. The first involves petrographic analysis of white ware temper to investigate the general organization of white ware production in southwestern

1. The research described here was aided and supported by a number of individuals and institutions. My dissertation committee (H.T. Wright, William Farrand, Richard Ford, Stephen Plog, and John Speth) provided important help and advice, particularly with the stylistic analysis. The Anasazi Heritage Center, Crow Canyon Archaeological Center, and the Center for Archaeological Investigations at Southern Illinois University provided me with access to collections and facilitated my research in many ways. Many DAP researchers helped me to understand the tremendous database they generated. I am particularly grateful to Eric Blinman and Dean Wilson for their advice on pottery. Jack Ellingson (Fort Lewis College) and William Melson (Smithsonian Institution) helped me with the petrographic analysis. The research was funded by the University of Michigan's Rackham School of Graduate Studies and James B. Griffin Fund (background work and stylistic analysis), by the Wenner-Gren Foundation (compositional analyses), and by a grant from the Karen S. Greiner Endowment of Colorado State University to Ricky Lightfoot and myself (petrographic analysis and research on temper sources). Much of the compositional analysis was done while I was a fellow at the Smithsonian Institution's Conservation Analytical Laboratory, under the guidance of Ronald Bishop and James Blackman.

Colorado. The second involves analysis of pottery chemical composition using Instrumental Neutron Activation Analyses (INAA). These data are used to understand white ware production at Duckfoot in more detail and to consider the production and distribution of red and white wares at a broader regional level. The third study involves stylistic analysis of painted white wares; it considers the role of pottery and pottery decoration in social communication and takes an interregional perspective.

Petrographic Analysis of White Ware Temper in Southwestern Colorado

Most Pueblo I pottery from the Mesa Verde region is tempered with crushed igneous rock. Igneous cobbles are available in many terraces and watercourses, as well as from the igneous features themselves. The rocks are quite variable, yet the same general suite of igneous rocks can be found across the area. These materials are derived from a series of intrusive geologic features, roughly late Cretaceous or early Tertiary in age (approximately 65 million years old), including the Abajo, La Plata, Ute, and Wilson mountains. Given this situation, it is unlikely that specific sources of prehistoric temper can be identified. However, if the pottery was produced at many locations across the region, it might be possible to identify temper differences among sites or areas, differences that reflect variation in the available rock or in the selections made by Pueblo I potters (see Shepard's [1939] analysis of igneous rock temper in La Plata–area pottery).

In order to investigate possible site-specific temper differences, petrographic analysis of white wares from Duckfoot and from a number of Dolores-area sites was undertaken. The research was conducted by Hegmon (1991) and by Garrett (1990). The analysis discussed here concerns 61 specimens of Pueblo I black-on-white pottery (either Piedra Black-on-white or untyped black-on-white technologically similar to Piedra), all tempered with crushed igneous rock. Although the same general type of rock (groundmass with porphyries of feldspars and ferromagnesium minerals [hornblende, biotite, and/or pyroxene]) was used as temper in most samples in both areas, there are a number of site- and area-specific differences in temper characteristics. These differences are most clearcut in the feldspar characteristics. Specifically, feldspars in the Duckfoot sample show more twinning and zoning than do those in the Dolores sample (twinning is a characteristic of plagioclase feldspars, zoning is a result of stress in an igneous environment). Duckfoot feldspars are also much less clouded than Dolores feldspars, suggesting that the rocks used for temper in the Duckfoot assemblage were less chemically altered than those used for the Dolores pottery. These results strongly suggest that different sources of temper were used in the white wares found at Duckfoot and at Dolores. Thus, they support the argument

that white wares were made at many locations across southwestern Colorado, although, from these data alone, it is not possible to determine whether white wares were actually made at the Duckfoot site.

Possible source material for igneous rock temper was also examined. The source closest to Duckfoot is Ute Mountain, particularly the Ute Creek dike, which extends north from the mountain and crosses McElmo Creek (Ekren and Houser 1965). Rocks from the dike and cobbles in McElmo Creek that derive from Ute Mountain are highly variable, but none appear to be a close match for the Duckfoot temper. A variety of igneous cobbles in the Dolores River was also sampled, and again, no close match for the white ware temper was found. Although failure to obtain a good match between source material and temper is frustrating, it leads to an important insight; that is, the analysis clearly shows that the prehistoric potters were selecting only certain kinds of rocks—from the wide variety available—to use as temper. Furthermore, although there are some differences in the temper used in the Duckfoot and Dolores samples, the same general selection criteria were applied in both areas. Thus, it appears that the technology of pottery production was widely shared at a very detailed level.

Chemical Composition and Distribution of Red and White Wares in the Mesa Verde Region

Evidence suggests that red and white wares were produced and distributed on very different scales in the Mesa Verde region. In order to better understand the relationship between pottery production and distribution and to understand production in more detail, a comparison of Pueblo I red and white wares from across the Mesa Verde region was undertaken (Hegmon et al. 1991). Samples from the Duckfoot site were the primary data point in southwestern Colorado. The analysis also included material from a number of sites in southeastern Utah, including Nancy Patterson Village in Montezuma Canyon and Edge of the Cedars (42SA700) further to the west, near Blanding.

Three pottery types were included in the comparison: Bluff Black-on-red, Piedra Black-on-white, and White Mesa Black-on-white (see Hurst et al. [1985] and Wilson and Blinman [1991a] for a description of White Mesa Black-on-white). Bluff Black-on-red, although most abundant in southeastern Utah, is common across the Mesa Verde region; 5 percent of the pottery found at Duckfoot is Bluff or Indeterminate San Juan Red (on the basis of sherd counts). San Juan red wares are also consistently present (though in lower frequencies) beyond the Mesa Verde region, including on Black Mesa in northeastern Arizona. In contrast, the two white wares have much more limited distributions. Piedra Black-on-white, although common in southwestern Colorado (5 percent of the Duckfoot assemblage is Piedra or other similar white ware), is rare in southeastern Utah. White Mesa Black-on-white, which differs from Piedra in terms of both technology and painted design style, is fairly common in southeastern Utah but is rare elsewhere. Much of the Duckfoot pottery analysis was done before White Mesa Black-on-white was formally described, so it would not have been recognized during standard analysis. However, an intensive reanalysis of at least 30 percent of the Duckfoot white wares revealed no examples of White Mesa Black-on-white.

The three types were compared in terms of chemical composition, firing effects, and morphological standardization, but the morphological data, which pertain primarily to red ware production in southeastern Utah, are not discussed here. The chemical (INAA) analysis included 17 specimens of White Mesa Black-on-white from Edge of the Cedars, 18 specimens of Bluff Black-on-red from Edge of the Cedars and 1 from Black Mesa Arizona, and 10 specimens of Piedra Black-on-white (including untyped black-on-white that is technologically similar to Piedra) from Duckfoot and 1 from Dolores. The distribution of the specimens based on the concentrations of iron and scandium is shown in Figure 3.25.

The two white ware types pattern very differently. They are different from each other chemically, and thus the compositional analysis supports the typological distinction. The sample of White Mesa Black-on-white clusters together quite tightly, suggesting that it was made at a single source (which is not unexpected, since the sample is from a single site). However, the sample of Piedra Black-on-white, although all from Duckfoot, does not form a chemically homogeneous group; instead, it shows extreme chemical heterogeneity. It appears that either the Duckfoot white wares were made at a very small scale (i.e., several production units at Duckfoot), or they were made in several locations in the Duckfoot area and exchanged, or both.

The red wares, including the specimen from Arizona, group together very tightly, suggesting that they were made at one or a few sources and distributed widely. Data on firing technology support this conclusion and suggest a possible core production area. Specifically, the analysis considered the presence of misfired red wares, that is, red wares that were not fully oxidized and therefore are gray in color. Misfired pieces are relatively scarce at the Duckfoot site (5 percent of the Duckfoot red wares in the sample were judged to be misfired), whereas they are much more common in southeastern Utah, particularly at Nancy Patterson Village in Montezuma Canyon, where 17 percent of the red wares were misfired. Red wares are also most abundant overall in southeastern Utah, particularly in Montezuma Canyon. It is likely that mistakes (i.e., misfired pieces) most often were left where they were produced and less often exported. Thus, firing information as well as compositional data suggest that red wares were produced in southeastern Utah and exported to southwestern Colorado. From this evidence, red ware production appears to have been specialized at the community level (see Costin [1991] for a discussion of community specialization). Furthermore, the abundance of misfired pieces at Nancy Patterson Village suggests that the Montezuma Canyon area may have been the core area of red ware production. Additional analysis of red wares from other sites, including Duckfoot, is underway (Hegmon and Allison 1990) in order to further examine this question.

Stylistic Analysis and Interregional Comparisons

Finally, the Duckfoot site data set was an important component of an analysis of Pueblo I pottery style (Hegmon 1992a, 1992b). The research considered variation within southwestern Colorado (using primarily Duckfoot and Dolores data) and compared patterns of stylistic variation between the Mesa Verde and the Kayenta regions (the Kayenta data were primarily from the Black Mesa Archaeological Project). The research included a study of rules of design style based on analyses of design structure on whole vessels and large fragments and a study of diversity and similarity based on analyses of design attributes on sherds and vessels.

When materials from the two regions were compared, it was found that designs on Mesa Verde pottery (including pottery from Duckfoot) were structured, although they were less rigidly rule-bound and much more diverse than designs on Kayenta pottery. The differences could be related to differences in social relations in the two regions. That is, Pueblo I settlement in the Mesa Verde region was much more intensive, with a larger population, larger and more permanent sites, and evidence of large-scale ritual activities. Kayenta settlement, with smaller and more short-term occupations and high mobility, is better characterized as extensive. In the Mesa Verde region, the stylistic diversity would have played an important role in establishing and marking social distinctions and differences, which would have been an important part of intensive social relationships. In contrast, pottery style in the Kayenta region was more likely an important means of expressing social similarity and maintaining extensive ties.

Although pottery style in general was highly diverse in the Mesa Verde region, apparently it was not used to maintain many rigid spatial boundaries. That is, Duckfoot white wares were stylistically very similar to those from Dolores, and samples from both areas were equally diverse. Thus, there is no evidence of stylistic differentiation among different drainages in southwestern Colorado, although differences between Piedra and White Mesa black-on-white suggest that spatial boundaries existed on a larger intraregional scale.

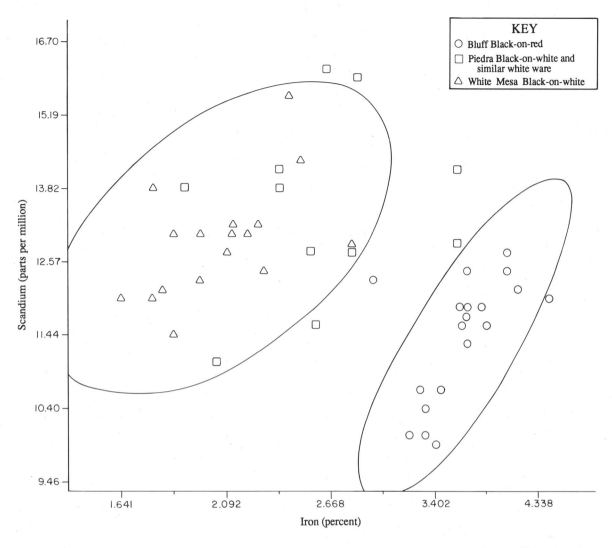

Figure 3.25. Logarithmic plot of the ratios of scandium to iron for White Mesa Black-on-white, Bluff Black-on-red, and Piedra Black-on-white (including specimens that are technologically similar to Piedra although they do not fit the strict type definition). The ellipses are 95 percent confidence intervals around Bluff and White Mesa. The Bluff ellipse was calculated excluding the outlier, which is from Edge of the Cedars. The one sample of Bluff from Arizona groups with the Utah sample.

Summary and Conclusions

Intensive analyses of pottery from Duckfoot and other sites in the northern Southwest reveal the kinds of information that can be gained about production and distribution systems and about social relationships. To conclude this section, the results of the three analyses are considered in relation to the role of pottery in Pueblo I social life.

The diversity of pottery designs at Duckfoot and elsewhere in the Mesa Verde region suggests that pottery played an important role in establishing and marking social distinctions. The finding that the designs were structured (although less rigidly structured than Kayenta designs) supports this interpretation. That is, the structure demonstrates that the diversity is not just random variation, a

result of what Morris (1939:177) suggested might be a "formative stage" in pottery decoration; instead, the structure indicates that the designs could have conveyed meaningful information.

The compositional analyses further support and enhance this conclusion with regard to the role of pottery style in social relations. Given the regularity of temper selection and other evidence of what appears to have been a well-established tradition of pottery technology in the Pueblo I period, it is unlikely that decorations on the same pottery would lack tradition. Rather, the technological consistency supports the argument that the design diversity relates to the social use of those designs. Furthermore, the chemical heterogeneity of Duckfoot white wares suggests that the pottery was produced on a small, local scale. Thus, it is likely that the Duckfoot residents would have had control over the designs on their white wares, either by making the pots themselves or through close social relations with the potters. Pottery designs could have easily conveyed information relating to both local and larger-scale social relations.

Finally, the white and red wares provide different, and complementary, lines of evidence. That is, the diversity of white ware designs, the evidence for small-scale local production, and the presence of different white ware types within the region indicate the expression of social differences at various levels. Comparable stylistic data on red wares are not available. However, concentrated red ware production in southeastern Utah and widespread distribution of the red wares (including to Duckfoot) are indicative of a broad regional social network. Thus, social differences were maintained at one level, although at another level, economic ties were far-reaching.

4

Stone and Mineral Artifacts

Mary C. Etzkorn

Stone and mineral objects make up the second-largest material category in the Duckfoot assemblage: approximately 25,390 individual items, including formal tools and ornaments, chipped-stone debris, and assorted unmodified materials, were recovered.[1] Formal tools and ornaments are stylized to greater or lesser degrees, and most were deliberately shaped before use, although they may also show evidence of wear or damage as a result of use. Flaking and grinding/polishing are the two most common forms of modification, and even though modification type is recorded during analysis, it is not used as a basis for broadly categorizing the tools in the assemblage (many items clearly were flaked *and* ground, which limits the usefulness of this observation as a point of separation). In this chapter, formal tools and ornaments are instead discussed in terms of broad functional categories: grinding tools, pounding/pecking tools, chopping tools, and so on. Because a study of use-wear patterns was beyond the scope of basic descriptive analysis, assignment of specific artifact types to broad functional categories is based entirely on morphology, ethnographic analogy, traditional assumptions about how certain objects were used, and common sense.

Chipped-stone debris consists primarily of the byproducts of stone tool and ornament manufacture. Most materials in this category are flakes, flake fragments, and pieces of shatter. However, also included in the assemblage are cores and informal, expedient flake tools; these items and the analytic system used to categorize them are described in more detail in the discussion of chipped-stone debris below.

A number of unmodified stone and mineral objects were recovered from Duckfoot and are also reported here because the contexts in which they were found or the improbability of their occurring naturally at the site suggest cultural significance.[2]

In this chapter, the artifacts in the Duckfoot assemblage are described in very general terms, with an emphasis on raw-material type and, for tools and ornaments, item condition. Broad patterns and exceptions to the patterns are noted. Refer to Appendix A for artifact summaries by major cultural unit and vertical subdivision within units.

Formal Tools and Ornaments

Analytic Criteria

The Duckfoot stone and mineral assemblage contains 1,645 formal tools and ornaments. After the items were cleaned and cataloged, each piece was examined individually, without magnification, and three observations were recorded: condition, material, and evidence of modification. On the basis of the last two observations and overall size and shape, each item was assigned to a specific artifact type category.

Condition

If an item was intact, its condition was recorded as *complete;* if broken so slightly that its final-use size and shape

---

1. This total is approximate because two artifact types—gizzard stones and pebbles—are not counted individually, but as occurrences within a given horizontal and vertical provenience, and a single occurrence may consist of one or multiple items.

2. Material types for most of the mineral objects were identified by Dr. John A. Campbell of Fort Lewis College, Department of Geology.

could be inferred with confidence, it was recorded as *incomplete;* and if a smaller portion of the specimen remained, making it impossible to determine its final-use size and shape, it was recorded as *fragmentary.* Occasionally it was not clear whether an object was whole or broken, and in such cases, *unknown* was recorded for item condition.

Material

The perennial problem of determining what constitutes a local versus a nonlocal resource was resolved, for purposes of this study, by the arbitrary designation of a 25-km (16-mi) radius around the Duckfoot site itself. Materials known to occur naturally within the circle so described are considered local; those found beyond are treated as nonlocal. Although this definition of local may or may not conform to the Anasazi view as to what constituted a reasonable hike, at least it has the advantage of being a generous estimate. Anything treated as nonlocal in this report probably could have been obtained only through intermediaries (trade) or by walking a relatively great distance. It should be noted that most materials that occur within a 25-km radius of Duckfoot are also known to occur at greater distances, and the designation of an item as "local" does not preclude the possibility that it was acquired elsewhere.

Local Materials. For stone artifacts, the type of stone and the geologic formation from which it came were recorded, when possible. Locally, the Dakota, Burro Canyon, and Morrison formations yield a variety of materials suitable for tool manufacture, and many of them would have been available to the inhabitants of Duckfoot at no great distance from the site (see discussion of geologic resources in Chapter 1). However, because the Duckfoot tools were examined without benefit of petrographic analysis or even simple magnification, the level of identification used was necessarily very general. Only two formations were recognized, the Dakota and the Morrison; items from the Burro Canyon Formation were included in tallies for the Morrison. Color and, to a lesser extent, texture were used to make formation assignments: Dakota Formation materials tend to be lighter—white, off-white, light gray, tan, pink, or buff—and coarser than Morrison Formation materials, which typically are green, purple, or brown and often have a finer texture.

Several types of stone yielded by the Dakota and Morrison formations were consolidated into just two classes during analysis. Claystones, mudstones, siltstones, and chert, all of which are characterized by relatively fine textures, were grouped into a single category called *chert/siltstone.* All silicified sandstones, ranging from very fine to very coarse in texture, were placed in the *orthoquartzite* category. There was a strong analytic bias in favor of orthoquartzites: specimens in which individual grains, however small, were discernible to the naked eye

and that were even slightly rough to the touch were classified as orthoquartzite; those for which individual grains could not be seen and that were smooth or slick to the touch were classified as chert/siltstone. This bias undoubtedly resulted in the misclassification of many siltstones as orthoquartzites.

Formation assignments were not made for three other types of stone materials assumed to be local. Sandstone and conglomerate materials most likely are from the Dakota Formation, and igneous rocks are assumed to have derived from nearby Ute Mountain or local stream and river beds, such as McElmo Creek and the Dolores River.

Most minerals in the Duckfoot assemblage could have been obtained locally. Azurite, though not abundant, is found in sparse copper deposits along the Ute Creek dike in McElmo Canyon (Ekren and Houser 1965:66), and calcite, gypsum, and ferruginous minerals such as hematite and limonite are available in numerous locations within 25 km of Duckfoot (Keane and Clay 1987:510–513; John A. Campbell, personal communication 1991).

Nonlocal Materials. Nonlocal stone types include obsidian, agate/chalcedony, and Washington Pass chert, all of which are distinctively finer grained than most materials found locally. Obsidian, or volcanic glass, usually is translucent light gray to black. It is known to occur naturally in, among other places, the San Francisco Peaks area near Flagstaff, Arizona, and the Jemez Mountains near Santa Fe, New Mexico. Both areas are among several believed to be represented in Dolores Pueblo I assemblages (Blinman 1986b:Figure 15.18), but because the Duckfoot specimens were not submitted for trace-element analysis, their place of origin is not known. Agate and chalcedony, cryptocrystalline silicates with excellent fracturing properties, are combined in the analytic system. The former is usually white or very light gray; the latter contains impurities that produce a wide range of colors, sometimes swirled through a lighter background matrix. Although Keane and Clay (1987:513–517) believe that these materials could represent both local and nonlocal sources, the nearest source of the high-quality agate and chalcedony seen in the Duckfoot collection is probably southeastern Utah, just outside the area defined as local (Bruce A. Bradley, personal communication 1991). Washington Pass chert is variable but, in most of its forms, distinctive. Also a high-quality silicate, it most often is a translucent or opaque pinkish brown, but occasionally it occurs in a pink, brown, gray, and white banded form. Washington Pass chert has a very restricted point of origin in the Washington Pass area of the Chuska Mountains in northwestern New Mexico, roughly 140 km south of the Duckfoot site (Bruce A. Bradley, personal communication 1991). Finally, any chert that does not fall within the texture and color ranges recognized in the local formations and that does not have the distinctive characteristics of Washington Pass chert is classified simply as nonlocal chert. Jasper is included in this category. The nonlocal chert classification is based more on visual, aesthetic considerations than on strict geologic criteria: extremely colorful, fine-textured, high-quality cherts are assumed to have originated somewhere other than the Montezuma Valley or environs.

The only mineral in the Duckfoot assemblage that is of nonlocal origin is turquoise, found in central and southern New Mexico and Arizona and in a few locations in central Colorado (Keane and Clay 1987:Table 21.2 and Figure 21.15). Although the closest documented natural source is roughly 260 km northeast of Duckfoot, near Villa Grove, Colorado (Keane and Clay 1987:Table 21.2 and Figure 21.15), the rugged mountains that separate the Montezuma Valley from this particular source may have served as an effective barrier. Sources to the south, though more distant, may have been more accessible. Keane and Clay (1987) provide an excellent summary of unusual rock and mineral sources for materials recovered from archaeological sites in the nearby Dolores River valley and should be consulted for additional information.

Modification

Evidence of modification was recorded for each tool or ornament. No attempt was made to distinguish among modification as a result of manufacture, modification as a result of use, and modification as a result of tool repair or maintenance. *Flaked* refers to the deliberate or incidental removal of flakes through either percussion- or pressure-flaking techniques. *Notched* and *serrated* describe two specialized forms of modification usually, though not always, created by flaking. Notches are distinct indentations frequently found in pairs on opposite edges of a tool; most often they probably served as hafting elements, that is, as anchors for shafts or handles that were tied to the tools. A serrated artifact is one whose edge is marked by a series of deep, regular indentations and sharp projections that give a jagged, sawtooth appearance. *Ground/polished* is used to describe an unnaturally smooth surface created by abrasion or rubbing; smoothing believed to be the result of natural weathering is not considered modification. *Grooved* and *incised* refer to the wearing of pronounced troughs in the surface of an artifact, and the two are distinguished by overall size and general appearance: grooves are deeper and broader; incisions are shallower and narrower, similar to cut marks. *Striated* describes the fine scratches that frequently are found on ground or polished artifact surfaces, although they also may be found unassociated with any other form of modification. Because artifacts were not examined under magnification, only striations pronounced enough to be visible to the naked eye were recorded, and many less obvious striations probably were not detected. Battered, pecked, and pitted are three forms of modification that result from similar processes, and certain assump-

tions about how the modification was acquired are used when describing specific artifacts. *Battered* refers to modification on a tool that is used to strike another object; *pecked* describes the modification that results from an item being struck. Thus, a peckingstone used to roughen the surface of a mano is said to be battered, but the surface of the mano is recorded as having been pecked. Pecking that is concentrated in one small spot, with the result that a pronounced depression is created in the surface of the rock, is described as *pitting*. *Perforated* artifacts are those through which a hole has been worn, punched, or drilled, and even incompletely perforated objects are included in this category.

None of the forms of modification described above is mutually exclusive of the others, and it is not uncommon for more than one form to be recorded for a given piece. For instance, a projectile point whose notches were created by flaking and whose edges are serrated would be described as flaked, notched, and serrated. It is hoped that such an approach to basic descriptive analysis will lend itself to future functional studies of artifacts that share similar attributes.

The Assemblage

In the following discussion, raw material and item condition data are reported, by artifact type, in Tables 4.1 and 4.2, respectively. Detailed modification data are on file at the Anasazi Heritage Center.

Grinding Tools

Metates. Metates are large rocks with distinctive shapes and wear patterns that clearly indicate they were used as grinding implements, specifically as the large, stationary bases against which the hand-held manos were rubbed. Together, these two tools form the basic Puebloan food-processing kit; their use in grinding corn and possibly other foodstuffs is well documented in the ethnographic and archaeological literature (e.g., Kidder 1932:66–69; Rohn 1971:201; Judd 1981:132–133; Raffensperger 1988:207). Depending on the characteristics of the grinding surface, it is possible to classify most metates as basin, trough, or slab, but in instances in which the metate is too fragmentary or the form too ambiguous, the item may be classified simply as a metate. A basin metate has a hollowed-out basin, usually a closed round or oval depression, which serves as the grinding surface; this type of metate is seen most often in Basketmaker assemblages, and none were recovered from Duckfoot. Trough metates were preferred during the Pueblo I period, and they are recognized by a distinct rectangular grinding trough that is open on one end. Trough depths range from shallow to quite deep, and occasionally a hole may have been worn through the bottom of the trough. A metate with a flat or nearly flat grinding

Table 4.1. Stone and Mineral Artifacts: Formal Tools and Ornaments by Raw Material Type

Artifact Type	Dakota Orthoquartzite N	%	Morrison Orthoquartzite N	%	Morrison Chert/Siltstone N	%	Sandstone N	%	Conglomerate N	%	Calcite N	%	Gypsum N	%	Quartz N	%	Igneous N	%	Other N	%
Grinding Tools																				
Metate (all types)							58	81.7	12	16.9							1	1.4		
Metate rest							5	100.0												
Mano (all types)	2	.9					139	65.6	55	25.9							14	6.6		
Abrader							71	92.2	6	7.8										
Pounding Tools																				
Hammerstone	1	3.0	1	3.0			3	9.1							5	15.2	9	27.3		
Peckingstone	151	32.2	287	61.2	10	2.1			1	.2					2	.4	5	1.1		
Maul	1	12.5	2	25.0													4	50.0		
Chopping Tools																				
Axe (all types)			5	33.3													10	66.7		
Axe/maul																	2	100.0		
Cutting, Piercing, Scraping Tools																				
Biface	12	40.0	2	6.7	6	20.0														
Projectile point	27	41.5	5	7.7	9	13.8														
Drill	4	36.4	2	18.2	2	18.2														
Modified flake	67	17.3	277	71.4	9	2.3											4	1.0		
Modified core			6	100.0																
Other chipped-stone tool	4	14.8	19	70.4																
Ornaments																				
Bead																				
Pendant											2	25.0								
Miscellaneous																				
Modified cobble			1	1.2			5	5.9							2	2.4	64	75.3		
Polishing stone	1	5.0															2	10.0		
Polishing/hammerstone																				
Stone disk							15	100.0												
Other modified stone/mineral	3	3.5	5	5.9			36	42.4	9	10.6			1	1.2	1	1.2	12	14.1	3	3.5
TOTAL	273	16.6	612	37.2	36	2.2	332	20.2	83	5.0	2	.1	1	.1	10	.6	127	7.7	3	.2

Table 4.1. Stone and Mineral Artifacts: Formal Tools and Ornaments by Raw Material Type (continued)

Artifact Type	Nonlocal												Unknown						Site Total	
	Agate/Chalcedony		Washington Pass Chert		Chert/Siltstone		Turquoise		Obsidian		Other		Orthoquartzite		Chert/Siltstone		Unknown			
	N	%	N	%	N	%	N	%	N	%	N	%	N	%	N	%	N	%	N	%
Grinding Tools																				
Metate (all types)																			71	4.3
Metate rest																			5	.3
Mano (all types)													2	.9					212	12.9
Abrader																			77	4.7
Pounding Tools																				
Hammerstone													7	21.2			7	21.2	33	2.0
Peckingstone					2	.4							3	.6	8	1.7			469	28.5
Maul																	1	12.5	8	.5
Chopping Tools																				
Axe (all types)																			15	.9
Axe/maul																			2	.1
Cutting, Piercing, Scraping Tools																				
Biface	3	10.0			6	20.0			4	6.2			1	1.5	1	3.3			30	1.8
Projectile point	8	12.3	1	1.5	8	12.3									2	3.1			65	4.0
Drill	1	9.1	1	9.1	1	9.1													11	.7
Modified flake	14	3.6			8	2.1			1	.3			1	.3	6	1.5			388	23.6
Modified core																			6	.4
Other chipped-stone tool	1	3.7							2	7.4					1	3.7			27	1.6
Ornaments																				
Bead							1	50.0									1	50.0	2	.1
Pendant							1	12.5							4	50.0	1	12.5	8	.5
Miscellaneous																				
Modified cobble					1	1.2					1	1.2	7	8.2			4	4.7	85	5.2
Polishing stone	1	5.0											4	20.0			12	60.0	20	1.2
Polishing/hammerstone													10	90.9			1	9.1	11	.7
Stone disk																			15	.9
Other modified stone/mineral									1	1.2	2	2.4	1	1.2	1	1.2	9	10.6	85	5.2
TOTAL	28	1.7	2	.1	26	1.6	3	.2	8	.5	4	.2	36	2.2	23	1.4	36	2.2	1,645	100.0

Table 4.2. Stone and Mineral Artifacts: Formal Tools and Ornaments by Condition

Artifact Type	Complete		Incomplete		Fragmentary		Unknown		Site Total	
	N	%	N	%	N	%	N	%	N	%
Grinding Tools										
Metate (all types)	31	43.7	4	5.6	36	50.7			71	4.3
Metate rest	2	40.0					3	60.0	5	.3
Mano (all types)	117	55.2	20	9.4	73	34.4	2	.9	212	12.9
Abrader	8	10.4	1	1.3	56	72.7	12	15.6	77	4.7
Pounding Tools										
Hammerstone	25	75.8	2	6.1	6	18.2			33	2.0
Peckingstone	398	84.9	6	1.3	35	7.5	30	6.4	469	28.5
Maul	7	87.5			1	12.5			8	.5
Chopping Tools										
Axe (all types)	12	80.0	1	6.7	2	13.3			15	.9
Axe/maul					2	100.0			2	.1
Cutting, Piercing, Scraping Tools										
Biface	4	13.3	2	6.7	24	80.0			30	1.8
Projectile point	11	16.9	36	55.4	18	27.7			65	4.0
Drill	3	27.3	5	45.5	1	9.1	2	18.2	11	.7
Modified flake	276	71.1	5	1.3	62	16.0	45	11.6	388	23.6
Modified core	5	83.3			1	16.7			6	.4
Other chipped-stone tool	14	51.9			6	22.2	7	25.9	27	1.6
Ornaments										
Bead	2	100.0							2	.1
Pendant	5	62.5	3	37.5					8	.5
Miscellaneous										
Modified cobble	39	45.9	6	7.1	37	43.5	3	3.5	85	5.2
Polishing stone	17	85.0	1	5.0	2	10.0			20	1.2
Polishing/hammerstone	9	81.8	1	9.1	1	9.1			11	.7
Stone disk	15	100.0							15	.9
Other modified stone/mineral	29	34.1	2	2.4	45	52.9	9	10.6	85	5.2
TOTAL	1,029	62.6	95	5.8	408	24.8	113	6.9	1,645	100.0

surface that spans or nearly spans the width of the metate is classified as a slab metate, a style seen frequently throughout the Pueblo II and III periods. The grinding surface may be slightly concave as the result of use, but it does not have a pronounced trough.

Of the 71 metates recovered from the Duckfoot site, 64 (90.1 percent) are trough metates, 3 (4.2 percent) are slab metates, and the remaining 4 (5.6 percent) are too fragmentary to assign to specific type. Selected specimens are shown in Figure 4.1. The Duckfoot assemblage is composed overwhelmingly of sandstone and conglomerate materials (Table 4.1), which should be expected given the availability of such resources in the nearby area and the suitability of these rough-textured rocks for grinding foodstuffs. The 35 metates that are complete or nearly so (Table 4.2) vary widely in their characteristics. Two types of sandstone were used—a thin, tabular variety and a massive, blocky stone. Textures range from fairly fine (some of the sandstones) to medium to very coarse (the conglomerates). This textural gradation may relate to the different stages of corn grinding, with metates made of very coarse materials having been used for the initial grind and metates made of

successively smaller grained materials having been used for progressively finer grinds (cf. Judd 1981:133). The metates range from oval to rectangular to triangular in shape, and most appear to have been at least lightly modified around their edges by flaking and/or pecking. Trough depths range from barely perceptible to very deep, and almost all the trough bases have been roughened by pecking. Roughening (sometimes called "sharpening") the surface of a metate that has grown too smooth through use to effectively grind corn extends the use life of the tool and is a well-documented Puebloan practice (Kidder 1932:69; Rohn 1971:202–203; Judd 1981:135). In the Duckfoot assemblage, surface pecking seems to have been carried to an extreme in at least one instance; the grinding surface of a trough metate recovered from the fill of Room 17 had been so deeply pitted that it is doubtful it could have continued to serve in its original capacity.

Metate Rests. Although technically not grinding implements, metate rests are discussed here because they serve as accessories to metates, which *are* used for grinding. As the name indicates, a metate rest is any object used to

Figure 4.1.   Metates. A slab metate is in the lower right corner; all others are trough metates. The deeply pitted metate from Room 17 is in the upper left corner. Length of metate in lower left corner = 47.5 cm, rest same scale.

support one end of the metate at an angle suitable for grinding. A metate rest may be another stone tool, an unmodified rock, or any object strong enough to support the weight of the metate. It is identified in the field on the basis of context; objects that may have served as metate rests but that were not discovered actually supporting a metate are not recognized as such. Five metate rests were recovered from the Duckfoot site (Tables 4.1 and 4.2). Of these, two are unmodified chunks of sandstone and three are pieces of sandstone that had been minimally shaped through grinding, pecking, and/or flaking.

Manos. Manos are the hand-held stone tools used in conjunction with metates to process grain; the smaller mano is drawn back and forth across the surface of the larger metate to reduce the corn kernels or other foodstuffs to a fine meal. Manos are round, oval, or rectangular in plan and oval, rectangular, or triangular in cross section. The edges may or may not have been deliberately shaped. At least one surface has been ground smooth as the result of use; however, the same surface frequently will have been roughened through pecking in order to provide or restore an effective grinding plane. Two types of mano are recognized in Crow Canyon's analytic system: one-hand and two-hand. As the name suggests, one-hand manos (sometimes called "biscuit manos") are small enough to be held comfortably in one hand, and although found on Anasazi sites dating to all time periods, they are most common on Basketmaker sites, where they were used in conjunction with basin metates. Two-hand manos, which predominate from Pueblo I times on, are large enough to require two hands to manipulate. Manos that are too fragmentary or

too ambiguous in terms of size and/or shape to be identified as one- or two-hand varieties are classified simply as manos.

Of the 212 manos recovered from the Duckfoot site, 195 (92.0 percent) are two-hand manos, 9 (4.2 percent) are one-hand, and 8 (3.8 percent) are too fragmentary to assign to either of the more specific categories. As with metates, the assemblage is composed overwhelmingly of sandstones and conglomerates (Table 4.1), and the range of textures again may indicate that different manos were used at different points in the grinding process. Of the 137 items that are complete or nearly complete (Table 4.2), the majority show signs of some deliberate shaping in the form of pecking or, very occasionally, flaking. Virtually all of the ground surfaces had also been roughened by pecking, and a few tools have pecked indentations that may have served as finger grips (Figures 4.2 and 4.3). A trace of pigment may have been present on the surface of one two-hand mano, which suggests that, at least occasionally, manos were used for grinding materials other than food.

Abraders. A piece of sandstone or conglomerate that has a ground surface, but that is not obviously a mano or a metate, is classified as an abrader. Included in this category are (1) complete or nearly complete tools that clearly never functioned as the aforementioned grinding implements and (2) broken items that may be mano or metate fragments but that are too small to be confidently identified as such. In all cases, at least some of the grinding is inferred to be the result of use and not a result of deliberate shaping of the tool as a part of manufacture. Although it is generally assumed that abraders were used to grind, smooth, or wear down other objects, this category

Figure 4.2. One-hand manos. Most are lightly pecked around their edges, and all are heavily ground on at least one surface. Several also show signs of battering on their edges, which suggests that they may have been used for pounding and pulverizing, as well as grinding, tasks.

Figure 4.3. Two-hand manos. The indentations pecked into the long sides of the three manos in the bottom row may have served as finger grips. Length of mano in upper left corner = 22.0 cm.

almost certainly encompasses a number of different tools used for a variety of specific purposes. Some abraders have uniformly ground, flat surfaces; others are ground on only parts of their available surfaces; a few are ground on their edges; and occasionally the grinding is confined to pronounced troughs or grooves in the rock face (although none of this last type was found at Duckfoot). Some may have been used as "active abraders," that is, as tools that were held in the hand and worked back and forth over other materials; others probably were used as "passive abraders," that is, as stationary bases against which other objects were ground. Food processing, tool manufacture and maintenance, and hide preparation are but a few of the possible uses to which various types of abraders may have been put, and studies of microscopic use-wear patterns may identify more specific tool types within this category. Because the term *abrader* is used in a very general sense in the Crow Canyon analytic system, it is difficult to draw exact parallels with other analytic systems and other site assemblages; however, ground sandstone artifacts have been inferred by various researchers in the Southwest to have served as files or whetstones or, if grooved, as arrow-shaft straighteners and smoothers (Kidder 1932:76–83; Rohn 1971:208; Hayes and Lancaster 1975:158; Wheeler 1980b:247–248; Judd 1981:118–124).

Seventy-seven abraders were recovered from the Duckfoot site. Selected specimens are shown in Figure 4.4. Most are fragments (Table 4.2), which leaves open the possibility that at least some are pieces of manos or metates that could not be identified as such. To the extent that it is possible to assess modification with such a fragmentary assemblage, it appears that many of the Duckfoot abraders had been only minimally shaped around their edges. By and large, abraders appear to be expediently produced tools that were modified more through use than through deliberate shaping during manufacture. As with the manos and metates, many abraders had been pecked in addition to having been ground, probably to extend the use life of the grinding surface.

Pounding/Pecking Tools

Hammerstones. A hammerstone is a naturally rounded stone tool that appears to have been battered through use. Hammerstones usually are small enough to be held in one hand but occasionally are large enough or heavy enough to require two hands to manipulate. In addition, a hammerstone must be made of a material of sufficient toughness and durability (for example, well-cemented orthoquartzite or igneous) to withstand being repeatedly struck against other

Figure 4.4. Abraders.

hard objects; a friable sandstone cobble, for instance, would not serve the purpose. In the Crow Canyon analytic system, hammerstones are distinguished from pecking-stones in that the former are naturally rounded before being battered, whereas the latter have ridges or remnants of sharp edges that were dulled through battering. It is as-sumed that rocks with different kinds of edges might have been used for different purposes. As defined here, hammer-stones are considered inherently better suited for use in stone tool manufacture (i.e., for removing flakes from cores and for flaking stone tools) and general hammering tasks. However, in many analytic systems, peckingstones and hammerstones are not differentiated (see Kidder 1932:60–61; Rohn 1971:210–211; Hayes and Lancaster 1975:149–150; Judd 1981:117–118).

Most of the 33 hammerstones in the Duckfoot assem-blage are made of small, water-worn river cobbles or pebbles (Figure 4.5). Igneous, quartz, and orthoquartzite are the predominant identifiable material types, but a fairly large number (7 of 33, or 21.2 percent) were classified simply as "unknown stone" (Table 4.1). The relatively high

percentage of unidentified materials reflects the difficulty with which heavily weathered specimens are identified: unless a fresh break exposes the rock interior, the type of material cannot be distinguished easily with the naked eye.

Peckingstones. A peckingstone is a rock that originally had sharp edges but that subsequently was battered, with the result that at least some of the edges became dulled or blunted. Most peckingstones probably were cores origi-nally, but large, sharp flakes and unmodified stones with naturally sharp edges occasionally were used as pecking-stones as well. In the literature, peckingstones are fre-quently referred to as hammerstones, and a wide range of uses has been suggested. Some may have been used to shape building stones or large stone tools such as metates; others may have been used to roughen the surfaces of manos and metates that had become too worn to grind corn effectively (Kidder 1932:61; Hayes and Lancaster 1975:149; Wheeler 1980b:243). Hayes and Lancaster (1975:149) state that some peckingstones could have been used to soften jerky. Clearly, such implements are suited for a number of

Figure 4.5. Hammerstones.

different purposes, and it is probably incorrect to assume a single predominant function for them.

Peckingstones are the most common formal stone tools in the Duckfoot assemblage: 469 individual specimens were recovered from all major cultural units at the site. Selected peckingstones are shown in Figure 4.6. The predominant type of material in the collection is Morrison orthoquartzite, followed by Dakota orthoquartzite and smaller quantities of igneous rock, chert/siltstone, and other materials (Table 4.1). The majority of peckingstones in the collection appear to have originated as cores, but some consist of river cobbles that were flaked to produce sharp edges, and several are simply large flakes with battered edges. Six specimens have distinctly ground facets

Figure 4.6. Peckingstones.

in addition to the battering requisite for identification as peckingstones. If the degree and extent of battering on a peckingstone is an indication of the duration of its use, the Duckfoot assemblage includes everything from minimally used peckingstones that are almost indistinguishable from cores to peckingstones that are so heavily battered that no sharp edges or ridges remain. Heavily used peckingstones that were still fairly large sometimes were resharpened by flaking so they could continue to be used as peckingstones, but smaller specimens probably were recycled as hammerstones or simply discarded.

Mauls. A maul is a heavy stone tool with two blunt ends and a grooved or notched hafting element located midsection or near the tool base. Suitable for pounding or pulverizing tasks, mauls may be thought of as hafted hammerstones, and like hammerstones, mauls must be made of resilient rock that can withstand repeated battering against other hard materials. Although this type of tool has been found in many Anasazi sites dating to various time periods, typically it is not found in great numbers (Kidder 1932:55; Rohn 1971:209–210; Hayes and Lancaster 1975:149; Judd 1981:240), which suggests that, for most pounding tasks, the smaller, unhafted hammerstones and peckingstones were the preferred implements.

Eight mauls, all made of river cobbles or massive stone nodules, were recovered from the Duckfoot site. Seven of the eight are made of tough igneous or orthoquartzite (Morrison and Dakota) materials (Table 4.1), and all but one are complete (Table 4.2). The hafting element on one item consists of a pecked groove that extends completely around the midsection of the tool; the remaining mauls are simply notched on their edges (Figure 4.7a–c). All but one show evidence of moderate to heavy battering on at least one and sometimes both ends, the exception being a notched river cobble that appears never to have been used. Several mauls in the Duckfoot collection may have originally served as axes, as evidenced by their tapered distal ends. These items apparently were recycled for use as mauls after their sharp edges became too dull to cut or chop.

Chopping Tools

Axes. An axe is similar to a maul except that it has a sharpened bit on one or both ends. It is possible to identify whole axes as either single- or double-bitted, but a fragment that does not include both ends is classified simply as an axe or an axe/maul. Eleven single-bitted and two double-bitted axes were recovered from Duckfoot; in addition, two bit fragments were too small to classify. Selected specimens are illustrated in Figure 4.7d–l. Igneous cobbles appear to have been the preferred raw material for axe manufacture, although five specimens were identified as Morrison orthoquartzite (Table 4.1). Most of the 13 complete or nearly complete specimens (Table 4.2) were only minimally shaped, which indicates that their makers had selected cobbles requiring relatively little effort to be modified into the desired end product. All had been roughly flaked and/or pecked into shape, and most had bits that had been ground to fine, sharp edges. Side notches are present on 12 specimens, and 1 unusual item has a basal notch in addition to side notches. A pronounced groove had been pecked three-quarters of the way around the proximal end of one axe. The ground surfaces of most of the Duckfoot axes are moderately to heavily striated, especially near their bits, and several show some signs of battering on their cutting edges, their butt ends, or both. Seven of the 15 axes were recovered from pit structure floors, 1 was

found on the floor of Room 19, and 1 was recovered from roof fall deposits in Room 11. The remaining six items were from the midden or postabandonment deposits in structures.

Typically, axes are thought to have been used primarily for chopping trees. However, in an experimental study in which stone axe replicas were used for a variety of purposes and the resulting use-wear patterns compared to those on Pueblo III specimens, Mills (1987:Table 7) concluded that, as a group, axes may have served more than one purpose. Wear patterns on most of the prehistoric axes (30 of 44, or 68.2 percent) were most consistent with those produced experimentally by chopping sagebrush at ground level and imbedding the axe bit in surrounding sediments. (Chopping, or "grubbing," sagebrush is one of the first steps in clearing land for building construction or agricultural-field preparation.) Only six axes (13.6 percent) had wear consistent with that seen on specimens used to chop trees, and four (10.0 percent) had damage indicative of hammering or pecking. The extent to which axe functions may have changed through time is not known, and therefore the Pueblo III analog may or may not apply to the Duckfoot axes.

Axes/mauls. A fragment of an axe or maul that does not include either end of the tool or one that includes only the butt, or proximal, end is classified in the Crow Canyon analytic system as an axe/maul because the portion of the tool needed to distinguish between the two is not present. Two such pieces were recovered from Duckfoot: one is a notched midsection, the other a notched butt fragment.

Cutting, Piercing, and Scraping Tools

Determining which stone implements were used for cutting, piercing, or scraping tasks is difficult in the absence of microwear studies, and in the Duckfoot study the assignment of specific artifact types to such broad functional classes is intended merely as a convenient organizational tool. Functional ambiguity and overlap are probably more the rule than the exception, and many tools may well have served multiple purposes.

Bifaces. Broadly defined, a biface is any flaked stone tool that has been bifacially thinned—that is, any item on which flake scars span both faces rather than be confined to the edges only. Under this broad definition, many "knives," most projectile points, and some drills would be classified as bifaces. In the Crow Canyon analytic system, however, biface is defined more narrowly to include only those bifacially thinned pieces that do not have hafting elements or other features that identify them as one of the more specialized forms such as drill or projectile point. Thus, included in this category are complete or nearly

Figure 4.7. Mauls and axes: *a–c,* mauls; *d–j,* single-bitted axes; *k–l,* double-bitted axes. Length of *l* = 19.8 cm.

complete items that clearly are not notched or stemmed and fragments (especially tip and midsection) that *could* be pieces of projectile points or drills but that do not include the part of the tool needed for more specific identification.

In the literature, bifaces ranging from small, crudely worked flakes to large, well-made items, with or without hafting elements, are often believed to have served as knives or general-purpose cutting implements (Kidder 1932:16–24; Rohn 1971:216–219; Hayes and Lancaster

Figure 4.8. Bifaces: *a–c,* possible projectile point fragments; *d,* "knife"; *e–h,* possible preforms ("blanks"); *i–m,* lanceolate forms.

1975:146; Wheeler 1980b:251–252). However, it is widely acknowledged that many bifacially thinned tools may have been used for purposes other than cutting (e.g., piercing or scraping) and, conversely, that some unifacially modified flakes may have been used as knives. Use-wear studies might identify specific functions of individual tools.

Thirty bifaces were recovered from the Duckfoot site, and two aspects of the collection are worth noting. First, as seen in Table 4.1, the percentage of nonlocal raw materials is unusually high—30.0 percent (six nonlocal chert and three agate/chalcedony)—and the percentage of items made of Morrison orthoquartzite, a common raw material in many other Duckfoot stone tool assemblages, is quite low—only 6.7 percent. Dakota orthoquartzite makes up 40.0 percent of the biface collection. Whether raw-material choice is a reflection of the suitability of certain material types for certain practical tasks or is suggestive of special ritual or other nonutilitarian functions for these items is not known. Second, the Duckfoot biface collection consists largely of fragmentary items (Table 4.2). Twenty-four of 30 bifaces (80.0 percent) are fragments, a reasonable figure given that this artifact category, by definition, includes pieces of other tools too fragmentary to identify to specific type. Certainly the number of fragmentary bifaces has been inflated by the inclusion of many projectile point tips and midsections.

Although it is difficult to characterize such a fragmentary assemblage, the size, shape, and overall level of workmanship indicate that several biface types or styles are present (Figure 4.8). The assemblage includes a number of short, broad, crudely worked pieces that may have served as tools but that could also be unfinished tool "blanks." At the other extreme are a number of very thin, very finely worked pieces that have such fragile edges and tips that it seems unlikely they could have survived any form of practical use (and, indeed, many of these are fragments). A third distinct type, represented only by fragments, is a relatively well shaped, relatively thin, but very large and broad biface; this type comes closest to the lanceolate variety most frequently interpreted to have functioned as a knife. Two small, crude bifaces with thick, pointed ends may be drills, but their characteristics are too ambiguous to assign them to that specific category. One tip and midsection fragment made of high-grade chalcedony was crudely serrated to produce very sharp, jagged edges. Perhaps the most distinctive piece in the collection is an item that technically should not be classified as a biface, because it has a constricted base, much like the hafting stem on some projectile points. However, the sides of this piece are almost perfectly parallel, and one edge has been worn or dulled, apparently through use, making this perhaps the most obvious example of a knife in the collection (Figure 4.8d).

Projectile Points. In the Crow Canyon analytic system, a projectile point is defined as a highly stylized tool with a sharp point on its distal end and a hafting element, such as a notch, stem, or flute, on its proximal end. Most projectile points have been bifacially thinned, but occasionally points are only lightly retouched around their edges. Although projectile points traditionally have been interpreted as hunting implements or weapons (attached to dart [spear] or arrow shafts) (Kidder 1932:14, 16–24; Hayes and Lancaster 1975:144–145; Wheeler 1980b:249–250; Judd 1981:253–254), it is also acknowledged that at least some may have served as knives (Kidder 1932:14, 16–24; Hayes and Lancaster 1975:144–145; Rohn 1971:218–219) and others as specialized ritual objects (see Stevenson [1894:78, 118, 122] for ethnographic accounts, and Judd [1981:254–255] and Bradley [1992] for archaeological examples).

Of the 65 projectile points recovered from Duckfoot, 21, or 32.3 percent, are made of nonlocal materials (agate/chalcedony, obsidian, Washington Pass chert, and nonlocal chert/siltstone, including jasper), a relatively high proportion similar to that noted for the more generalized biface category (Table 4.1). Of the local materials in the assemblage, many are exceptionally high quality cherts and orthoquartzites. Almost three-quarters of the collection is made up of items that are complete or nearly complete (Table 4.2), and by definition even the fragments include the hafting element. As a result, the Duckfoot projectile point assemblage lends itself particularly well to typological evaluation, and nine different projectile point styles recognized throughout the West and Southwest were identified (see Irwin-Williams 1973; Christenson 1987; Morris and Burgh 1954:56–57, Fig. 106; Hayes and Lancaster 1975:144–145).

Twenty points (30.8 percent) are classified as Archaic, with specific type breakdowns as follows: one Jay, one San Jose, two Mallory, seven late Archaic, and nine unidentifiable Archaic (Figure 4.9). All are believed to date to before A.D. 1, and the earliest type (Jay) is believed to date to between 5500 and 4800 B.C. (Irwin-Williams 1973:5–6). Jay and San Jose styles are marked by stemmed hafting elements and sloping shoulders. Jay points are larger and have slightly convex bases; San Jose points usually are smaller and frequently have concave bases. Mallory points are very distinctive, with large square or slightly concave bases and side notches. Fourteen of the Archaic projectile points in the Duckfoot assemblage are made of readily obtained Dakota and Morrison orthoquartzites or cherts. Four are made of cherts not believed to be available locally, one is made of agate/chalcedony, and one is made of orthoquartzite of unknown origin. Two points have serrated edges.

Fifteen of the early points were recovered from midden or other extramural deposits; four were found in collapsed structural debris in Rooms 3, 7, 8, and 10; and one was found on the floor of Pit Structure 1. The most plausible explanation for the recovery of Archaic points from Pueblo I cultural deposits is that the Duckfoot inhabitants collected these specimens and brought them to the site. Whether the

Figure 4.9. Archaic projectile points: *top row* (*left to right*), Jay, San Jose, Mallory, Mallory; *middle row,* late Archaic; *bottom row,* Archaic, not further specified.

points were actually used as tools or simply collected as curios is not known.

Fourteen points (21.5 percent) are classified as Basketmaker (A.D. 50–750) in origin: five as Basketmaker II and nine as Basketmaker III (Figure 4.10). Basketmaker II projectile points tend to be larger than Basketmaker III points, and they have corner notches, convex bases, and relatively straight to slightly convex sides. Four of the five Duckfoot specimens are made of Dakota orthoquartzite, and one is made of chert of nonlocal origin. Basketmaker III points have straight to slightly convex sides, corner notches, and straight to slightly convex bases. The material source composition of the Duckfoot Basketmaker III projectile point collection is noteworthy. Five of the nine items are made of nonlocal materials (one nonlocal chert, two agate/chalcedony, and two obsidian), and those made of local materials are relatively fine grained Dakota orthoquartzite and Morrison chert. Seven of the 14 Basketmaker points were recovered from the midden or courtyard areas, 6 were found on floors or in collapsed wall and roof debris within structures, and 1 was recovered from post-abandonment pit structure fill. The presence of Basketmaker points is not surprising, given that numerous Basket-maker sites are located in the immediate vicinity and the nearest is only 30 m distant. Prehistoric scavenging could account for the recovery of these earlier materials.

Twenty-three of the Duckfoot projectile points (35.4 percent) are classified as Pueblo I types (A.D. 750–900). These points are intermediate in size between the larger Archaic and Basketmaker points and the smaller Pueblo II and III types. The sides are straight to slightly concave, and characteristically the hafting element consists of a simple straight or contracting stem (Figure 4.11). In marked contrast to the Basketmaker III points, most of the Pueblo I points (16 of 23) are made of locally available Dakota and Morrison orthoquartzites and cherts. Nonlocal materials include chert (three points), agate/chalcedony (two), and Washington Pass chert (one). A single specimen is made of a chert of unknown origin. One point with a single notch on one side and a straight edge on the other appears to have been broken during manufacture.

Like the distribution of Basketmaker points, that of Pueblo I points at Duckfoot encompasses a wide variety of contexts, ranging from midden, courtyard, and other ex-tramural areas (17) to collapsed structural debris (4) and features within structures (2). One particularly well made point was found in the fill of the capped sipapu in Pit Structure 1; the point appeared to have been deliberately placed, perhaps as part of a ritual dedication.

One small, corner-notched projectile point was identi-fied as a Pueblo II type (A.D. 900–1150); one small, side-notched point was classified as Pueblo III (A.D. 1150–1300); and six points were classified as indeterminate types. The Pueblo II and III points are shown in Figure 4.11.

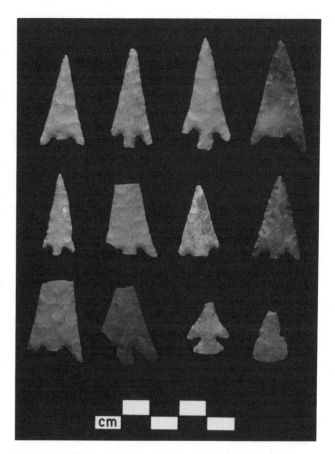

Figure 4.11. Pueblo I, Pueblo II, and Pueblo III projectile points. Pueblo II and III points are in the bottom row, third and fourth from left, respectively. All others are Pueblo I points.

The presence of two points that postdate the occupation of Duckfoot may be explained by the presence of Pueblo II and III sites in the vicinity. One of the indeterminate projectile points has a basal notch in addition to two side notches, and another point has a worn and beveled tip that suggests use as a drill.

Drills. A drill is a stone tool with a deliberately shaped or use-modified projection suitable for drilling holes with a rotary motion. Some drills are very informal and incon-spicuous, retaining most of the original characteristics of the flakes from which they were made and having only a few small flake scars and perhaps some light polish around their bits. Others appear to be reworked, or recycled, projectile points whose tips were modified by flaking into narrow bits. For all drill types, bits tend to be bevelled in opposite directions. Both hafted and unhafted drills are widely reported for Puebloan sites dating to all time periods (Kidder 1932:24–29; Rohn 1971:222, Figure 254; Hayes and Lancaster 1975:147, Figure 181; Judd 1981:132).

Eleven drills were recovered from Duckfoot (Figure 4.12). Eight of the 11 are made of locally available materials (Dakota orthoquartzite and Morrison ortho-

Figure 4.10. Basketmaker projectile points: *top row,* Basketmaker II; *bottom row,* Basketmaker III.

Figure 4.12. Drills: *a–c,* reworked projectile points; *d–f,* heavily modified flakes; *g–k,* minimally modified flakes. All specimens with intact bits showed signs of wear in the form of polish. Item *f* is unusual in that two of its three corners appear to have been used for drilling, as evidenced by the distinctive rounding and wearing of the projections.

quartzite and chert); the remaining 3 items are made of agate/chalcedony, Washington Pass chert, and other non-local chert (Table 4.1). A cursory examination of the collection suggests that three different styles of drill are present. The most common style (five items) is also the least formal and consists of flakes that have been only lightly retouched along their edges, particularly (and sometimes solely) in the area of the bit. These drills retain most of the original features of the flakes or spalls from which they were manufactured and were recognized as drills because one small tip or projection had been bevelled through unifacial or bifacial flaking. These informal drills probably were held in the hand rather than hafted to handles or shafts. The second drill style in the Duckfoot collection consists of moderately to heavily modified items that appear more stylized than the expedient variety just described. The three drills of this type were bifacially thinned or retouched along all their edges to produce very regular, formal shapes. Two have broad bases tapering to narrow bits; these could have been hafted to handles, but there is no direct evidence indicating that they actually were. The third drill of this style is a small, triangular piece on which two of the three corners appear to have served as bits; given the size and shape of this drill, hafting probably was not possible. The last type of drill in the Duckfoot collection consists of reworked projectile points. All three in the assemblage are highly stylized, bifacially thinned, notched tools that could have been easily hafted to pump-drill shafts or other handles.

Modified Flakes. By definition, a modified flake is one that has been deliberately modified prior to use but that does not fall into one of the other more formal tool categories; it is a very general category that encompasses a wide range of possible functions. Modification usually

consists of flaking along one or more edges but not across the surface of the tool. The flake scars are usually perpendicular to the edge of the flake, and they must be large enough to suggest deliberate modification rather than incidental damage or wear through use (although use wear may be evident as well). In addition, the scars must clearly represent flakes that were removed after the flake itself was detached from the core. A modified flake still shows at least some of the characteristic features of the flake from which it was made (platform, bulb, ripples on the ventral face). A chipped-stone tool not made from a flake, or one made from a flake on which most of the original flake features have been obliterated by subsequent modification, is classified as one of the other stone tool types, such as biface, projectile point, drill, or other chipped-stone tool.

Modified flakes constitute the second-largest stone tool category in the Duckfoot assemblage: 388 individual items were recovered. The assemblage is composed almost entirely of local materials, with Dakota and Morrison orthoquartzites predominating (Table 4.1). Most are unifacially flaked along one or more edges, and many of these tools were probably used as scrapers. Several items are bifacially retouched, and a few are lightly ground or polished along an edge or around a tip or projection. The former could have been used as knives or scrapers, the latter as gouges or drills. Given the enormous morphological and technological variability of the modified flake assemblage, detailed studies of use wear are needed to identify the specific tool types and functions represented. Selected modified flakes are shown in Figure 4.13.

Modified Cores. A modified core is a rock that appears to have served originally as a core (refer to the discussion of chipped-stone debris later in this chapter) but that later was modified, either incidentally as the result of use or

Figure 4.13. Modified flakes *(top and middle rows)* and "other chipped-stone tools" *(bottom row)*. The modified flakes have at least one deliberately shaped edge, but flake scars do not span the tool surfaces. Similarly, two of the three "other chipped-stone tools" are simply shaped along an edge but were placed in the more general tool category because they were made from naturally angular pieces of stone rather than from true flakes. The third specimen *(bottom right)* is a more extensively modified piece that lacks the morphological characteristics necessary for assignment to one of the more specific tool categories.

deliberately in order to shape the tool for some specific function, such as scraping. Modification usually consists of flaking, grinding, or polishing along one or more edges or ridges. Flake scars along a modified edge should be small enough and regular enough to suggest use damage or deliberate shaping rather than removal of large flakes for the purpose of furnishing stone for tool manufacture. A core that has been modified specifically by battering is classified as a peckingstone.

Six modified cores, all made of Morrison orthoquartzite and all but one complete, were identified in the Duckfoot assemblage (Tables 4.1 and 4.2). Four of the six had been unifacially flaked along one or more edges, which suggests use as scrapers, and two had been bifacially flaked, possibly indicating that they were used as choppers or combination scrapers-choppers.

Other Chipped-Stone Tools. Stone tools that were at least primarily shaped by flaking but that do not fall into any of the stone tool categories already listed are classified as "other chipped-stone tools." Examples include (1) "chunks" or spalls that have at least one edge that was modified by flaking, (2) flakes that have been modified more extensively than modified flakes but less extensively than some of the more formal tool types, and (3) tool fragments that clearly have been modified primarily by flaking but that are too small to be identified as a specific tool type.

Twenty-seven artifacts were classified as other chipped-stone tools. Many of these items are chunks or spalls that

were unifacially retouched to produce a working edge suitable for scraping tasks, and in this respect they are similar to the majority of modified flakes in the assemblage (Figure 4.13). Others are fragments or complete but unusual items that simply do not lend themselves to standard description or classification. One very unusual piece probably is not a tool at all but, rather, an eccentric or effigy; this tiny item, found in the midden area, is made of high-quality chalcedony and has multiple finely worked projections that possibly represent the limbs of an animal (Figure 4.14).

Ornaments

Only two types of ornament were recognized in the Duckfoot stone and mineral assemblage: beads and pendants. Both are objects designed to be strung on a cord, but they are distinguished by the placement of the holes used for threading. Beads typically have centrally located holes that allow them to be threaded side by side with other similar objects to form a beaded strand or necklace. Pendants usually have holes located close to one of their edges, which causes them to dangle once suspended.

Beads. One turquoise bead and one bead of unknown material (but presumed to be stone or mineral) were recovered from Duckfoot (Table 4.1; Figure 4.15). The turquoise bead, found in the fill of a Courtyard 2 feature (Feature 3), is a tiny (5.7 mm, maximum diameter), flat,

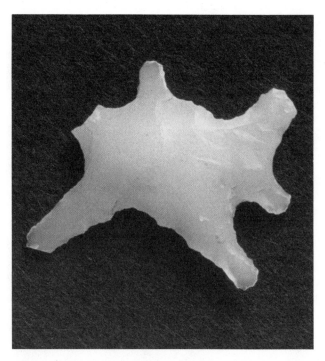

Figure 4.14.   Chipped-stone artifact, possible effigy. Maximum dimension = 18.2 mm.

asymmetrical piece that was polished to a high sheen. The other specimen, recovered from midden deposits, is an almost perfectly round, flat disk that has been ground but not polished. This bead is even smaller (4.8 mm, maximum diameter) than the turquoise specimen, which is the main reason that the type of material was not identified.

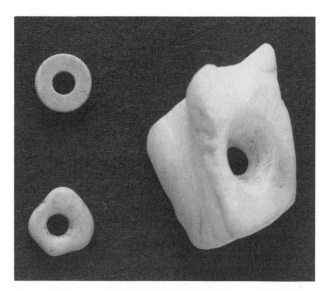

Figure 4.15.   Stone and mineral beads or similar objects: *top left*, bead of unknown material; *lower left*, turquoise bead; *right,* modified gypsum. Maximum dimension, turquoise bead = 5.7 mm.

Pendants.   Eight stone or mineral pendants were recovered from a variety of contexts at Duckfoot: two from pit structure floors, two from collapsed structural debris, one from other postabandonment deposits inside a room, and three from midden and courtyard deposits. Four are made of colorful, banded siltstone of unknown, but likely local, origin; two are calcite; one is turquoise; and one is made of a very soft, almost chalky, unknown stone, probably a decomposing mudstone or siltstone (Table 4.1; Figure 4.16). The pendants vary widely in size and shape. The turquoise pendant is the smallest in the assemblage (13.3 by 10.3 mm) and has a very irregular shape that appears to conform closely to the natural shape of the stone. The pendant made of unknown stone and the four pendants made of siltstone were much more heavily modified and range from rectangular to oval to leaf-shaped in outline; these are small- to medium-size pieces. The two calcite pendants are the most unusual in the collection. One consists of a very thin, slightly curved disk, 30.9 mm in diameter, with a lustrous, polished surface; the other, the largest in the collection, is a thick, flat, well-shaped rectangle that measures 62.2 by 38.6 mm.

Miscellaneous Objects

A number of other stone and mineral objects that do not clearly fall within one of the broad functional categories described above were recovered from the Duckfoot site.

Modified Cobbles.   A cobble that appears to have been modified, either deliberately or incidentally through use, but that does not conform to any of the more specific stone tool type definitions is classified as a modified cobble. Eighty-five such items, of many different sizes and shapes and undoubtedly representative of a wide variety of uses, were recovered from Duckfoot. Three-quarters of the assemblage is made of igneous river cobbles; orthoquartz-ites, sandstones, and other materials make up the remainder of the assemblage (Table 4.1). Almost all the cobbles appear to have been at least lightly ground or polished beyond what would be expected as the result of natural weathering, and many are battered and/or pecked as well. They range in size from relatively small cobbles that can be grasped easily in one hand, to medium-size rocks probably requiring two hands to manipulate, to large, heavy, flat cobbles that probably were used as stationary objects on surfaces. Most of the large, flat specimens are at least lightly ground or polished and have apparently random peck marks on at least one of their flat surfaces, similar to items described by Hayes and Lancaster (1975:159–160) as "lapstones." Eight such tools were recovered from Duckfoot: four from pit structure floors, two from front-room floors, and two from collapsed structural debris in rooms.

Figure 4.16. Stone and mineral pendants. Length of largest pendant (*bottom row, left*) = 62.2 mm.

Two cobbles had single notches, and two had double notches; one of the latter might be a maul fragment. Six small- to medium-size cobbles were pitted on at least one, and sometimes both, of their surfaces. In addition, the faces of these pitted cobbles most often were uniformly ground or polished, and all had been battered on at least their ends and sometimes around their entire circumferences. In three of the six cases, single deep pits were located on one surface only. In another case, a discrete pit had been pecked into one surface, and no fewer than four had been pecked into the other. On two cobbles, the pitting was more diffuse. Selected modified cobbles are shown in Figure 4.17.

Artifacts similar to some of those classified as modified cobbles in the Duckfoot assemblage have been described for many other Puebloan sites. Wheeler (1980b:248–249) classifies polished and battered cobbles recovered from Long House on Mesa Verde as "rubbing stones" and believes that they may have been used in the final smoothing or buffing of certain tools or pieces of household furniture (e.g., door and niche covers). Hayes and Lancaster (1975:159)

refer to several large, polished or pecked alluvial cobbles from Badger House Community as "lapstones" and suggest they may have been used in hide working. A number of battered and/or pitted cobbles without hafting elements and small enough to be held in the hand were recovered from Mug House and are described by Rohn (1971:210–211) as "pitted pounding stones"; some of these were also ground or polished, like several of the Duckfoot specimens.

Polishing Stones. In the Crow Canyon analytic system, a pebble or small cobble that appears to have been ground or polished as the result of use is classified as a polishing stone. Traditionally these artifacts are inferred to have been used in the manufacture of pottery; even today Pueblo potters use small pebbles to smooth the walls of unfired vessels. Although some polishing stones almost certainly functioned in this capacity, others may have been used for polishing other materials such as stone, bone, and wood. It is also possible that some polishing stones served in a ritual or combined utilitarian-ritual capacity. Their primary

Figure 4.17. Modified cobbles. The cobbles in the top row are pitted or pitted and notched. The cobbles in the bottom row are similar to "lapstones" (cf. Hayes and Lancaster 1975:159-160). Length of cobble in upper right corner = 13.1 cm.

practical function may have been to smooth pottery, but they apparently were also viewed as objects of particular social significance, at least in historic times. Guthe (1925:28), as quoted in Kidder (1932:63–64), states that, at San Ildefonso Pueblo, polishing stones usually

> are distinctly the personal property of the potter, and apparently have a semi-sacred significance. For the most part they are heirlooms, handed down from mother to daughter, but additional stones are picked up from ruins which the potters have visited. . . . Stones are sometimes lent by one potter to another, but they very seldom find their way outside the family group.

Twenty pebbles recovered from the Duckfoot site had sufficient use wear to warrant classification as polishing stones. Because pebbles and cobbles were not examined under magnification, it is likely that some polishing stones that were only slightly used escaped detection. Over half of the Duckfoot polishing stones are of unknown material type (Table 4.1). As with hammerstones, specific identification of highly weathered specimens is difficult in the absence of fresh breaks that expose the rock interiors. All are small enough to be grasped easily with the fingers. Shapes range from round to oval to irregular, and polished surfaces from flat to slightly concave to slightly convex. Hayes and Lancaster (1975:157) state that the stones with

polished convex surfaces were used to smooth bowl interiors, and those with concave surfaces were used for vessel exteriors. One unusual Duckfoot specimen is lightly flaked on some of its edges and has several highly polished angular facets. The contexts in which the Duckfoot polishing stones were found do not suggest any special function for these particular specimens. Thirteen of the 20 were recovered from midden deposits, 6 were found in collapsed structural debris inside structures, and 1 was found in general courtyard deposits.

Polishing/hammerstones. As the type name implies, a polishing/hammerstone is a pebble or small cobble with distinctive use wear patterns that suggest it was used to both rub and strike other objects. The main difference between a polishing/hammerstone and some of the modified cobbles that are both ground and battered is size: polishing/hammerstones are small enough to be grasped easily with the fingertips rather than require the whole hand to manipulate. Eleven such objects were recovered from Duckfoot. Most are pebbles identified only as some type of orthoquartzite (Table 4.1), and although most would have been suitable as pottery polishers, the presence of battering suggests a different or additional, but unknown, function. The amount and degree of battering ranges from very minimal damage on one or both ends to more pronounced battering around the entire tool circumference.

Stone Disks. A flat, more or less round stone that has been deliberately shaped, usually by flaking, grinding, or pecking, is classified as a stone disk. Fifteen stone disks, all complete and all made of sandstone, were recovered from Duckfoot (Tables 4.1 and 4.2). Modification and workmanship vary widely. Several pieces were simply snapped around their edges, others were bifacially flaked, and still others had been ground and/or flaked into shape. Grinding on the flat stone faces is rare, but it does occur. Most of the disks are round, but one well-shaped piece is closer to rectangular, and some of the more crudely worked "snapped" pieces are fairly angular. Similar objects have been documented for many prehistoric Puebloan sites throughout the Southwest and are sometimes inferred to have served as jar or pot lids, niche covers, or pit/cist covers (Kidder 1932:75–76; Rohn 1971:197–199; Hayes and Lancaster 1975:159; Judd 1981:127). Two of the Duckfoot stone disks are large enough and heavy enough to have been unsuitable for use as jar lids, unless the jars were buried and had extra support from the surrounding dirt (cf. Kidder 1932:75–76); both disks would have made better pit or cist covers. One disk of intermediate size could have served as either a jar or a pit cover, and the remaining items were too small to have effectively served either function (the smallest measures only 4.5 cm in diameter). It is possible that the very smallest disks could have been used as gaming pieces.

Other Modified Stone/Mineral. Any stone or mineral that has been modified by a human being but that does not conform to any of the more specific tool or ornament categories defined above is classified as "other modified stone/mineral." Frequently used to classify extremely small artifact fragments that cannot be identified to specific type, this term may also be used to describe complete or nearly complete objects that, on the basis of morphological attributes, simply do not fit handily into any other category. This category encompasses an enormous variety of material type and modification classifications, and use-wear studies are needed to identify specific functions represented by the 85 pieces in the Duckfoot collection. However, five items are distinctive enough to warrant individual description here. An oddly shaped piece of ironstone that had been partly drilled may be a pendant that was never completed, and a tiny piece of polished turquoise may be a pendant fragment. A third item consists of a small, heavily ground chunk of iron-rich siltstone that could have supplied deep red pigment. The only piece of worked gypsum in the Duckfoot assemblage is a very irregular chunk with a pronounced groove in one side and a relatively large hole through the middle; this item is too irregular to be confidently classified as a bead or a pendant (Figure 4.15). Technically, the fifth item probably should not be considered a mineral, but it was treated as such as a matter of expediency. An unfired clay ball, obviously deliberately rolled or molded into shape, was recovered cracked from the floor of Pit Structure 1 and later broke into numerous pieces. Because the item is so fragmented, its actual size is difficult to estimate, but given the curvature of some of the larger chunks, the ball probably had a diameter of 3.5 to 4.5 cm.

Chipped-Stone Debris

Chipped-stone debris consists of the byproducts of stone tool manufacture and maintenance, and this category constitutes the bulk of the Duckfoot stone artifact assemblage (23,412 pieces, or 92.2 percent). Although most of the chipped-stone debris consists of waste flakes and flake fragments, the collection also includes cores, use-damaged flakes, and spalls that may or may not be related to tool manufacture. Basic descriptive analysis was designed to provide general assemblage information and an organizational format that would facilitate future studies of resource procurement and tool-manufacturing technology.

Analytic Criteria

Chipped-stone debris was recovered from the field in bulk lots by specific provenience, and a single lot could consist of one or many (several hundred) items. After each lot was cleaned and assigned a catalog number, the individual pieces were examined, without magnification, and classified on the basis of raw material and morphological type.

Material

The gradual development of Crow Canyon's laboratory system led to some inconsistencies in the treatment of chipped-stone debris as compared to formal tools and ornaments. Unlike formal tools and ornaments, for which both geologic formation and specific stone type were recorded, chipped-stone debris (with the exception of igneous material) was identified only to formation or, if the material was believed to have been imported, simply as nonlocal. Thus, only five chipped-stone debris material options were used to describe the assemblage: Dakota, Morrison, igneous, nonlocal, and unknown. All locally available sandstones, conglomerates, orthoquartzites, and chert/siltstones were assigned to either the Dakota or the Morrison category, but the specific rock types were not distinguished. All igneous rock was assigned to a third, presumed local, category. Nonlocal materials include agate/chalcedony, obsidian, Washington Pass chert, petrified wood, other chert (including jasper), and any other rocks not available locally, but again, specific types were not distinguished. The "unknown" category was reserved for items of unknown origin; these materials could represent either local or nonlocal sources. The level of identification used in chipped-stone analysis limits specific, direct comparisons between this data set and that generated for formal tools and ornaments; however, general comparisons are possible and are attempted here.

Type

Five morphological categories were used to describe the chipped-stone debris in the Duckfoot assemblage: cores, flakes with platforms, flakes without platforms, edge-damaged flakes, and "other." A core is any rock from which at least one flake is inferred to have been removed either for direct use or for purposes of supplying stone for tool manufacture. Cores usually are not themselves flakes, although occasionally a very large flake may serve as a core. All cores are recognized by the presence of flake scars, curved indentations that indicate where flakes were removed. A flake is identified on the basis of several distinctive characteristics that result from deliberate removal from the core stone: (1) a platform, which is the point or facet where the force (pressure or percussion) was applied to remove the flake; (2) a bulb of applied force, located on the ventral flake surface just below the platform; (3) concentric ripples radiating out from the platform on the ventral face, a result of the "stress waves" rippling through the stone as the fracture is propagated; and (4) cortex, flake scars, or a combination of the two on the dorsal surface (see Crabtree [1972] for definitions of commonly used flintknapping terms). Because the platform contains valuable information with regard to flaking technology (for instance, platform preparation and type of force applied), flakes that retain this feature potentially are of greater interest than those that do not; therefore, the two were differentiated during basic analysis. Additional analysis of striking platform characteristics has not been undertaken, however.

Any flake, whether it has a platform or not, that shows signs of incidental damage through use, but *not* deliberate modification prior to use, is classified as an edge-damaged flake. In other systems, such tools are frequently referred to as "used" or "utilized" flakes. Edge-damaged flakes usually have very small, regular flake scars along an edge, but occasionally modification takes the form of grinding, polishing, or battering. Flakes that appear to have been damaged as a result of prehistoric or recent "accidents" (for example, modification as a result of being stepped on, dropped, or scraped with a trowel) are not included in the edge-damaged category. It is worth emphasizing that edge-damaged flakes are tools, albeit an informal, expedient variety. The only difference between an edge-damaged flake and a modified flake (see definition of modified flake in discussion of stone tools) is the degree of effort expended on manufacture: modified flakes were deliberately shaped prior to use, whereas edge-damaged flakes were simply used.

The "other" category was created to accommodate angular, blocky pieces of stone made of the same knappable materials seen throughout the chipped-stone assemblage but lacking the diagnostic characteristics of cores or flakes. Sometimes called "shatter" or "angular debris," such items could be byproducts of tool manufacture or, alternatively, may be naturally angular pieces of rock that were incidentally incorporated into site deposits. Because their origins and associations are ambiguous, they are tallied separately from the known cultural materials.

The Assemblage

Table 4.3 summarizes material and typological data, by grouped provenience, for the Duckfoot chipped-stone debris assemblage; the same data are provided by individual study unit in Appendix A. Materials from the Morrison Formation, which include both orthoquartzites and chert/siltstones, dominate the assemblage; other local materials are present but in much smaller quantities. Nonlocal stone constitutes 4.4 percent of the assemblage. Table 4.4 compares the raw-material profile for chipped-stone debris with that for selected tool types; the tools chosen for this comparison are those that were manufactured primarily by flaking or that could reasonably be expected to be damaged by flaking during normal use (hammerstones, peckingstones, mauls, axes [all types], axe/mauls, bifaces, projectile points, drills, modified flakes, modified cores, and other chipped-stone tools). Although the two profiles are generally similar, the difference in the relative proportions of Dakota and Morrison formation materials is worth noting: the chipped-stone debris assemblage is made up of 75.9 percent Morrison and 14.1 percent Dakota materials; corresponding values for the tool assemblage are 60.9 and 25.3 percent, respectively. The difference may be due at least in part to analytic bias: most of the debris was analyzed very early in the project, and the tools much later, after the lab staff was better able to recognize the full range of variation and distinguish between the two sources. However, it also is possible that, compared to tools made of Morrison materials, tools made of Dakota materials were more likely to be completely or partly shaped away from the site and brought to Duckfoot in a finished or nearly finished state.

Unmodified flakes make up 92.8 percent of the chipped-stone debris assemblage, and there are nearly twice as many items with platforms as without (Table 4.3). Only 340 cores were recovered, but if peckingstones made from cores were included in the count, this number would at least double. Most of the 843 edge-damaged flakes in the assemblage were flaked, although a few had ground, polished, or battered edges. Many probably served as scrapers or knives, and microwear studies are needed to identify the full range of functions for these possibly multipurpose items. It is important to note that, because

Table 4.3. Chipped-Stone Debris Data Summary

	Room				Pit Structure				Midden		Modern Ground Surface		Other		Site Total	
	Surfaces		Roofs		Surfaces		Roofs									
	N	%	N	%	N	%	N	%	N	%	N	%	N	%	N	%
MATERIAL																
Local																
Dakota	128	23.7	60	12.3	257	37.7	17	10.4	2,325	13.7	122	13.0	399	11.1	3,308	14.1
Morrison	375	69.6	376	77.0	358	52.6	119	73.0	13,038	76.6	728	77.8	2,787	77.7	17,781	75.9
Igneous	9	1.7	11	2.2	18	2.6	15	9.2	518	3.0	21	2.2	78	2.2	670	2.9
Nonlocal	17	3.1	33	6.8	23	3.4	7	4.3	718	4.2	21	2.2	203	5.6	1,022	4.4
Unknown	10	1.8	8	1.6	25	3.7	5	3.1	418	2.4	44	4.7	121	3.4	631	2.7
TYPE																
Core	7	1.3	15	3.1	10	1.5	9	5.5	238	1.4	7	.7	54	1.5	340	1.4
Flake with platform	233	43.2	307	62.9	178	26.1	74	45.4	10,281	60.4	573	61.2	2,247	62.6	13,893	59.3
Flake without platform	279	51.8	130	26.6	412	60.5	65	39.9	5,569	32.7	307	32.8	1,081	30.1	7,843	33.5
Edge-damaged flake	14	2.6	23	4.7	16	2.3	11	6.7	608	3.6	32	3.4	139	3.9	843	3.6
Other	6	1.1	13	2.7	65	9.5	4	2.4	321	1.9	17	1.8	67	1.9	493	2.1
TOTAL	539	2.3	488	2.1	681	2.9	163	.7	17,017	72.7	936	4.0	3,588	15.3	23,412	100.0

NOTE: Structure surfaces include surfaces and surface-associated features containing cultural or constructional fill. Roofs include roof fall and mixed roof and wall fall deposits. Midden includes modern ground surface, fill, and all midden features. Modern ground surface includes all modern surface proveniences except for the midden. "Other" includes all other site deposits (e.g., structure and feature fills not included elsewhere, courtyard deposits, arbitrary units).

Table 4.4. Material Type Comparison, Chipped-Stone
Debris and Selected Stone Tools

Material	Debris (N = 23,412) %	Tools[a] (N = 1,054) %
Local		
Dakota	14.1	25.3
Morrison	75.9	60.9
Igneous	2.9	3.2
Nonlocal	4.4	5.9

[a] Tool types used in comparison are hammerstone, peckingstone, maul, axe (all types), axe/maul, biface, projectile point, drill, modified flake, modified core, and other chipped-stone tool.

individual items were not examined under magnification, it is likely that many edge-damaged flakes were not recognized, and therefore the number reported here probably greatly underestimates the actual number in the assemblage. Also included in the Duckfoot assemblage, though not specifically identified in Table 4.3, are numerous flakes that clearly were removed from peckingstones and groundstone tools such as axes; these flakes probably are the result of tool breakage and attempts to refurbish or repair dulled or slightly broken tools.

Pit structure floors yielded somewhat more chippedstone debris than did room floors, but roof deposits in rooms yielded three times the number of items as similar deposits in pit structures. This could indicate that, although stone tools were made or refurbished inside pit structures as often as inside rooms, room rooftops were preferred over pit structure rooftops when the same activity was conducted out of doors. Alternatively, room rooftops, which were probably not routinely subjected to heavy foot traffic, may have been a preferred discard location for chipped-stone debris.

Unmodified Stone and Minerals

The majority of unmodified stone and mineral objects were not collected during excavation of the Duckfoot site. However, when the natural physical attributes of a piece or the context in which an item was found led the excavator to suspect the piece was of potential cultural significance, or when the item most likely would not occur naturally in the immediate site vicinity, the specimen was retained as part of the artifact assemblage. Four categories of unmodified materials were recognized: gizzard stone, pebble, unmodified cobble, and unmodified stone/mineral.

Gizzard stones, sometimes called gastroliths, are small stones carried in the gizzards of birds to aid in digestion. When first consumed by the bird, gizzard stones may be quite angular, but eventually their sharp edges are worn smooth. Although the definition of a gizzard stone may be relatively unambiguous, recognizing one in an archaeological assemblage cluttered with miscellaneous pebbles and gravels is not a straightforward task. In the Duckfoot assemblage, very small, flat stones with angular facets but rounded edges are classified as gizzard stones. It is acknowledged that some gizzard stones probably were not recognized during excavation and analysis and, conversely, that many items classified as gizzard stones probably are nothing more than very small, weathered pebbles. The archaeological importance of gizzard stones is not clear, as it usually is not possible to relate their presence to a specific type of bird—for instance, turkey—that may have been of particular cultural significance. Fifty-three occurrences of gizzard stones are recorded for the Duckfoot site, primarily from extramural midden and courtyard deposits (an occurrence is defined as the recovery of one or multiple stones from the same vertical and horizontal provenience within a given study unit).

Unmodified pebbles and cobbles were collected because they do not occur naturally at the site and therefore are assumed to have been deliberately transported. Sources of small, water-worn rocks are located not far from Duckfoot in the many natural drainages in the area, the closest of which is Alkali Canyon. Larger cobbles are available in quantity in the Dolores River valley, roughly 24 km distant. It is possible that at least some of these items were used as tools but do not show signs of modification when viewed without magnification. For example, a pebble used only briefly as a polishing stone may not show the polishing or striations indicative of such use and would be classified simply as a pebble. Thirty-three unmodified cobbles were recovered from Duckfoot. Like gizzard stones, pebbles recovered from the site are recorded as occurrences, of which 89 are recorded for Duckfoot.

Unmodified stones and minerals other than pebbles and cobbles were collected only if they had unusual physical characteristics or were found in contexts that indicated possible cultural significance. The Duckfoot collection is made up of 158 items, which may be broadly subdivided as follows: 47 specimens that would have been suitable for use as pigment (hematite, ironstone, iron oxides, carnotite, limonite, azurite, and various ochres), 24 concretions, 23 fossil shells, 18 pieces of calcite, 7 clay samples, 7 pieces of caliche, 4 pieces of sandstone, 4 pieces of igneous rock, 2 pieces of turquoise, 1 conglomerate rock, 1 piece of gypsum, and 20 miscellaneous rocks (iron-rich mudstones, siltstones, sandstones) or materials of unknown type or origin. Among the unidentified materials are several pieces of a very friable, light gray, powdery material, originally believed to be clay or gypsum. Subsequent laboratory analysis eliminated both of these possibilities but failed to establish an alternative identification.

5

Perishable Artifacts

Mary C. Etzkorn, G. Timothy Gross, and Karen R. Adams

Artifacts made of animal bone, shell, and plant material were recovered from a variety of contexts at Duckfoot, and although they make up a relatively small proportion of the total material assemblage, they include some of the more unusual individual pieces in the collection. In this chapter, the discussions of vegetal and animal bone artifacts focus on descriptions of formal tools and ornaments; unmodified plant and bone materials are discussed in Chapters 6 and 8, respectively. The shell assemblage, however, is so small that separate discussion of modified and unmodified items is not warranted, and this material class is described in its entirety here, with an emphasis on those items that were fashioned into formal artifacts.

Bone Tools and Ornaments

Mary C. Etzkorn

Of the 5,710 animal bones recovered from Duckfoot, 92 (or 1.6 percent) were modified in ways that suggest their use as tools or ornaments. By the definition of modification used here, bone that appears to have been altered in the course of food preparation is not treated as modified. Thus, breakage, burning, and cut marks indicative of butchering, skinning, or cooking are not considered in this analysis. Two bone artifacts, a gaming piece from the floor of Room 12 and a tube from the floor of Room 15 (neither assigned PL numbers), were misplaced before analysis and are not included in the following discussion.

Analytic Criteria

All bone tools and ornaments were identified to the most specific taxonomic level possible, using criteria described

in Chapter 8. The results of this phase of analysis are included in the overall results reported in that chapter but are repeated here to facilitate comparisons by artifact type. In the second phase of analysis, two observations—condition and type of modification—were recorded to describe the physical attributes of each item and provide information relevant to artifact manufacture and use. On the basis of these observations, an artifact type assignment was made for each piece. Complete analytic data for the entire modified bone assemblage are on file at the Anasazi Heritage Center.

Condition was recorded as *complete, incomplete* (broken, but enough present to infer final-use size and shape), *fragmentary* (not enough present to infer final-use size and shape), and *unknown.* As mentioned above, *modification* refers specifically to alteration believed to be the result of tool manufacture and/or use, and several types are recognized in the modified bone assemblage: cutting, striating, grooving, perforating, and grinding/polishing. Because bone was examined without magnification, it is likely that some tools were not recognized, and for those tools that were recognized, some forms of modification may have gone undetected.

The artifact types recognized in the Duckfoot assemblage are similar to those reported for other Anasazi sites dating to all time periods (see Kidder 1932; Brew 1946; Rohn 1971; Hayes and Lancaster 1975), but specific definitions may differ. Awl, bead, gaming piece, needle, pendant, and tube are the identifiable artifact types in the Duckfoot assemblage. When it was not possible to assign an item to one of these types, usually because it was too fragmentary, the piece was described as "other modified bone." Implicit in the type categories used are assumptions and inferences with regard to artifact use. These are discussed below under "Artifact Types."

The Modified Bone Assemblage

Taxonomic Composition

The taxonomic composition of the Duckfoot modified bone assemblage in most ways does not parallel that of the total bone assemblage from the site (see Chapter 8). For instance, mule deer (*Odocoileus hemionus*) accounts for only 1.0 percent of the total bone assemblage from Duckfoot but constitutes 39.1 percent (36 of 92 analyzed items) of all modified bone (Table 5.1). Medium mammal, at 18.5 percent, and indeterminate artiodactyl (most of which probably is *Odocoileus*), at 11.9 percent, are the next most common taxa represented in the modified assemblage. By contrast, these same taxa account for only 4.9 percent and 4.7 percent, respectively, of the total Duckfoot bone assemblage. The number of modified items made of bone from larger animals reflects the suitability of heavier, denser bone for certain types of tools (e.g., awls) that must withstand greater stress during use. Bones of smaller animals, including small mammals and birds, make up relatively small percentages of the total Duckfoot modified bone collection, and they are often used for items inferred to have served nonutilitarian purposes (for example, beads and possibly tubes). It is interesting, however, that relative proportions of mammal and bird bone used for certain types of tools appear to change through time. For instance, in Pueblo III modified bone assemblages, many awls are made of bird (specifically, turkey) bone (Rohn 1971:224; Wheeler 1980a:309; Bradley 1992) rather than of deer bone. Whether this signals a functional change (that is, awls were used for different purposes during Pueblo I and

Pueblo III times), a change in animal availability, or simply a shift in material preference is not known.

The taxonomic composition of the Duckfoot modified bone assemblage is similar to that reported for other Pueblo I sites in the Mesa Verde area. At Site 1676 in the Badger House community at Mesa Verde National Park, 12 of 37 bone tools (32.4 percent) recovered from structures that date to roughly the same time period as Duckfoot were identified as mule deer, and an additional 7 (18.9 percent) were classified as indeterminate artiodactyl (Hayes and Lancaster 1975:34–63). The use of slightly different analytic and taxonomic systems by the Dolores Archaeological Program (DAP) hinders direct comparison of Duckfoot and Dolores assemblages, but in general it appears that, in the identifiable tool assemblage, artiodactyl bones are much more common than those of other animals (see, for example, Neusius and Gould 1988:Tables 15.15 and 15.20; Wilshusen 1986a:Table 2.38, 1986b:Tables 3A.19–3A.22).

Artifact Types

The Duckfoot modified bone assemblage is composed overwhelmingly of pointed, sharpened perforating tools. The absence of bevel-edged tools (e.g., scrapers) suitable for smoothing or scraping hides or plant fibers is notable, particularly in view of the fact that such items have been reported, though not in large numbers, for other Pueblo I sites in the area (Brew 1946:Figure 186; Hayes and Lancaster 1975:169–170; Neusius and Gould 1988:Table 15.19; Wilshusen 1986a:Table 2.38, 1986b:Tables 3A.19–3A.22). Modified bone from Duckfoot is described below, by

Table 5.1. Modified Bone by Artifact Type and Taxonomic Category

Taxonomic Category	Awl N	Awl %	Bead N	Bead %	Gaming Piece N	Gaming Piece %	Needle N	Needle %	Pendant N	Pendant %	Tube N	Tube %	Other[a] N	Other[a] %	Site Total N	Site Total %
Class Aves																
Avian, indeterminate	1	2.0									2	66.7	1	3.8	4	4.3
Class Mammalia																
Order Lagomorpha																
Lepus cf. *californicus*	1	2.0											4	15.4	5	5.4
Leporid, indeterminate			1	100.0									2	7.7	3	3.3
Order Carnivora																
Canis sp.											1	33.3			1	1.1
Vulpes cf. *macrotis*	1	2.0													1	1.1
Medium carnivore									1	50.0					1	1.1
Order Artiodactyla																
Odocoileus hemionus	32	62.7											4	15.4	36	39.1
Artiodactyl, ind	8	15.7											3	11.5	11	11.9
Small mammal, ind	1	2.0											2	7.7	3	3.3
Medium mammal, ind	6	11.8			3	37.5			1	50.0			7	26.9	17	18.5
Mammal, indeterminate	1	2.0			5	62.5	1	100.0					3	11.5	10	10.9
TOTAL	51	100.0	1	100.0	8	100.0	1	100.0	2	100.0	3	100.0	26	100.0	92	100.0

ind = indeterminate.

[a] Other modified bone.

artifact type, and suggestions are offered with regard to the possible use or uses to which the various artifacts may have been put. These suggestions are based largely on interpretations presented in the ethnographic and archaeological literature, and use-wear studies are needed to corroborate or refute current interpretations of function.

Awls. An awl is a bone that has been deliberately shaped into a point on at least one end; occasionally awls are pointed on both ends, but none of this type was recovered from Duckfoot. Awls may or may not have deliberately drilled holes in their butt ends, but when they do, they are distinguished from needles on the basis of larger holes, larger overall size, and blunter tips (needles usually are made of thinner slivers of bone and have more-tapered tips and smaller holes).

Awls make up over half (51 items, or 55.4 percent) of the modified bone assemblage from Duckfoot, a proportion that is similar to that noted for the combined late Pueblo I components at Site 1676 at Mesa Verde (56.8 percent) (Hayes and Lancaster 1975:34–63). Lower percentages are recorded for a number of Dolores-area sites, for example, 21.9 percent at Grass Mesa Village (Neusius and Gould 1988:Table 15.19) and 31.4 percent at Periman Hamlet (Wilshusen 1986a:Table 2.38). The lower percentages for some Dolores sites may reflect differences in the types of activities represented but may also be attributed, at least in part, to the large numbers of tool fragments that were not identified to specific artifact type.

The majority of awls in the Duckfoot assemblage are made of mule deer (62.7 percent) or indeterminate artiodactyl (15.7 percent) bone, although bird (avian), black-tailed jackrabbit (*Lepus* cf. *californicus*), kit fox (*Vulpes* cf. *macrotis*), and indeterminate mammal are represented as well (Table 5.1). As seen in Table 5.2, over half of the Duckfoot awls are fragmentary. One very short awl appears to be the reworked tip of a larger awl that broke; the broken end was lightly worn, either deliberately to make the grip more comfortable or incidentally through the continued use of the fragment as if it were the whole tool. Although a detailed typological analysis was not undertaken, it is clear from even a cursory glance at the collection that several "styles" of awl are represented, the most common being a short, thick variety, compact enough to fit easily in the palm

of the hand (Figure 5.1). Longer and more slender awls fall into intermediate and large size categories, although the latter is relatively uncommon in the Duckfoot assemblage. The butt ends of the awls presumably served as handles, and a variety of styles was noted, ranging from unmodified articular ends, to lightly worn or polished ends, to heavily ground handles on which most or all of the joint had been removed. Type and degree of modification vary widely. All items show signs of grinding and/or polishing, and many are heavily striated. The extent to which these forms of modification are the result of tool manufacture or tool use is not known, but it is likely that both types of modification are represented. Originally, many bones were probably split or cut to roughly shape them into points, but subsequently they were so heavily ground that no evidence of the initial modification remains. A hole had been drilled through the butt end of one short awl, and a deep groove had been carved or worn around the butt end of a second. A small indentation or nick was observed near the tip of a broken awl.

Traditionally, awls are assumed to have been used as perforators—for example, to punch holes in hides to make clothes—and this seems a plausible function for at least the compact variety noted in the Duckfoot assemblage. Nearly all of these tools are made of deer, indeterminate artiodactyl, or other robust mammal bone, and with their short, thick shafts, they appear able to withstand being pushed and twisted through a resilient material such as animal skin. Other possible functions for awls have been suggested in the literature. Hayes and Lancaster (1975:168–169) speculate that exceptionally slender and fragile awls may have served as clothing fasteners or hair pins. Kidder (1932:225–227) and Rohn (1971:226–227) believe that awls with transverse grooves and highly polished shafts may have been used in weaving to beat down the weft; the repeated rubbing of the relatively soft bone against the fibers eventually would wear a groove and produce a lustrous polish on the bone surface. Although this seems a reasonable inference, recent experiments in which replicas of prehistoric bone awls have been used as weaving aids have failed to produce the anticipated wear pattern (Bullock 1991:4). However, awls *have* been grooved experimentally as the result of being used to split yucca fibers to prepare them for basket making (Bruce A. Bradley, personal com-

Table 5.2. Modified Bone by Artifact Type and Condition

Condition	Awl		Bead		Gaming Piece		Needle		Pendant		Tube		Other[a]		Site Total	
	N	%	N	%	N	%	N	%	N	%	N	%	N	%	N	%
Complete	14	27.4			7	87.5	1	100.0	2	100.0	2	66.7	1	3.8	27	29.3
Incomplete	11	21.6	1	100.0	1	12.5							2	7.7	15	16.3
Fragmentary	26	51.0									1	33.3	22	84.6	49	53.3
Unknown													1	3.8	1	1.1
TOTAL	51	100.0	1	100.0	8	100.0	1	100.0	2	100.0	3	100.0	26	100.0	92	100.0

[a]Other modified bone.

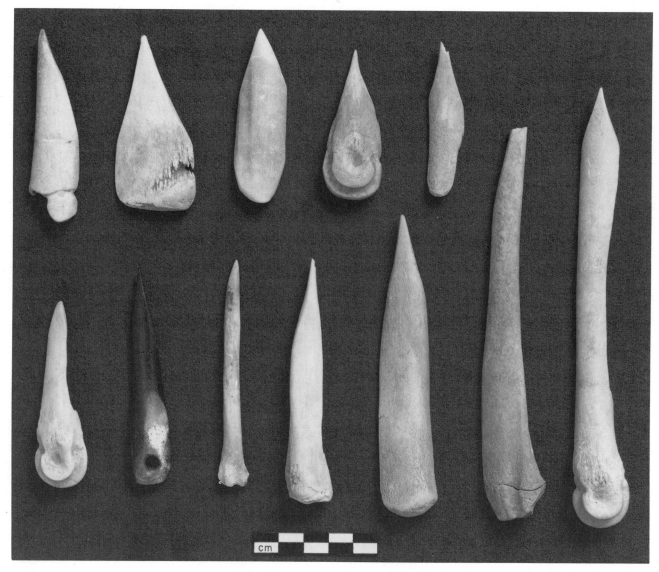

Figure 5.1. Bone awls.

munication 1991). The one grooved Duckfoot specimen is puzzling because the groove wraps around the butt end, a modification form and location not accounted for by any of these scenarios.

Bead. A bead is a small object with a hole that would have allowed it to be threaded on a cord or string, possibly with other similar objects to form a beaded strand or necklace. The hole was often made deliberately by drilling, but occasionally it was a natural feature that was deliberately enlarged or made more regular. Beads can be made of a variety of materials, including bone, stone, and shell. Only the one bead made of bone is discussed here; the remaining Duckfoot specimens are described in Chapter 4 and in the discussion of shell artifacts later in this chapter. A bead differs from a pendant in the placement of the hole: beads usually have central holes, whereas pendants have

holes near their top edge, so that the piece dangles when suspended. Refer to the discussion of "Tubes," below, for an explanation of the difference between a bone tube and a bead cut from a long-bone shaft.

The one bone bead from Duckfoot was cut from the long-bone shaft of an indeterminate leporid (rabbit) (Tables 5.1 and 5.2; Figure 5.2d). Although the bead is broken, enough of it is present to measure its length (25.3 mm) and diameter (7.0 mm). The shaft surface is noticeably better polished than the shaft surfaces of the bone tubes described below.

Gaming Pieces. Gaming pieces are small, deliberately shaped pieces of bone with incised patterns on one or both faces. Gaming pieces usually are round or oval and often have one flat and one convex surface. Not counting the one missing specimen from Room 12, eight gaming pieces,

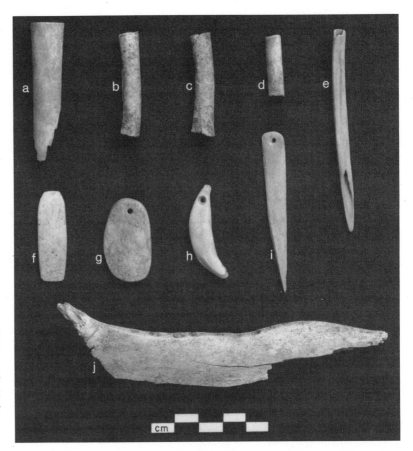

Figure 5.2. Miscellaneous bone artifacts: *a–c*, tubes; *d*, bead; *e–f*, other modified bone; *g–h*, pendants; *i*, needle; *j*, heavily used deer mandible (other modified bone). Item *f* may be an unfinished pendant or gaming piece. Item *j* is heavily ground and striated and may have been used as a scraper or "saw."

seven complete and one incomplete, were recovered from Duckfoot (Tables 5.1 and 5.2; Figure 5.3). All are made of indeterminate mammal bone, although three could be identified as medium mammal. Two of the eight pieces are round (diameter of the complete specimen is 13.2 mm); the remaining items are ovals that range from 20.6 to 30.0 mm in length. All had been heavily ground into shape, and all are at least lightly polished. Incised designs are simple and range from crude slashes in apparently random patterns to much more carefully executed hatch and crosshatch designs. In all cases, designs are confined to the flat surface only. The two round gaming pieces in the collection have center holes, and it might be argued that they are actually disk beads. However, the designs etched into their faces would not have been visible had they been strung with other beads on, say, a necklace strand.

As the name implies, gaming pieces, sometimes called "dice," are often inferred to have been used in games, particularly games of chance. Culin (1975:46–49) describes the ethnographic use of incised wooden dice ("stick dice") in Zuni games and believes that shaped and incised pieces of bone found in prehistoric sites may have served a similar function. Bone gaming pieces have been recovered from many prehistoric sites from Pueblo I through Pueblo III (Judd 1981:279), and little variation is seen among specimens from different time periods. The Duck-

foot gaming pieces are very similar to specimens from Pueblo Bonito in Chaco Canyon (Judd 1981:279–280); Site 13 at Alkali Ridge (Brew 1946:Figure 188); Badger House community at Mesa Verde National Park (Hayes and Lancaster 1975:170–171); and several Dolores-area sites, most notably Weasel Pueblo (5MT5106), where 40 bone gaming pieces were found in association with a multiple burial on a pit structure floor (Morris 1988a:724–725). Judd (1981:279) notes that although games involving dice historically have been used for recreation by many Native American groups, such games are often incorporated into important religious ceremonies as well. The context in which the gaming pieces at Weasel Pueblo were found certainly suggests that these artifacts at times may have served more than recreational purposes.

Needle. Unlike modern needles, which simultaneously pierce and sew, prehistoric bone needles are believed to have been used primarily to draw sinew or cordage through a preexisting hole created by another tool, such as an awl. The one bone needle recovered from Duckfoot was made from a very thin long-bone shaft fragment (indeterminate mammal) that had been ground into a fine point on its distal end and drilled at its proximal end (Tables 5.1 and 5.2; Figure 5.2*i*). The surface is heavily striated (probably the result of manufacture) and well polished (possibly the result

Figure 5.3. Gaming pieces. Diameter of complete round specimen (*top row, left*) = 13.2 mm.

of use), and the tip is beveled in opposite directions, which suggests that this particular needle may have been used for perforating as well as for sewing through a preexisting hole. The needle shaft measures 65.9 mm in length, and the hole measures 1.0 mm in diameter. The presence of needles, or "bodkins," as the larger specimens sometimes are called, has been documented at many prehistoric Puebloan sites (Kidder 1932:222–225; Rohn 1971:225–226; Hayes and Lancaster 1975:168; Neusius and Gould 1988:Table 15.20).

Pendants. Two bone pendants were recovered from Duckfoot (Tables 5.1 and 5.2; Figure 5.2*g* and *h*). One, made of medium mammal bone, is a well-shaped oval measuring 34.2 mm in length and 20.0 mm in width. Its edges are heavily ground, and fine striations cover both surfaces. The second pendant consists of the canine tooth of a medium carnivore. The tooth, including the root, is 38.6 mm long. Both pendants are complete.

Tubes. Tubes are deliberately modified cylindrical objects with hollow centers. They are made of long bones, usually bird but also occasionally mammal, whose articular

ends have been cut off and natural marrow cavities cleaned and enlarged. Most appear to have been lightly ground and/or polished on their cut ends, the shaft, or both. Occasionally bone tubes will have one or more holes drilled through one surface, but no tubes of this type were recovered from Duckfoot. The only other artifact with which tubes may be confused is a type of bone bead, also cut from the shaft of a long bone. Beads are distinguished by their relatively smaller size (length and diameter), as compared to tubes. At this point, the tube and bead sample sizes are too small to quantify the distinction, and artifact type assignments are made somewhat intuitively.

Not counting the missing specimen from Room 15, three bone tubes were identified in the Duckfoot assemblage (Tables 5.1 and 5.2; Figure 5.2*a–c*). Two of these, made of indeterminate bird bone, are almost identical, although they were recovered from two separate Courtyard 1 features. One is 42.4 mm long and 10.1 mm in diameter; the other is 42.7 mm long and 9.2 mm in diameter. Both are complete, and both were cut and very lightly polished. The third tube is fragmentary; its diameter measures 14.0 mm, and its existing length is 56.2 mm, although its original length cannot be determined. This piece was identified as *Canis* sp. (canid), and with its thicker walls and greater weight, it differs markedly from the other two tubes in the collection.

Bone tubes as defined in the Crow Canyon analytic system subsume a wide variety of artifacts described in the literature on prehistoric bone implements. Objects similar to the two Duckfoot specimens made of bird bone are variously referred to as tubes, beads, or tubular beads; and tubes with single or multiple drill holes have been called flageolets, pipes, whistles, perforated tibias, tibia tinklers, and rattles (Kidder 1932:248–268; Brew 1946:Figure 187; Rohn 1971:127–128, 249–250; Hayes and Lancaster 1975:170–171; Neusius and Gould 1988:Table 15.19). Today, the Zia and other Pueblo Indians use simple, unperforated tubes made of turkey long-bone shafts as "turkey calls" or noisemakers (Bruce A. Bradley, personal communication 1991). The exact functions of the Duckfoot specimens remain unknown. Perhaps they are bird calls, large beads, or "bead stock," that is, long tubes from which pieces were cut to make beads.

Other Modified Bone. The "other modified bone" category was created to accommodate modified items that could not be assigned to one of the more specific and descriptive categories listed above, usually because they were too fragmentary. In some cases, a specific assignment could not be made because the item simply did not conform to the established artifact type definitions. Twenty-six items, constituting 28.3 percent of the Duckfoot modified bone assemblage, were assigned to this category. It is likely that the assemblage includes tool and ornament fragments as well as manufacturing debris. Twenty-two of the 26 items

(84.6 percent) are fragments, 2 (7.7 percent) are incomplete, 1 (3.8 percent) is complete, and 1 (3.8 percent) is of unknown condition (Table 5.2). The one complete item is a small, well-shaped piece that appears to be either an unfinished pendant or an unfinished gaming piece (Figure 5.2f). One of the two incomplete items is made of a black-tailed jackrabbit tibia, the end of which had been sheared off. The other incomplete piece is made of a jackrabbit radius shaft; one end of the long, slender bone had been ground into a rounded tip, unlike the tip of an awl or needle (Figure 5.2e). The item of unknown condition is made of a bird long-bone shaft, and cut marks on one end suggest that it may have served as a source of bone for beads.

Perhaps the most unusual tool in the "other modified bone" category is a deer mandible that shows signs of heavy postmortem grinding and abrasion (Figure 5.2j). The tooth row apparently served as the working edge of the tool, and longitudinal striations on both the inside and the outside surfaces suggest that this item may have been used as a scraper or "saw."

Artifact Distribution

In this discussion, items recovered from floor, roof fall, and mixed roof and wall fall contexts are given primary consideration, because their specific cultural associations are believed to be less ambiguous than those of artifacts found in other contexts. Table 5.3 compares the number of bone tools recovered from surface rooms and pit structures, and it is clear that bone implements are far more common in the latter. Of the 46 bone tools and ornaments recovered from structure floors and roof fall deposits, 36 (78.3 percent) were found in pit structures, 6 (13.0 percent) were recovered from domestic rooms, and 4 (8.7 percent) came from storage rooms. Awls, the single largest bone tool category, were found overwhelmingly in pit structures, with 29 of 33 being recovered from floor and roof deposits in these structures. On the basis of the distribution of bone implements, it appears that sewing, and possibly weaving

and basket making, took place most often in pit structures, although some probably occurred at least occasionally in domestic rooms. Very few, if any, activities requiring the use of bone implements were conducted in storage rooms, and it is possible that the few items recovered from these structures were stored or abandoned, rather than used, on the floors or roofs.

Midden data are not reported in tabular form, but the quantity and diversity of bone implements from this area of the site are similar to those reported for pit structures. Including artifacts found in features, 21 modified bone items were recovered: 5 awls, 5 gaming pieces, and 11 specimens classified as "other modified bone." The relatively high proportion of items that could not be identified to a specific artifact type is not surprising for a midden; all but two of these pieces are fragmentary, and it is reasonable to assume that many broken implements and pieces of bone-tool manufacturing debris would be discarded in the trash area. The number of gaming pieces recovered from the midden is not as easily explained. Five of the eight (or nine, counting the one missing item) gaming pieces recovered from Duckfoot are from this portion of the site, and all but one are complete. Why usable artifacts, and particularly those that may have served in a ritual capacity, would be discarded in this fashion is not known.

Shell

G. Timothy Gross

Methods

Excavations at the Duckfoot site yielded 36 shell items, including 18 worked pieces. Each item was cleaned, cataloged, and then analyzed using procedures patterned after those developed at the DAP (Neusius and Canaday 1985; Phagan and Hruby 1984). Variables recorded include taxon, modification, number of items, weight, completeness, shell structures present, condition, and evidence of

Table 5.3. Modified Bone Distribution, Combined Floor and Roof Contexts

Artifact Type	Domestic Rooms		Storage Rooms		Pit Structures		Total	
	N	%	N	%	N	%	N	%
Awl	3	9.1	1	3.0	29	87.9	33	71.7
Bead	1	100.0					1	2.2
Gaming piece					1	100.0	1	2.2
Needle								
Pendant			1	100.0			1	2.2
Tube								
Other modified bone	2	20.0	2	20.0	6	60.0	10	21.7
TOTAL	6	13.0	4	8.7	36	78.3	46	100.0

NOTE: Floor contexts do not include features. Roof contexts include roof fall and mixed roof and wall fall deposits. Room 20 is excluded from consideration because it is not located in the roomblock, cannot be dated with confidence, and is ambiguous in terms of function.

burning. Taxon, number of items, weight, and whether or not the shell had burned need no further explanation. The variable *completeness* indicates how much of the original shell is present, expressed as a percentage. *Shell structures present* allows for the recording of which parts of the shell (such as the hinge, aperture, margins, or body) are present, and *condition* refers to whether the shell is hard and fresh-looking or whether and how it has weathered.

All items were examined for evidence of modification, and a 10× hand lens or a variable-power binocular microscope was used when necessary. The shell tool typology developed by Haury (1937:137) for Snaketown was used to describe worked, or modified, shell items in the Duckfoot assemblage. For all modified shell, observations of a series of technological variables were recorded, including the type of worked item, evidence of production (such as grinding or chipping), presence or absence of specialized treatments (such as drilling or grooving), and the completeness of the item as a tool or an ornament. Length, width, and thickness of modified items were measured with sliding calipers.

Results

In general, the Duckfoot shell artifacts were in fairly good condition (over half of them were classified as fresh looking or only slightly chalky), although over half were fragments. None were burned. Selected analysis results are summarized in Tables 5.4, 5.5, and 5.6; complete data records are on file at the Anasazi Heritage Center. Table 5.4 summarizes the taxonomic composition of the shell assemblage from the Duckfoot site. Identification was aided by the use of standard references (Keen 1963, 1971; Pennak 1978; Pilsbry 1939, 1948) and by comparison with reference specimens.[1] The level of taxonomic identification possible varied considerably as a result of breakage, weathering, and modification. Each taxonomic class is discussed briefly below and is correlated with artifact type, when applicable. Refer to Table 5.5 for a breakdown, by artifact type, of all worked pieces.

The single disk bead recovered from the site was so heavily modified that it was identifiable only as shell (indeterminate mollusk). Given the thickness and structure of the shell, it is probably marine in origin. This disk bead, which is 4.1 mm in diameter and 1.2 mm thick, was recovered from the midden.

All five examples of *Olivella* recovered from Duckfoot had been modified into beads. In each case, the spire had been deliberately removed to create a hole through which

Table 5.4. Taxonomic Composition of the Mollusk Collection

Taxon	N	%
Marine		
Gastropod		
Olivella cf. *dama* Wood	5	13.9
Pelecypod		
Glycymeris sp. Da Costa	11	30.5
Nonmarine		
Gastropod	18	50.0
Indeterminate		
Indeterminate gastropod	1	2.8
Indeterminate mollusk	1	2.8
TOTAL	36	100.0

a cord or string could be passed. This modification, along with wear and breakage in the aperture area, makes a positive identification difficult, but the beads appear to be *O. dama,* a species available in the Gulf of California (especially at the head of the gulf) but not along the Pacific coast of California (Keen 1971:628). *Olivella* beads were recovered from Room 8, Pit Structures 1 and 2, and two excavation units in the midden; selected specimens are shown in Figure 5.4.

Glycymeris was the most common type of marine shell recovered from the Duckfoot site. All the *Glycymeris* items are modified: 10 are bracelet fragments, and 1 is a whole pendant that was made from a piece of broken bracelet (Figure 5.4). The pieces are not identifiable to species, but they probably are *G. maculata* and/or *G. gigantea,* species identified from Hohokam sites in Arizona (Haury 1976:307; Craig 1982). These species are found in the Gulf of California (Keen 1971:55).

The bracelet fragments are parts of thin shell bands. To make a bracelet, the central area of a *Glycymeris* valve is removed and the hole enlarged until only a relatively thin ring of shell remains. The margins are also ground to

Table 5.5. Worked Shell

Artifact Type[a]	N	%
Worked fragment (other modified shell)	1	5.5
Bead, disc	1	5.5
Bead, whole shell	5	27.8
Bracelet fragment	10	55.5
Pendant (reworked bracelet fragment)	1	5.5
TOTAL	18	100.0

[a]Based on Haury (1937:137).

1. Dr. Richard C. Brusca and Regina Wetzer allowed me to use the comparative collections of the Department of Marine Invertebrates, San Diego Museum of Natural History. Their assistance is greatly appreciated.

Table 5.6. Distribution of Shell Taxa

Taxon	Room		Pit Structure		Courtyard		Midden		Other		Total	
	N	%	N	%	N	%	N	%	N	%	N	%
Marine												
Gastropod												
Olivella cf. *dama* Wood	1	20.0	2	40.0			2	40.0			5	100.0
Pelecypod												
Glycymeris sp. Da Costa	7	63.6	1	9.1	2	18.2	1	9.1			11	100.0
Nonmarine												
Gastropod	12	66.7					2	11.1	4	22.2	18	100.0
Indeterminate												
Indeterminate gastropod							1	100.0			1	100.0
Indeterminate mollusk							1	100.0			1	100.0
TOTAL	20	55.5	3	8.3	2	5.5	7	19.4	4	11.1	36	100.0

varying degrees, and, depending on the amount of grinding and the degree of thinness, the umbo and hinge areas are modified to greater or lesser degrees. No umbo fragments were recovered from Duckfoot.

The width, thickness, and shape of plain-band bracelets have been used to classify these artifacts into types that Haury (1976:313) suggests are temporally sensitive. The temporal correlations of the types have been questioned (e.g., Howard 1987), but the classification is a useful descriptive device. Few of the Duckfoot bracelets fall within the size ranges presented by Haury for both width and thickness, but the four that do are in the Type 2 range. Haury (1976:314) reports that this type of bracelet occurs throughout the occupation at Snaketown, which he dates to between 300 B.C. and A.D. 1100. Recent research, how-

ever, suggests that the Hohokam occupation probably began several centuries later than previously thought (cf. McGuire and Schiffer 1982).

Of all the types of shell recovered, *Glycymeris* fragments were found in the widest variety of contexts (Table 5.6). They were recovered from Rooms 6, 15, and 19 (five pieces from the same bracelet in Room 19), Pit Structure 1, Courtyard 2 (two pieces), and the midden. A comparison of the fragments indicates that at least six bracelets are represented.

The shell collection from the Duckfoot site includes two different types of nonmarine gastropods: one that appears to be *Succinea sp.* and one that is probably *Oreohelix strigosa depressa*. These remains were usually quite fragmentary, and most specimens could be identified only to

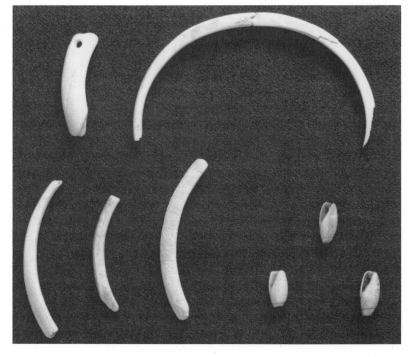

Figure 5.4. Shell artifacts: pendant made from a fragment of a *Glycymeris* shell bracelet (*upper left*), *Glycymeris* bracelet fragments (*upper right, lower left*), and *Olivella* beads (*lower right*). Pendant length = 33.7 mm.

the level of nonmarine shell. One piece of *Oreohelix* recovered from Feature 13 (bell-shaped cist) in Room 19 may have been a bead. The spire of the shell had been removed, leaving an opening through which the shell could have been strung. A concentration of similar *Oreohelix* shells was reported by Rohn (1971:128, Figure 161) from Mug House on Wetherill Mesa at Mesa Verde National Park.

Both taxa of nonmarine mollusks recovered from the site have been collected from their natural habitats in southwestern Colorado. The genus *Succinea* includes a number of widely distributed species that occur in a variety of habitats (Pilsbry 1948). Drake (1959:150) says that members of the family Succineidae in the Southwest are "almost amphibious and are usually considered so because they have generally been found alive in the damp, wet zone on the substratum at the edge of bodies of water." Other sources indicate that a wider range of conditions is suitable for members of the genus *Succinea,* including drier areas (Baker 1939:121–127; Pilsbry 1948:839). The type locality for *Oreohelix strigosa depressa* is a canyon near Durango, Colorado (Pilsbry 1939:431). *Oreohelix* occurs in talus or duff in areas that are rich in lime, where they are found near the surface, concealed by rock, bark, or leaves (Pilsbry 1939:415).

Summary and Discussion

Shell artifacts are not common in sites of the Mesa Verde region, especially compared to the types and quantities of shell recovered from sites in Chaco Canyon (Jernigan 1978:153; Morris 1939). For example, excavations at Badger House community at Mesa Verde National Park resulted in the recovery of only two shell beads from the sites investigated (Hayes and Lancaster 1975:73).

The distribution of shell on sites investigated by the DAP provides some useful comparisons. Blinman (1986b:Table 15.3) reports the occurrences of shell by time period for Dolores-area sites. An occurrence is defined as all the items of a similar artifact type from the same general provenience. This measure was provided in an attempt to allow comparisons of the number of discrete artifacts that may have been present, rather than the number of constituent pieces such as beads, since several hundred beads may be found in a single necklace. For the Dolores shell assemblage, a total of 87 occurrences (347 artifacts) is reported from 18 sites, for an average of 4.8 occurrences (19.3 artifacts) per site. One burial from Tres Bobos Hamlet (5MT4545) yielded 172 beads, counted as one occurrence, which inflates the total number of items considerably. Two large, complex sites, Grass Mesa Village (5MT23) and McPhee Pueblo (5MT4475), had over 20 occurrences each, and these two sites account for an additional 102 shell artifacts. Shell occurrence values for the remaining 15 sites ranged from 1 to 6, with artifact counts of 1 to 38 (the mean

count is 4.9). Beads, bracelets, and pendants are the shell artifact types identified in the Dolores collections.

Dominguez Ruin (5MT2148) has been noted as being unusual for the region in the types of shell artifacts recovered (Nickens and Hull 1982:211; Reed 1979:53). This site yielded shell beads, a shell-and-turquoise frog effigy, and two round shell ornaments, all associated with a burial. The elaborateness of the ornaments has been attributed to the burial being found in association with a Chacoan outlier (Nickens and Hull 1982:211). Burials may contain large quantities of shell as compared with other features at sites, because necklaces and bracelets were sometimes included in the grave goods.

The Duckfoot shell assemblage is well within the range observed for other sites in the region, then. The taxa of marine shell reported for Duckfoot are present in the Dolores inventory (Blinman 1986b:687; Gross 1985), as are the specific artifact types. If the five bracelet fragments from Room 19 are counted as a single occurrence, the Duckfoot site yielded 18 shell artifacts in 14 occurrences. This is a large number compared to sites of similar size in the Dolores area, but differences in percent of site excavated may account in part for this, as Duckfoot was more completely excavated than most Dolores sites.

The nonmarine shell probably is present at the site as a result of natural, rather than cultural, processes, although one of the specimens of *Oreohelix strigosa depressa* may have been modified into a bead. Both types of snail identified at the site have been recorded in recent times from southwestern Colorado.

Modified Vegetal Artifacts

Karen R. Adams

Approximately 20 charred pieces of coiled-basketry fragments were recovered from the floor of Pit Structure 3 (PL 160), and all could belong to a single basket burned in place. The largest piece is 15 × 5 cm, and the remaining pieces are generally less than 5 × 2 cm. All fragments are .6 cm thick.

The basket fragments consist of a bunched, two-rod and welt foundation with a noninterlocking stitch; rings, or coils, were added as the basket was constructed concentrically outward from the center (terminology according to Adovasio [1977:64]; see Blinman [1986a:57] for diagram of foundation composition). The 2- to 3-mm-diameter rods appear to be an entire single-season's growth of *Rhus aromatica* stems, each stem containing a relatively large pith area and a single woody ring. Although the branched growth habit of *Rhus* makes it difficult to imagine that anyone would use it for constructing baskets, apparently one can promote the growth of long, supple "sucker" shoots in the spring by cutting or burning the plant back in the fall (Bohrer 1983). The welts (flat pieces laid opposite one another, alongside the two rods) are composed of unretted Monocotyledon tissue with abundant fibrovascular bundles; these pieces probably are *Yucca* leaves. The sewing elements (stitches) have such a limited cross section that they are difficult to identify, but after comparison with modern charred specimens of all the tree and shrub taxa present in the vicinity, the Duckfoot stitching material appears to best match the first-season's growth of *Rhus aromatica* stems as well. These young stems have been split into more than two lengthwise strips, since the cross section of each reveals two flat facets rather than a hemispheric shape. Stitch width ranges from 1.0 to 2.0 mm (mean = 1.7 mm), and stitch frequency ranges from 4 to 5 stitches per centimeter (mean = 4.7). Coil height, essentially reflecting basket thickness, averages .6 cm.

With some exceptions, the Duckfoot basket fragments are similar in basic construction to the majority of those recovered from Dolores-area sites and described by Blinman (1986a). Although the materials and basic construction technique are predominantly the same in both assemblages, the Dolores fragments are composed of split *Rhus* rods and unsplit *Rhus* sewing elements, just the reverse of the Duckfoot materials described here.

The charred basket fragments in Pit Structure 3 were intermixed with a number of other plant remains. *Juniperus* charcoal, a number of small juniper twig ends, and a *Juniperus* seed with attached fruit coat were identified. In addition, a single *Stipa hymenoides*–type caryopsis and three burned *Atriplex/Sarcobatus*-type twigs preserved. The context in which these materials were recovered sheds no light on whether any represent the actual basket contents.

Two charred examples of twining were recovered from the roof fall stratum in Room 3. The largest piece measures 3 × 1.2 cm and is 4 mm thick. Both fragments are closely twined, with the warp and weft composed entirely of Monocotyledon material, most likely *Yucca*. The warp has been fashioned from three 1.2-mm-diameter plies of retted fibers wrapped together in an S-twist (Adovasio 1977:20) to form a 2-mm-diameter foundation cord. The weft appears not to have been entirely stripped of epidermis, displaying at minimum a two-ply S-twist for the filling material. The tight weave has at least 20 weft cords per centimeter.

Six charred fragments of cordage were found on the floor of Pit Structure 4 (PL 170). All are constructed of retted Monocotyledon fiber bundles, again likely to have been acquired from *Yucca* leaves. A single 3-cm-long piece of cordage measuring 4 mm in diameter is composed of two two-ply strands. Each two-ply strand has an S-twist, and although the overall cordage twist is difficult to recognize, it is probably a Z-twist (Adovasio 1977:20). The remaining five two-ply fragments are between 1 and 2 mm in diameter, and all are less than 2 cm in length. The individual strands are Z-twisted but are joined with an S-twist to form the cordage.

6

Carbonized Plant Remains

Karen R. Adams

Introduction

Plant remains were recovered from the Duckfoot site using two techniques. Excavators hand-picked from site deposits a limited number of larger, easily recognized plant parts, such as *Zea mays* (corn) cob segments and kernels, large pieces of charred wood, and basketry and cordage fragments. These individually hand collected items were retrieved as vegetal specimens, of which only the recognizable corn remains and worked (modified) items were submitted for analysis. Results of the *Zea* analysis are included here; however, most of the plant remains discussed in this chapter, including the wood charcoal, were recovered from flotation samples, that is, samples of dirt processed with a water-separation technique that concentrates organic remains. Of 262 variable-size samples collected, 103 from a variety of cultural contexts were processed and analyzed, and the greater part of this chapter is devoted to reporting those results.[1]

Methods

Flotation Sample Processing

Two different methods were used to process the Duckfoot flotation samples. Samples collected before 1986 were processed with water twice. First, the sediments were washed through a $\frac{1}{16}$-in-mesh screen. The residue remaining in the screen, which consisted of both organic and inorganic debris, was then placed in a container of water, and any floating organic material was scooped from the water's surface using a tea strainer. When dried, this material was sieved through a series of graduated screens and divided into five groups based on particle size (larger than 4.0 mm, 2.0 to 4.0 mm, 1.0 to 2.0 mm, .5 to 1.0 mm, and smaller than .5 mm) to facilitate analysis.

The second group of Duckfoot flotation samples, collected during the 1986 and 1987 field seasons, was water-processed only once. The method used (Schaff 1981) concentrates charred debris while sorting material by particle size. For the Duckfoot samples, a small container of water was placed on top of a nested set of graduated U.S. Geological Survey sediment-sorting screens—in this case, of 4.75-, 2.80-, 1.40-, .71-, and .25-mm mesh sizes. The flotation sample was poured into the water and gently stirred to loosen organic material, which floated to the surface. Then more water was added to cause the container to overflow. Because the diameter of the water container was smaller than that of the screens, the overflow passed through the screens, and organic material was trapped in one of the meshes depending on the sizes of individual particles. This procedure was continued until most or all of the organic material was freed from the sediment

---

1. Several individuals and institutions contributed to the study reported in this chapter. The Crow Canyon Program staff extended to me, as an outside consultant, every courtesy in yearly visits to their campus to develop modern comparative collections and confer with archaeologists. Ricky R. Lightfoot established clear archaeobotanical priorities for the Duckfoot site and was extremely conscientious about providing archaeological data to enhance interpretation. His skills in plant collecting are also appreciated; he gathered big-sagebrush flowering stalks over a four-month period and collected a suite of modern wood samples with corresponding herbarium specimens that became the foundation of the wood comparative collection. Rex K. Adams accompanied me on a number of trips to the Cortez area, serving as an able and observant field assistant. He also assisted in laboratory analysis of the botanical material. The role of the Colorado Historical Society in partially funding this analysis is gratefully acknowledged.

and rinsed clean, presorted by size in one of the five screens.

It should be noted that certain sampling and method biases may have influenced the Duckfoot flotation sample data. For example, some fragile organic material may have been lost in those samples processed twice with water. Also, variability in original sample size would be expected to influence the record of uncommon items, with decreased chances of recovery in samples of small volume. Finally, items smaller than the smallest meshes employed (i.e., tea strainer and .25-mm screen) would be consistently lost from samples. These might include seeds of such taxa as *Nicotiana* and *Descurainia*.

Subsampling for Seeds and Other Microremains

Examination of the entire volume of concentrated organic material from each processed sample for seeds and other plant parts (excluding wood charcoal, discussed below) was beyond the practical scope of this project. Therefore, a method was developed to maximize the number of samples examined without compromising the potential to assess the taxonomic diversity of any one sample. This approach has the advantage of taking into consideration actual recovered data in deciding how much effort should be spent sorting a given sample. The recovery of separate taxa is believed to provide the most insight into prehistoric plant use, and the underlying assumption is that taxa are randomly distributed within each sample.

First, the total sample was divided into five particle-size categories, as described above for the different processing methods. Samples that are grouped by similar particle size require less focusing of eyes as well as microscope. Also, if one begins analysis with the largest-particle-size group, taxa recognized as whole items early in sorting can often be identified as fragments in smaller-particle-size groups. Usually all items in the largest-particle-size group (i.e., larger than 4.75 or 4.0 mm) are examined, because sorting large items proceeds relatively rapidly and unique specimens often occur in this group. For the remaining particle-size groups, a number of standard subsample volumes are sorted for each group until no new taxa are encountered, at which time one can move on to sort the next particle-size group. (A standard subsample volume is defined as "the volume of material that can be packed, but not piled, contiguously under a field-of-view for each particle size" [Bohrer and Adams 1977:40].) Although the particle-size groups differed slightly for Duckfoot samples processed in different years, the subsample volumes used are as follows: 1.8 ml for particles larger than 2.8 or 2.0 mm, .9 ml for particles larger than 1.4 or 1.0 mm, .5 ml for particles larger than .71 or .50 mm, and .3 ml for particles smaller than .71 or .50 mm. There are two advantages to this approach: particle-size groups rich in taxonomic diversity

receive more attention than those low in diversity, and the samples themselves provide data helpful in deciding whether to continue or curtail sampling. This approach is similar to the one developed by ecologists (known as the "species area curve") to evaluate when field sampling of plant diversity may be reasonably curtailed (Mueller-Dombois and Ellenberg 1974:52–53).

The amount of material examined from Duckfoot flotation samples varied widely. Samples less than 50 ml in volume were sorted completely. In samples of larger volume, the amount examined varied but was never below the minimum required by the method described above. In cases in which materials of a specific particle size were only partly sorted, the potential total number of items of a taxon in the entire particle-size group was estimated using the following formula:

$$\text{estimated no. of items of a taxon} = \frac{\text{actual no. recovered} \times \text{total particle-size volume}}{\text{volume of particle size sorted}}$$

This estimation technique assumes that the rate of recovery would remain the same in the unsorted portion. When a taxon was recovered from more than one particle-size group, data from all particle-size groups were summed to obtain the number actually recovered from, and the total number estimated in, the sample.

Wood Charcoal Analysis

Water processing of flotation samples provides unbiased recovery of charcoal in terms of size, shape, and species composition (Willcox 1974:123). Therefore, wood charcoal was examined from 77 of the same flotation samples sorted for seeds and other small plant parts. The goal of identifying 20 pieces of burned wood per sample was usually accomplished by spreading out all specimens and randomly selecting a subsample of 20 large enough (larger than 4 mm^2) to be manipulated effectively by hand and likely to retain the anatomical features necessary for identification (Minnis 1987:122). Samples containing fewer than 20 specimens were examined in their entirety, when possible. The charred wood was examined under a binocular microscope, with magnifications ranging up to 50×. Each piece was oriented under the microscope, then snapped to produce a clean cross (transverse) section. When cross section provides the only source of evidence for identification, the level of reliability is lower than when the radial and tangential sections also are examined, especially under higher magnifications (Bohrer 1986:34).

A collection of modern wood specimens representing 19 common tree and shrub species in a 10-km radius around the Duckfoot site provided modern material for comparison. This collection, carbonized (burned) by the author, is backed by specimens deposited in the University of Arizona Herbarium (ARIZ). The modern collection is made up of a broad size range of material, including smaller stems and twigs of each species. Many of the ancient Duckfoot charcoal specimens are twigs of very small diameter, and cross sections of small twigs do not always display the same characteristics as older, larger specimens. Direct comparison of like-size modern and ancient specimens is crucial to identification. A number of references were also consulted for anatomical descriptions of genera (Friedman 1978; Barefoot and Hankins 1982; Schweingruber 1982) and for identification criteria for other species present in southwestern Colorado (Dale 1968; Minnis 1987).

Results

At least 30 taxa are represented in the plant materials recovered from flotation samples from the Duckfoot site (Table 6.1). All items discussed here are carbonized and are assumed to have become so because of prehistoric activities. Noncarbonized specimens, some clearly modern, have been excluded from consideration. Most taxonomic identifications are followed by the word "type" in both the text and the tables of this chapter. This word should convey to the reader that, even though the Duckfoot specimen resembles the taxon named, it may also resemble other related (or occasionally unrelated) taxa. This conservative approach is used because of the similarity in appearance of parts of various southwestern taxa, especially when an ancient specimen has been carbonized and likely altered in appearance. The specific criteria for the identification of the parts of each taxon, including wood charcoal, can be found in Appendix C.

In the following section, the distribution of each taxon at Duckfoot is examined. The number and variety of contexts in which a particular taxon was found provide the basis for evaluating whether it was intentionally gathered and what its relative cultural importance might have been. With such a site-wide perspective for all taxa, one can better interpret the significance of plant remains in specific structures or features.

Domesticated Plants

Phaseolus vulgaris-type cotyledon (Leguminosae)

A single cotyledon (half bean) fragment from the Pit Structure 1 hearth was identified on the basis of size as domesticated *Phaseolus vulgaris,* or the "common bean" cultivated in the prehistoric Southwest (Kaplan 1956). In the Anasazi area, the first bean records coincide with the introduction of pottery, and beans were identified in a number of sites in the nearby Dolores River area (Griffitts 1987). Scarcity of bean remains at Duckfoot could relate to poor preservation of a normally boiled resource, or to low incidence of use, or both.

Table 6.1. Carbonized Plants from the Duckfoot Site

Taxon	Part
DOMESTICATED PLANTS	
Phaseolus vulgaris–type	cotyledon
Zea mays	cob, kernel, cupule
WILD PLANTS	
Gymnosperms	
Juniperus osteosperma–type	fruit, seed, twig, scale leaf
Juniperus–type	wood charcoal
Pinus edulis–type	needle fragment
Pinus–type	bark scale, wood charcoal
Angiosperms	
Artemisia tridentata–type	flowering head, leaf
Artemisia–type	wood charcoal
Atriplex canescens–type	fruit core
Atriplex–type	wood charcoal
Capparaceae–type	seed
Cercocarpus/Artemisia–type	axillary bud
Cercocarpus–type	wood charcoal
Cheno-am	seed
Compositae–type	achene
Descurainia–type	seed
Echinocereus fendleri–type	seed
Galium–type	fruit
Gramineae	caryopsis
Helianthus/Viguiera–type	achene
Leguminosae–type	seed
Lycium–type	wood charcoal
Malvaceae–type	seed
Mentzelia albicaulis–type	seed
Monocotyledon	fibrovascular bundle
Opuntia–type	seed
Peraphyllum–type	seed
Peraphyllum/Amelanchier–type	wood charcoal
Phragmites australis–type	stem
Physalis longifolia–type	seed
Populus angustifolia–type	wood charcoal
Populus/Salix–type	wood charcoal
Portulaca retusa–type	seed
Purshia–type	wood charcoal
Rhus aromatica–type	seed
Scirpus–type	achene
Stipa hymenoides–type	floret, caryopsis
Yucca baccata–type	seed
Type 1 unknown	
Type 2 unknown	

NOTE: The word "type" following a family, genus, or species designation indicates that the ancient Duckfoot items are similar to the taxon named but that other taxa may have parts whose characteristics are within the range of the taxon cited.

Zea mays cob, kernel, cupule (Gramineae)

Numerous charred *Zea mays* cupules and kernel fragments were recovered from Duckfoot flotation samples, and a limited number of larger charred cob segments and whole kernels were hand-collected as vegetal specimens during regular excavation. The larger cob segments and whole kernels provide general information on the characteristics of prehistoric corn, and the distribution of cupules indicates the ubiquity of corn remains at Duckfoot.

Cobs: Excavators hand-collected a number of cob segments, none over 3 cm in length (Table 6.2). With the exception of two cob base specimens discussed below, the specimens lack attached kernels. All cobs appear to have straight rows, although the diminutive nature of some cupules suggests that kernels occasionally did not completely develop. Measurements made with an eyepiece micrometer at $8 \times$ magnification include the commonly recorded features of cob rachis diameter (Wellhausen et al. 1952), cupule width (Nickerson 1953), and cob shank diameter (Nickerson 1953), when possible.

The cob apex and base segments provide additional data on the characteristics of the population. Both apex segments taper abruptly. The two cob bases from Pit Structure 3 retain kernels and husks but differ noticeably from each other. One has a round to pentagonal rachis; this cob with kernels still attached measures 26 mm at the base, then tapers abruptly to a very narrow, 5.1-mm round shank. The other cob base with kernels has an elliptical rachis measuring 22 mm, but this specimen tapers quite gradually to a much wider, elliptical, 12.8-mm shank.

This small sample of cob segments varies in characteristics recorded. The number of cob rows ranges from 8 to 14, with 10- and 12-row cobs predominating. The average row number (N = 12) is 10.8. Cobs are generally round in cross section, although four specimens appear elliptical. Considerable variation in rachis diameter is also noted. This variation may be due in part to the fact that both apex and base segments are present, as well as a number of incomplete specimens whose exact positions along the cob cannot be determined. Average cupule width

varies from 4.2 to 8.3 mm, with an overall mean of 6.2 mm; on the 5 to 10 cupules examined, cupule width varies by as much as 2.0 mm on a single cob segment. The fragmentary condition of the cob segments made it impossible to standardize the location of cupule measurements at mid-cob.

Row number and cupule width data are plotted in Figure 6.1. Three Courtyard 3 cobs and three Pit Structure 3 cobs tend to have diverse characteristics, whereas three Pit Structure 1 cobs are more similar to one another, with generally fewer rows and smaller cupule width. Single cobs from Rooms 15 and 16 and Pit Structure 4 are characterized by intermediate row numbers and cupule widths.

Kernels: Charred complete kernels, some single and others in adherent, orderly stacks, were recovered from two pit structures and two courtyards (Table 6.3). Up to 10 whole, randomly selected kernels from seven proveniences were scrutinized under $8 \times$ magnification for presence of denting and husk striations across the apex (if there were 10 or fewer kernels, all were examined; if there were 11 or more, 10 were randomly selected). At the same time, thickness (Nickerson 1953:83) and width (at right angles to thickness) were measured. Broken kernels that accompanied the whole kernels were examined to determine the general type (fine grained or porous) of endosperm (Doebley and Bohrer 1983:32–34).

The 53 whole kernels examined are variable. Mean thickness is 4.4 mm and mean width is 8.1 mm, which suggests a general rectangular shape; however, the whole kernels actually range from isodiametric to elongate. A dented apex was observed on a single kernel; husk striations, too, were noted on only a single specimen. Endosperm ranges from fine grained to porous, which suggests that both pop/flint and flour types may be represented.

Table 6.2. Observations on Selected *Zea mays* Cob Segments

Plant Part	Location	Cob Shape	Rachis Diameter (mm)	Shank Diameter (mm)	No. of Rows	Cupule Width (mm) Mean	Cupule Width (mm) Range
Cob apex	Room 16	elliptical	9.0 × 12.8		12	5.5	5.1–5.8
	Pit Structure 4	elliptical	9.0 × 12.8		10	6.4	5.8–7.0
Cob base[a]	Pit Structure 3	elliptical	7.7 × 12.8	12.8	10	8.3	7.6–9.6
		round	15.4	5.1	10	7.0	
Cob segment	Room 15	round	6.4		12	5.1	
	Pit Structure 1	round	6.4		10	4.2	3.8–4.5
		round	7.0		10	5.3	4.5–6.0
		round	7.7		8	5.1	
	Pit Structure 3	elliptical	8.3 × 12.2		14	5.5	5.1–5.8
	Courtyard 3	round	11.5		10	7.2	6.4–7.9
		round	14.1		12	8.3	
		round	7.7		12	4.2	3.8–4.5

NOTE: Table summarizes cobs recovered by hand during regular excavation. No cob segments were recovered from analyzed flotation samples.
[a] Kernels and leaves (husks) still attached.

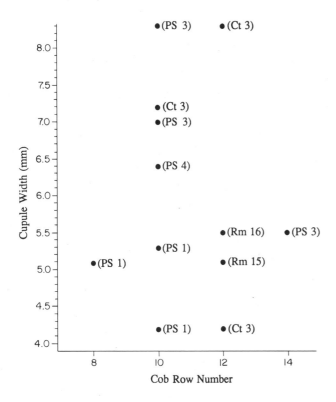

Figure 6.1. Distribution of Duckfoot cob segments by row number and cupule width. Cobs were recovered by hand during regular excavation. No cob segments were recovered from analyzed flotation samples. Ct = Courtyard, PS = Pit Structure, Rm = Room.

Two pieces of evidence suggest that a number of the Duckfoot kernels were charred while on the cob. More than one group of neatly stacked kernels adhered to one another, mimicking the arrangement of kernels in rows on a cob. Also, these kernels displayed virtually no swelling or extruding surfaces of any kind, as if charring had occurred while they were held tightly in place, limiting expansion or shape alteration. Slow charring in a reducing atmosphere may also produce kernels with little distortion.

Table 6.3. Measurements of Selected Whole
Zea mays Kernels

Location	N	Thickness (mm)	Width (mm)
Pit Structure 3	2	6.7	8.9
Pit Structure 4	10	3.8	8.2
	10	4.1	8.5
Courtyard 1	5	5.4	8.2
	8	4.7	8.2
Courtyard 3	8	4.7	6.7
	10	3.8	8.4
	N = 53	MEAN = 4.4	MEAN = 8.1

NOTE: Table summarizes kernels recovered by hand during regular excavation. Kernels from flotation samples are not included.

The arrangement of the fused kernels suggests that at least some of the Duckfoot corn had regular rows, as opposed to having kernels in a mosaic, or tessellate, arrangement. The arrangement of kernels attached to the two cob base segments was irregular, but kernel placement is often irregular near the cob base.

Cupules: Numerous *Zea* remains, including cob fragments, kernel fragments, and cupules, were also found in flotation samples from the Duckfoot site (Table 6.4). A cupule is a cup-like structure that holds two spikelets, and each spikelet, in turn, holds a single kernel, or grain. The cupule is the most durable part of the cob and often preserves after all other parts have degraded.

Recovery of cupules and cob fragments from hearths and ash pits in many locations suggests secondary use of the cobs as fuel. Their presence in various pits, cists, bins, and courtyard features may relate to processing or storage of corn or possibly to general mixing of site debris. Limited numbers of charred *Zea* cupules/cobs with vessels and metates may represent trash or derive from roof collapse. The presence of charred *Zea* parts in upper control samples reveals that some mixing of plant remains occurred at Duckfoot.

Wild Plants: Gymnosperms

Juniperus osteosperma-type   (Cupressaceae)
fruit, seed, twig, scale leaf

Juniperus-type wood charcoal

Charred juniper fruit, seeds, twigs, and twig scale leaves are known from a variety of locations at Duckfoot (Table 6.5). These parts are morphologically similar to those of *Juniperus osteosperma*, the common juniper species that grows near the site today. This record complements the widespread recovery of burned juniper wood in flotation samples (see "Discussion" section at the end of the chapter), which suggests that juniper resources were often carried into the Duckfoot site.

The presence of twigs and individual twig scale leaves in many samples from the floors of Pit Structures 1 and 3 suggests that branches may have been carried in to provide fuel to deliberately burn both structures. The remaining distribution of juniper twigs and scale leaves attests to regular use of the boughs as fuel and possibly as a material to line the sides and bottoms of bins and cists. Occasional recovery of whole juniper fruit and seeds may relate to use as a food and/or to incidental inclusion of the reproductive parts on branches carried in for some other purpose. The fruit of numerous juniper species has been gathered by historic groups for food, and wood and branches have been sought for construction material and fuel (Adams 1988).

Pinus edulis–type cone scale,   (Pinaceae)
needle fragment

Pinus edulis (pinyon) is known at Duckfoot from single charred needle fragments in the Room 13 hearth and a Room 15 cist. Seven pinyon needle fragments were also excavated from the floor beneath a metate and in the upper fill of Pit Structure 3. A single pinyon cone scale was recovered from the hearth in Room 11.

Pinus-type bark scale,   (Pinaceae)
wood charcoal

Charred *Pinus*-type bark scales were present in all four pit structures, eight rooms, and the midden (Table 6.6), often in features where burning had occurred. Although the *Pinus*-type bark scales and wood charcoal probably represent pinyon trees that grew in the surrounding pinyon-juniper woodland, these parts could have also

Table 6.4. Carbonized *Zea mays* Remains from Flotation Samples

Location[a]	Number Recovered	Number Estimated[b]	Location[a]	Number Recovered	Number Estimated[b]
Pit Structure 1			Feature 3 (cist)	4c	4c
Feature 8 (hearth)	4c, 3kf	4c, 38kf	Feature 6 (hearth)	10cf, 18c	10cf, 18c
Feature 9 (ash pit)	1c, 9kf	7c, 56kf	Feature 10 (warming pit)	8c	8c
Floor fill above Features 8 and 9 (hearth and ash pit)	5kf	63kf	Room 12		
Vessel (PL 64) contents	1cf, 2c, 3kf	1cf, 4c, 5kf	Feature 6 (hearth)	78c, 7kf	78c, 7kf
Roof fall, upper control	2c, 2kf	3c, 2kf	Wall and roof fall, metate trough contents	10c, 1kf	10c, 1kf
Pit Structure 2			Wall and roof fall, upper control	53c	64c
Feature 3 (hearth)	3cf, 41c, 3kf	3cf, 41c, 3kf	Room 13		
Floor	2c	2c	Feature 1 (hearth)	17c, 4kf	125c, 29kf
Roof and wall fall, upper control	1c	1c	Feature 2 (slab-lined pit)	2c	2c
Pit Structure 3			Room 15		
Feature 36 (ash pit)	2cf, 24c, 14kf	2cf, 27c, 14kf	Feature 1 (bin)	1c, 7kf	1c, 7kf
Floor beneath metate (PL 72)	3c, 4kf	4c, 4kf	Feature 4 (ash-filled pit)	6cf, 23c, 2kf	6cf, 46c, 4kf
Vessel (PL 50) contents	1cf, 1kf	1cf, 1kf	Feature 5 (cist)	3cf, 18c, 1kf	3cf, 18c, 1kf
Closing/fuel on floor	1kf	1kf	Room 16		
Roof fall, upper control	4kf	4kf	Feature 3 (hearth)	14c, 3kf	25c, 3kf
Pit Structure 4			Feature 7 (hearth)	18c, 5kf	19c, 5kf
Vessel (PL 35) contents	1c, 2kf	1c, 2kf	Room 17		
Vessel (PL 47) contents	1cf, 2c	1cf, 2c	Upper fill, metate (PL 2) trough contents	2c	2c
Room 2			Room 19		
Upper fill, control under vessels (PL 1)	1c	1c	Feature 5 (pit)	5c	5c
Room 3			Feature 13 (bell-shaped cist)	2c, 1kf	2c, 1kf
Floor, ash concentration	1kf	1kf	Feature 19 (hearth)	3c	3c
Room 4			Courtyard 2		
Feature 2 (pit)	16cf, 52c	16cf, 52c	Feature 8 (fire pit)	17cf, 79c, 1kf	17cf, 79c, 1kf
Room 5			Feature 15 (slab-lined pit)	9c, 2kf	9c, 2kf
Feature 1 (fire pit)	1kf	1kf	Courtyard 3		
Feature 3 (pit)	1c	1c	Feature 18 (pit)	5c	5c
Room 6			Feature 24 (roasting pit)	1c	1c
Feature 1 (cist)	2c	2c	Arbitrary Unit 4		
Room 10			Feature 1 (fire pit)	2c, 1kf	2c, 1kf
Feature 3 (corner bin)	2c	2c	Midden		
Room 11			Feature 2 (pit)	63c	113c
Feature 1 (corner bin)	4c	4c	General midden	1c, 1kf	1c, 1kf

c = cupule; *kf* = kernel fragment; *cf* = cob fragment.
[a] All artifacts and features are associated with floors/surfaces, unless noted otherwise.
[b] When only a portion of the material in a flotation sample was examined, a proportional estimate of the total present was calculated, assuming a continued constant rate of recovery.

derived from *Pinus ponderosa* and *Pseudotsuga*, trees that today grow on the north face of the Mesa Verde and on the Sleeping Ute Mountain less than 16 km (10 mi) from Duckfoot.

Bark scales might become incorporated into ancient deposits for a number of reasons. Trunks or older branches gathered for hearth fuel could leave bark scales in ashy deposits; bark removed from large trunks used as roof beams might provide a handy source of tinder; and layers of bark also might offer soft material for bedding.

The bark scales in flotation samples from the floor and the fill above the hearth and ash pit in Pit Structure 1 could be the remains of fuel carried in to burn the structure or might reflect postabandonment mixing of site deposits. The distribution of bark scales parallels the widespread occurrence of *Pinus*-type wood charcoal in flotation samples, supporting a suggestion that pine-type wood was regularly sought as hearth fuel (see "Discussion" section) and roof-construction material.

Pines have provided southwestern historic groups with many resources (Adams 1988). Seeds ("nuts") of the nut-pines, including *Pinus edulis*, have been sought as an important food. Wood has been used in construction and gathered as a hearth fuel. Other parts collected include the inner bark for food, pollen for various needs, and pitch and needles for medicinal treatments. Prehistoric southwestern groups, too, have left occasional records of ancient *Pinus* bark use (Adams 1988).

Wild Plants: Angiosperms

Artemisia tridentata-type (Compositae)
flowering head, leaf segment

Artemisia-type wood charcoal

Artemisia tridentata (big sagebrush) is represented at the Duckfoot site by charred immature flowering heads and leaf segments in a number of locations (Table 6.7). The presence of immature flowering heads and leaf segments in so many floor and above-floor samples in Pit Structures 1 and 3 is used to support the argument that sagebrush

Table 6.5. Carbonized *Juniperus*-type Remains from Flotation Samples

Location[a]	Number Recovered	Number Estimated[b]	Location[a]	Number Recovered	Number Estimated[b]
Pit Structure 1			Upper fill, vessel (PL 1) contents	16*t*	16*t*
Feature 8 (hearth)	4*s*, 1*sl*	4*s*, 34*sl*	Room 6		
Floor fill above Features 8 and 9 (hearth and ash pit)	1*t*	8*t*	Feature 1 (cist)	3*sl*	9*sl*
Vessel (PL 8) contents	1*t*	3*t*	Room 10		
Vessel (PL 64) contents	1*s*	1*s*	Feature 1 (corner bin)	1*t*	1*t*
Roof fall, vessel (PL 15) contents	1*sl*	7*sl*	Room 11		
Roof fall, upper control	1*t*	1*t*	Feature 3 (cist)	1*t*	1*t*
Pit Structure 2			Room 12		
Feature 19 (sipapu)	1*t*	1*t*	Feature 6 (hearth)	3*sl*, 1*s*, 9*sf*	3*sl*, 1*s*, 9*sf*
Floor	1*t*	1*t*	Wall and roof fall, upper control	1*t*	2*t*
Pit Structure 3			Room 13		
Feature 34 (hearth)	1*t*, 2*s*	2*t*, 2*s*	Feature 1 (hearth)	1*t*	1*t*
Feature 36 (ash pit)	3*t*	3*t*	Room 15		
Floor, basket (PL 160) and possible contents	1*s*, 78*t*	1*s*, 190*t*	Feature 5 (cist)	2*t*	2*t*
Floor beneath metates (PLs 19 and 72)	47*t*	95*t*	Room 16		
			Feature 3 (hearth)	3*s*	5*s*
Floor near Feature 33 (sipapu)	2*t*	2*t*	Room 17		
Vessel (PL 50) contents	103*t*	138*t*	Upper fill, metate (PL 2) trough contents	1*sl*	1*sl*
Vessel (PL 52) contents	55*t*	90*t*	Room 19		
Closing/fuel on floor	1*f*, 1*s*, 515*t*	1*f*, 1*s*, 990*t*	Feature 13 (bell-shaped cist)	1*t*	1*t*
Roof fall, upper control	82*t*	97*t*	Feature 19 (hearth)	5*t*	5*t*
Pit Structure 4			Courtyard 2		
Feature 1 (hearth)	1*f*, 1*s*, 1*t*	1*f*, 1*s*, 1*t*	Feature 15 (slab-lined pit)	5*t*	5*t*
Vessel (PL 47) contents	2*s*, 10*t*	2*s*, 22*t*	Midden		
Room 2			General midden	1*t*	1*t*
Upper fill, control under vessels (PL 1)	7*t*, 4*sl*	7*t*, 4*sl*			

f = fruit; *s* = seed; *sf* = seed fragment; *t* = twig; *sl* = twig scale leaf.
[a] All artifacts and features are associated with floors/surfaces, unless noted otherwise.
[b] When only a portion of the material in a flotation sample was examined, a proportional estimate of the total present was calculated, assuming a continued constant rate of recovery.

Table 6.6. Carbonized *Pinus*-type Bark Scales from Flotation Samples

Location[a]	Number Recovered	Number Estimated[b]
Pit Structure 1		
Feature 8 (hearth)	5	194
Floor fill above Features 8 and 9 (hearth and ash pit)	8	89
Floor scrape	1	1
Vessel (PL 12) contents	2	2
Vessel (PL 18) contents	5	5
Vessel (PL 64) contents	1	2
Roof fall, vessel (PL 11) contents	5	8
Roof fall, vessel (PL 15) contents	16	134
Pit Structure 2		
Feature 3 (hearth)	20	20
Feature 4 (ash pit)	2	2
Pit Structure 3		
Feature 36 (ash pit)	4	4
Closing/fuel on floor	1	1
Pit Structure 4		
Feature 1 (hearth)	3	3
Floor fill	1	1
Room 2		
Upper fill, vessel (PL 1) contents	1	1
Room 4		
Feature 1 (burned spot)	1	1
Feature 2 (pit)	1	1
Room 10		
Feature 1 (corner bin)	1	1
Floor	4	4
Room 11		
Feature 6 (hearth)	1	1
Feature 10 (warming pit)	2	2
Room 12		
Feature 6 (hearth)	93	93
Wall and roof fall, metate trough contents	8	8
Wall and roof fall, upper control	3	7
Room 13		
Feature 1 (hearth)	47	347
Room 15		
Feature 1 (bin)	1	1
Feature 4 (ash-filled pit)	2	6
Feature 5 (cist)	3	3
Room 16		
Feature 3 (hearth)	9	20
Feature 7 (hearth)	15	21
Midden		
Feature 2 (pit)	6	12
General midden	1	1

[a] All artifacts and features are associated with floors/surfaces, unless noted otherwise.
[b] When only a portion of the material in a flotation sample was examined, a proportional estimate of the total present was calculated, assuming a continued constant rate of recovery.

branches were carried into these structures to provide fuel for intentional burning of the roofs.

The state of flowering head maturity (described in Appendix C) provides an important clue about the season when this resource was gathered. Flowering stalks of modern *Artemisia tridentata* plants were collected at one- to two-week intervals from early August until late November 1987 to document big-sagebrush flower development in the Duckfoot region. The charred Duckfoot specimens best match the September collections of modern flowering heads with fully swollen flower buds just ready to open. Modern heads collected in August contained buds that had only just begun to swell, appearing less mature than the Duckfoot specimens, whereas heads gathered toward the end of September had passed out of bud form and begun full flowering. The mature achenes begin to disperse in October and can continue to do so for weeks afterward.

Artemisia leaves in the Duckfoot assemblage provide no seasonality clues, because the fragrant, three-toothed leaves remain attached throughout the year. Sagebrush wood, also available year-round, was often recovered from hearths and roof fall debris, which suggests that the Duckfoot occupants routinely gathered the wood and branches for use as fuel and as roofing material (see "Discussion" section).

The bulk of the historic references to sagebrush document the use of this plant for medicinal purposes. However, a few historic groups gathered *Artemisia tridentata* as fuel when no other firewood was available (Robbins et al. 1916:45; Bye 1972:93). The Navajo used sagebrush as a floor covering for a sweat lodge (Bohrer 1964). The Duckfoot record suggests that the role of big sagebrush for southwestern groups was formerly broader.

Atriplex canescens-type fruit core   (Chenopodiaceae)

Atriplex-type wood charcoal

Charred *Atriplex canescens*-type (four-wing saltbush) fruit cores were found in two areas at Duckfoot. A single fruit core was recovered from the fire pit in Arbitrary Unit 4. Twenty-nine specimens were recovered from various Pit Structure 3 contexts, including the floor beneath a metate, floor debris thought to represent either roof-closing material or fuel, the hearth, and an upper control sample above the floor. Four-wing saltbush grows in the area today and was observed in full bloom during the latter part of June (ARIZ 271498). By fall, the plants produce a characteristic hard fruit with four papery wings, which can cling to branches for months after maturing.

The fruits recovered from Duckfoot may have been attached to *Atriplex* branches used for fuel or to smaller branches used in roof construction, as *Atriplex*-type wood charcoal was often found in hearths, fire pits, and roof fall debris (see "Discussion" section). The hardness of the fruit core greatly increases its chances of preservation.

Table 6.7. Carbonized *Artemisia tridentata*–type Remains from Flotation Samples

Location[a]	Number Recovered	Number Estimated[b]
Pit Structure 1		
Feature 9 (ash pit)	2*ls*	2*ls*
Floor scrape	20*fh*, 13*ls*	42*fh*, 26*ls*
Vessel (PL 1) contents	5*fh*, 8*ls*	47*fh*, 76*ls*
Vessel (PL 3) contents	5*fh*, 10*ls*	25*fh*, 34*ls*
Vessel (PL 8) contents	18*fh*, 8*ls*	36*fh*, 14*ls*
Vessel (PL 14) contents	11*fh*, 20*ls*	91*fh*, 148*ls*
Vessel (PL 18) contents	15*fh*, 12*ls*	15*fh*, 12*ls*
Vessel (PL 64) contents	16*fh*, 31*ls*	128*fh*, 157*ls*
Vessel (PL 78) contents	63*fh*, 21*ls*	137*fh*, 56*ls*
Roof fall, vessel (PL 11) contents	11*fh*, 1*ls*	67*fh*, 4*ls*
Roof fall, vessel (PL 15) contents	67*fh*, 20*ls*	535*fh*, 150*ls*
Roof fall, upper control	5*fh*, 16*ls*	6*fh*, 25*ls*
Pit Structure 3		
Feature 34 (hearth)	1*ls*	2*ls*
Floor, basket (PL 160) and possible contents	6*ls*	10*ls*
Floor beneath metate (PL 19)	5*ls*	13*ls*
Vessel (PL 50) contents	4*fh*, 34*ls*	5*fh*, 40*ls*
Vessel (PL 52) contents	1*fh*, 8*ls*	9*fh*, 43*ls*
Closing/fuel on floor	25*ls*	51*ls*
Roof fall, upper control	6*fh*, 7*ls*	9*fh*, 7*ls*
Pit Structure 4		
Vessel (PL 47) contents	1*ls*	3*ls*
Floor fill	2*ls*	4*ls*
Room 2		
Upper fill, control under vessels (PL 1)	1*ls*	1*ls*
Room 12		
Wall and roof fall, metate trough contents	4*fh*, 1*ls*	4*fh*, 1*ls*
Midden		
General midden	1*ls*	1*ls*

fh = flowering head; *ls* = leaf segment.
[a] All artifacts and features are associated with floors/surfaces, unless noted otherwise.
[b] When only a portion of the material in a flotation sample was examined, a proportional estimate of the total present was calculated, assuming a continued constant rate of recovery.

The historic record of *Atriplex canescens* is varied and includes documentation of use as firewood and kiva fuel (Whiting 1966:23, 73), as arrowhead material (Jones 1931:24), and as a medicine and material for religious ceremonies (Stevenson 1915:44, 88; Whiting 1966:73). The Hopi add the alkaline ashes of this plant to blue cornmeal to maintain the blue color in piki bread (Whiting 1966:73).

Capparaceae-type seed

A single charred seed considered to derive from the caper family was recovered from a pit in the midden. Taxa that have seeds of this type and are likely to be in the area are *Cleome* and *Polanisia* (Harrington 1964:255). The evidence is too meager to suggest use of any of these plants by the Duckfoot occupants.

Cercocarpus/Artemisia-type   (Rosaceae/Compositae)
axillary bud

Charred specimens believed to be the base of foliar buds that would have developed into twigs with leaves were identified as *Cercocarpus/Artemisia*-type axillary buds in a number of Duckfoot flotation samples (Table 6.8). These items probably were attached to branches of *Cercocarpus* (mountain mahogany) and/or *Artemisia* (sagebrush). Examination of *Cercocarpus* herbarium specimens collected at various times during the year reveals that these axillary bud bases can always be found on older stems, precluding the use of this material as a seasonality indicator.

The occurrence of *Cercocarpus/Artemisia* axillary buds in hearths suggests that the branches were used for fuel. Axillary buds in a number of samples from and above the floor of Pit Structures 1 and 3 may derive from branches used to burn the structures intentionally. A similar pattern of recovery of *Artemisia tridentata*-type flowering heads and leaf segments corroborates this suggestion.

Cercocarpus-type wood charcoal   (Rosaceae)

Cercocarpus montanus (mountain mahogany) grows in the region today. The hard, dense wood of various species of *Cercocarpus* has provided historic groups with loom tools, digging sticks, pipe heads, rabbit sticks, dice, and arrow points; wood charcoal has been used to treat burns (Adams 1988). The Navajo called one species "heavy as a stone" because of its compact, fine-grained structure (Matthews 1886:772).

Most historic references to mountain mahogany describe its use in artifact manufacture, yet *Cercocarpus*-type wood was recovered at Duckfoot from numerous hearths, as well as from bins and pits, room fill, roof fall, and the midden (see "Discussion" section). This distribution suggests that the Duckfoot occupants regularly carried mountain mahogany into the site for fuel, roof construction, and possibly artifact needs.

Cheno-am seed   (Chenopodiaceae/Amaranthaceae)

Seeds from *Chenopodium*, in the goosefoot family, and *Amaranthus*, in the pigweed family, can look alike (Isely 1947:339), especially when carbonized and degraded. For that reason, the name "cheno-am" (a combination of the two names) is assigned to ancient seeds that could represent either genus. Charred cheno-am seeds were recovered from a wide variety of structures and features at Duckfoot (Table 6.9), and representatives of both taxa grow in the region today.

Table 6.8. Carbonized *Cercocarpus/Artemisia*-type
Axillary Buds from Flotation Samples

Location[a]	Number Recovered	Number Estimated[b]
Pit Structure 1		
Floor under metate (PL 41)	1	1
Floor scrape	4	13
Vessel (PL 1) contents	5	47
Vessel (PL 3) contents	2	10
Vessel (PL 8) contents	5	9
Vessel (PL 9) contents	8	8
Vessel (PL 14) contents	5	45
Vessel (PL 18) contents	2	2
Vessel (PL 64) contents	2	14
Vessel (PL 78) contents	12	12
Roof fall, vessel (PL 11) contents	12	30
Roof fall, vessel (PL 15) contents	13	122
Roof fall near Burial 2, upper control	25	57
Roof fall, upper control	2	3
Pit Structure 2		
Feature 3 (hearth)	3	3
Pit Structure 3		
Vessel (PL 50) contents	3	3
Vessel (PL 52) contents	1	1
Roof fall, upper control	1	1
Room 2		
Upper fill, control under vessels (PL 1)	27	27
Room 3		
Floor, ash concentration	7	7
Room 4		
Floor near Feature 3 (pit)	1	1
Room 12		
Feature 6 (hearth)	5	5
Wall and roof fall, upper control	12	18
Room 13		
Feature 1 (hearth)	13	96
Feature 2 (slab-lined pit)	23	23
Room 16		
Feature 7 (hearth)	3	4
Room 19		
Feature 13 (bell-shaped cist)	18	19
Feature 19 (hearth)	4	4
Courtyard 3		
Feature 5 (pit)	65	65
Feature 18 (pit)	7	7
Midden		
General midden	1	1

[a] All artifacts and features are associated with floors/surfaces, unless noted otherwise.
[b] When only a portion of the material in a flotation sample was examined, a proportional estimate of the total present was calculated, assuming a continued constant rate of recovery.

Many of the charred cheno-am seeds occurred in contexts suggestive of food processing. Seeds on floors and in hearths, vessels, pits (including one that may have been a roasting pit), bins, cists, and the midden suggest food preparation, storage, and discard. The recovery of cheno-am seeds in roof fall and occasionally in upper-level room debris may relate to general postabandonment mixing of deposits.

Historic records of various southwestern groups harvesting *Chenopodium* and *Amaranthus* plants and seeds are extensive (Adams 1980b, 1988). Leaves were cooked like spinach during summer and fall, as new plants germinated with each rain. Seeds, easily harvested and processed, provided the basis for flour or mush.

Compositae-type achene

A charred specimen believed to be an achene (fruit) of the sunflower family preserved in a vessel on the floor of Pit Structure 1. This limited evidence may represent nothing more than incidental introduction of the item into the site.

***Descurainia*-type seed**   (Cruciferae)

Charred seeds identified as *Descurainia*-type (tansy-mustard) were recovered from three structures at Duckfoot. The specific contexts that yielded seeds were a vessel in the fill of Room 2 (one seed), the fill of the cist in Room 6 (two seeds), and the hearth of Room 13 (one seed). *Descurainia pinnata* plants have been observed growing near the site today, with mature seeds in mid-June (ARIZ 271481).

The contextual data weakly suggest that *Descurainia* seeds may have been a limited food resource at Duckfoot. Possibly the seeds were stored and then parched before being consumed. Many southwestern groups have used the seeds of tansymustard as food (Adams 1988).

***Echinocereus fendleri*-type seed**   (Cactaceae)

A single charred hedgehog cactus seed was recovered in association with a vessel on the floor of Pit Structure 4. This species is a common member of the regional vegetation (Turner 1982:145). The limited evidence only hints that the ripe fruit was carried into the structure. The evidence for direct consumption of fleshy cactus fruit (including seeds) without preparation might be recovered only in the coprolite record, and no coprolites were found at the Duckfoot site.

***Galium*-type fruit**   (Rubiaceae)

A single charred specimen from hearth fill (Feature 7) in Room 16 most closely matches the fruit of *Galium*, in the madder family. This sparse record suggests accidental

Table 6.9. Carbonized Cheno-am Seeds from Flotation Samples

Location[a]	Number Recovered	Number Estimated[b]
Pit Structure 1		
Feature 8 (hearth)	2	68
Floor under metate (PL 41)	2	2
Vessel (PL 1) contents	1	9
Vessel (PL 8) contents	1	5
Vessel (PL 12) contents	1	1
Pit Structure 2		
Feature 3 (hearth)	1	1
Feature 5 (sand pile)	1	1
Pit Structure 3		
Feature 34 (hearth)	6	12
Roof fall, upper control	1	1
Pit Structure 4		
Feature 1 (hearth)	1	1
Floor fill	52	156
Vessel (PL 35) contents	36	36
Vessel (PL 47) contents	40	100
Room 2		
Upper fill, vessel (PL 1) contents	1	1
Room 4		
Feature 1 (burned spot)	8	24
Feature 2 (pit)	2	2
Room 6		
Feature 1 (cist)	17	53
Room 11		
Feature 1 (corner bin)	1	1
Feature 6 (hearth)	2	2
Room 12		
Feature 6 (hearth), Level 1	21	21
Feature 6 (hearth), Level 2	20	20
Wall and roof fall, upper control	1	2
Room 13		
Feature 1 (hearth)	11	50
Room 15		
Feature 1 (bin)	3	3
Feature 5 (cist)	5	5
Room 16		
Feature 3 (hearth)	2	12
Room 19		
Feature 19 (hearth)	3	3
Courtyard 2		
Feature 15 (slab-lined pit)	3	3
Courtyard 3		
Feature 18 (pit)	1	3
Feature 24 (roasting pit)	1	1
Arbitrary Unit 4		
Feature 1 (fire pit)	1	6
Midden		
Feature 2 (pit)	2	10

[a] All artifacts and features are associated with floors/surfaces, unless noted otherwise.
[b] When only a portion of the material in a flotation sample was examined, a proportional estimate of the total present was calculated, assuming a continued constant rate of recovery.

introduction of the fruit, especially because the closely spaced tubercles on the fruit coat would enhance transport on sandals and clothing.

Gramineae caryopsis

Charred items recognized as the caryopses, or "grains," of various members of the grass family were identified in samples from potential food-processing and storage features in four rooms and in the midden (Table 6.10). These grains appear to derive from more than one grass taxon. It seems likely that the Duckfoot inhabitants gathered them for food, at least in limited quantities. Although grinding may have been a preparation step, the specimens associated with metates could owe their presence to trash discarded into rooms. Wild grasses are known to have been a major, often staple, food of southwestern groups (Doebley 1984:53).

Helianthus/Viguiera-type achene   (Compositae)

A single charred *Helianthus/Viguiera*-type fruit from a Room 13 pit feature is weakly suggestive of the use of sunflower achenes as food at Duckfoot. Although *Helianthus* grows in disturbed habitats in the region today and the achenes have provided food for people throughout the centuries (Adams 1988), its limited occurrence could represent nothing more than accidental entry of the item into Room 13.

Leguminosae-type seed

A single charred seed identified as a member of the legume (pea) family was recovered from a shallow pit in Room 15. This item probably represents a wild member of the legume family, but gathering of wild legume seeds prehistorically cannot be inferred from a single specimen.

Lycium-type wood charcoal   (Solanaceae)

The intricately branched, spiny shrub *Lycium pallidum* produces fleshy berries that ripen in the summer in the Duckfoot area (ARIZ 271500). Wolfberry is one of the few members of the potato family in Colorado that is woody. *Lycium*-type charcoal was recognized in a single Courtyard 2 fire pit (see "Discussion" section).

Malvaceae-type seed

Carbonized Malvaceae-type (mallow) seeds were found in three locations at the Duckfoot site. Single specimens recovered from under a vessel in the fill of Room 2, from the hearth in Room 12, and from floor fill above the hearth and ash pit in Pit Structure 1 vary in size and may represent different taxa. The distribution of these seeds does not

Table 6.10. Carbonized Gramineae Caryopses from
Flotation Samples

Location[a]	Number Recovered	Number Estimated[b]
Room 4		
Wall fall, metate (PL 4) trough contents	1	1
Room 6		
Feature 1 (cist)	5	25
Room 12		
Feature 6 (hearth)	1	1
Wall and roof fall, metate trough contents	1	1
Room 15		
Feature 1 (bin)	2	2
Midden		
Feature 2 (pit)	1	15

[a] All artifacts and features are associated with floors/surfaces, unless noted otherwise.
[b] When only a portion of the material in a flotation sample was examined, a proportional estimate of the total present was calculated, assuming a continued constant rate of recovery.

clearly explain their presence at the site. At least one species of *Sphaeralcea* (in the Malvaceae family) grows in the area today, producing ripe fruit in mid-June.

Mentzelia albicaulis-type seed   (Loasaceae)

A total of five charred *Mentzelia albicaulis*-type (stickleaf) seeds came from two locations at Duckfoot. One seed was recovered from the contents of a vessel in the fill of Room 2, and the others were found in the fill of the cist in Room 6. This limited distribution hints that, on occasion, stickleaf seeds were brought to the site, possibly as food. *Mentzelia albicaulis* seeds have been eaten by historic groups and are presumed to represent a food at a number of prehistoric southwestern archaeological sites (Bohrer 1978).

Monocotyledon fibrovascular bundle cordage

Monocotyledons are a major group of angiosperms (flowering plants) that are characterized by parallel veining, scattered vascular bundles within pithy tissue, fibrous root systems, a single seed leaf (cotyledon), and flower parts that usually occur in threes. Refer to the discussion of modified vegetal materials in Chapter 5 for a description of items composed at least partly of Monocotyledon fibers.

Opuntia sp. (prickly pear)-type seed   (Cactaceae)

Small numbers of charred prickly pear-type seeds were present in Pit Structures 1, 3, and 4, and in Rooms 10, 11, and 16 (Table 6.11). Seeds found in association with vessels, a cist, and three hearths may reflect storage of the

fruit and consumption of the seeds after parching. Seeds recovered from roof fall deposits may have been stored in containers suspended from the rafters or may have become mixed with these deposits during or after abandonment.

At least three species of prickly pear are listed as present in the area today (Turner 1982:145). *Opuntia* offers a variety of parts that humans have sought for food, including flower buds, joints (stems), and fruit (Adams 1980b). The ability of ripe fruit to cling to stems for months after maturing means that harvest is not restricted to a particular season of the year. People may find this an appealing source of food during winter months when access to other plant products is limited.

Peraphyllum-type seed   (Rosaceae)

Peraphyllum/Amelanchier-type wood charcoal

Four charred specimens from a cist in Room 6 and one from the floor of Pit Structure 4 closely match the seed morphology of *Peraphyllum ramosissimum* (squaw apple), a shrubby member of the rose family that grows near the Duckfoot site today (ARIZ 271463, 271471). *Peraphyllum* seeds are produced in small, fleshy, apple-like (pome) fruits that ripen at or after the end of June.

There are no known references to *Peraphyllum* seed use in the ethnographic literature. *Amelanchier* (service berry) fruits, very similar in appearance and timing of maturity,

Table 6.11. Carbonized *Opuntia* (Prickly Pear)-type Seeds
from Flotation Samples

Location[a]	Number Recovered	Number Estimated[b]
Pit Structure 1		
Vessel (PL 8) contents	1	1
Vessel (PL 14) contents	3	19
Vessel (PL 64) contents	3	6
Roof fall near Burial 2, upper control	4	8
Roof fall, upper control	1	1
Pit Structure 3		
Closing/fuel on floor	1	1
Pit Structure 4		
Floor fill	2	2
Vessel (PL 47) contents	1	1
Room 10		
Feature 5 (cist)	1	1
Room 11		
Feature 6 (hearth)	8	8
Room 16		
Feature 3 (hearth)	1	1
Feature 7 (hearth)	1	1

[a] All artifacts and features are associated with floors/surfaces.
[b] When only a portion of the material in a flotation sample was examined, a proportional estimate of the total present was calculated, assuming a continued constant rate of recovery.

were once eaten at Isleta Pueblo (Castetter 1935:16) and were consumed fresh by the Navajo (Franciscan Fathers 1910) and Hopi (Whiting 1966:79). The Duckfoot distribution of squaw apple remains hints that *Peraphyllum* was an occasional prehistoric food source.

Duckfoot occupants infrequently carried *Peraphyllum* and/or *Amelanchier* wood into the site (see "Discussion" section). Wood of these taxa was not commonly sought by historic groups, although the Hopi made bows, arrows, prayer sticks, and other ceremonial equipment from *Amelanchier* (Whiting 1966:41, 44, 79).

Phragmites australis-type   (Gramineae)
stem fragment

A single charred *Phragmites australis*-type (reedgrass) stem fragment was preserved in the floor fill above the hearth and ash pit of Pit Structure 1. Although this generally robust grass does not seem to be common in the area today, one plant was seen 3 km south of the site, and a thick stand is located 16 km to the northeast.

An interpretation of the single Duckfoot specimen is difficult. Perhaps it was used in one of the smaller layers of vegetation in the roof. Alternatively, the hollow stem makes this grass a good candidate for cigarette material. Modern groups have used *Phragmites australis* stems for sleeping mats (Russell 1975: 134); for roofing, pipestems, and weaving rods (Whiting 1966:66); and for arrows, game sticks, and many other items (Robbins et al. 1916:66). Apparently many southwestern groups collected a sweet exudate from the stems as a food (summarized in Yanovsky 1936:8; Bye 1972:91).

Physalis longifolia-type seed   (Solanaceae)

Physalis longifolia-type (groundcherry) seeds were found in a variety of locations and feature types (Table 6.12). The fruit of the groundcherry is fleshy, similar to that of a small tomato, and is usually available by mid- to late summer. *Physalis longifolia* (ARIZ 271477) grows very near the Duckfoot site today.

The association of groundcherry seeds with a vessel and two bin/pit features suggests storage (possibly after drying) and cooking of the fruits. The recovery of a burned seed from a Courtyard 3 feature that may have been a roasting pit could indicate that the small, fleshy tomatoes were processed in courtyard areas. Presence of groundcherry seeds in the fill of a metate trough could be the result of general mixing of debris.

Recovery of charred groundcherry seeds from five hearths is not clearly explained by the modern ethnographic record of *Physalis* use. In historic times, the fruit has been gathered and eaten raw, dried for winter consumption, or boiled with other ingredients as a condiment (Adams 1988). These references do not really explain how *Physalis*

Table 6.12. Carbonized *Physalis longifolia*-type Seeds from Flotation Samples

Location[a]	Number Recovered	Number Estimated[b]
Pit Structure 1		
Floor fill above Features 8 and 9 (hearth and ash pit)	1	12
Vessel (PL 18) contents	2	2
Pit Structure 3		
Feature 34 (hearth)	1	2
Pit Structure 4		
Feature 1 (hearth)	3	3
Vessel (PL 35) contents	1	1
Room 4		
Wall fall, metate (PL 4) trough contents	1	1
Room 11		
Feature 1 (corner bin)	2	2
Feature 6 (hearth)	2	2
Floor	1	1
Room 12		
Feature 6 (hearth), Level 1	9	9
Feature 6 (hearth), Level 2	4	4
Room 13		
Feature 1 (hearth)	29	171
Room 15		
Feature 1 (bin)	1	1
Feature 4 (ash-filled pit)	3	10
Feature 5 (cist)	3	3
Room 16		
Feature 3 (hearth)	11	35
Courtyard 2		
Feature 15 (slab-lined pit)	2	2
Courtyard 3		
Feature 24 (roasting pit)	1	1

[a] All artifacts and features are associated with floors/surfaces, unless noted otherwise.

[b] When only a portion of the material in a flotation sample was examined, a proportional estimate of the total present was calculated, assuming a continued constant rate of recovery.

seeds might routinely become charred. Regardless, charred *Physalis* seeds have been recovered from prehistoric contexts, including in human coprolites, in New Mexico, Arizona, and Colorado (Adams 1988).

Populus angustifolia-type wood charcoal   (Salicaceae)

Populus/Salix-type wood charcoal

Various cottonwood and willow trees grow in the general region of the Duckfoot site today and are usually confined to mesic areas along streams in canyon bottoms, but today one would have to walk at least 15 to 20 minutes to reach some of these trees in Alkali Canyon northwest of the site. *Populus/Salix* and *Populus angustifolia* wood charcoal was recovered from a few contexts at Duckfoot (see "Discussion" section).

The extensive use of *Populus* and *Salix* in historic times has been summarized elsewhere (Adams 1988). Parts that have been collected by historic groups include the wood, wood knots, branches/young shoots, leaves and leaf buds, bark, inner bark, catkins, fruit, and roots. The wood has been used extensively in ceremonies, for construction, as firewood, and in the manufacture of household tools. The branches have provided prime material for baskets, and the roots have been carved into rafts and dolls.

Portulaca retusa–type seed   (Portulacaceae)

Portulaca (purslane) seeds preserved in a limited number of contexts at the Duckfoot site. Small, round seeds with tough seed coats accidentally spilled into a hearth during parching have a good chance of surviving in the archaeological record. Single seeds recovered from hearths in Rooms 13 and 16 hint at food preparation (parching), and possibly the seeds were stored in one of the Pit Structure 1 vessels.

Prehistoric use of purslane seeds was extensive, and historically *Portulaca* seeds have been gathered and ground into flour by numerous southwestern groups (Adams 1988). In addition to using the seeds, humans have used the succulent plants as potherbs and medicines.

Purshia–type wood charcoal   (Rosaceae)

Purshia tridentata (bitterbrush) and *Purshia mexicana* (cliffrose) are two shrubby members of the rose family in the Duckfoot area today. A small amount of wood charcoal thought to represent either or both of these taxa was recovered from the Duckfoot site (see "Discussion" section). Historic references to *Purshia* use are difficult to locate in the literature. *Purshia* (as *Cowania*) has been used by modern groups, not for fuel or construction material, but for arrow shafts (Swank 1932:38), medicines, and cigarettes (Reagan 1929:156, 159; Whiting 1966:78). The bark was once an important source of fiber for clothing, belts, sandals, mats, and ropes (Bye 1972:95; Whiting 1966:78).

Rhus aromatica–type seed,   (Anacardiaceae)
wood charcoal

Four charred seeds from the cist in Room 6 and one from roof fall debris on the floor of Pit Structure 4 display characteristics typical of *Rhus aromatica* (squawbush, lemonade berry) in the cashew family. This intricately branched shrub, which bears ripe sour fruit in June, grows in the region today (ARIZ 271473). The limited evidence hints that the fruits were carried into the site. Historic use of a variety of *Rhus* parts (bark, leaves, twigs, roots, buds,

pollen, and wood) has been documented, but it is the use of the edible fruit that has been most widely recorded (Adams 1988). A few pieces of *Rhus*–type wood charcoal were recovered from Duckfoot deposits. The reader is referred to the discussion of modified vegetal materials in Chapter 5 for a description of modified items incorporating *Rhus* parts in their construction.

Scirpus–type achene   (Cyperaceae)

A single charred *Scirpus*–type (sedge) achene was identified in the roof fall debris on the floor in Pit Structure 4. *Scirpus acutus* grows not far from the Duckfoot site, in a wet canyon bottom (ARIZ 271511).

Stipa hymenoides–type floret,   (Gramineae)
caryopsis

A variety of charred *Stipa hymenoides*–type (Indian rice grass) reproductive parts was identified in flotation samples from seven rooms, a courtyard feature, and Pit Structure 4 (Table 6.13). Complete florets (the mature grain and chaffy enclosing parts) or their component parts were recovered. This cool-season grass ripens in the Duckfoot area in mid-June (ARIZ 271483).

Table 6.13.   Carbonized *Stipa*-type Remains from Flotation Samples

Location[a]	Number Recovered	Number Estimated[b]
Pit Structure 4		
Floor fill	1*f*	1*f*
Vessel (PL 47) contents	4*f*	10*f*
Room 4		
Feature 1 (burned spot)	7*c*, 1*f*	13*c*, 3*f*
Room 6		
Feature 1 (cist)	120*c*, 80*cf*, 31*f*, 12*l*, 18*p*	206*c*, 138*cf*, 47*f*, 29*l*, 51*p*
Room 11		
Feature 10 (warming pit)	1*c*	1*c*
Room 12		
Feature 6 (hearth), Level 1	1*f*	1*f*
Room 13		
Feature 1 (hearth)	6*f*	6*f*
Room 16		
Feature 3 (hearth)	2*f*	9*f*
Room 19		
Feature 13 (bell-shaped cist)	1*c*	2*c*
Courtyard 2		
Feature 15 (slab-lined pit)	2*c*, 1*p*	2*c*, 1*p*

c = caryopsis; *cf* = caryopsis fragment; *f* = floret; *l* = lemma fragment; *p* = palea fragment.

[a] All artifacts and features are associated with floors/surfaces.

[b] When only a portion of the material in a flotation sample was examined, a proportional estimate of the total present was calculated, assuming a continued constant rate of recovery.

Most of the Indian rice grass remains at Duckfoot are associated with potential food-processing features. The variety of Indian rice grass parts in the Room 6 cist may indicate that the florets were processed nearby and the processing debris was later incorporated into the cist. The grains of this grass have been used extensively by historic groups and are found frequently in ancient southwestern sites (Bohrer 1975).

Yucca baccata-type seed   (Liliaceae)

Charred broad-leaf *Yucca baccata*-type seeds were recovered from two locations at Duckfoot. Single seeds were found in one of the Room 16 hearths and in the Room 6 cist. The limited recovery of these prehistoric remains only weakly suggests that yucca fruits were gathered by the Duckfoot occupants, possibly to be prepared for consumption by cooking. *Yucca baccata* plants were recently observed to produce fleshy immature fruit in mid-June in the region (ARIZ 271501).

In a survey of food plants of North American Indian groups, Yanovsky (1936:15) wrote that southwestern people ate the fleshy, banana-like fruit of *Yucca baccata* fresh or cooked, roasted the unripe fruit before eating it, ground and pressed the fruit into cakes and dried it in the sun for winter use, and roasted the flower buds as food. In New Mexico, a fermented beverage was made from the fruit.

Type 1 Unknown

The Type 1 unknown was recovered from three pit structures and Room 12. Remains were collected from ash pit fill (Pit Structures 1 and 3), hearth contents (Pit Structure 3 and Room 12), structure and floor fill (Room 12 and Pit Structure 1), and a vessel on a structure floor (Pit Structure 4). Possibly these remains represent some type of overwintering bud, a gall, or a plant parasite brought in on firewood.

Type 2 Unknown

A single charred Type 2 unknown was recovered from the fire pit in Room 5. It is not certain that the Type 2 unknown is plant material. No interpretation of this specimen is offered.

Distribution of Plant Remains

In this section, the data are examined to gain insights into plant use in individual structures and outdoor activity areas. Remains from a variety of contexts are discussed; however, those from features retaining primary deposits provide an especially valuable source of information on activities in specific locations. A complete inventory of analyzed plant remains is on file at the Anasazi Heritage Center.

Pit Structure 1

Pit Structure 1 burned with its contents still inside, which resulted in the preservation of a rich assemblage of household tools, containers, and utensils. Certain features of the carbonized plant remains reveal that the Pit Structure 1 fire was extremely hot. Many vegetal items had vitrified, becoming shiny and glasslike in appearance. Sides of rounded stems were swollen and had bubbled out in blisters. Some specimens believed to be resin globs had become so hot that they turned liquid, then resolidified with drip tails pointing in one direction.

The plant remains recovered from Pit Structure 1 provide insight into material culture needs, food products, and possibly the destruction of the structure. On the basis of the recovery of wood charcoal, *Pinus* bark scales, and a *Juniperus* scale leaf, it seems that *Pinus, Juniperus, Artemisia,* and *Cercocarpus* were used to fuel the hearth. *Zea mays* cobs also fed the hearth fire, leaving cupule remains.

Charred roof beam fragments reveal construction timber choice. The roof was composed of larger wooden beams (vigas) overlain by smaller branches and topped with 15 to 20 cm of mud. Choice of timber included *Juniperus, Pinus ponderosa,* and *Pinus edulis,* a pattern confirmed by the charcoal from the burned roof stratum. That sagebrush (cf. *Artemisia*) may have formed part of the roof is suggested by the recovery of *Artemisia*-type charcoal from roof fall debris. The presence of *Cercocarpus/Artemisia*-type axillary buds and charcoal suggests that the occupants of Pit Structure 1 may have used *Cercocarpus* as one of the layers in roof construction.

Some of the plant remains in Pit Structure 1 could relate to either roof construction or the destruction of the structure. For example, *Pinus, Juniperus,* and *Cercocarpus* wood charcoal on the floor could derive from fuel placed in the structure to intentionally burn it, or these materials may have fallen there as the roof collapsed while burning. The same explanations could apply to *Artemisia tridentata* parts, *Pinus* bark scales, and *Juniperus* parts in floor and vessel samples and to pine, juniper, and sagebrush charcoal present in floor fill above the hearth and ash pit. A single *Phragmites* (reedgrass) stem segment from ashy floor fill above the hearth and ash pit could represent material employed in roof construction, although this robust grass could have been used for other purposes as well.

Some charred seeds of wild plants from Pit Structure 1 were found in several other locations at the site. Cheno-am seeds in three vessels (PLs 1, 8, and 12) and the hearth, as well as on the floor beneath a metate (PL 41), probably represent a wild food product stored, parched, and ground in preparation for consumption. *Opuntia* (prickly pear) seeds recovered from three different vessels (PLs 8, 14, and 64) may be the remains of stored cactus fruit. *Opuntia* seeds in roof fall could indicate that this plant was suspended from the ceiling rafters; alternatively, the seeds might have been incidentally mixed with the roofing debris. *Physalis longifolia*-type seeds in a vessel (PL 18) and in floor fill above the hearth and ash pit probably were food items.

Other seeds in Pit Structure 1, generally rare at the site, hint at occasional use of certain wild resources. *Juniperus* seeds from a vessel (PL 64) and in the hearth fill may have been gathered as food or inadvertently transported on branches used for fuel. Although it is possible that the *Portulaca* seed and Compositae achene in separate vessels (PLs 11 and 78, respectively) and the Malvaceae-type seed in floor fill above the hearth and ash pit represent ancient food, the limited evidence is only suggestive of such use.

Cultivated plants in Pit Structure 1 include beans (*Phaseolus vulgaris*-type) and corn (*Zea mays*). The single bean cotyledon is the only evidence of domesticated *Phaseolus* recovered from the Duckfoot site. Pit Structure 1 definitely is a locus for *Zea* use, because remains were recovered from the hearth, ash pit, and floor fill associated with these two features. Although the cupules probably derive from use of the cobs as fuel, the presence of kernels could reflect accidental charring of a food resource. The presence of *Zea mays* parts in association with only one olla (PL 64) suggests that little corn was being stored in pottery vessels when the structure burned. The presence of *Zea* parts in roof fall may derive from dry cobs suspended from the beams or could reflect postabandonment mixing of debris.

Pit Structure 2

Compared with the plant assemblages in the burned pit structures, the quantity and variety of charred plant remains in Pit Structure 2 are limited. Occupants gathered a variety of wood for the hearth (Feature 3), including *Juniperus, Pinus, Artemisia, Atriplex, Cercocarpus,* and possibly even *Purshia*—an interpretation supported by the presence of *Pinus*-type bark scales and *Cercocarpus/Artemisia*-type axillary buds. They also used *Zea mays* cobs as fuel.

A cheno-am seed and some *Zea mays* kernel fragments in the hearth are the only items that potentially represent ancient foods. A cheno-am seed in a pile of clean sand on the floor (Feature 5) possibly was a ritual offering, although it could also have been incorporated into deposits during abandonment. The latter explanation could also account for the presence of *Zea* cupules and a *Juniperus* twig on the floor and a *Juniperus* twig in the sipapu (Feature 19).

Pit Structure 3

Wood charcoal in the hearth (Feature 34) and ash pit (Feature 36) reveals that the occupants used *Juniperus, Pinus,* and *Atriplex* to fuel their hearth fire. The recovery of *Juniperus* twig, *Pinus*-type bark scale, and *Atriplex* fruit core remains supports this interpretation. *Zea mays* cupules and cob fragments suggest that corncobs served as a fuel

source as well. Possibly *Artemisia* branches were burned, leaving charred leaves as evidence.

Material suspected of representing fuel brought in to intentionally burn the structure includes *Juniperus, Pinus, Artemisia,* and *Atriplex* charcoal. Burned juniper wood was found in all three of the roof-closing/fuel-on-floor samples examined. Supporting evidence for the taxa used is provided by the presence of *Artemisia tridentata*-type leaves and flowering heads; *Atriplex canescens*-type fruit cores; *Juniperus* twigs, one fruit, and a seed; *Pinus* needle and bark scale remains; and a *Cercocarpus/Artemisia*-type axillary bud in the upper control and closing-material/fuel-on-floor samples. Since both juniper and pinyon pine wood were identified frequently in the tree-ring samples collected from Pit Structure 3, the burning roof could well have contributed some of the juniper and pine-type charcoal to the debris on and above the floor.

Items recovered from the floor beneath a metate (PL 19), in association with two floor-related vessels (PLs 50 and 52), and from the area around the sipapu (Feature 33) include many of the taxa and parts already discussed. These are *Artemisia tridentata*-type flowering head and leaf fragments, *Atriplex canescens*-type fruit cores, *Juniperus* twigs, *Pinus edulis* needle fragments, *Cercocarpus/Artemisia*-type axillary buds, and some *Zea mays* remains. Most of this material probably relates to the destruction of the structure.

Foodstuffs in hearth and ash pit fill include *Zea mays* kernels, cheno-am seeds, *Juniperus* seeds (although these may have come in on branches used for fuel), and one *Physalis longifolia*-type seed. Food remains that probably were deposited after abandonment include a *Zea mays* kernel and an *Opuntia* seed in the roof-closing-material/fuel-on-floor sample, as well as a cheno-am seed and several *Zea* kernel fragments in an upper-level control sample.

Pit Structure 4

Juniperus, Pinus, Cercocarpus, and *Atriplex* wood charcoal were recovered from the hearth (Feature 1) in Pit Structure 4. A *Juniperus* twig and some *Pinus*-type bark scales support the record of fuel choice. The *Juniperus* fruit and seed remains in the hearth could represent branches used as fuel or food that accidentally fell into the fire.

Artemisia, Juniperus, and *Pinus*-type charcoal in a roof fall sample above the floor, along with *Artemisia tridentata*-type leaf fragments, a *Pinus*-type bark scale, and *Juniperus* twigs on or above the floor, could derive from a combination of roof-construction material and fuel brought in to intentionally burn the pit structure. Both juniper and pinyon were identified in tree-ring samples from the collapsed roof.

Food remains clearly associated with Pit Structure 4 activities are limited to cheno-am and *Physalis longifolia*-type seeds in the hearth. Items on the floor or associated with vessels on the floor may relate either to food use or to debris deposited during abandonment and burning. These items include cheno-am seeds, an *Echinocereus fendleri*-type seed, a *Physalis longifolia*-type seed, *Juniperus* seeds, an *Opuntia* seed, *Stipa hymenoides*-type florets, a *Peraphyllum*-type seed, and some *Zea mays* kernel, cob, and cupule fragments.

Reproductive parts recovered from floor samples containing roof fall debris include cheno-am and *Opuntia* seeds, both considered to represent food remains elsewhere at the site. *Rhus aromatica* and *Scirpus*-type achenes, although edible, are too rare to more than hint at food use at Duckfoot. All these items may have been on the roof when it collapsed.

Room 2

Flotation samples taken from the contents of two partial vessels (assigned a single point-location number, PL 1) and from the sediment beneath them yielded a cheno-am seed, probably a food item at Duckfoot, and *Descurainia* and *Mentzelia albicaulis*-type seeds, too rare at Duckfoot to indicate common harvest as food. In addition, the vessel samples contained *Juniperus* twigs and a *Pinus*-type bark scale, which may represent fuel and/or construction materials.

With the possible exception of a Malvaceae seed, most of the taxa recovered from beneath the PL 1 vessels do not represent potential foods. A *Zea mays* cupule, an *Artemisia tridentata*-type leaf fragment, *Cercocarpus/Artemisia*-type axillary buds, and *Juniperus osteosperma*-type leaves and twigs are believed to be the remains of hearth fuel and roof-construction materials.

Room 3

An ash concentration on the floor of Room 3 was sampled for plant materials. The only remains present were some *Cercocarpus/Artemisia*-type axillary bud segments, possibly representing one of the smaller layers of roof construction, and a *Zea mays* kernel fragment. Because, in sharp contrast to the floor, the roof stratum was relatively full of artifacts, this kernel fragment could have been on the roof surface.

Room 4

Originally Room 4 was used as a storeroom. Later, a fire or fires were built directly on the surface of the floor (Feature 1), perhaps to cook small animals (the floor of the room was littered with over 300 burned and unburned small-mammal bones and bone fragments). A shallow pit in the room (Feature 2), although filled with charcoal, revealed no evidence of in situ burning.

Juniperus wood charcoal, a *Pinus*-type bark scale, and *Zea mays* cob remains suggestive of fuel use were pre-

served within the shallow pit. *Juniperus* and *Cercocarpus* charcoal and a *Pinus*-type bark scale were recovered from the burned area on the floor. Elsewhere on the floor, the charcoal of these three taxa preserved, as did a *Cercocarpus/Artemisia*-type axillary bud. This same pattern of charcoal preservation is repeated in the fill on the floor and in the trough of a metate located in the fill above the floor. *Rhus*-type charcoal in the shallow pit may derive from some specialized activity in this room, such as basket making.

Two charred cheno-am seeds in the shallow pit might be the remains of food processed elsewhere. Charred cheno-am seeds and some *Stipa hymenoides*-type remains from the burned area on the floor perhaps represent foods that were prepared in the fire(s) on the floor surface. A Gramineae caryopsis and a *Physalis longifolia*-type seed in the metate trough may relate to rooftop activities or to general mixing of debris after abandonment of the room.

Room 5

Like Room 4, Room 5 served first as a storage room, then as a locus for domestic activities. The floor contained a variety of artifacts and a number of features. The fire pit (Feature 1) yielded juniper wood (fuel) and a *Zea mays* kernel (possible food). The corner bin (Feature 2), filled with collapsed structural debris, retained *Artemisia*, *Atriplex*, *Juniperus*, and *Pinus* charcoal. Since this room did not burn at abandonment, the material in the bin probably derives from mixed postabandonment deposits. Of two shallow pits (Features 3 and 4) located in the room, only one yielded charred plant remains: a single *Zea mays* cupule, perhaps representing use of cobs as fuel, was recovered from Feature 3 fill.

Room 6

The diversity of plant taxa recovered from Feature 1, a cist in the southeast corner of Room 6, is impressive. The assemblage includes cheno-am seeds and *Stipa hymenoides*-type parts, both common food items. Other plant parts possibly used for food include Gramineae grains and *Descurainia*, *Mentzelia albicaulis*, *Peraphyllum*, *Rhus aromatica*, and *Yucca baccata* seeds. The *Juniperus osteosperma*-type leaves and *Zea mays* cupules probably derive from fuel use.

Room 10

Charcoal recovered from the floor includes *Cercocarpus*, *Juniperus*, and *Pinus* wood, in addition to several *Pinus*-type bark scales. Whether this material represents debris deposited in the structure after abandonment or is part of the burned roof is unknown. The tree-ring record documents the use of both juniper and pinyon as roof beams.

Cercocarpus branches might have been used in one of the smaller layers of the roof.

Two bins and one cist contained plant remains. The cist in the northwest quadrant of the room (Feature 5), thought to have been a storage cist open at the time of abandonment, contained burned roof fall debris, including *Juniperus*, *Artemisia*, and *Cercocarpus* wood charcoal. The cist also yielded a single *Opuntia* (prickly pear)–type seed, possibly relating to the original function of the cist as a storage feature. The two bins (Features 1 and 3) in the northeast and southeast quadrants of the room contained postabandonment or mixed deposits. Feature 1 yielded *Cercocarpus* and *Juniperus* charcoal; Feature 3, *Artemisia* and *Pinus*. These items, in addition to a *Juniperus* twig, a *Pinus*-type bark scale, and two *Zea mays* cupules, probably represent fuel use.

Room 11

Several different fuels were used in Room 11. The charcoal preserved in the hearth (Feature 6) includes *Juniperus*, *Pinus*, *Artemisia*, *Atriplex*, *Cercocarpus*, and the only example of *Peraphyllum/Amelanchier*-type wood recovered from the site. This record is supplemented by the recovery of *Pinus*-type bark and cone scales, as well as *Zea mays* cob and cupule fragments. The warming pit (Feature 10) retained specimens of *Juniperus*, *Atriplex*, and *Artemisia*.

The tree-ring record suggests that the roof of Room 11 was constructed of both juniper and pinyon wood. This inference is supported by the fact that *Juniperus* and *Pinus* charcoal was found in one of the bins (Feature 1) filled with burned roof fall. *Cercocarpus* and *Atriplex* charcoal was recovered from this same location, and *Cercocarpus* charcoal was found on the room floor. These shrub remains may represent some of the smaller elements employed in roof construction.

The storage cist (Feature 3) in Room 11 contained thin sandstone slabs and stone tools. It also yielded some *Zea mays* cupules; a *Juniperus* twig; and *Juniperus*, *Artemisia*, and *Atriplex* charcoal. These plant remains generally represent fuel and roofing resources at Duckfoot and apparently became mixed with the stone artifacts actually stored in the cist.

Plant remains that suggest food use in Room 11 are primarily from the hearth. These include cheno-am seeds, *Opuntia* (prickly pear)–type seeds, and *Physalis longifolia*-type seeds. *Stipa hymenoides* remains were recovered from the warming pit associated with the hearth, a *Physalis* seed was found on the floor, and a cheno-am seed was recovered from the bin. Collectively, the evidence suggests that food was both stored and prepared in Room 11.

Room 12

The ash-filled hearth (Feature 6) contained evidence of both food and fuel resource use. Both levels of the hearth had cheno-am and *Physalis* seeds, grass remains (including *Stipa hymenoides* florets), and *Zea mays* kernel fragments. The single Malvaceae-type seed, a rare item at Duckfoot, could represent an infrequent food choice. The *Juniperus* seeds in the hearth were either gathered for food or brought in on juniper branches used for fuel.

Fuel evidence recovered from the hearth includes *Juniperus*, *Pinus*, *Artemisia*, *Atriplex*, and *Cercocarpus* wood charcoal, as well as *Juniperus* scale leaves, *Pinus*-type bark scales, and *Cercocarpus/Artemisia*-type axillary buds. Presence of *Zea mays* cupules suggests that corncobs were also burned.

Other plant remains from Room 12 are not as easily interpreted. A Gramineae caryopsis and a *Zea mays* kernel fragment may be only incidentally associated with the two metates in which they were found and are not necessarily indicative of food preparation. Incidental association is likely, because nonfood items (*Artemisia tridentata*-type flowering heads and a leaf, *Pinus*-type bark scales, and *Zea mays* cupules) are also present in the metate sample, along with charcoal of *Populus/Salix* and the same five wood types found in the hearth. The roof of Room 12 was constructed in part of *Juniperus* and *Pinus edulis;* when the roof burned and collapsed, charcoal probably was deposited in the metates. *Juniperus*, *Pinus*, *Artemisia*, and *Cercocarpus* charcoal, in addition to *Cercocarpus/Artemisia* axillary buds, a *Juniperus* twig, *Pinus*-type bark scales, *Zea mays* cupules, and a cheno-am seed from an upper-control flotation sample, may derive from an isolated pocket of charred debris and should not necessarily be viewed as representative of general room fill.

Room 13

Plant remains from the hearth (Feature 1) in Room 13 reveal that foodstuffs common in other Duckfoot contexts, such as cheno-am and *Physalis* seeds, *Zea mays* kernels, and *Stipa hymenoides*-type florets, were prepared in this room as well. The hearth also contained *Descurainia* and *Portulaca retusa*-type seeds, potential food resources occasionally recovered elsewhere at Duckfoot.

Wood charcoal types in the two strata of the hearth were identical and include *Cercocarpus*, *Juniperus*, and *Pinus*. *Pinus*-type bark scales and a single needle fragment, a *Juniperus* twig, and *Cercocarpus/Artemisia*-type axillary bud remains provide additional evidence of fuel use. Presence of *Zea mays* cupules signals use of the cobs as fuel.

The plant remains in the slab-lined pit (Feature 2) are similar to those nonfood items recovered from the hearth, with the exception of a rare *Helianthus/Viguiera*-type achene. Other remains include *Cercocarpus/Artemisia*-type axillary buds, *Zea mays* cupules, and charcoal of *Cercocarpus*, *Juniperus*, and *Pinus*, all apparently representative of general site debris.

Room 15

Room 15 originally may have served as a habitation structure but later was converted to a storage area. A shallow pit filled with ash (Feature 4), but with no evidence of in situ burning, is the only indication that a hearth may have existed at one time; if so, construction of a cist (Feature 5) destroyed it. A slab-lined bin (Feature 1) was built over both the pit and the cist sometime after the site was abandoned. Feature 4 fill consisted of primary debris related to its function as a receptacle for ash; Feature 5 fill was made up of postabandonment deposits mixed with cultural refuse; and Feature 1 was filled with postabandonment sediments.

Plant remains recovered from Features 1, 4, and 5 are similar. For example, all three retained some *Pinus*-type bark scale and *Zea mays* cob/cupule evidence, which probably indicates fuel choice. In addition, the cist and the ash-filled pit had four wood charcoal types in common—*Atriplex, Artemisia, Juniperus,* and *Pinus*—suggesting that at least some of the contents of these two features represent similar activities. *Cercocarpus* and *Juniperus* charcoal fragments in the postabandonment fill of the slab-lined bin shed no light on feature function.

Likely food items discarded into the postabandonment fill of Feature 1 include cheno-am seeds, Gramineae caryopses, a *Physalis longifolia*-type seed, and *Zea mays* kernel fragments. The ash-filled pit also had *Physalis longifolia*-type seeds and *Zea mays* kernels, and it contained a Leguminosae seed. The cist contained cheno-am seeds, *Physalis longifolia*-type seeds, and a *Zea mays* kernel. Potential food remains in postabandonment fill may indicate that similar activities were conducted in nearby rooms after Room 15 ceased to be used for its primary function and refuse from other areas of the site began to accumulate in it.

Room 16

Room 16 apparently was serving as a domestic structure when it was intentionally filled with brush and burned, a scenario suggested by the presence of a layer of charred twigs on the floor. Two basin-shaped hearths were excavated, only one of which (Feature 3) was in use at the time of abandonment. The other (Feature 7) had charcoal and ash in its lower fill, but the top of the pit had been deliberately capped with clean sediment. According to the tree-ring record, the roof of this room was constructed partly of juniper wood.

The hearth in use when Room 16 burned (Feature 3) contained *Artemisia, Atriplex, Juniperus,* and *Pinus* charcoal in addition to *Pinus*-type bark scales and *Zea mays* cupules. All are evidence of fuel choice. Juniper and pine charcoal, *Pinus*-type bark scales, and *Zea mays* cupules were also recovered from both the upper (clean fill) and

lower (use-related) strata of the capped hearth (Feature 7). The *Populus angustifolia*-type charcoal recovered from the lower stratum of Feature 7 represents one of only four occurrences of this taxon at Duckfoot and may relate to special needs for a wood type normally found in mesic (moderately moist) habitats.

Plant remains representing food resources were identified from both hearths. Feature 3 had cheno-am, *Opuntia, Physalis longifolia,* and *Yucca baccata*-type seeds; *Stipa hymenoides*-type florets; and *Zea mays* kernels. The presence of *Juniperus* seeds could be attributed to use as food or to the gathering of juniper branches for fuel. The upper layer of Feature 7 contained an *Opuntia* seed and several *Zea mays* kernel fragments; the lower stratum, associated with the use of the hearth, yielded *Portulaca retusa*-type seeds and a *Zea mays* kernel fragment. There is no satisfactory explanation for the presence of *Galium*-type fruit in the lower layer other than chance occurrence.

Room 17

Room 17 appears to have served as a storage facility. A large pit below the floor of Room 17 (Feature 18 of Courtyard 2) was actually associated with courtyard activities that took place before the construction of Room 17. Two metates in the fill of the room may have fallen inside when the roof on which food-preparation equipment was kept deteriorated and collapsed.

Since both metates are in fill, plant remains recovered from their troughs may not relate specifically to metate use but, rather, to general site activities. *Artemisia, Juniperus,* and *Pinus* wood charcoal, a *Juniperus* scale leaf, and *Zea mays* cupule fragments are believed to derive from postabandonment debris.

Room 19

Two features on the floor of Room 19 (Feature 19, a hearth, and Feature 13, a large bell-shaped cist) contained wood charcoal and ash. The bell-shaped cist contained structural debris, and any burned material inside was probably deposited when the roof burned and fell into the room.

Evidence of fuel choice from the hearth includes *Juniperus, Pinus, Cercocarpus,* and the rarely occurring *Populus angustifolia*-type charcoal, supplemented by *Cercocarpus/Artemisia*-type axillary buds, *Juniperus* twigs, and *Zea mays* cupules. The bell-shaped cist contained an identical assemblage of resources. *Cercocarpus, Juniperus,* and *Pinus* charcoal excavated from beneath the metate may derive from roof fall. According to tree-ring records, the charred roof debris contained both pinyon and juniper wood.

Food remains in the room are sparse. Cheno-am seeds are the only items in the hearth suggesting that foods were prepared in Room 19. One *Stipa hymenoides*–type caryopsis and one *Zea mays* kernel fragment in the bell-shaped cist could be stored food items or materials that fell into the room with the burning roof.

Extramural Features

Six pit features excavated in three courtyard areas contained both food and fuel remains: Courtyard 1, Feature 8; Courtyard 2, Features 8 and 15; and Courtyard 3, Features 5, 18, and 24. Only one, Feature 8 in Courtyard 2, contained material related to primary use; the remaining features contained mixed deposits. The food remains recovered, including cheno-am and *Physalis longifolia*–type seeds, *Stipa hymenoides* parts, and *Zea mays* kernels, are seen in many Duckfoot deposits. *Juniperus* was the most common type of wood charcoal identified in the courtyard feature samples. *Artemisia, Cercocarpus,* and *Pinus* charcoal occurred occasionally, and *Atriplex*-type charcoal was recovered from a single pit (Courtyard 2, Feature 15). All are common fuelwoods at the site. The sole occurrence of *Lycium*-type charcoal, in a Courtyard 2 fire pit (Feature 8), hints that wood of this shrub was brought in; perhaps its use in an open-air feature is explained by the undesirable odor and acrid smoke emitted when it burns. *Juniperus* twigs, *Cercocarpus/Artemisia*-type axillary buds, and a *Pinus*-type needle fragment provide additional evidence of the materials brought in for fuel. The presence of *Zea mays* cob and cupule fragments in four courtyard pits suggests use of cobs as fuel in these outdoor features.

An outdoor fire pit in Arbitrary Unit 4, east of Pit Structure 3, contained evidence of a variety of fuels, including *Zea mays* cupules, an *Atriplex canescens*–type fruit core, and *Artemisia, Juniperus, Pinus,* and *Atriplex*-type charcoal. Only a cheno-am seed and a *Zea mays* kernel fragment suggest the foods that may have been prepared there.

Midden

Charcoal in the midden should be of wood routinely burned in hearths and then discarded. *Artemisia, Atriplex, Juniperus,* and *Pinus* charcoal were recovered in two or more midden deposits; *Cercocarpus* and *Populus/Salix*-type charcoal were each found in single locations. This record of fuel choice is supported by the presence of *Pinus*-type bark scales, a *Juniperus* twig, an *Artemisia tridentata*–type leaf, and a *Cercocarpus/Artemisia*-type axillary bud in the same samples. Additional fuel probably was provided by *Zea mays* cobs, as evidenced by the presence of cupules in some of the midden deposit samples.

Remains of possible foodstuffs in the midden include a Capparaceae-type seed, cheno-am seeds, a Gramineae caryopsis fragment, and a *Zea mays* kernel. With the exception of the Capparaceae seed, the other taxa were

routinely recovered in Duckfoot rooms and pit structures and are believed to represent common foods.

Discussion

Carbonized plant remains recovered from Duckfoot flotation samples offer a view of the resources used by the site's inhabitants. In the following discussion, features containing primary, de facto, and mixed refuse are assessed to establish which plants were routinely chosen for food, fuel, and roof construction and to see if any plant remains clearly suggest ceremonial or medicinal use. Then distributions of plant taxa within the site are examined to shed light on the activities conducted inside different types of structures, and architectural suites are compared to reveal potential differences in intensity of use. Plant data are examined to gain insights into seasons of occupation and abandonment, at least for some of the structures. Finally, plant materials from Duckfoot are compared with those recovered from Dolores Archaeological Program (DAP) sites of the same age to assess similarities and differences in the acquisition of foods and woody taxa.

Food Resources

One can learn about common foods by examining the plant remains recovered from vessels, in-place metates, hearths, roasting pits, ash pits, and other subsistence-related features containing at least some primary or de facto refuse that derives from artifact or feature use (Table 6.14). Charred plant remains from floors and postabandonment structure and feature fills are generally excluded from this attempt to understand routine food-related activities within structures; however, contexts that yielded a mixture of primary and nonprimary debris *are* considered.

Cheno-am seeds, *Zea mays* kernels, and *Physalis longifolia* seeds were the most commonly recovered foods in features associated with food processing, cooking, or storage (Table 6.15). The two wild resources would be available first, in mid- to late summer, with the corn ready for harvest later in the season. The cool-season (June) grass *Stipa hymenoides* is also common, although it was not recovered from any of the pit structures. Less common or rare foods include *Opuntia* seeds, *Portulaca retusa* seeds, *Descurainia* seeds, Gramineae parts, Malvaceae seeds, *Mentzelia albicaulis* seeds, *Yucca baccata* seeds, a Compositae achene, and a *Phaseolus* cotyledon. *Juniperus* seeds and *Atriplex canescens* fruit cores in hearths and other contexts could signify use of branches for fuel or acquisition of the fruit as food.

The distribution of the plant remains does not reveal if domesticated corn contributed more to the ancient diet than did wild plants. Both *Zea mays* kernels and cheno-am seeds are especially ubiquitous in features and in the midden.

Table 6.14. Selected Contexts Used in Intrasite
Comparisons

Burned or Partly Burned Structures	
Pit Structure 1	Room 11
Feature 8 (hearth)	Feature 6 (hearth)
Feature 9 (ash pit)	Feature 10 (warming pit)
Metate (PL 41)	Room 12
Vessels (PLs 1, 3, 8, 9, 11,	Feature 6 (hearth)*
12, 14, 15, 18, 64, 78)*	Room 13
Pit Structure 3	Feature 1 (hearth)
Feature 34 (hearth)	Room 16
Feature 36 (ash pit)	Feature 3 (hearth)
Pit Structure 4	Feature 7 (hearth)*
Feature 1 (hearth)	Room 19
Room 3	Feature 19 (hearth)
Floor	

Unburned Structures	
Pit Structure 2	Room 5
Feature 3 (hearth)	Feature 1 (fire pit)
Feature 4 (ash pit)	Room 15
Room 2	Feature 4 (ash-filled pit)
Vessels (PL 1)*	
Room 4	
Feature 1 (burned spot)	
Feature 2 (pit)	

NOTE: Features and artifacts listed here are interpreted to have contained primary or de facto cultural refuse or to have been filled with mixed deposits containing at least some primary or de facto refuse. Features and artifacts containing mixed deposits are marked with an asterisk (*).

Because frequent harvest of *Physalis longifolia* and *Stipa hymenoides* is also suggested, the role of wild plants at Duckfoot is assumed to have been at least as important as that of corn.

Fuel Resources

One can learn of fuel choice by examining the tree and shrub charcoal recovered from all floor-related hearths, roasting pits, ash pits, and other thermal features filled with primary, de facto, or mixed deposits (Table 6.16). Included in this study are features in which a fire was actually built, as well as those in which no evidence of in situ burning exists but into which someone may have placed hot coals for food processing.

It appears that the Duckfoot inhabitants routinely used the wood of *Juniperus* and *Pinus* and the cobs of *Zea mays* as fuel. Commonly used fuel also includes wood from three shrubs—*Cercocarpus, Atriplex,* and *Artemisia*— each identified from at least six structures. These shrubs probably were available locally, as they are today, providing material for tinder or to supplement fires of *Juniperus* and *Pinus*.

Four woody taxa occur only rarely in burned features. Charcoal identified as *Populus angustifolia*-type was preserved in only two domestic rooms (Rooms 13 and 19). *Peraphyllum/Amelanchier*-type and *Purshia*-type wood charcoal were each found in single hearths, and *Rhus*-type charcoal was recovered from a shallow pit. More than one factor may be responsible for the sparse recovery of these types. Certain wood makes poor firewood or may grow too far from a dwelling to be gathered routinely (e.g., *Populus*). Other wood might be better suited to the manufacture of tools and implements (*Purshia*) or baskets (*Rhus*). Also, people might avoid depleting local food-bearing taxa (*Peraphyllum/Amelanchier, Rhus*).

Other Wood Resources

Wood from contexts other than hearths and presumed cooking/heating features provides evidence for the acquisition of trees and shrubs for reasons unrelated to fuel use. Charcoal recovered from roof fall, room fill/upper control, floor surface, and floor-associated feature samples may shed light on choices of material for roof construction and other purposes. Since the burned structures might have contained more than just a routine assemblage of remains (Hally 1981), patterns of wood preservation have been tallied separately for burned (Table 6.17) and unburned (Table 6.18) structures.

When one compares features in burned structures with those in unburned structures, some patterns emerge. For example, the ubiquitous charcoal type is *Juniperus*, regardless of whether the rooms were destroyed by fire. *Pinus* charcoal is second in occurrence, a pattern that fits the predominant pattern of charcoal preserved in cooking and heating features. Likewise, *Artemisia, Cercocarpus,* and *Atriplex* were recovered frequently.

A preservation bias is suggested, in that the burned rooms retained evidence of three charred woods (*Populus angustifolia*-type, *Populus/Salix*-type, and *Rhus*-type) not recovered in the unburned structures. The cottonwood charcoal recovered from the troughs of two Room 12 metates might be the remains of wooden implements or could represent smaller roof beams that collapsed as the rooms burned. The *Rhus*-type charcoal and Monocotyledon fibrovascular bundles found in association with a burned basket on the floor of Pit Structure 3 are materials used in basket manufacture.

Apparently, a number of the structures at Duckfoot were ritually abandoned. For example, deliberate destruction of the four pit structures (three of them by burning) appears to have followed the placement of one or more human bodies on their floors. The presence of cottonwood in material used to intentionally burn Pit Structure 3 suggests that this wood was one of the fuels used to destroy the structure. Similarly, the floor of Room 16 was covered with a thick layer of charred twigs, which suggests that the room had been filled with brush before being destroyed by fire. Charcoal types in fill and on the floors of most of the

Table 6.15. Summary of Food Taxa from Selected Contexts

Taxon and Part	Architectural Suite 1					Architectural Suite 2							Architectural Suite 3	
	HR 16	SR 2	SR 3	SR 15	PS 1	HR 4	HR 5	HR 11	HR 12	HR 13	PS 2	PS 4	HR 19	PS 3
Cheno-am seed	+	+			+	+		+	+	+	+	+	+	+
Zea mays kernel	+		+	+	+		+		+	+	+			+
Physalis longifolia seed	+			+	+			+	+	+		+		+
Stipa hymenoides parts	+					+		+	+	+				
Juniperus seed*	+				+			+						+
Opuntia seed	+				+		+							
Portulaca retusa seed	+				+					+				
Descurainia seed		+								+				
Gramineae parts									+					
Mentzelia albicaulis seed		+												
Yucca baccata seed	+													
Atriplex canescens fruit core*														+
Compositae achene					+									
Malvaceae seed									+					
Phaseolus cotyledon					+									

NOTE: Plant materials summarized here were recovered from contexts interpreted to have contained primary or de facto cultural refuse or to have been filled with mixed deposits containing at least some primary or de facto refuse (see Table 6.14). Taxa are listed in order of ubiquity.
HR = habitation room; SR = storage room; PS = pit structure.
*Could represent food or fuel resource use.

burned rooms include *Artemisia, Atriplex, Cercocarpus, Juniperus, Pinus,* and the rarely occurring *Populus.* Some of this wood may represent fuel placed in these structures to promote burning.

Pinus and *Juniperus* charcoal in roof fall debris supports the tree-ring record of pine and juniper as commonly used roof beam materials. *Artemisia* and *Cercocarpus* identified in roof fall hint that some of the smaller layers of the roof in Pit Structures 1 and 4 and in Room 10 were of sagebrush and mountain mahogany, although these two shrubs are not known to routinely provide roofing material. The intended use of *Atriplex,* the other shrub commonly identified in the charcoal remains at Duckfoot, is not known.

Ceremonial or Medicinal Plant Use

There is no clear evidence of ceremonial or medicinal plant use at the Duckfoot site. Such evidence might include items thought to have been carried long distances or parts not generally known in the ethnographic record as being food resources or as serving daily needs. However, it is also possible that plants routinely used as food or for ordinary

Table 6.16. Summary of Fuel Taxa from Selected Thermal Features

Taxon and Part	Architectural Suite 1			Architectural Suite 2							Architectural Suite 3	
	HR 16	SR 15	PS 1	HR 4	HR 5	HR 11	HR 12	HR 13	PS 2	PS 4	HR 19	PS 3
Juniperus wood	+	+	+	+	+	+	+	+	+	+	+	+
Pinus wood	+	+	+	+		+	+	+	+	+	+	+
Zea mays cob/cupule	+	+	+	+		+	+	+			+	+
Cercocarpus wood			+	+		+	+	+	+	+	+	
Atriplex wood	+	+				+	+		+	+		+
Artemisia wood	+	+	+			+	+		+			
Populus angustifolia wood	+										+	
Peraphyllum/Amelanchier wood							+					
Purshia wood									+			
Rhus wood					+							

NOTE: Plant materials summarized here were recovered from contexts interpreted to have contained primary or de facto cultural refuse or to have been filled with mixed deposits containing at least some primary or de facto refuse (see Table 6.14). Taxa are listed in order of ubiquity.
HR = habitation room; SR = storage room; PS = pit structure.

Table 6.17. Tree and Shrub Taxa Not Related to Fuel Use, Burned Structures

Location	Taxon											
	Arte	Atri	Cerc	Juni	Pine	PiEd	PiPo	Abie	Pseu	PoAn	Pop/Sal	Rhus
Pit Structure 1				⊕		⊕	⊕					
Floor fill above Features 8 and 9 (hearth and ash pit)	+			+	+							
Floor scrape			+	+	+							
Roof fall, upper control	+		+	+	+							
Pit Structure 3				⊕		⊕	⊕	⊕	⊕			
Floor, basket (PL 160) and possible contents	+	+		+								+
Closing/fuel on floor	+	+		+	+					+		
Roof fall, upper control	+	+		+	+							
Pit Structure 4				⊕		⊕						
Roof fall	+			+	+							
Room 10				⊕		⊕						
Feature 1 (corner bin)			+	+								
Feature 3 (corner bin)	+				+							
Feature 5 (cist)	+		+	+								
Floor			+	+	+							
Room 11				⊕		⊕						
Feature 1 (corner bin) fill			+	+	+							
Feature 1 (corner bin) floor		+		+								
Floor			+	+								
Room 12				⊕		⊕						
Wall and roof fall, metate trough contents	+	+		+	+						+	
Wall and roof fall, metate trough contents	+		+	+	+						+	
Wall and roof fall, upper control	+		+	+	+							
Room 13				⊕		⊕	⊕					
Feature 2 (slab-lined pit)			+	+	+							
Room 16				⊕								
Room 19				⊕		⊕						
Floor fill under metate			+	+	+							

NOTE: Table summarizes charred wood from roof fall, room fill/upper control, floor surface, and selected floor-associated feature samples.

\+ = taxon identified in flotation sample.

⊕ = taxon identified for tree-ring sample.

Arte = *Artemisia*-type; Atri = *Atriplex*-type; Cerc = *Cercocarpus*-type; Juni = *Juniperus*-type; Pine = *Pinus*-type; PiEd = *Pinus edulis*–type; PiPo = *Pinus ponderosa*–type; Abie = *Abies*-type; Pseu = *Pseudotsuga*; PoAn = *Populus angustifolia*-type; Pop/Sal = *Populus/Salix*-type; Rhus = *Rhus*-type.

household requirements were also used occasionally for medical/religious purposes. The problem is that similarity in both the condition and the distribution of remains used for different purposes makes it impossible to distinguish secular events from medical/religious ones.

Structure Type Comparisons

A comparison of the food and fuel remains recovered from habitation rooms, storage rooms, and pit structures (Tables 6.15 and 6.16) reveals patterns of plant use associated with different types of structures. The presence of temporary storage facilities, hearths or other thermal features, and a wide variety of pottery and stone artifacts is used to infer a domestic (habitation) function for some surface rooms. Storage rooms, on the other hand, generally have fewer features and artifacts. Pit structures at Duckfoot contain the most abundant and diverse sets of features and artifacts, including those related to mealing, cooking, and storage. Pit structures at Duckfoot are considered to be primarily domestic structures; however, the presence of sand-filled pits, including sipapus, and the use of all four pit structures as burial chambers suggest at least some amount of ritual activity (cf. Wilshusen 1989).

Seven domestic surface rooms contained a collective average of four taxa likely to represent food products, and the same seven contained *Juniperus, Pinus,* and *Zea* remains indicative of fuel use. Three domestic rooms with

Table 6.18. Tree and Shrub Taxa Not Related to Fuel Use,
Unburned Structures

Location	Taxon					
	Arte	Atri	Cerc	Juni	Pine	PiEd
Pit Structure 2				⊕		
Floor				+	+	
Roof and wall fall, upper control	+	+		+		
Room 4				⊕		
Feature 1 (burned spot)			+	+	+	
Floor near Feature 3 (pit)			+	+	+	
Wall fall, metate (PL 4) trough contents			+	+		
Room 5						
Feature 2 (corner bin)	+	+		+	+	
Room 15						⊕
Feature 5 (cist)	+	+	+	+	+	
Room 17						
Upper fill, metate (PL 2) trough contents	+			+	+	

NOTE: Table summarizes charred wood from roof fall, room fill/upper control, floor surface, and selected floor-associated feature samples.
+ = taxon identified in flotation sample.
⊕ = taxon identified for tree-ring sample.
Arte = *Artemisia*-type; Atri = *Atriplex*-type; Cerc = *Cercocarpus*-type; Juni = *Juniperus*-type; Pine = *Pinus*-type; PiEd = *Pinus edulis*-type.

complete sets of mealing tools (Rooms 12, 13, and 16) were represented by six to eight possible food taxa, including the commonly recovered cheno-am and *Physalis longifolia* seeds, *Zea mays* kernels, and *Stipa hymenoides* parts. In general, domestic surface rooms yielded the greatest diversity of food remains at Duckfoot; the diversity seen in these rooms is exceeded only by that in the Pit Structure 1 assemblage. Three storage rooms also have food taxa, although common foods are seen less frequently. These rooms generally retain fewer foods than do domestic structures, and the remains recovered from them are more variable.

Pit Structure 1 contained a total of eight seed or fruit taxa also recovered in domestic and/or storage rooms, with the exception of a Compositae achene and a *Phaseolus* cotyledon. On the other hand, the other three pit structures have five or fewer taxa, in all cases including the commonly occurring cheno-am seeds and in two cases including *Physalis longifolia*-type seeds. Despite the variability among pit structures, the common taxa recovered do not differ from those recorded for domestic and storage rooms, except for a lack of *Stipa hymenoides* remains in the pit structure samples (N = 44 samples). For example, the ample Pit Structure 1 assemblage is not unusual for the taxa that are present or notable for those that are missing.

Although there are some differences in the distribution of food remains among the separate architectural types,

fuels carried into domestic structures and pit structures are similar. The ubiquitous and common wood charcoal types have been identified in most fire-related features, regardless of location at the site.

Comparisons of Architectural Suites

The archaeobotanical record from the three architectural suites at Duckfoot includes materials from seven rooms serving domestic needs, three rooms with storage or other nonhabitation functions, and four pit structures. The following discussion focuses on the abandonment assemblages associated with each architectural suite.

Common foods recovered from all three suites include cheno-am seeds, *Zea mays* kernels, and *Physalis longifolia* seeds (Table 6.15). The remaining likely food products are so restricted in distribution that they shed no light on room-suite associations. More possible food items were recovered from the western and central architectural suites than from the eastern suite. Since rooms in all suites burned, preservation may not account for this distribution. Perhaps fewer people used the eastern rooms as the time of abandonment approached, or those who did were not processing or consuming much food there. It is also possible that sample size is at least partly responsible for the observed pattern.

Comparison of the fuels put into hearths and fire pits before abandonment does little to delineate different patterns of use in the three architectural suites (Table 6.16). *Juniperus* wood, *Pinus* wood, *Zea mays* cobs, and, to a lesser extent, *Cercocarpus* and *Atriplex* wood were brought into most habitation rooms and pit structures, regardless of location. *Artemisia* wood appears to be restricted to hearths in the western and central suites. Charcoal of other taxa occurs so rarely in these features that no association with architectural suites is apparent.

Seasonality Considerations

Seasons of Occupation

Plant remains recovered from Duckfoot indicate that the site was occupied from at least early summer (June) through late fall (October). As with most site assemblages, the plant remains are not helpful in providing evidence for or against winter occupation. However, since most plant parts can be stored for long periods, winter occupation cannot be ruled out.

The Duckfoot occupants most likely gathered mature *Descurainia* seeds, *Stipa hymenoides* florets, and *Rhus aromatica* fruit at least by early summer (mid-June), which is when these resources ripen in the region. Harvest probably lasted at least a month, as different plants of each species matured their fruit in succession. Cheno-am, *Physalis, Portulaca, Peraphyllum,* and *Mentzelia albicaulis* fruits would ripen in midsummer or later, as would achenes of plants in the sunflower family. Some of these later-ripening resources would be available until the first severe fall frost. Except for *Peraphyllum*, these plants could easily grow in the disturbed conditions of a cultivated cornfield or along the edges of paths and areas frequently disturbed by humans.

Use of *Artemisia tridentata*-type branches in the month of September is suggested by the presence of immature flowering heads in three structures. Presumably, harvest of the domesticated corn and bean crops would take place during this time as well.

A number of plant resources could have been gathered at any time during the year. For example, branches of local juniper, pine, big sagebrush, four-wing saltbush (with mature fruit capable of clinging for months), cottonwood/willow, and members of the rose family would always be available for fuel. *Phragmites australis* stems, once fully grown, remain standing in mesic settings for months. Fruits of *Juniperus osteosperma, Opuntia,* and *Yucca baccata* can cling to plants for long periods following maturity, sometimes into the next reproductive season.

Season of Abandonment

Various lines of evidence suggest that at least some of the Duckfoot structures were abandoned after the fall harvest but before deep winter set in, although the evidence is equivocal. A case for spring or summer abandonment is weaker and requires more complex assumptions to support it. Both hypotheses are examined here.

A strong indication of fall abandonment is the presence of *Artemisia tridentata*-type immature flowering heads on the floors of burned Pit Structures 1 and 3 and in limited amounts in Room 12. Modern big-sagebrush flowering heads of comparable maturity are available in September in the Duckfoot region. It is assumed that the ancient flowering heads are from branches brought in as fuel to intentionally destroy the structures, although they might also derive from roof construction.

The second major line of evidence for fall abandonment concerns the recovery of cheno-am seeds, *Zea mays* kernels, and *Physalis longifolia*-type seeds from deposits in food-related features in many structures. Assuming that these are the remains of the last foods prepared in these rooms, they could represent a complex of recently harvested summer and fall resources. The common occurrence of *Stipa hymenoides*-type parts weakens the case for fall abandonment, as this grass matures in early summer; however, Indian rice grass florets could have been stored from an earlier harvest.

If the site was abandoned in the fall, the Duckfoot occupants took the harvest of all their domesticated crops with them. No temporary or permanent storage area contained *Zea* kernels in quantities suggesting that a good harvest was abandoned. In fact, the entire kernel assemblage from all areas of the site could easily be stored in a small shoe box. An alternative explanation is that a meager *Zea* harvest provided some of the impetus for leaving.

The case for spring or summer abandonment is more difficult to build. A major piece of evidence arguing against this scenario is the presence of immature flowering heads of big sagebrush. One would have to assume that people had gathered the branches the previous fall (in September) and piled them in structures for burning some months later. One would have to further assume that the delicate flowering structures remained attached and, although probably tinder dry, were not completely consumed during combustion. One would also have to assume that the *Zea,* cheno-am, and *Physalis* resources harvested the previous year were abundant and successfully stored for some months. Although this notion is not unreasonable, it is easier to imagine the Duckfoot inhabitants harvesting and storing a single resource (the Indian rice grass of the fall-abandonment scenario) than to imagine the same for the three separate resources of the spring/summer-abandonment model.

A Comparison Between Duckfoot and Dolores Sites

The DAP documented plant use at a number of other Pueblo I sites in the region. Patterns of plant use at Duckfoot generally compare well with those inferred for chronologically similar Dolores sites, although there are some differences. Unless noted otherwise, all DAP comparative data cited below are from contexts that date between A.D. 840 and 880, a time when the Dolores area had the highest population of the A.D. 600 to 1250 period studied (Matthews 1986:166). In both the Duckfoot and the Dolores assemblages, plant remains most likely represent relatively limited periods of time (days or weeks) as abandonment of a location approached, and they cannot be assumed to represent plant use over the entire occupation of a site or group of sites.

The roles played by domesticates at Duckfoot and at the Dolores sites (Matthews 1986:158) appear to have been similar. In both assemblages, *Zea mays* kernels are common, indicating that corn was a ubiquitous domesticate in both places. On the basis of small samples at both locations, it appears that mean cob row number is nearly identical: 11.0 rows at Dolores sites (N = 26) and 10.8 rows at Duckfoot (N = 12). The small number of *Phaseolus vulgaris* remains from Dolores sites (Griffitts 1987) was thought to be related to preparation, preservation, and sampling biases and, therefore, to underrepresent the prehistoric importance of beans in the diet. This explanation might also apply to the recovery of only a single bean cotyledon at Duckfoot.

Use of wild plants as food differed somewhat in the two projects. The following comparison is based on slightly different databases. For Dolores sites, only fire-related features such as hearths (with primary, secondary, de facto, or mixed cultural debris), ash pits, and other pits with evidence of burning were considered to provide data of sufficiently high quality (Matthews 1986:173). At Duckfoot, food-related contexts with primary, de facto, and mixed cultural refuse include not only hearths and other thermal features but also vessels, a pit feature, and an in situ metate.

The two most commonly recovered wild foods in the Duckfoot samples were cheno-am and *Physalis longifolia*-type seeds, in that order. Likewise, at Dolores sites, *Chenopodium/Amaranthus/*cheno-am and *Physalis* are the most ubiquitous taxa in fire-related features (Matthews 1986:178). At this point the data diverge. For example, the common recovery of *Stipa hymenoides*–type parts at Duckfoot is not duplicated at Dolores sites. Instead, it appears that the Dolores populations favored *Portulaca* and *Descurainia*. Potential food taxa that occur infrequently in both Duckfoot and Dolores contexts include Gramineae, *Mentzelia albicaulis*, *Yucca baccata*, and *Opuntia*.

A number of woody plants commonly provided hearth fuel at Duckfoot and at Dolores sites. Evidence from charcoal- and ash-filled features assumed to retain primary deposits from fires indicates that the Duckfoot occupants routinely used *Juniperus*-type and *Pinus*-type (including *Pinus*, *Pseudotsuga*, and/or *Picea*) wood as fuel. In addition, they used *Cercocarpus*, *Atriplex*, and *Artemisia* branches. *Zea mays* cobs provided an important nonwoody combustible resource. Similarly, juniper and pine charcoal was frequently retrieved from hearths and middens of Dolores sites of all time periods (Matthews 1985:44, 45), along with *Artemisia* and *Cercocarpus* wood (Benz 1984:204–205). Both groups infrequently added *Rhus* and rose family

(e.g., *Amelanchier*, *Peraphyllum*, *Purshia*) wood to hearths or other fire-related features.

A few differences are apparent in fuel selection at Duckfoot and at the Dolores sites. For example, the notable role of *Atriplex* as a fuel at Duckfoot seems not to have been duplicated at Dolores (Matthews 1986:162–163). Nor is the acquisition of *Quercus* and *Populus/Salix* for fuel by Dolores groups repeated at Duckfoot. Elevation differences and access to varied plant communities could partly explain differences in fuel use, as the Dolores sites generally are above 2073 m (6800 ft) in elevation, and many are located nearer to riparian habitats and more extensive oak (*Quercus*) stands.

Vegetal materials used in roof construction were similar, although not identical, at Duckfoot and the Dolores sites. Pueblo I occupants of the Dolores area used *Juniperus*, ponderosa pine (*Pinus ponderosa*), and Douglas fir (*Pseudotsuga*) from A.D. 800 to 850, then increased their reliance on pinyon (*Pinus edulis*) and cottonwood (*Populus*) as they decreased use of juniper from A.D. 850 to 900 (Matthews 1986:180). At Duckfoot, on the other hand, *Juniperus* and pine (*Pinus edulis*, *P. ponderosa*) beams predominated during both the original building in the A.D. 850s and the final construction and remodeling in the A.D. 870s. Whereas more and more cottonwood was incorporated into roofs in the Dolores area, cottonwood/willow remained a rarely used roofing material at Duckfoot. A combination of resource depletion and changing regional population densities may have provided the impetus for changing patterns of wood use at Dolores sites (Kohler and Matthews 1988).

Although the archaeobotanical record of roof-closing material may be incomplete, both Duckfoot and Dolores groups (Matthews 1986:164, 176) gathered *Artemisia* for this purpose. Charcoal recovered from roof fall suggests that the Duckfoot occupants used *Cercocarpus* as well.

7

Fossil Pollen

Jannifer W. Gish

Research Objectives

The primary objective of the Duckfoot pollen study was to provide an assessment of ethnobotanic plant use at the site. It was anticipated that distributions of pollen types within pit structures and rooms could be used to enhance interpretations of activity-area and structure function and provide clues about season of structure use. The apparently deliberate destruction of several structures with their contents in place resulted in excellent preservation of last-use floor assemblages, with obvious indications of activity-area specialization. Since much of the destruction is inferred to have had ritual connotations, a specific goal of the pollen study was to identify evidence of ceremonial plant use, if possible.

Most of the 40 analyzed pollen samples were collected from sediments beneath artifacts (vessels, manos, metates, and slabs) on structure floors (Table 7.1). The purpose of this collection strategy was to focus on sealed contexts that had been protected from postabandonment pollen rain and that, therefore, might be expected to more accurately reflect pollen deposition during the occupation of the site. Rapid fill accumulation, particularly in the pit structure depressions, further contributed to the good preservation of contextual evidence, and there were few indications of postabandonment disturbance to site deposits. The pollen samples selected for analysis were from all four pit structures, five front-row rooms (Rooms 10, 12, 13, 16, and 19), and two back-row rooms (Rooms 3 and 5). Over half of the samples analyzed came from Architectural Suite 3.

Reconstruction of past environmental conditions was not of primary concern in this study. The discussion of environmental issues is oriented more toward an assessment of the cultural effects on the background pollen rain, and only a tentative reconstruction of past vegetation is presented.

Chemical Processing and Data Recovery

All samples were processed chemically by Dr. John Jones at the Texas A & M University Palynology Laboratory. A standard zinc bromide flotation procedure was used to extract the pollen, and several samples with large amounts of cellulose were also subjected to acetolysis. One tablet of *Lycopodium* spores, calculated at $11,300 \pm 400$ grains, was added to each sample to calculate pollen concentration in the sediment. The addition of *Lycopodium* spores allowed evaluation of pollen deposition at the site and provided a means of assessing the effectiveness of pollen recovery during processing. An estimate of the number of grains per slide (number of pollen grains per transect × number of possible transects per slide) was calculated for all samples. This, and the proportion of *Lycopodium* spores to prehistoric (fossil) pollen in the initial transects, provided the basis for deciding whether a 100- or 200-grain count (or no count) would be tabulated for a given sample. In all, 32 of the 40 samples produced sufficient pollen for counts of at least 100 grains of combined arboreal and nonarboreal pollen.

The estimated number of grains per slide and the pollen concentration ranges for each sample with sufficient pollen for analysis are listed in Table 7.2. Pollen concentration is related to the number of *Lycopodium* tracer spores counted while recording the 100 or 200 grains of prehistoric pollen, as indicated by the following formula (in which the number of *Lycopodium* spores per tablet = $11,300 \pm 400$):

$$\text{pollen concentration} = \frac{\text{no. of Lycopodium spores added} \times \text{no. of fossil pollen counted}}{\text{no. of Lycopodium spores counted} \times \text{amount of sediment processed}}$$

Table 7.1. Pollen Sample Contexts

Structure	Sample Code[a]	Context
Pit Structure 1	68-6	burned roof fall
	72-6	floor under mano (PL 34)
	72-95	floor under gray ware vessel, southeast corner
Pit Structure 2	223-8	unburned roof fall
	332-12	floor under slab (no PL), north half of structure
	362-2	floor, trough metate (PL 108, Surface 1) fill
	373-9	floor, trough metate (PL 1, Surface 2) fill
Pit Structure 3	569-9	roof fall under slab (no PL), center of structure
	588-10	floor under mano (PL 14)
	589-2	floor under mano (PL 4)
	590-14	floor under mano (PL 238)
	594-5	floor under abrader (PL 217)
	595-3	floor under vessels (PL 140)
	600-1	floor under modified cobble (PL 189)
	602-1	floor under slab (no PL) near tool cluster, east wall
	604-3	floor under jar (PL 129)
	606-11	floor around sipapu (Feature 33)
	612-3	floor under slab (no PL), northwest corner
Pit Structure 4	981-3	roof fall
	986-14	floor south of wing wall, southwest corner
	986-15	floor under mano (PL 138)
	986-17	floor under modified cobble (PL 43)
	986-19	floor under mano (PL 45)
	986-23	floor under mano (PL 166)
Room 3	91-6	floor
Room 5	129-6	floor under slab (no PL)
Room 10	582-4	corner bin (Feature 1) fill
	582-5	corner bin (Feature 1) base
	583-1	bin (Feature 2) base
	583-2	bin (Feature 2) fill
	614-3	corner bin (Feature 3) base
Room 12	47-4	floor northeast of hearth (Feature 6)
Room 13	139-5	floor, slab metate (PL 41) fill
Room 16	143-4	fill above floor (Surface 2)
	143-45	floor (Surface 2), trough metate fill
Room 19	559-8	floor beneath pollen sample (559-9), north of hearth
	559-9	floor fill under inverted trough metate (PL 52, Stratum 3), north of hearth
	559-13	floor under mano (PL 51), north of hearth
	680-4	burned roof fall in cist (Feature 13)
	680-5	floor under mano (no PL) at base of cist (Feature 13)

[a] The first number in the two-number sample code is a provenience indicator, the second is a laboratory catalog number.

Generally, in circumstances in which there is good pollen preservation, one would expect a positive relationship between pollen concentration per milliliter and the estimated number of grains per slide. In the case of Duckfoot, samples with the highest concentrations (more than 7,200 grains per ml) tend to have the highest estimated abundance of pollen (more than 1,000 grains per slide). There are, however, some exceptions to this pattern. In three samples (559-8, 559-13, and 680-4, all from Room 19) with low estimates of pollen abundance, low numbers of *Lycopodium* spores resulted in the calculation of a high concentration of pollen per milliliter of sediment. In these cases, the recovery of both the tracer spores and the prehistoric pollen grains seems to have been inhibited. All the Duckfoot samples contained particulate charcoal as well as substantial quantities of decayed organic detritus.

Table 7.2. Concentration of Pollen in Duckfoot Samples

Structure	Sample Code[a]	*Lycopodium* Count	Amount of Sediment Processed (ml)	Fossil Pollen Count	Estimated Grains per Slide	Concentration (grains per ml)
Pit Structure 1*	72-6†	31	10	100	270	3,516–3,774
Pit Structure 2	223-8	53	20	200	923	2,056–2,207
	332-12	8	20	200	2,250	13,625–14,625
	362-2†	15	20	200	1,035	7,266–7,800
	373-9†	21	20	200	675	5,190–5,571
Pit Structure 3*	569-9	40	20	100	225	1,362–1,462
	588-10	30	20	200	1,710	3,633–3,900
	589-2	62	20	200	1,350	1,758–1,887
	594-5	36	20	100	315	1,513–1,625
	600-1	16	20	100	225	3,406–3,656
	602-1†	15	20	200	1,530	7,266–7,800
	604-3†	57	20	200	720	1,912–2,052
	612-3	28	20	200	1,080	3,892–4,178
Pit Structure 4*	981-3	36	20	100	315	1,513–1,625
	986-14†	10	20	200	2,205	10,900–11,700
	986-15†	4	20	200	2,430	27,250–29,250
	986-19	51	20	200	675	2,137–2,294
Room 3*	91-6†	50	20	200	720	2,180–2,340
Room 5	129-6†	32	20	200	1,080	3,406–3,656
Room 10*	582-4	93	10	100	270	1,172–1,258
	582-5	30	10	200	1,665	7,266–7,800
	583-1	79	10	100	405	1,379–1,481
	583-2	97	20	200	720	1,123–1,206
Room 12*	47-4†	10	20	100	315	5,450–5,850
Room 13*	139-5	12	20	100	225	4,541–4,875
Room 16*	143-4	33	20	100	270	1,651–1,772
	143-45	54	20	100	450	1,009–1,083
Room 19*	559-8	7	20	100	450	7,785–8,357
	559-9	20	20	200	855	5,450–5,850
	559-13	7	20	100	315	7,785–8,357
	680-4	4	20	100	405	13,625–14,625
	680-5†	15	20	200	3,150	7,266–7,800
Pit Structure 1	68-6	not counted	20	IP	135	not calculated
	72-95	not counted	20	IP	45	not calculated
Pit Structure 3	590-14	not counted	20	IP	45	not calculated
	595-3	not counted	20	IP	135	not calculated
	606-11	not counted	20	IP	135	not calculated
Pit Structure 4	986-17	not counted	10	IP	90	not calculated
	986-23	not counted	20	IP	45	not calculated
Room 10	614-3	not counted	5	IP	22	not calculated

IP = insufficient pollen for a 100-grain count.

[a] The first number in the two-number sample code is a provenience indicator, the second is a laboratory catalog number. Samples marked with † were subjected to large-fraction scanning.

*Structures that burned at abandonment.

In some cases, the debris was large, which may have interfered with efficient pollen and spore recovery during processing of some samples.

Macrobotanical (flotation) remains were generally better preserved in structures that burned at abandonment than in those that did not. However, the relationship between burning and pollen preservation is not as clear. There are high concentrations of pollen in both burned (Pit Structures 3 and 4, Room 10) and unburned (Pit Structure 2) structures (Table 7.2). There are also eight samples from burned structures (Pit Structures 1, 3, and 4, and Room 10) that did not have sufficient pollen to produce a 100-grain count. Thus, there does not appear to be a significant relationship between quality of pollen preservation and presence or absence of burning at abandonment. Other factors, such as sample context and how well that context was sealed prehistorically, probably had a greater effect on pollen preservation at Duckfoot.

After the 100- or 200-grain combined arboreal and non-arboreal pollen counts were made at 400× magnification, the entire slide for each sample was scanned at 100× to detect additional economic pollen. Pollen identified as the result of this type of scanning is noted with a "+" (for presence) in the base data tables.

Pollen aggregates were tabulated systematically as well. Aggregates are clumps of the same type of pollen. They are generally indicative of short-distance pollen dispersal or of actual introduction of plant parts into a particular context. In wind-pollinated plants, aggregates are not an effective means of pollen dispersal. In insect-pollinated plants, too, aggregates have disadvantages. Large or abundant aggregates are rare in modern pollen surface control studies of many community types (Gish 1982), although in certain depositional environments (for instance, in flood-plain situations) there may be exceptions to this general pattern. Large aggregates in cultural contexts are presumed to be ethnobotanically significant. Aggregates are not abundant in the Duckfoot samples, although in several instances they provide apparent confirmation of ethno-botanic plant use in structures.

As a final step in the evaluation of the Duckfoot site pollen, 11 samples were subjected to large-fraction scanning (Table 7.2). In this procedure, sample residues, which were preserved in glycerin and stored in 1-dram vials, were thinned by the addition of reagent alcohol. The thinned liquid was decanted through a 45-μ-mesh screen, and the larger material was pipetted off and transferred to a slide for microscopic evaluation. This technique, based on a method developed by Clary (1988), allows examination of the entire polliniferous residue extracted from a sample rather than just a droplet of the residue, which is generally the case during routine analysis. Because cultigens and other ethnobotanic taxa, such as cacti, have large pollen grains, this procedure greatly increases the chances of detecting ethnobotanic taxa in a sample. Pollen identified during large-fraction scanning is noted with the symbol "⊕" in the base data tables. Since the results of large-fraction scanning can greatly modify the picture of plant use and subsistence activities at a site, it is important to emphasize that this study includes the large-fraction scanning results and therefore is not strictly comparable to other pollen studies that used standard procedures only.

Pollen Taxa

Botanical nomenclature follows Kearney and Peebles (1960) and Lehr (1978). Pollen taxa designations follow those in common usage (Table 7.3). The two pine categories were separated on the basis of body breadth: grains up to 45μ in breadth were identified as pinyon-type pollen; larger grains were classified as ponderosa-type. Body breadth was generally less than 40μ for the pinyon-type grains, and there were few borderline cases. Although size separations are not always reliable (depending on the geographic area of study and the plant populations involved), the distinctions were clear at the Duckfoot site, and no attempt was made to examine other haploxylon/diploxylon points of separation.

The overall variety of pollen taxa in the Duckfoot samples is moderate (Table 7.3), with a total of 33 taxa recorded and a maximum of 22 identifiable taxa observed in a single sample. Most of the pollen could be identified only to the level of family or genus, but Table 7.4 lists some of the species that could be represented. Seasonality information, pollination mode, and some possible plant uses are also included in Table 7.4.

Assessment of Ethnobotanic Significance

The term *ethnobotanic plant* is used in a broad sense to refer to plants inferred to have been used by humans for food, fuel, construction, medicine, clothing, ritual, or other purposes. The term *subsistence* is used specifically to refer to a food plant or to a plant that is used primarily for food. The term *subsistence activity* refers to an activity related to the procurement, processing, consumption, or storage of a food item. Correspondingly, the term *non-subsistence plant* subsumes all plants used for nonfood purposes.

It is important to emphasize that the interpretations of plant use presented in this chapter are not absolute. Few pollen taxa are unequivocal indices of cultural manipulation of plants; maize is the main exception. Certain limitations apply to other taxa to greater or lesser degrees. Squash pollen in archaeological contexts, for example, probably represents cultivated squash but may also represent gathered gourds. The more equivocal taxa in terms of interpretation include Cheno-am, composites, and grasses. For example, high Cheno-am percentages may reflect food use of edible seeds or greens, nonfood use of herbs or shrubs, weedy proliferations of opportunistic species in disturbed areas around sites, or natural occurrences of herbs and shrubs. Given the multiple possibilities, it might be argued that no attempt at interpretation should be made. However, by establishing specific criteria for inferring ethnobotanic significance and applying these criteria consistently to the sample set, the data can be evaluated more fully.

In this chapter, assessments of ethnobotanic plant use are based on the following criteria: (1) relatively high

Table 7.3. Taxa Represented in Duckfoot Pollen Samples

Scientific Name	Common Name	Abbreviation[a]
Arboreal Pollen		
Picea	spruce	
Pseudotsuga	Douglas fir	
Pinus edulis–type	pinyon-type pine	Ped
P. ponderosa–type	ponderosa-type pine	Ppond
Pinus	pine (fragments)	
Juniperus	juniper	Ju
Quercus	oak	
Nonarboreal Pollen		
Low-spine Compositae	composite family; spines less than 2μ, including *Ambrosia* (ragweed)	Lo
High-spine Compositae	composite family; spines over 2μ, including *Helianthus* (sunflower)	Hi
Artemisia	sagebrush	Ar
Liguliflorae	Liguliflorae tribe, including *Lactuca* (wire lettuce)	
Cheno-am (Chenopodiaceae-*Amaranthus*)	goosefoot family and amaranth	Ch
Sarcobatus	greasewood	
Gramineae	grass family	Gr
Cruciferae	mustard family	
Ephedra nevadensis–type	joint-fir	Ephn
E. torreyana–type	joint-fir	
Leguminosae	pea family, including *Astragalus* (loco weed)	Le
Sphaeralcea-type	mallow-type	Sph
Boerhaavia-type	spiderling-type	
Gaura-type	gaura-type	
Oenothera-type	evening primrose–type	
Eriogonum	wild buckwheat	Er
Portulaca	purslane	Por
Rosaceae	rose family	
Solanaceae	potato family	
Umbelliferae	parsley family	Umb
Cyperaceae	sedge family	Cyp
Cleome	beeweed	Cl
Cylindropuntia	cholla	Cy
Platyopuntia	prickly pear	Pl
Zea	maize	Z
Cucurbita	squash	Cu

[a]Abbreviations used for pollen taxa in other tables in this chapter.

pollen percentage (determined individually for each taxon), (2) large or abundant pollen aggregates, (3) large or abundant pollen aggregates associated with high pollen percentages, (4) relatively high percentage, as compared with other samples from the same study unit, (5) unique occurrence, as compared with other samples from the same study unit, (6) sample context, and (7) complementary flotation sample evidence. For several pollen taxa, any occurrence is presumed to have ethnobotanic significance. This is not limited to pollen from such an obvious economic plant as maize but also includes squash, cholla, prickly pear, purslane (*Portulaca*), beeweed (*Cleome*), and parsley family (Umbelliferae) pollen, which are extremely rare in any natural background pollen rain. When the significance criteria are applied to the Duckfoot pollen results, 15 pollen

categories are interpreted to have been of subsistence significance in one or more samples at the site (Table 7.5), and an additional 5 are interpreted to have been of non-subsistence significance.

In inferring use, ethnographic accounts, such as Whiting (1939) and Elmore (1943), were also taken into consideration. Ethnographic accounts, however, do not always reflect the full range of uses of a plant. An example is prickly pear. Ethnographic accounts (Elmore 1943:64; Whiting 1939:85) document use of prickly pear fruit but not of buds or flowers. Prickly pear fruits normally do not retain remnant flower parts, and experimental pollen washes by this author have resulted in no pollen recovery. Therefore, the common occurrence of prickly pear pollen in cultural contexts at Duckfoot indicates bud/flower use, which in

Table 7.4. Seasonality, Pollination Mode, and Uses of Selected Plant Taxa Likely Represented in the Duckfoot Pollen Assemblage

Taxon	Seasonality[a]	Mode[b]	Use[c]
HIGHER ELEVATION ARBOREALS:			
Pinaceae			
Picea sp. (spruce)			
flowers	May–Jun	w	
branches	Jan–Dec		R
Pseudotsuga menziesii (Douglas fir)			
flowers	May–Jul	w	
branches	Jan–Dec		R
Pinus edulis (pinyon)			
flowers	May–Jul	w	
pollen	May–Jul		R
nuts	Sep–Oct		E
gum	Jan–Dec		M,R,U,O
wood	Jan–Dec		B,F,R
P. ponderosa (ponderosa)			
flowers	May	w	
nuts	Sep–Oct		E
branches/needles	Jan–Dec		R
wood	Jan–Dec		B,F
Cupressaceae			
Juniperus sp. (juniper)			
flowers	Mar–Jul	w	
berries	Jan–Dec		E,O
bark	Jan–Dec		O
branches/twigs/leaves	Jan–Dec		M,R
wood	Jan–Dec		B,F,R
Fagaceae			
Quercus sp. (oak)			
flowers	Apr–May	w	
nuts (acorns)*	Aug–Sep		E
wood	Jan–Dec		B,F,R,U
CACTI:			
Cactaceae			
Opuntia sp. (cholla) (*Cylindropuntia*)			
buds/flowers	May–Jun	a	E
fruits	Jun–Jul		E
young shoots (joints)	May–Jun		E
Opuntia sp. (prickly pear) (*Platyopuntia*)			
buds/flowers	Apr–Jun	a	E
fruits	Apr–Jul		E
young pads	May–Jun		E
HERBS AND SHRUBS:			
Amaranthaceae			
Amaranthus sp. (amaranth)			
flowers	Jun–Jul	w	
greens (whole plant)	Jun–Aug		E
seeds	Jul–Sep		E
Capparidaceae			
Cleome serrulata (beeweed)			
flowers	Jun–Aug	a	
greens	Jun		E,O
seeds	Jul–Aug		E

Table 7.4. Seasonality, Pollination Mode, and Uses of Selected Plant Taxa *(continued)*

Taxon	Seasonality[a] (Jan–Dec)	Mode[b]	Use[c]
Chenopodiaceae			
Chenopodium sp. (goosefoot)			
flowers	▬	w	
greens (leaves)	▬		E
seeds	▬		E
Atriplex canescens (four-wing saltbush)			
flowers	▬	w	
greens (young branches)	▬		E
seeds	▬		E
branches	▬		F,O
Sarcobatus vermiculatus (greasewood)			
flowers	▬	w	
wood	▬		R, O
Compositae			
Artemisia sp. (sagebrush)			
flowers	▬	w	
seeds (achenes)	▬		E
leaves (young leaves)	▬		E,M
Low-spine Compositae			
Ambrosia confertiflora (slimleaf bursage)			
flowers	▬	w	
Dicoria brandegei (dicoria)			
flowers	▬	a	E
seeds	▬		E
High-spine Compositae			
Erigeron sp. (fleabane)			
flowers	▬	a	
Helianthus sp. (sunflower)			
flowers	▬	a	
leaves	▬		M
seeds (achenes)	▬		E, R, O
Cruciferae			R
Descurainia pinnata (tansymustard)			
flowers	▬	a	E
greens	▬		E
seeds	▬		
Cyperaceae			
Scĭrpus sp. (bulrush)			
pollen	▬		E
flowers	▬	w	E
seeds	▬		E
stems, roots	? ▬ ?		
Ephedraceae			
Ephedra viridis (N-type) (joint-fir)			
flowers	▬	w	
branches	▬		M
E. torreyana (T-type) (joint-fir)			
flowers	▬	w	
branches	▬		M
Gramineae (grass)			
flowers	▬	w	
seeds	▬		E

Table 7.4. Seasonality, Pollination Mode, and Uses of Selected Plant Taxa *(continued)*

Taxon	Seasonality[a]												Mode[b]	Use[c]
	Jan	Feb	Mar	Apr	May	Jun	Jul	Aug	Sep	Oct	Nov	Dec		
Leguminosae														
Astragalus sp. (loco weed)														
flowers				―――	━━━	―――							a	
young roots				―――――										E
Malvaceae														
Sphaeralcea sp. (globemallow)														
flowers					―――――――――								a	
seeds						―――――――― – –								E
roots		? ―――――――――――――― ?												M
Polygonaceae														
Eriogonum alatum (wild buckwheat)														
flowers							――――――――						w	
roots		? – – – ――――――――――― – ?												E,M
Portulacaceae														
Portulaca retusa (purslane)														
flowers						―――――――							a	
greens						―――――――								E
seeds							―――――							E
Rosaceae														
Cowania mexicana (cliffrose)														
flowers					――――――― – –								a	
wood/bark	―――――――――――――――――――――――													Fi,O
Solanaceae														
Lycium pallidum (wolfberry)														
flowers				―――――― – – – –									a	
fruits				―――――― – – – –										E
Physalis longifolia (groundcherry)														
flowers				――――――――									a	
fruits						―――――――								E
Solanum triflorum (nightshade)														R
flowers					――――――――								a	
Umbelliferae														
Cymopterus purpurascens (camote)														
flowers				――――									a	
roots		? – – ―――――――――― – – – ?												E
CULTIGENS:														
Gramineae														
Zea mays (maize)														
planting				――――										
pollen						――――――							w	R
harvest (fruits)							– – – ・ ―――――							E
Cucurbitaceae														
Cucurbita sp. (squash)														
planting				――――										
flowers						―――――							a	E
harvest (fruits)							– – – ――――							E

[a] Season of availability: single solid line = normal period; double solid line = peak period; dashed line = extended period; ? = period uncertain. Seasonality information is based on Kearney and Peebles (1960), field observation, and study of herbarium specimens. Information generally is relevant to areas above 1524 m (5000 ft) in elevation.

[b] Mode of pollination: a = animal- or self-pollinated; w = wind-pollinated.

[c] Ethnographically documented and potential uses of plants: B = building material; E = edible food; F = fuel; Fi = fiber; M = medicine; R = ritual; U = utensils; O = other. The information on cultigens is based on Whiting (1939). Other plant use information is from Whiting (1939), Kearney and Peebles (1960), and Kirk (1970).

* Mature in one season.

Table 7.5. Frequency of Occurrence of Subsistence Plant Pollen, by Structure

							Taxon								
	Hi	Ch	Gr	Ephn	Le	Sph	Er	Por	Umb	Cyp	Cl	Cy	Pl	Z	Cu
Architectural Suite 1															
Pit Structure 1 (N = 2)†		1										1	2	2	
Room 3 (N = 1)†												1	1	1	
Room 16 (N = 2)												1	1		1
Architectural Suite 2															
Pit Structure 2 (N = 4)†		3							1		1	2	2	2	
Pit Structure 4 (N = 4)†		1							1	1	1	3	2	3	
Room 5 (N = 1)		1											1	1	
Room 12 (N = 1)†		1							1			1	1	1	1
Room 13 (N = 1)		1							1				1	1	
Architectural Suite 3															
Pit Structure 3 (N = 8)		3				1		1	2	1		6	4	5	1
Room 10 (N = 4)		1?									1		2	1	
Room 19 (N = 5)†	2?	2	1	1	1		1	2			1	2	2	4	2
Pit Structures															
N = 17		8				1		1	4	2	2	12	10	12	2
%		47.1				5.9		5.9	23.5	11.8	11.8	70.6	58.8	70.6	11.8
Rooms															
N = 15	2	6	1	1	1		1	2	2		2	5	9	9	3
%	13.3	40.0	6.7	6.7	6.7		6.7	13.3	13.3		13.3	33.3	60.0	60.0	20.0
Pit Structures (floors)															
N = 14		8				1		1	4	2	2	11	8	10	2
%		57.1				7.1		7.1	28.6	14.3	14.3	78.6	57.1	71.4	14.3
Rooms (floors)															
N = 9	1	4	1	1				2	2			4	6	6	1
%	11.1	44.4	11.1	11.1				22.2	22.2			44.4	66.6	66.6	11.1

NOTES: Table lists the number of samples in which pollen of a specific type was found in percentages or aggregates relatively high enough or large enough to indicate that the taxon was of subsistence significance (includes results of standard and large-fraction scanning). See Table 7.3 for explanation of pollen taxa abbreviations.
N = number of samples.
† Structures for which large-fraction-scanning data are included.

turn suggests that this plant was used for more diverse purposes in the past than it is in the present. In this chapter, both known and potential uses are considered in discussions of plant use. Kirk (1970) is a particularly good source of information in this regard.

Determinations of ethnobotanic significance in this chapter also take into account individual sample context and the results of Duckfoot flotation sample analysis. Like all archaeological data, pollen data cannot be interpreted independently of context. When samples are collected from contexts with high integrity (well-sealed contexts with low potential for later sediment/pollen influx and dilution of the initially preserved pollen assemblage), very specific information can be obtained. Even high contextual integrity, however, cannot eliminate the possibility of noncultural or incidental cultural influences on pollen assemblages. Therefore, comparisons with complementary flotation sample results are very important. At Duckfoot, for example, abundant remains (but no seeds) of sagebrush in flotation samples from structures point to nonsubsistence rather than subsistence use of plants. Thus, unusual values in the pollen results, such as a 74 percent sagebrush value in the roof fall sample from Pit Structure 4, are also interpreted as indicating a nonsubsistence role. Table 7.6 presents a comparison of Duckfoot flotation and pollen sample analysis results.

Environmental Considerations

Local and Nonlocal Pollen Rain

At Duckfoot, most of the pollen taxa probably represent plants that grew locally in the past. A comparison of the pollen taxa list provided in Table 7.3 with the modern floristic description of the site vicinity provided in Chapter 1 illustrates this point. In most regards, the past and present floras were probably similar. In the arboreal pollen categories, pinyon, juniper, and oak are considered local trees. Nonlocal taxa represented in the pollen rain include spruce (*Picea*), Douglas fir (*Pseudotsuga*), and ponderosa-type pine (*Pinus ponderosa*-type). None of these are sufficiently represented in the pollen rain to suggest local plant presence at the site in the past. However, the use of ponderosa pine as a construction material at the site and also its apparent use as fuel, as suggested by the pollen results (Room 19), seem to indicate that ponderosa pine was available nearby.

Most of the nonarboreal pollen taxa, too, probably represent plants that grew locally. These include the two cacti, cholla (*Opuntia* or, archaically, *Cylindropuntia*) and prickly pear (*Opuntia* or, archaically, *Platyopuntia*). Both are presumed to be ethnobotanically significant when they occur in the Duckfoot pollen records. An example of a possibly nonlocal nonarboreal taxon is Cyperaceae. Plants of this family may have been available at no great distance, however, because sedges are found at the base of Alkali Canyon today. Cattail was notably absent in the pollen record. Although it is risky to place much emphasis on the absence of a certain type of pollen, the tetrad type of cattail (*Typha latifolia*), which grows at this elevation, has a readily recognizable pollen form that tends to preserve fairly well. Absence here suggests that cattail was not used, possibly because it was not available nearby. Cattail is more dependent on permanent water for survival than are many members of the sedge family. Perhaps the riparian conditions at the base of the nearby canyons were no more favorable in the past than they are today.

Human Effects on the Pollen Record and Duckfoot Vegetation Reconstruction

If the predominant plant community at Duckfoot in the past was pinyon-juniper woodland with a sagebrush understory, as it is today, one might expect these three taxa to dominate the 32 Duckfoot samples that yielded 100- or 200-grain pollen counts. The actual situation, however, is more complex. All the Duckfoot samples are from archaeological contexts and therefore do not represent natural pollen rain in an undisturbed habitat. The pollen content of each sample would have been affected by land disturbance, clearing, and introduction of plants into the site for economic uses. The three most abundantly represented categories in the Duckfoot samples are pinyon-type pine, sagebrush, and Cheno-am. Most of the records are dominated by Cheno-am (18 samples) or sagebrush (11 samples) pollen, with only 2 samples dominated by pinyon-type pine pollen and 1 sample (Pit Structure 3, 602-1) showing codominance of juniper and sagebrush. The results of tree-ring analysis (Appendix B) make it clear that pinyon and juniper were the most common construction materials. The flotation sample results support this conclusion and document the importance of these plants, as well as of sagebrush, as fuel sources. Flotation sample results also point to food use of Cheno-am species. In using pollen to reconstruct past vegetation at Duckfoot, the problem is to recognize when human use of pine, juniper, sagebrush, and Cheno-am has magnified (skewed) the natural expressions of these wind-pollinated types.

For example, high percentages of pinyon-type pine pollen were recorded in Pit Structure 3 (sample 594-5), Pit Structure 4 (three samples excluding roof fall), Room 10 (583-1), Room 16 (143-45), and Room 19 (559-13). Pinyon wood was commonly identified during tree-ring sample analysis (Appendix B). In the flotation sample analysis (Chapter 6), it was generally not possible to separate pine parts by species. Pine remains in the various structures included wood, bark scales, and needle fragments. Remains of pinyon nuts were not noted, but the shells are thin and perishable and may be underrepresented in the flotation

Table 7.6. Comparison of Flotation and Pollen Sample Results

| Structure | Flotation[a] | | Pollen[b] | |
	Floor and Features	Roof Fall	Floor and Features	Roof Fall
Pit Structure 1†	Op, Ch, Comp, Ar, Ju, Phr, Mal, Pi, Por, Cerc/Ar, Cerc, Phy, Z, Pha	Op, Ar, Ju, Pi, Cerc/Ar, Cerc, Z	Subsistence: Cy, Pl, Ch, Z Nonsubsistence: none	Subsistence: Pl, Z Nonsubsistence: none
Pit Structure 2	Atr, Ch, Ar, Ju, Pi, Cerc/Ar, Cerc, Pur, Z		Subsistence: Cy, Pl, Cl, Ch, Umb, Z Nonsubsistence: Ju	Subsistence: Ch Nonsubsistence: Ar, Lo, Ephn
Pit Structure 3	Rh, Op, Atr, Ch, Ar, Ju, Ped, Pi, Cerc/Ar, Pop, Phy, Z	Atr, Ch, Ar, Ju, Ped, Pi, Cerc/Ar, Z	Subsistence: Cy, Pl, Ch, Cyp, Sph, Por, Umb, Z, Cu Nonsubsistence: Ju, Ped	Subsistence: Cy, Pl, Z Nonsubsistence: Ju
Pit Structure 4†	Rh, Ech, Op, Atr, Ch, Ar, Ju, Scir, Sti, Pi, Cerc, Per/Am, Phy, Z	Ar, Ju, Pi	Subsistence: Cy, Pl, Cl, Ch, Cyp, Umb, Z, Cu Nonsubsistence: Ju, Ped	Subsistence: none Nonsubsistence: Ar
Room 3†	Cerc/Ar, Z		Subsistence: Cy, Pl, Z Nonsubsistence: none	no sample
Room 5	Atr, Ar, Ju, Pi, Z	no sample	Subsistence: Pl, Ch, Z Nonsubsistence: none	no sample
Room 10	Op, Ar, Ju, Pi, Cerc, Z		Subsistence: Pl, Cl, Ch, Z Nonsubsistence: Ar, Ju, Ped	no sample
Room 12†	Atr, Ch, Ar, Ju, Gr, Sti, Mal, Pi, Cerc/Ar, Cerc, Phy, Z		Subsistence: Cy, Pl, Ch, Umb, Z, Cu Nonsubsistence: none	no sample
Room 13	Ch, Ar, He/Vi, Des, Ju, Sti, Ped, Pi, Por, Cerc/Ar, Phy, Z		Subsistence: Pl, Ch, Umb, Z Nonsubsistence: none	no sample
Room 16	Op, Atr, Ch, Ar, Ju, Sti, Yuc, Pi, Por, Cerc/Ar, Rub/Gal, Pop, Phy, Z		Subsistence: Cy, Pl Nonsubsistence: Ar, Ped	no sample
Room 19†	Ch, Ju, Sti, Pi, Cerc/Ar, Cerc, Pop, Z		Subsistence: Cy, Pl, Ch, Hi, Ephn, Gr, Por, Er, Z, Cu Nonsubsistence: Ped, Ppond	Subsistence: Cl, Ch, Le, Z, Cu Nonsubsistence: none

Flotation taxa abbreviations: Ar = *Artemisia*; Atr = *Atriplex*; Cerc = *Cercocarpus*; Cerc/Ar = *Cercocarpus/Artemisia*; Ch = cheno-am; Comp = Compositae; Des = *Descurainia*; Ech = *Echinocereus*; Gr = Gramineae; He/Vi = *Helianthus/Viguiera*; Ju = *Juniperus*; Mal = Malvaceae; Op = *Opuntia* (prickly pear); Ped = *Pinus edulis*; Per/Am = *Peraphyllum/Amelanchier*; Pha = *Phaseolus*; Phr = *Phragmites australis*; Phy = *Physalis longifolia*; Pi = *Pinus*; Pop = *Populus angustifolia*; Por = *Portulaca retusa*; Pur = *Purshia*; Rh = *Rhus*; Rub/Gal = *Rubiaceae/Galium*; Scir = *Scirpus*; Sti = *Stipa hymenoides*; Yuc = *Yucca*; Z = *Zea*.

For explanation of pollen taxa abbreviations, see Table 7.3.

[a] Taxa listed for flotation samples are for general comparative purposes and do not reflect the level of identification actually used in analysis. For precise taxonomic categories, refer to Chapter 6.

[b] Taxa listed for pollen samples were found in percentages or aggregates relatively high enough or large enough to indicate that the occurrence was of ethnobotanical significance. Includes results of standard and large-fraction scanning. Structures for which large-fraction-scanning data are included are indicated by the symbol †.

record. Much of the flotation evidence is believed to reflect hearth fuel use (Chapter 6), and higher pinyon-type pollen percentages probably also reflect such uses. Because pollen washes of mature pinyon cones and nuts by this author have resulted in almost no pollen recovery, it is unlikely that high pinyon percentages would be contributed by nut processing. Pollen, however, could easily become trapped in the rough bark of branches and be transported to sites through fuelwood gathering. The wood may have been collected when the pine was pollinating, or the wood gathered may have included branches with male cones. In Room 19, a high pinyon-type pollen percentage (42.0 percent) and a high ponderosa-type value (13.0 percent) were recorded for a sample collected from near the hearth (sample 559-13). Thus, the highest pine pollen percentages in the Duckfoot records may not be an aspect of the natural past pollen rain but, rather, a result of human activity such as fuelwood collection and storage.

Juniper remains recovered from flotation samples from various structures at Duckfoot include wood, twigs, scale leaves, fruits, and seeds. Evidently, juniper branches were used for roof construction and fuel, and the fruits were used for food (Chapter 6). Fruits can stay on trees for extended periods of time, even through the next reproductive season, and fruit harvesting when male cones are present may result in the introduction of pollen into cultural contexts. Thus, either subsistence or nonsubsistence use could be indicated by high juniper pollen percentages, and sample context and flotation sample results are crucial to the interpretation of the pollen results. Given the quantities of nonseed parts recovered from flotation samples collected from structures, it is probable that the highest pollen percentages predominantly reflect cultural use of plant materials for nonsubsistence purposes.

In Pit Structure 3, high percentages of juniper pollen tend to be associated with larger aggregates of juniper pollen (sample 569-9, roof fall; sample 602-1, under slab). This pattern is not as strong elsewhere at the site, but high percentages tend to be more common in pit structures (Pit Structure 2, sample 373-9; Pit Structure 4, sample 986-19) than in the above-ground rooms (only Room 10, sample 582-5, has a fairly high percentage of juniper pollen). This trend probably relates in part to differences in abandonment activities; Pit Structures 3 and 4 and Room 10 were burned, and the introduction of juniper as fuel to promote the burning of the structures could have contributed to the pollen evidence. In Pit Structure 2, which was not burned, the sample with the highest percentage of juniper pollen was from the hearth area, and this sample may reflect use of juniper for hearth fuel. The highest values of juniper, therefore, probably were not entirely the result of natural pollen rain in the past but, instead, the result of human activity.

A similar conclusion can be drawn about the highest sagebrush percentages. In the flotation samples, common elements of big sagebrush include flower heads, leaves, and axillary buds (*Cercocarpus/Artemisia*) (Chapter 6). Use of sagebrush as fuel is likely. An evaluation of the archaeological evidence suggests that in some structures sagebrush may have been used as a fuel to intentionally burn the structures at abandonment. High sagebrush pollen values occur in Pit Structure 2 (223-8), Pit Structure 4 (981-3), Room 10 (582-5), and Room 16 (143-4). According to ethnographic accounts, sagebrush was used as a flavoring agent (Kearney and Peebles 1960:938), and the seeds of Wright sagebrush (*Artemisia wrightii*) were used for food (Elmore 1943:82). Use of sagebrush for medicinal purposes has also been documented (Elmore 1943:81). Nonetheless, the high values of sagebrush pollen at Duckfoot, considered in conjunction with flotation sample results, suggest nonsubsistence fuel use of sagebrush, perhaps for purposes related more to abandonment activities than to typical preabandonment activities. The average sagebrush value at Duckfoot is 26.1 percent, with individual values occasionally dropping below 20 percent. Although the highest values probably are not completely natural, local presence of the plants is indicated.

The average Cheno-am value for all Duckfoot samples is 32.7 percent, and values rarely drop below 20 percent. Although cultural enhancement almost certainly is a factor, local presence of source plants also seems necessary to account for such a high percentage. Saltbush is a major contributor of Cheno-am pollen when saltbush is present in the vicinity. Currently it is found within 2.4 km (1.5 mi) of the Duckfoot site. The recovery of four-wing saltbush fruit cores from two Pit Structure 3 flotation samples and common use of saltbush as a fuelwood (Chapter 6) point to the probability of local saltbush presence in the past, or at least of transport into the site. Saltbush, of course, is not the only possible contributor of the Cheno-am pollen that is so well represented in the Duckfoot samples. Numerous herbaceous weedy species such as amaranth, or goosefoot, also could have occurred at the site. Cheno-am seeds of either *Chenopodium* (goosefoot) or *Amaranthus* (amaranth) are common items in the flotation samples, and local presence of these weedy types seems highly likely.

Considering the manifold evidence for cultural effects on the pollen record, it is difficult to reconstruct the predominant vegetation surrounding the Duckfoot site during its occupation. Both the pollen and the flotation sample results suggest that pinyon, juniper, sagebrush, and chenopodiaceous plants were important components of the local and regional vegetation. At Duckfoot, pinyon-type pine pollen values generally exceed juniper values, with only a few exceptions. This does not necessarily mean, however, that pinyon was the more dominant arboreal, since numerous factors affect pollen dominance expressions (Gish 1982:106). Overall, a mix of arboreals, understory shrubs, and herbaceous plants is indicated for the site area in the past.

Ethnobotanic Plant Use, by Structure

In the following discussion, the Duckfoot pollen data are used to enhance interpretations of structure and activity-area function that were based on archaeological observation (for example, presence/absence of certain features and artifacts). Complete pollen analysis results are provided in Tables 7.7 and 7.8; provenience details are provided in Table 7.1.

Pit Structure 1

Three pollen samples from Pit Structure 1 were analyzed: one from roof fall, one from the floor beneath a vessel, and one from the floor beneath a mano (PL 34). The sample beneath the mano produced a 100-grain count; the other two had insufficient pollen for analysis. The Cheno-am value in the sample from beneath the mano is enhanced (that is, higher than would be expected if it were the result of natural pollen rain), and maize pollen was present (Table 7.7). Cholla and prickly pear were noted during large-fraction scanning. Prickly pear and maize pollen were observed in the standard scan of the roof fall sample (68-6). The presence of these pollen types in roof fall could be related to the drying of food items on top of the roof or to the suspension of food items from roof rafters inside the structure. Use of Cheno-am species for food suggests spring-summer-fall plant gathering; the presence of maize implies summer-fall planting, crop maintenance, and harvesting; and the recovery of cholla and prickly pear pollen suggests spring collecting of buds or flowers (Table 7.4). On the basis of pollen evidence alone, one might not infer year-round occupation; the limited variety of ethnobotanic taxa (Tables 7.5 and 7.6) suggests spring-through-summer or -fall use. However, these results are based on a single Pit Structure 1 sample that produced a 100-grain pollen count. The diversity of subsistence items in the flotation record (Table 7.6) *is* suggestive of year-round occupation.

The pollen and flotation evidence together support an interpretation of domiciliary use for Pit Structure 1. Human skeletons on the floor of the structure suggest the possibility of funerary rituals at abandonment, but no unusual pollen types or percentages are apparent in the record. Ritual plant (flower or pollen) use may be inferred on the basis of high percentages or large or abundant aggregates of unusual types, particularly those not normally used for subsistence purposes. Ritual use may also be inferred from the presence of types that have ethnographically documented ritual uses. The Pit Structure 1 pollen profile reveals no evidence of ritual plant use. It should be noted that none of the analyzed samples from this or any other structure were directly associated with on-floor skeletal remains. Analysis of samples from such contexts perhaps would provide additional insights into the nature of funerary rituals conducted at the site.

Pit Structure 2

Four Pit Structure 2 samples were analyzed, and all produced 200-grain counts (Table 7.7). The ethnobotanic variety (Tables 7.5 and 7.6) was moderate. Two samples from the troughs of metates yielded the greatest variety of ethnobotanic pollen types. The metate near the hearth (PL 1, Surface 2) seems to have served as a good pollen trap, and analysis results suggest that such food items as Cheno-am, cholla, prickly pear, and maize might have been processed in this area (sample 373-9). The juniper value for this same sample is also fairly high, which may indicate processing of juniper seeds. The other metate sample (362-2, from PL 108, Surface 1) yielded Cheno-am, cholla, prickly pear, maize, and Umbelliferae pollen. The two metates may have been the foci of two fairly similar activity areas.

Beeweed and Cheno-am pollen were notable in the floor sample from the north part of the structure (332-12). The presence of these pollen types could relate to generalized food use within the house. Beeweed also has other known uses—as an organic paint, for example (Whiting 1939:78). Analysis of additional contrasting floor samples would be required to determine if this part of the structure was used for subsistence or nonsubsistence activities.

The unburned roof fall sample (223-8) produced the highest values of low-spine Compositae, *Artemisia,* and *Ephedra nevadensis*-type pollen of the four Pit Structure 2 samples. The Cheno-am value for this sample also may be enhanced (note aggregate). Perhaps shrubs grew around the structure or were used in roof construction. The absence of beeweed, cholla, and prickly pear pollen in the roof fall sample emphasizes the probable ethnobotanic significance of these plants in the floor samples.

Overall, the pollen evidence suggests that gathered plants, in addition to maize, were processed or used in Pit Structure 2. Spring-through-summer or -fall food procurement is indicated. There is a greater variety of ethnobotanic plants than in Pit Structure 1, in part because of the larger number of productive samples. The greater ethnobotanic diversity in Pit Structure 2 (six taxa that might have been used for food [Tables 7.5 and 7.6]) is suggestive of fairly intense occupation and, hence, probably year-round use of the structure.

Pit Structure 3

Eight of the 11 pollen samples from Pit Structure 3 produced 100- or 200-grain counts (Table 7.7). The largest number of taxa inferred to be of ethnobotanic significance in any one sample is four, and a total of nine subsistence categories is represented in the structure overall (Tables 7.5 and 7.6). A floor sample (604-3) beneath a jar (PL 129) yielded squash pollen. No additional taxa were observed in the large-fraction scan of two Pit Structure 3 samples (602-1 and 604-3). Spring-through-fall plant acquisition is

Table 7.7. Pollen Type Percentages, Pit Structures

	Pit Structure 1[a]	Pit Structure 2				Pit Structure 3								Pit Structure 4			
Sample Code:	72	223	332	362	373	569	588	589	594	600	602	604	612	981	986	986	986
	6	8	12	2	9	9	10	2	5	1	1	3	3	3	14	15	19
Arboreal Pollen																	
Picea															.5		
Pseudotsuga															⊕	.5	+
Pinus edulis-type	4.0	13.5	5.5	10.5	13.0	14.0	11.0	8.0	22.0	11.0	10.5	7.5	8.0	6.0	25.0	23.0	15.5
P. ponderosa-type	3.0	1.5	5.0	2.5	1.5	1.0	6.5	1.0	1.0	2.0	2.0	.5	1.0	.5	2.0	3.5	2.0
Pinus	1.0	.5	.5	.5	.5	1.0	.5	.5		1.0		1.0			1.5	1.0	1.5
Juniperus		3.0	3.0	2.5	18.0	19.0	6.0	5.0	6.0	4.0	26.5	2.0	2.0		2.0	1.0	7.0
Quercus						1.0	1.0				.5	.5	.5		2.0	1.5	
Nonarboreal Pollen																	
Low-spine Compositae	4.0	13.0	6.0	2.5	3.0	4.0	4.5	12.5	9.0	5.0	8.5	6.5	11.0	7.0	3.5	1.5	4.5
High-spine Compositae		1.0	1.5		.5	1.0	1.5	1.0	1.0		2.0	.5	1.0	1.0	1.0	2.5	1.5
Artemisia	26.0	35.5	14.5	17.5	18.5	21.0	20.0	29.5	30.0	37.0	26.5	35.0	29.5	74.0	34.5	28.5	20.5
Cheno-am	58.0	25.5	59.0	58.0	39.0	28.0	46.0	33.5	27.0	29.0	16.0	26.5	41.0	9.0	24.0	29.5	41.0
Sarcobatus											1.0						
Gramineae			2.0	1.5	1.0	2.0		2.0		3.0	2.5	.5	1.5	1.0	1.0	2.0	3.0
Ephedra nevadensis-type	1.0	3.0							1.0					1.0	2.0	⊕	
E. torreyana-type		.5				1.0		.5			.5	.5	.5		.5	⊕	
Sphaeralcea-type						1.0						16.0	.5			⊕	
Boerhaavia-type															⊕	⊕	
Gaura-type											+				⊕		
Oenothera-type															⊕		
Portulaca										2.0							
Rosaceae												.5					
Umbelliferae								1.0					1.0				
Cyperaceae			.5														
Cleome				.5													
Cylindropuntia	⊕					+			2.0						.5	.5	1.5
Platyopuntia	⊕		⊕	⊕		1.0						.5			1.0	.5	
Zea	2.0			1.0		+	.5		1.0			.5			+	2.5	.5
Cucurbita																⊕	
Unknown	1.0	3.0	2.0	1.0	2.5	5.0	2.0	3.5		5.0	2.0	1.0	3.0	1.0	.5	1.5	1.0
Total count	100	200	200	200	200	100	200	200	100	100	200	200	200	100	200	200	200
Aggregates[b]	Ch-12b	Ch-50b	Ch-10b	Ch-10b Z-4a	Ch-6b Lo-5a (Z-2) [Ch-12]	Ju-8b Ch-2a (Z-2)	(Ch-12) Ar-3a	Ch-12b Ar-3a	Ar-2a Lo-2a	none	Ju-10d Ar-2a Ch-6b	Ju-2a Lo-3b Ar-5b Ch-2b	Ch-4b	Lo-2a Ar-3c	Ju-2a Ar-5b Ch-13b (Ch-15) [CI-8]	Ped-3a Ar-15a	Ch-4b

NOTES: The first number in the two-number sample code is a provenience indicator, the second is a laboratory catalog number. Highlighted percentages denote occurrences interpreted to be of ethnobotanic significance. + = observed in standard scan; ⊕ = observed in large-fraction scan.

[a] Platyopuntia and Zea pollen were noted in the standard scan of one Pit Structure 1 sample with insufficient pollen (68-6).

[b] Aggregate code explanation: The letter or letters preceding the dash refer to the plant taxon (see Table 7.3); the number after the dash refers to the estimated number of grains in the largest aggregate; and the letter following the number refers to the maximum number of aggregates observed (a = 1, b = up to 5, c = up to 10, d = up to 15, e = up to 25, f = up to 50). Codes in parentheses indicate aggregates observed during the standard scan; codes in brackets indicate aggregates observed during the large-fraction scan. Maximum number of aggregates observed was not recorded for scans.

Table 7.8. Pollen Type Percentages, Rooms

Sample Code:	Room 3 — 91 / 6	Room 5 — 129 / 6	Room 10 — 582 / 4	Room 10 — 582 / 5	Room 10 — 583 / 1	Room 10 — 583 / 2	Room 12 — 47 / 4	Room 13 — 139 / 5	Room 16 — 143 / 4	Room 16 — 143 / 45	Room 19 — 559 / 8	Room 19 — 559 / 9	Room 19 — 559 / 13	Room 19 — 680 / 4	Room 19 — 680 / 5
Arboreal Pollen															
Picea	⊕		1.0	.5				2.0	1.0	2.0			1.0	1.0	
Pinus edulis-type	15.5	16.5	14.0	20.5	28.0	15.5	7.0	19.0	8.0	44.0	20.0	24.5	42.0	8.0	14.5
P. ponderosa-type	2.5	1.0	1.0	1.0		.5	⊕	1.0	2.0	4.0	4.0	5.0	13.0	3.0	1.0
Pinus		.5		.5	1.0	.5		2.0		2.0	1.0	1.5	2.0		.5
Juniperus	2.0	1.5	2.0	13.0	3.0	4.5	3.0	8.0	2.0	6.0	4.0	5.0	1.0	4.0	1.0
Quercus	.5					1.0	1.0					1.0			2.5
Nonarboreal Pollen															
Low-spine Compositae	9.0	8.0	11.0	6.5	8.0	12.5	6.0	2.0	9.0	6.0	9.0	3.5	1.0	7.0	6.5
High-spine Compositae	.5				1.0	2.0						.5			1.5
Artemisia	33.5	25.0	24.0	45.0	31.0	34.0	13.0	15.0	40.0	13.0	6.0	13.0	4.0	18.0	23.0
Liguliflorae	.5		2.0			.5									
Cheno-am	30.0	42.0	32.0	7.0	15.0	22.5	45.0	34.0	32.0	16.0	46.0	32.5	28.0	44.0	30.5
Sarcobatus	3.0	.5	3.0	3.5	4.0	2.0	2.0	3.0	1.0	2.0	3.0	4.5	1.0	3.0	4.0
Gramineae	.5		1.0	1.0	1.0							.5			.5
Cruciferae															.5
Ephedra nevadensis-type	.5	1.5		1.0	1.0	.5	1.0	2.0	1.0	1.0	1.0	1.5	4.0	1.0	.5
E. torreyana-type		.5						2.0			1.0				.5
Leguminosae													1.0	2.0	
Sphaeralcea-type	.5	1.5	2.0		1.0						2.0	1.0			⊕
Oenothera-type															+
Eriogonum			1.0					1.0							3.0
Portulaca						.5					+	.5			
Rosaceae															
Solanaceae		.5													
Umbelliferae						.5	4.0	1.0							
Cleome	⊕						1.0					.5		1.0	
Cylindropuntia							⊕			1.0					1.0
Platyopuntia	.5	.5		+			14.0	1.0		1.0		.5			3.0
Zea	.5	.5	1.0		1.0		⊕	3.0			1.0	2.0	3.0	3.0	3.5
Cucurbita														+	⊕
Unknown	1.0		5.0	2.5	6.0	3.0	3.0	5.0	4.0	2.0	2.0	2.5	3.0	4.0	3.0
Total count	200	200	100	100	100	200	100	100	100	100	100	200	100	100	200
Aggregates[a]	Ar-3b, Ch-12b, Gr-2a	Ar-6b, Ch-15b	Ch-4a	Ar-3b, Lo-2a, Ch-2a	Ar-3a, Ch-2a, Sph-2a	Ar-3a	Umb-2a, Z-2b(12), [Ch-65]	Ch-2a	Ar-20b, Ch-8b	Ch-2a	Ch-70b, (Por-2)	Ju-4a, Hi-2a, Ch-10b, Gr-3a, (Z-2)	Ch-5b	Ped-2a, Ar-2a, Le-2a, [Ch-25]	Ch-12b, Er-2a, [Ch-25], [Pl-2]

NOTES: The first number in the two-number sample code is a provenience indicator, the second is a laboratory catalog number. Highlighted percentages denote occurrences interpreted to be of ethnobotanical significance. + = observed in standard scan; ⊕ = observed in large-fraction scan.

[a] Aggregate code explanation: The letter or letters preceding the dash refer to the plant taxon (see Table 7.3); the number after the dash refers to the estimated number of grains in the largest aggregate; and the letter following the number refers to the maximum number of aggregates observed (a = 1, b = up to 5, c = up to 10, d = up to 15, e = up to 25, f = up to 50). Codes in parentheses indicate aggregates observed during the standard scan; codes in brackets indicate aggregates observed during the large-fraction scan. Maximum number of aggregates observed was not recorded for scans.

suggested by the nine subsistence taxa. Squash in particular indicates fall activities, because it often is a slightly later harvest than maize.

The variety of ethnobotanic plants differs from sample to sample, which may reflect differences in the types of food prepared and stored in the various parts of the structure. The results from the roof fall sample (569-9) raise the possibility that some subsistence items (cholla, prickly pear, and maize) may have been temporarily placed on the roof for drying or suspended from the ceiling as a means of storage. Such materials could have become incorporated in the roof layers as the roof collapsed when the structure burned. Alternatively, later reuse of the pit structure depression, as evidenced by an intrusive fire pit (Feature 2) in postabandonment fill, could have contributed this evidence. Thus, postabandonment disturbance could also be a factor here.

It is interesting that in this structure the greater concentrations of ethnobotanic taxa are away from the hearth area, unlike the situation in Pit Structure 2. This is an unusual distribution because the hearth is often seen as the focus of food preparation, and it may reflect storage of food items around the periphery of the structure. Overall, the pollen evidence is consistent with a view of the structure as a domiciliary unit, and there are no obvious indications of ritual plant use, despite the evidence for ceremonial abandonment, including on-floor human skeletal remains.

Pit Structure 4

Four of the six samples from this structure yielded 100- or 200-grain pollen counts (Table 7.7). The roof fall sample (981-3) had a very high percentage of sagebrush pollen (74 percent), which may indicate use of sagebrush in roof construction or its introduction as fuel to promote the burning of the structure at abandonment. The overall variety of ethnobotanic categories (Tables 7.5 and 7.6) is fairly diverse (eight taxa, including squash in the large-fraction scan). Samples from the portion of the pit structure north of the wing wall (986-19 and especially 986-15, from under manos PL 45 and PL 138, respectively) yielded a broad range of ethnobotanic pollen types. In the sample from south of the wing wall (986-14), the ethnobotanic taxa are all staple food types: cholla, prickly pear, and maize. Ground, pecked, and/or polished stone was common on the floor in this area, and this space could have been used for processing or storing the major plant resources. Alternatively, the lower economic diversity may reflect differences in sampling contexts, since, in contrast to the other samples, sample 986-14 sediments were not sealed by an artifact.

The pollen evidence is consistent with an interpretation of domiciliary use for Pit Structure 4, and the diversity of economic taxa suggests year-round occupation. Flotation samples also yielded a diverse complement of ethnobotanic

types (Table 7.6). There are no pollen indicators of ritual activity in the form of unusually high percentages or large pollen aggregates of plants that might have been used in ceremonies.

Pit Structure 4 differs from the other pit structures in that the diversity of pollen taxa is not directly related to the number of samples collected and analyzed. Only four Pit Structure 4 samples yielded sufficient pollen for analysis, yet the diversity of ethnobotanic pollen types is high. One Pit Structure 4 sample (986-15) yielded a greater diversity of subsistence taxa (six items, or seven, if large-fraction scanning results are included) than any other single sample from a pit structure. This Pit Structure 4 sample may simply have been collected from a particularly well protected context, and the results do not necessarily signify any difference in function for this structure.

Room 3

One pollen sample from the floor was analyzed, and it yielded a 200-grain count. Ethnobotanic taxa are limited to cholla (in the large-fraction scan), prickly pear, and maize (Tables 7.5, 7.6, and 7.8). The limited ethnobotanic diversity could be related to several factors: the kinds of items stored, the "state" of food items being stored, preservation, and the single-sample database. Room 3 may have been a specialized structure used specifically for the storage of maize and cacti, or it may have been used for storage of a variety of processed foods that were in refined stages of reduction, which would have resulted in sparse pollen deposition. Alternatively, the conditions under which the room was abandoned may have contributed to poorer pollen preservation than seen elsewhere at the site. Because so few artifacts were left on the floor, Room 3 lacked the well-sealed contexts that characterized the pit structures, and exposed contexts are less conducive to preservation of intact ethnobotanic pollen assemblages. Sample size could be affecting the Room 3 pollen profile as well, given that only one sample from this room was analyzed. Despite the limitations of the Room 3 pollen record, the possibility remains that the low diversity of pollen types is related to structure function. Perhaps surface rooms, or at least some surface rooms, were used primarily for "passive" storage of foods, in contrast to pit structures, which may have been used for daily living and "active" food processing.

Room 5

The one pollen sample from the floor of Room 5 yielded a 200-grain count, and Cheno-am, prickly pear, and maize are interpreted as being ethnobotanically significant (Tables 7.5, 7.6, and 7.8). No additional taxa were recorded in the large-fraction scan of this sample. The sample was collected from under a slab in the northwest corner of the structure, away from the central fire pit, and therefore could have been from a food-storage area rather than from a food-processing area. As in Room 3, the single-sample database may be a factor in the low ethnobotanic diversity.

Room 10

Four of the five samples from Room 10 processing bins (Features 1 and 2) yielded 100- or 200-grain counts. A limited variety of ethnobotanic taxa was recorded (Tables 7.5, 7.6, and 7.8). Pollen types identified include prickly pear, Cheno-am, maize, and beeweed. The functions of the bins apparently were different, since the prickly pear and high Cheno-am value were recorded for Feature 1 (582-4), whereas the beeweed and maize were recorded for Feature 2 (583-2, 583-1). The ethnobotanic diversity in Room 10 is not as great as in Pit Structures 2 and 4, which, like Room 10, had four samples with sufficient pollen. This suggests a more specific or restricted use of this room at the time of abandonment, which is consistent with the artifact and feature evidence for this room.

Room 12

The one pollen sample analyzed from this structure was collected from the floor in the vicinity of the hearth. Ethnobotanic taxa include Cheno-am, Umbelliferae, cholla, and maize (Tables 7.5, 7.6, and 7.8). Prickly pear and squash were observed in the large-fraction scan. This assemblage is more diverse than the assemblages from the three previous rooms and more similar to some of the pit structure assemblages. The pollen record suggests activities consistent with a domestic, or habitation, function for this room. The presence of both spring-harvested cacti and summer- or fall-harvested maize and squash indicates that activities were at least multiseasonal.

Room 13

The one analyzed pollen sample from Room 13 was collected from the surface of a slab metate (PL 41) on the floor. The ethnobotanic plant variety is similar to that described for Room 12, except that cholla and squash pollen were not identified in Room 13 (Tables 7.5, 7.6, and 7.8). Multiseasonal activities are again represented.

Room 16

Two pollen samples associated with Surface 2, the floor of Room 16, produced markedly contrasting results (Table 7.8). Sagebrush pollen (40 percent) dominates the above-floor sample (143-4), whereas pinyon-type pollen (44 percent) is outstanding in the trough metate sample (143-45). Remains of both pine and sagebrush were recovered from Room 16 flotation samples (Table 7.6), which suggests that

the pollen values are related to construction or fuelwood use. If both plants were used when they were pollinating, the results would suggest different seasons of activity (Table 7.4). Cholla and prickly pear were identified in the metate fill sample. This further suggests multiseasonal activities encompassing spring (pinyon and cacti)-through-summer or -fall (sagebrush) collection.

The limited ethnobotanic variety (Tables 7.5 and 7.6) could be a result of the small size of the database. It is not clear where the results of Room 16 pollen analysis should be placed in comparison with pollen results from the other structures. The flotation data reveal a more detailed picture of plant use (Chapter 6).

Room 19

The five analyzed Room 19 pollen samples (Tables 7.1 and 7.8) were collected from two locations: the fill (roof fall) and floor of a large bell-shaped cist (Feature 13) and the floor north of the hearth (Feature 19). Feature 13 had a layer of roof fall at its base. The variety of ethnobotanic taxa—maize, squash, beeweed, Cheno-am, and a member of the pea family (Leguminosae)—in the sample collected from this roof fall layer (680-4) suggests that food items might have been on, or suspended from, the structure roof. Alternatively, the roof may have fallen on top of foods stored in the cist. Even greater ethnobotanic diversity was evident in the sample (680-5) collected from beneath a mano on the floor of the cist: maize, squash, cholla, prickly pear (3 percent), wild buckwheat (*Eriogonum*), and probably high-spine composite pollen were the ethnobotanically significant taxa present in this sample. The diversity may relate to cist function rather than roof fall; if so, many different types of plants possibly were stored throughout the use of this feature.

The sample (559-13) beneath the mano (PL 51) near the hearth is the only one that revealed evidence of nonsubsistence items; both pinyon-type and ponderosa-type values appear enhanced. This may be the result of fuel items having been placed near the hearth for easy access during cooking. Ethnobotanic taxa in sample 559-13 and in the remaining two samples north of the hearth (559-8 and 559-9) include Cheno-am, purslane, cholla, prickly pear, and maize, and probably high-spine composites, joint-fir, and grasses (note aggregate). The purslane (*Portulaca*) associated with the metate near the hearth is particularly noteworthy. Although several subsistence plants are common to both the hearth area and Feature 13, the presence of purslane pollen near the hearth and the recovery of squash pollen from Feature 13 suggest that there may have been at least two separate activity areas for processing and storing food items. Multiseasonal food procurement is evident. The accumulation of a wide variety of food items in the room (Tables 7.5 and 7.6) raises the possibility of year-round use of the structure.

Summary

Many ethnobotanic taxa are common to pit structures and rooms at Duckfoot, although there are a few points of contrast. Some of these involve taxa (for example, globemallow, purslane, wild buckwheat, and sedge) that occur too infrequently to infer functional differences between structure types. In the case of cholla, however, there is a definite difference between pit structure and surface room pollen records. Cholla pollen is far more common in pit structures than in other sampled contexts. Active food processing in the pit structures as opposed to passive food storage in rooms might have resulted in greater pollen accumulation on pit structure floors and could account for this difference.

Overall, the pollen evidence at Duckfoot supports archaeological interpretations of structure function. All pollen-based inferences with regard to plant use and structure function at Duckfoot are tendered with caution, however, because the data patterns may be due at least in part to the small numbers of samples that were evaluated for some contexts, differences in sample contexts within structures, and differences in abandonment activities among the various structures.

8

Faunal Remains

Danny N. Walker

Introduction

The primary goal of the Duckfoot faunal analysis was to provide a basic descriptive summary of the non-human animal bone assemblage recovered from the site. Over 5,700 individual bones or bone fragments from every major cultural unit at the site were analyzed (Table 8.1). This total includes bone tools and ornaments, as well as the large assemblage of unmodified bone that reflects both prehistoric cultural activity (for example, food consumption and discard) and noncultural, or natural, phenomena (for example, natural animal deaths and postabandonment introduction into site deposits). Because over 90 percent of the Duckfoot site was excavated and over 99 percent of the recovered bone was analyzed, the data reported here are considered representative of the local fauna (Hibbard 1958:3) deposited during and after the occupation of the site. In this chapter, tabular data presentations are organized by major cultural unit (structure, courtyard, and midden area); the accompanying text discussion is organized by taxonomic category. Although vertical provenience within individual structures is given general consideration in the text discussion, a detailed comparison of specific contexts was not a primary focus of this study. The interested reader is referred to Appendix D for tabular presentations of data organized by horizontal and vertical provenience.

This chapter treats all faunal materials, modified and unmodified, equally, and the emphasis is on basic zoological identification. Evidence of cultural modification of faunal material was recorded during analysis and is described in general terms. However, more-detailed descriptions of bone tools and ornaments, and interpretations based on the distributions of these items, are provided in Chapter 5.

Methods

A number of recovery techniques were used to collect animal bone from the Duckfoot site: conventional trowel excavation, dry screening through ¼-in mesh, water screening through ¹⁄₁₆-in mesh, and flotation sampling of feature contents. After having been cleaned, bones were identified to the most specific taxonomic category possible; taxonomy follows Jones et al. (1982). Categories range from class level (e.g., mammal, indeterminate) to species level (e.g., *Odocoileus hemionus*), with all possible taxonomic levels in between. If specimens could not be identified to genus, general size groupings within the next-highest taxonomic level possible were used—for example, small mammal (jackrabbit and smaller), medium mammal (larger than jackrabbit to deer), small rodent (vole to small ground squirrel), large rodent (large ground squirrel to prairie dog), small carnivore (mink or smaller), and medium carnivore (fox to coyote). This type of description provides additional information that would not be available were such specimens identified only to class. No attempts were made to reconstruct specimens that appeared to be from the same bone.

The specific skeletal element present and evidence of cultural modification were also recorded during analysis, and these data are on file at the Anasazi Heritage Center. Modifications noted include burning, tool modification, cut marks from skinning or defleshing, and impact fractures from bone breakage. No tool manufacturing debris per se was observed in the collection.

Identifications were made by comparison with known osteological materials in three collections: the personal osteological collection of the author, the comparative osteological collection of the University of Wyoming Department of Anthropology, and the collection of the University

Table 8.1. Faunal Remains Recovered from
the Duckfoot Site

Taxon	NISP	%
Class Reptilia		
cf. Colubridae (colubrid snake)	8	.1
Class Aves		
Large raptor, eagle (?)	2	.0
Meleagris gallopavo (turkey)	12	.2
Grouse	1	.0
Zenaida macroura (mourning dove)	1	.0
Thrush	1	.0
Avian, indeterminate	42	.7
Class Mammalia		
Order Lagomorpha		
Sylvilagus sp. (cottontail)	1,043	18.3
Lepus cf. *californicus* (black-tailed jackrabbit)	334	5.8
Leporid, indeterminate	425	7.4
Order Rodentia		
Spermophilus cf. *lateralis* (golden-mantled ground squirrel)	5	.1
Spermophilus sp. (ground squirrel)	9	.2
Cynomys gunnisoni (Gunnison's prairie dog)	126	2.2
Sciurid, indeterminate	10	.2
Thomomys bottae (Botta's pocket gopher)	11	.2
Dipodomys ordii (Ord's kangaroo rat)	34	.6
Peromyscus sp. (small mouse)	15	.3
Neotoma albigula (white-throated woodrat)	28	.5
Neotoma cf. *mexicana* (Mexican woodrat)	1	.0
Neotoma sp. (woodrat)	15	.3
Cricetid, indeterminate	37	.6
Rodent, indeterminate	644	11.3
Order Carnivora		
Canis familiaris (domestic dog)	39	.7
Canis sp. (canid)	32	.6
Vulpes cf. *macrotis* (kit fox)	1	.0
Canid, indeterminate	18	.3
Felis rufus (bobcat)	1	.0
Small carnivore	2	.0
Medium carnivore	3	.0
Order Artiodactyla		
Odocoileus hemionus (mule deer)	59	1.0
Artiodactyl, indeterminate	267	4.7
Small mammal, indeterminate	2,074	36.3
Medium mammal, indeterminate	278	4.9
Mammal, indeterminate	132	2.3
TOTAL	5,710	100.0

NISP = number of identified specimens.

of Wyoming Department of Zoology and Physiology Vertebrate Museum.[1] These collections are composed primarily of Wyoming mammals, and some indigenous southern Colorado species are not represented, which may have prevented specific assignment of some specimens identified only to genus.

Several abbreviations used throughout this chapter warrant explanation: "cf." is used to describe materials that show greatest similarity to the taxon listed; "sp." denotes that the specimen is not identifiable to species on the basis of the available evidence; and NISP refers to the number of individual specimens counted.

The Duckfoot Site Faunal Assemblage

Remains of 1 reptilian, 5 avian, and at least 16 mammalian taxa were recovered from the Duckfoot site (Tables 8.1–8.5). All taxa are extant in the region (Anderson 1961; Armstrong 1972; Hammerson and Langlois 1981; Bissell and Dillon 1982; Chase et al. 1982; Neusius 1985b, 1985c; also see Chapter 1, Table 1.2, this volume) or are domesticated species (e.g., *Canis familiaris*) associated with the prehistoric occupation of the Duckfoot site. In the following discussion, each taxon at Duckfoot is discussed according to the structure or other excavation unit in which its presence was noted; more specific provenience groupings by vertical subdivision are used in the data presentations in Appendix D. Observations about possible cultural use or modification are also noted in this chapter, and comparisons with the Dolores Archaeological Program (DAP) faunal assemblage are made when appropriate.

CLASS REPTILIA

Family cf. Colubridae

Colubrid snake: Five thoracic vertebrae identified as colubrid snake were recovered from Room 5 (Table 8.3), and three were recovered from the midden (Table 8.5). Two large colubrids are listed by Hammerson and Langlois (1981) and Neusius (1985b, 1985c) as occurring in southwestern Colorado: *Pituophis melanoleucus* (gopher snake) and *Thamnophis elegans* (garter snake). Only the gopher snake attains a size commensurate with the Duckfoot vertebrae. No direct evidence of cultural use or modification of these specimens was noted. The specimens from Room 5 were recovered from the layer of collapsed wall debris, and the specimens from the midden area were recovered from the main cultural stratum in two adjoining

---

1. Permission to use the University of Wyoming Department of Anthropology comparative osteology collection was given by Dr. George C. Frison, Department Head. Dr. Mark Boyce provided permission to use the University of Wyoming Zoology Museum collections.

Table 8.2. Faunal Remains, Architectural Suite 1

Taxon	Room 1		Room 2		Room 3		Room 15		Room 16		Room 20		Pit Structure 1		Courtyard 1		Total	
	N	%	N	%	N	%	N	%	N	%	N	%	N	%	N	%	N	%
Class Aves																		
Large raptor, eagle (?)			1	20.0									1	.1			1	.1
Meleagris gallopavo							1	.4									2	.2
Grouse													1	.1			1	.1
Avian, indeterminate					1	2.4							19	2.2	4	10.8	24	1.9
Class Mammalia																		
Order Lagomorpha																		
Sylvilagus sp.					7	16.7	81	33.8	3	5.6			531	61.5	4	10.8	626	50.3
Lepus cf. californicus			2	40.0	1	2.4	6	2.5	1	1.9	1	33.3	43	5.0	5	13.5	59	4.7
Leporid, indeterminate					12	28.6	5	2.1	7	13.0			56	6.5	6	16.2	86	6.9
Order Rodentia																		
Spermophilus cf. lateralis							1	.4	1	1.9							2	.2
Cynomys gunnisoni			1	20.0			5	2.1									6	.5
Thomomys bottae													1	.1			1	.1
Peromyscus sp.					1	2.4	1	.4					8	.9			10	.8
Neotoma sp.													9	1.0			9	.7
Cricetid, indeterminate									1	1.9			2	.2			3	.2
Rodent, indeterminate					1	2.4	56	23.3					98	11.4			155	12.4
Order Carnivora																		
Canis familiaris							19	7.9					1	.1	2	5.4	21	1.7
Canis sp.					1	2.4	2	.8							1	2.7	5	.4
Vulpes cf. macrotis													1	.1			1	.1
Order Artiodactyla																		
Odocoileus hemionus	1	100.0					1	.4	2	3.7			6	.7	1	2.7	11	.9
Artiodactyl, indeterminate							2	.8	19	35.2			1	.1	7	18.9	29	2.3
Small mammal, indeterminate					7	16.7	42	17.5	20	37.0	1	33.3	27	3.1	4	10.8	101	8.1
Medium mammal, indeterminate					11	26.2	3	1.3			1	33.3	50	5.8	2	5.4	67	5.4
Mammal, indeterminate			1	20.0			15	6.3					8	.9	1	2.7	25	2.0
TOTAL	1	100.0	5	100.0	42	100.0	240	100.0	54	100.0	3	100.0	863	100.0	37	100.0	1,245	100.0

NOTE: Includes modified and unmodified bone.

Table 8.3. Faunal Remains, Architectural Suite 2

Taxon	Room 4 N	Room 4 %	Room 5 N	Room 5 %	Room 6 N	Room 6 %	Room 7 N	Room 7 %	Room 11 N	Room 11 %	Room 12 N	Room 12 %	Room 13 N	Room 13 %
Class Reptilia														
cf. Colubridae			5	2.9										
Class Aves														
Large raptor, eagle (?)														
Zenaida macroura														
Thrush														
Avian, indeterminate					1	.9	1	3.4	1	.3				
Class Mammalia														
Order Lagomorpha														
Sylvilagus sp.	26	4.0	2	1.2	1	.9	9	31.0	10	2.5	21	11.5	4	1.1
Lepus cf. californicus	112	17.4			7	6.0	6	20.7	8	2.0	7	3.8	19	5.1
Leporid, indeterminate	119	18.5	20	11.8	1	.9	2	6.9	42	10.5	9	4.9	5	1.3
Order Rodentia														
Spermophilus sp.					1	.9	1	3.4						
Cynomys gunnisoni	7	1.1			67	57.8	2	6.9	3	.8	2	1.1	1	.3
Sciurid, indeterminate									4	1.0	1	.5		
Thomomys bottae														
Dipodomys ordii														
Peromyscus sp.					2	1.7			1	.3				
Neotoma albigula			1	.6	4	3.4					4	2.2	1	.3
Neotoma cf. mexicana											1	.5		
Neotoma sp.			1	.6					2	.5				
Cricetid, indeterminate	1	.2											1	.3
Rodent, indeterminate	1	.2	122	71.8							5	2.7	317	84.3
Order Carnivora														
Canis familiaris					1	.9			1	.3				
Canis sp.					1	.9								
Canid, indeterminate									4	1.0	1	.5		
Small carnivore									1	.3				
Medium carnivore							1	3.4						
Order Artiodactyla														
Odocoileus hemionus	2	.3			1	.9	4	13.8	1	.3	1	.5		
Artiodactyl, indeterminate	1	.2	6	3.5	4	3.4			3	.8	3	1.6	2	.5
Small mammal, indeterminate	363	56.5	5	2.9	16	13.8	2	6.9	319	79.8	127	69.4	14	3.7
Medium mammal, indeterminate	1	.2	5	2.9	6	5.2	1	3.4			1	.5	1	.3
Mammal, indeterminate	9	1.4	3	1.8	3	2.6							11	2.9
Total	642	100.0	170	100.0	116	100.0	29	100.0	400	100.0	183	100.0	376	100.0

NOTE: Includes modified and unmodified bone.

Table 8.3. Faunal Remains, Architectural Suite 2 (continued)

Taxon	Room 14 N	%	Room 17 N	%	Pit Structure 2 N	%	Pit Structure 4 N	%	Courtyard 2 N	%	Total N	%
Class Reptilia												
cf. Colubridae											5	.2
Class Aves												
Large raptor, eagle (?)					1	.1					1	.0
Zenaida macroura			1	3.7							1	.0
Thrush											1	.0
Avian, indeterminate					4	.4					6	.2
Class Mammalia												
Order Lagomorpha												
Sylvilagus sp.	8	29.6	1	3.7	23	2.6	61	25.6	8	10.8	174	5.5
Lepus cf. californicus					17	1.9	3	1.3	9	12.2	188	5.9
Leporid, indeterminate	1	3.7			4	.4	3	1.3	7	9.5	213	6.7
Order Rodentia												
Spermophilus sp.											2	.1
Cynomys gunnisoni			1	3.7	2	.2	1	.4	2	2.7	88	2.8
Sciurid, indeterminate									2	2.7	7	.2
Thomomys bottae									6	8.1	6	.2
Dipodomys ordii					14[a]	1.6					14	.4
Peromyscus sp.											3	.1
Neotoma albigula									1	1.4	11	.3
Neotoma cf. mexicana											1	.0
Neotoma sp.					1	.1					4	.1
Cricetid, indeterminate					1	.1					3	.1
Rodent, indeterminate			22	81.5	1	.1					468	14.7
Order Carnivora												
Canis familiaris							1	.4	2	2.7	4	.1
Canis sp.					1	.1					3	.1
Canid, indeterminate					2	.2					7	.2
Small carnivore											1	.0
Medium carnivore											1	.0
Order Artiodactyla												
Odocoileus hemionus	2	7.4			6	.7	9	3.8			26	.8
Artiodactyl, indeterminate			1	3.7	9	1.0	36	15.1	3	4.1	68	2.1
Small mammal, indeterminate	7	25.9	1	3.7	805	90.0	2	.8	14	18.9	1,675	52.7
Medium mammal, indeterminate	9	33.3			1	.1	118	49.6	9	12.2	152	4.8
Mammal, indeterminate					2	.2	4	1.7	11	14.9	43	1.4
TOTAL	27	100.0	27	100.0	894	100.0	238	100.0	74	100.0	3,176	100.0

[a] Includes two nearly complete skeletons; each skeleton was counted as three bones.

Table 8.4. Faunal Remains, Architectural Suite 3

Taxon	Room 8 N	Room 8 %	Room 9 N	Room 9 %	Room 10 N	Room 10 %	Room 18 N	Room 18 %	Room 19 N	Room 19 %	Pit Structure 3 N	Pit Structure 3 %	Courtyard 3 N	Courtyard 3 %	Total N	Total %
Class Aves																
Meleagris gallopavo													1	1.7	1	.2
Avian, indeterminate	1	.6									1	.5			2	.4
Class Mammalia																
Order Lagomorpha																
Sylvilagus sp.	170	95.5	16	84.2	6	22.2			4	7.3	6	2.9	2	3.4	204	37.0
Lepus cf. californicus	2	1.1			3	11.1	1	10.0			6	2.9	6	10.3	18	3.3
Leporid, indeterminate	1	.6			4	14.8	5	50.0	3	5.5	9	4.4	8	13.8	30	5.4
Order Rodentia																
Spermophilus cf. lateralis									3	5.5			1	1.7	3	.5
Spermophilus sp.									4	7.3					5	.9
Cynomys gunnisoni	1	.6	1	5.3	2	7.4			3	5.5	13	6.4	2	3.4	22	4.0
Thomomys bottae					1	3.7									1	.2
Dipodomys ordii											17[a]	8.3			17	3.1
Peromyscus sp.											1	.5			1	.2
Neotoma albigula	1	.6							1	1.8	13	6.4			14	2.5
Cricetid, indeterminate											30	14.7			31	5.6
Rodent, indeterminate													4	6.9	4	.7
Order Carnivora																
Canis familiaris											4	2.0	2	3.4	6	1.1
Canis sp.					1	3.7			1	1.8			3	5.2	5	.9
Canid, indeterminate													3	5.2	3	.5
Small carnivore											1	.5			1	.2
Medium carnivore													1	1.7	1	.2
Order Artiodactyla																
Odocoileus hemionus	1	.6			1	3.7			2	3.6	5	2.5	1	1.7	10	1.8
Artiodactyl, indeterminate											45	22.1	2	3.4	47	8.5
Small mammal, indeterminate					8	29.6	4	40.0	21	38.2	44	21.6	18	31.0	95	17.2
Medium mammal, indeterminate	1	.6	2	10.5	1	3.7			13	23.6	8	3.9	4	6.9	29	5.3
Mammal, indeterminate											1	.5			1	.2
TOTAL	178	100.0	19	100.0	27	100.0	10	100.0	55	100.0	204	100.0	58	100.0	551	100.0

NOTE: Includes modified and unmodified bone.

[a] Includes one nearly complete and two partial skeletons; each skeleton was counted as three bones.

Table 8.5. Faunal Remains, Midden and Arbitrary Units

Taxon	Midden		Arbitrary Units	
	N	%	N	%
Class Reptilia				
cf. Colubridae	3	.4		
Class Aves				
Meleagris gallopavo	9	1.3		
Avian, indeterminate	9	1.3	1	2.1
Class Mammalia				
Order Lagomorpha				
Sylvilagus sp.	32	4.6	7	14.6
Lepus cf. *californicus*	58	8.4	11	22.9
Leporid, indeterminate	95	13.8	1	2.1
Order Rodentia				
Spermophilus sp.	2	.3		
Cynomys gunnisoni	10	1.4		
Sciurid, indeterminate	3	.4		
Thomomys bottae	3	.4		
Dipodomys ordii	3	.4		
Peromyscus sp.	1	.1		
Neotoma albigula	1	.1	2	4.2
Neotoma sp.	2	.3		
Rodent, indeterminate	14	2.0	3	6.3
Order Carnivora				
Canis familiaris	8	1.2		
Canis sp.	19	2.8		
Canid, indeterminate	8	1.2		
Felis rufus	1	.1		
Medium carnivore	1	.1		
Order Artiodactyla				
Odocoileus hemionus	11	1.6	1	2.1
Artiodactyl, indeterminate	120	17.4	3	6.3
Small mammal, indeterminate	192	27.8	11	22.9
Medium mammal, indeterminate	25	3.6	5	10.4
Mammal, indeterminate	60	8.7	3	6.3
TOTAL	690	100.0	48	100.0

NOTE: Includes modified and unmodified bone.

units. Neusius (1986) reported a total of two gopher snake bones from all Dolores-area Pueblo I sites.

CLASS AVES

Order Falconiformes
Family cf. Accipitridae

Large raptor, eagle(?): Terminal phalanges of a large raptor were recovered from Room 2 fill (Table 8.2) and Pit Structure 2 roof fall (Table 8.3) at Duckfoot. The specimen from Room 2 was burned and calcined. Comparison with numerous raptor bones in the University of Wyoming Anthropology Department collection showed that the specimens compared well in size and general morphology with *Aquila chrysaetos* (golden eagle). However, because of general similarities with many other raptor terminal phalanges, positive identification was not made. Other than their presence in site deposits and the burned condition of one bone, there are no indications of cultural use or modification of these bones.

Order Galliformes
Family Meleagrididae

Meleagris gallopavo **(turkey):** Twelve specimens identified as turkey were recovered from various locations at Duckfoot: Room 15 (Feature 2, cist), Pit Structure 1 (roof fall), Courtyard 3 (Feature 39, pit), and six excavation units in the midden (Tables 8.2, 8.4, and 8.5). The presence of turkey bones in Pit Structure 1 roof fall is suggestive of some form of cultural use of this species, although no direct evidence of cultural modification or use was seen on any of the specimens. In comparison with Pueblo I sites at Mesa Verde, turkey bones were not recovered in large numbers from Dolores-area Pueblo I sites (Benz 1984; Neusius 1985a, 1986). Benz (1984:211) speculated that this might have been the result of habitat differences between the two areas. Neusius (1986) reviewed the use of turkey at Dolores Pueblo I sites and concluded that the species was used for food; no evidence suggested the keeping of turkeys for feather use alone. The small number of turkey bones in the Duckfoot collection suggests that the site inhabitants did not keep or use turkeys to any large degree, at least at the site itself.

Family Tetraonidae

Grouse: A single distal tibiotarsus identified as a member of the grouse family was recovered from postabandonment fill in Pit Structure 1 (Table 8.2), but comparative osteological material necessary for specific identification was not available. Three grouse species are reported by Chase et al. (1982) and Neusius (1985b, 1985c) for southwestern Colorado. Comparisons were made with two of

these species, *Centrocercus urophasianus* (sage grouse) and *Pedioecetes phasianellus* (sharp-tailed grouse), but dissimilarities beyond species level were noted. The Duckfoot specimen probably represents the third species, *Dendragapus obscurus* (blue grouse). No signs of cultural modification are apparent on the Duckfoot specimen. *Dendragapus obscurus* may have been used by inhabitants of Pueblo I sites in the Dolores area (Benz 1984). Neusius (1986) identified only sage grouse from Dolores sites but, as was the case in the Duckfoot analysis, was unable to compare the specimens with known blue grouse skeletal material.

Order Columbiformes
Family Columbidae

Zenaida macroura **(mourning dove):** A single right humerus identified as mourning dove was recovered from postabandonment fill in Room 17 (Table 8.3). This specimen compared well with a series of mourning dove humeri, but it is also possible that the specimen represents *Columba fasciata,* the band-tailed pigeon (not examined during this study). Although the band-tailed pigeon is known to nest in southwestern Colorado today (Chase et al. 1982; Robbins et al. 1966), its prehistoric distribution is not known. Osteological differences between the two genera should be greater than those noted during comparison of the Duckfoot specimen with the series of mourning dove skeletons. Mourning doves are known to occupy abandoned buildings and other structures. No definite evidence for cultural use or modification was seen on the specimen. Neusius (1986) reported this species for Dolores sites and discussed its ethnographic use for both food and ceremonial purposes (Neusius 1985b).

Order Passeriformes
Family Turdidae

Thrush: A complete humerus of a small passerine bird was recovered from the layer of collapsed wall debris in Room 6 (Table 8.3). This specimen compared closely with several members of the family Turdidae in the University of Wyoming Department of Anthropology comparative collection. Five members of this passerine family are found in southwestern Colorado (Chase et al. 1982; Neusius 1985b, 1985c). Two of these are *Sialia* spp. (bluebirds) and are smaller than the Duckfoot specimen; *Turdus migratorius* (robin) is larger than the Duckfoot example. Skeletons of the remaining two birds, *Catharus guttatus* (hermit thrush) and *Myadestes townsendi* (Townsend's solitaire), were not available for comparison. Other thrushes were examined (*Catharus ustulata* [Swainson's thrush] and *Catharus fuscescens* [verry]) but were specifically dissimilar to the Duckfoot specimen. The specimen's presence in the upper fill of Room 2 may indicate that it is intrusive into site deposits and not associated with the cultural occupation. Neusius (1986) reported both mountain bluebird (*S. currucoides*) and robin from the Dolores sites.

Avian, indeterminate

Fragmentary avian material was recovered from Rooms 3, 6, 7, 8, and 11; Pit Structures 1, 2, and 3; Courtyard 1; and the midden (Tables 8.2–8.5). Roof fall, floor, feature (hearth and bin), and postabandonment contexts are represented in the assemblage. For the most part, the specimens are broken long-bone shaft fragments; none of the material retains morphological characteristics necessary for specific, or even ordinal, identification. Cut marks were observed on one bone found in the mixed wall and roof fall stratum in Room 7. A bird bone awl was recovered from the floor of Pit Structure 1, and two bone tubes were found in two pit features in Courtyard 1.

CLASS MAMMALIA

Order Lagomorpha

Sylvilagus sp. (cottontail): Cottontail is the most common taxon in the identifiable Duckfoot assemblage (Table 8.1). Cottontail bones were recovered from 16 surface rooms, all 4 pit structures, 4 features and 3 excavation units in the courtyard areas, and 23 of the 70 (32.8 percent) midden-area excavation units that yielded bone (Tables 8.2–8.5). In addition, when considering NISP, cottontails occurred in greater numbers than jackrabbits throughout the site (Table 8.1). This suggests that cottontails may have been actively hunted by the prehistoric inhabitants of Duckfoot and that they served as an important subsistence item. Floor, roof fall, feature, and postabandonment contexts are represented. No definite cut marks were seen on any of the cottontail material, which may reflect the ease with which cottontails can be skinned and dismembered without using tools of any kind (personal observation). It is also possible that cut marks simply were not preserved on the bones.

The large number of cottontail bones recovered from Duckfoot may reflect, at least partly, a hunting strategy designed to take advantage of the various "pest" animals that invade disturbed land such as agricultural fields. Benz (1984) proposed that animals attracted to agricultural plots may have been hunted by humans as a means of eliminating competition for food crops and providing dietary protein in the form of meat. Such exploitation of animal pests is referred to as "garden-hunting," and Benz (1984) proposed that cottontails, jackrabbits, prairie dogs, and turkeys may have been among the species taken in this manner at Dolores.

Some of the Duckfoot cottontail specimens are clearly from animals that died naturally after the site had been abandoned. Although there is no direct evidence in the form of articulated remains, numerous cottontail specimens were recovered from upper fill in various structures, which suggests natural mortalities following abandonment. An especially convincing case of postabandonment inclusion in site deposits may be seen in Pit Structure 1, where over 500 cottontail bones were recovered from the fill of the vent shaft (Feature 10). The interior opening of the vent tunnel had been sealed by a large sandstone slab, so animals that fell into the vertical shaft would not have been able to gain access to the structure interior, where they might have been able to crawl to safety over the collapsed roof debris.

Two species of cottontail have been reported in southwestern Colorado today (Armstrong 1972): *Sylvilagus audobonii* (desert cottontail) and *Sylvilagus nuttallii* (Nuttall's cottontail). These two species, although sympatric in their ranges, occur in different local habitats (Hall 1981). Nuttall's cottontail is most commonly found in riparian or densely wooded areas. The desert cottontail prefers the uplands and is found most commonly in parkland savannas. The two species have been identified only on the basis of external characteristics, complete skulls, or mandibular measurements (Flint and Neusius 1987; Neusius and Flint 1985). Characteristics allowing separation of the two species on the basis of postcranial osteological material have not been defined in the literature.

Given the upland location and environment of Duckfoot, it is possible that the cottontail specimens are *Sylvilagus audobonii*. However, prehistoric inhabitants of Duckfoot may have hunted in a wide variety of environmental zones, including the riparian and wooded habitats of *Sylvilagus nuttallii*. *Sylvilagus audobonii* was the most common species reported for Dolores Pueblo I sites (Neusius 1986; Flint and Neusius 1987). Flint and Neusius (1987) proposed that cottontails were procured close to the location where they were ultimately used. If so, this would suggest that most cottontails recovered from Duckfoot are *S. audobonii,* assuming that the prehistoric environment was similar to the modern environment. However, disturbed agricultural areas near Duckfoot might have attracted *Sylvilagus nuttallii* into the area (see Benz 1984; Flint and Neusius 1987), and it seems likely that both species are present in the collection.

Lepus cf. *californicus* (black-tailed jackrabbit): Bones of black-tailed jackrabbit were recovered from 18 of 24 structures, 3 features and 9 excavation units in the courtyards, and 30 of 70 (42.8 percent) excavation units in the midden that yielded bone (Tables 8.2–8.5). As with cottontail, the relatively large number of jackrabbit bones may reflect, at least in part, a garden-hunting strategy. Remains were recovered from floors, roof fall strata, features, and postabandonment deposits. Five tools or tool fragments were recognized—one awl and four small fragments that had been cut, perforated, or polished. In the general bone assemblage, cut marks were observed on only two specimens. For both jackrabbit and cottontail, indications of stylized breakage, such as that reported from Rabbit Bone Cave, Wyoming (Walker 1988), were present. However, the sample was not large enough to state definitely that such breakage had occurred. Because of the fragility of leporid bones, similar breaks could have been caused by trampling or structural collapse.

Both *Lepus californicus* and *Lepus townsendii* (white-tailed jackrabbit) are extant in southwestern Colorado (Armstrong 1972), but only the black-tailed jackrabbit is recorded in the immediate vicinity of Duckfoot. As with cottontails, complete skulls are necessary to osteologically separate the two jackrabbit species. It is possible that the prehistoric range of the white-tailed jackrabbit was wider than the modern range or that prehistoric hunters traveled north and east to hunt white-tailed jackrabbit. Neusius (1986) reported the two species to be about equally represented in Dolores Pueblo I sites, but those sites were dispersed over a more varied habitat than that surrounding Duckfoot.

Leporid, indeterminate: Numerous specimens identified simply as members of the leporid family were recovered from a variety of contexts at Duckfoot (Tables 8.2–8.5). This material was too fragmented to be identified as either cottontail or jackrabbit, although both animals are probably represented. Two small fragments showed signs of modification in the form of light polish, and one long-bone shaft fragment had been shaped into a bead. Snowshoe hares (*Lepus americanus*) are also extant in southwestern Colorado and can be distinguished from cottontails and jackrabbits by their intermediate size (adult specimens). No mature leporid material from Duckfoot was of this intermediate size. Neusius (1986) reported snowshoes to be common in the Dolores Pueblo I lagomorph material. Again, the apparent differences between the Duckfoot and Dolores assemblages may reflect habitat differences between the two areas.

Order Rodentia
Family Sciuridae

***Spermophilus* cf. *lateralis* (golden-mantled ground squirrel):** Five unmodified bones of a small spermophilid were recovered from Duckfoot: one each from Rooms 15 (wall fall) and 16 (floor) and three from Room 19 (postabandonment fill) (Tables 8.2 and 8.4). These specimens are tentatively identified as golden-mantled ground squirrel (*Spermophilus lateralis*), whose preferred habitat is found near Duckfoot today. *Spermophilus lateralis* is known to burrow in rock fall areas, as might have been found at Duckfoot after abandonment. The Duckfoot specimens may be intrusive. This interpretation is supported by the predominance of *Spermophilus lateralis* remains in the upper fill of various rooms and the rarity of remains in the

midden and on structure floors. Neusius (1986) also reported the presence of *S. lateralis* in the Dolores Pueblo I faunal assemblages.

Another species of small ground sciurid, *Spermophilus spilosoma* (spotted ground squirrel), has been recorded for western Montezuma County (Armstrong 1972) but is extremely rare in the region. Comparative osteological material for *Spermophilus spilosoma* was not available for this analysis. The specimen from Room 15 has complete dentition and closely resembles known *Spermophilus lateralis* material examined; however, comparisons with known *Spermophilus spilosoma* remains are needed for definite assignment of all of the Duckfoot specimens.

***Spermophilus* sp. (ground squirrel):** Specimens from three rooms (Rooms 6, 7, and 19), Courtyard 3, and the midden could not be identified beyond the genus level (Tables 8.3–8.5). Several species of *Spermophilus* occur in the area, but the bones in question do not exhibit the morphological characteristics necessary for specific identification. No definite evidence of cultural modification or use of these specimens was observed.

***Cynomys gunnisoni* (Gunnison's prairie dog):** This species was the most common ground-dwelling sciurid recovered from Duckfoot. Prairie dog bones were found in most structures, several midden excavation units, and two courtyard pit features (Feature 16 of Courtyard 2 and Feature 18 of Courtyard 3) (Tables 8.2–8.5). In general, prairie dog remains were more common in the various structures than in the midden. Floor, roof/wall fall, feature, and postabandonment fill contexts are represented. Most of the individual specimens were recovered from Room 6, and most of them were found in Feature 1, a cist. The number of left mandibles indicates that a minimum of six individuals is represented by the remains found in this room. No signs of patterned cultural modification were seen on the specimens; however, as said earlier for the cottontails, this could be explained by the relative ease with which these animals may be skinned and dismembered without the use of tools.

Prairie dogs dig their own burrows and are not known to live in habitats like those seen around abandoned buildings. No large burrows that could be attributed to animals of this size were observed during excavations at Duckfoot (Ricky R. Lightfoot, personal communication 1988). Thus, the presence of prairie dog remains in the numbers seen here is suggestive of cultural activity. Large numbers of *Cynomys gunnisoni* were also recorded for many Dolores-area sites (Benz 1984; Neusius 1986), and Benz (1984) proposed that some of these animals may have been garden-hunted by humans for food. As with other prairie dog species, Gunnison's prairie dog hibernates during the winter months and is active above ground only from April through October (Pizzimenti and Hoffmann 1973).

The species is reported to have been extirpated from the McElmo Creek area near Cortez as the result of modern agricultural practices (Pizzimenti and Hoffmann 1973:3). However, prairie dogs are still found in great numbers around Cortez and within 2 km of the Duckfoot site itself (Ricky R. Lightfoot, personal communication 1988).

Sciurid, indeterminate: Ten sciurid bones did not exhibit the morphological characteristics necessary for assignment to genus, much less species. On the basis of size only, it is assumed that most probably represent some species of *Spermophilus,* or ground squirrel.

Family Geomyidae

Thomomys bottae (Botta's pocket gopher): Pocket gophers are well known for their intrusive burrowing into archaeological sites (Wood and Johnson 1978), and this is one possible explanation for the presence of gopher bones in the Duckfoot faunal assemblage. There is no direct evidence of cultural modification on any of the specimens. Pocket gopher remains were recovered from Pit Structure 1 (Feature 10, vent), Courtyard 2 (mixed courtyard deposits and construction fill of Feature 18), Room 10 (roof fall), and the midden (Tables 8.2–8.5). The contextual data do not strongly suggest that pocket gophers were used by the site inhabitants. Neusius (1986) reported this species to be common in Dolores Pueblo I sites but concluded that, although pocket gophers could have been a species that was garden-hunted, no direct evidence was found for food use.

Family Heteromyidae

Dipodomys ordii (Ord's kangaroo rat): Remains of several kangaroo rats were recovered from Pit Structures 2 (postabandonment fill, roof fall, and floor) and 3 (roof fall, floor, and hearth) and from the midden area (Tables 8.3–8.5). Edentulous skulls of kangaroo rats are distinctive, and the edentulous Duckfoot specimens were easily identified. Postcranial material from Pit Structure 2 was compared with several other heteromyids and cricetids but was almost identical to a series of known kangaroo rats. These are burrowing animals (Wood and Johnson 1978), and the remains recovered from postabandonment deposits are probably intrusive. However, the remaining *Dipodomys* bones, or at least those recovered from the two structures, most likely were related to the prehistoric occupation of the site. Bones that may reflect cultural use were recovered from floor and roof fall deposits in Pit Structures 2 and 3; three partial or nearly complete skeletons were recovered from the floor and hearth (Feature 34) in the latter structure. As none of the specimens had cut marks or signs of

modification, the assessment of cultural significance is based solely on an evaluation of the contexts from which the bones were recovered. Neusius (1986) noted that this species was rare in the Dolores Pueblo I faunal assemblage.

Family Cricetidae

Peromyscus sp. (small mouse): Five species of *Peromyscus* are known to be present today in southwestern Colorado (Armstrong 1972). Specimens from Room 3 (roof fall), Room 6 (wall fall and cist [Feature 1]), Room 11 (Feature 6, hearth), Room 15 (Feature 4, pit), Pit Structure 1 (Feature 10, vent), Pit Structure 3 (floor), and the midden could represent any of the five species (Tables 8.2–8.5). Series of dentitions are necessary to separate sympatric species such as these, and the remains recovered from Duckfoot could not be assigned to species. No definite evidence indicating cultural use or modification was detected on any of these specimens, although their presence in floor and roof fall deposits and in two features with cultural fill (Feature 1 in Room 6 and Feature 4 in Room 15) is suggestive of at least occasional use. Neusius (1986) also recorded *Peromyscus* in the Dolores faunal assemblage but was unable to separate the individual species or determine any possible cultural use.

Neotoma albigula (white-throated woodrat): Three species of woodrat have been recorded in southwestern Colorado today (Armstrong 1972): *Neotoma cinerea* (bushy-tailed woodrat), *Neotoma albigula* (white-throated woodrat), and *Neotoma mexicana* (Mexican woodrat). On the basis of tooth morphology, it is evident that *Neotoma cinerea* is not present in the Duckfoot assemblage. *Neotoma albigula* is the most common woodrat species recorded for Duckfoot, followed by *Neotoma* cf. *mexicana*. Neusius (1986) did not separate the three species in the Dolores faunal collections. *Neotoma* was common in those collections, and possibly all three species are represented, given modern habitat requirements and paleoenvironmental reconstructions.

At Duckfoot, white-throated woodrat bones were recovered from floor or lower fill in Rooms 5 and 12, from wall fall in Room 6, and from the courtyard and midden areas (Tables 8.3–8.5). Remains also were recovered from Feature 1 (cist) in Room 6, Feature 6 (hearth) in Room 12, Feature 1 (hearth) in Room 13, Feature 1 (pit) in Room 19, and Feature 1 (vent shaft) in Pit Structure 3.

There was no indication of patterned cultural modification on any of these specimens. Presence of white-throated woodrats on a structure floor and in cultural deposits in several features is suggestive of at least occasional use of this species as a food resource. However, abandoned and partly filled structures would provide the ideal habitat for most woodrat species, and some of the Duckfoot specimens, particularly those found in upper fill deposits, probably are the remains of animals that invaded after the site was abandoned.

***Neotoma* cf. *mexicana* (Mexican woodrat):** This species of woodrat favors pinyon-juniper woodlands (Armstrong 1972; Cornely and Baker 1986), and it nests in horizontal rock shelters. *Neotoma albigula* prefers more open areas, and it often nests away from rock ledges and overhangs (Armstrong 1972). In the Duckfoot assemblage, only a single mandible was identified as *Neotoma* cf. *mexicana;* it was recovered from the hearth (Feature 6) in Room 12. Given the preferred habitats of the two species, the paucity of *Neotoma mexicana* remains is puzzling, although the species may be represented among the woodrat material not identified to species.

***Neotoma* sp. (woodrat):** Additional *Neotoma* specimens from Duckfoot could not be identified to species. They were recovered from the floor in Room 5, roof fall in Room 11, Feature 6 (hearth) in Room 11, Feature 10 (vent) in Pit Structure 1, Feature 8 (pit) in Pit Structure 2, and the midden (Tables 8.2, 8.3, and 8.5). Because all the features were filled with collapsed structural debris or other postabandonment deposits, the presence of *Neotoma* sp. specimens in these contexts does not necessarily suggest cultural use. The recovery of *Neotoma* sp. remains from floor and roof fall contexts may be significant, although none of the bones had cut marks or other signs of modification.

Rodent, indeterminate

Additional rodent material was recovered from numerous structures and other excavated areas at Duckfoot (Tables 8.2–8.5). This material did not exhibit the morphological characteristics necessary for speciation, although some could be identified as cricetid rodent. For the most part, these specimens were fragmentary postcranial elements, mainly long bones.

Order Carnivora
Family Canidae

***Canis familiaris* (domestic dog):** Except for one specimen, all canid material identified to species is domestic dog. Dog remains were recovered from Rooms 6 and 15 (wall fall strata); Courtyards 1, 2, and 3; Pit Structure 3 (roof fall); Pit Structure 4 (postabandonment fill); and the midden (Tables 8.2–8.5). Most of the specimens are cranial elements, and mandibles are especially common. At least eight individual dogs are represented: one old, five mature, and two puppies. Cut or skinning marks could be seen only on paired adult mandibles from the midden and on a tibia

from Pit Structure 4. The two mandibles from the midden had also been chewed by carnivores, which is to be expected on specimens recovered from a trash area. A single anatomical unit consisting of a distal humerus and proximal radius/ulna was recovered from Pit Structure 3 roof fall. This is suggestive of a small butchering unit, although no distinct cut marks were seen on these bones. Extensive use of domestic dogs for food was recorded for Dolores sites (Benz 1984; Clark et al. 1987; Neusius 1985a). Like the dog remains recovered from Dolores sites (Clark et al. 1987), those from Duckfoot appear to be of a short-faced variety, often called the "Small Pueblo Dog" (Allen 1920; Colton 1970).

***Canis* sp. (canid):** Thirty-two additional *Canis* sp. specimens were identified from Duckfoot (Tables 8.2–8.5). Several of these specimens are postcranial bones, which are not easily identified to species, especially considering the extreme variation seen in postcranial elements of domestic dogs (Olsen 1979a, 1985). None of the specimens have characteristics indicating that they represent any species other than domestic dog, but because they lack the necessary morphological characteristics, they were not assigned to species. One long-bone shaft fragment from Feature 7 (pit) in Room 19 had been fashioned into a bone tube.

***Vulpes* cf. *macrotis* (kit fox):** A single distal radius shaft fragment modified into an awl was recovered from the roof fall stratum in Pit Structure 1 (Table 8.2). This specimen was compared with a series of canid radii in the University of Wyoming Department of Anthropology comparative collection and was found to be most similar to *Vulpes velox* (swift fox), although it still was not referable to that species. *Vulpes macrotis* is closely related to *Vulpes velox* of the northern Plains and closely resembles that species in appearance and habits (McGrew 1979). Kurtén and Anderson (1980) state that the two species cannot be separated on the basis of fossil material and believe that they may be only subspecifically distinct. The specimen needs to be compared with known osteological material of *Vulpes macrotis* before definite species assignment can be made. If the specimen can be referred to *Vulpes macrotis,* it will be only the fourth prehistoric record of the species from North America (McGrew 1979). Neusius (1986) did not record the kit fox in any of the Dolores Pueblo I faunal assemblages, although the red fox (*Vulpes vulpes*) was well represented.

Canid, indeterminate: Eighteen canid bones could not be referred to genus. These are elements such as phalanges and ribs that, even on the basis of size, could not be assigned to *Canis* or *Vulpes*. Because there are so few *Vulpes* bones in the Duckfoot assemblage, most of the canid material probably represents *Canis*. If so, it is probably domestic dog. However, because the bones did not exhibit identifying morphological characteristics, they were not assigned to genus, much less species.

Family Felidae

***Felis rufus* (bobcat):** A single distal left mandible with canine, third and fourth premolars, and first molar was recovered from the midden (Table 8.5). Bobcats are extant in southwestern Colorado today (Armstrong 1972; Bissell and Dillon 1982). No cultural modifications could be seen on the Duckfoot specimen, and therefore its cultural use cannot be ascertained. The specimen has carnivore chew marks, possibly canid. Neusius (1986) considered the bobcat (and not the lynx, *Felis canadensis*) to be the felid probably represented in the Dolores collections.

Small carnivore

Two mandibular fragments identified only as indeterminate small carnivore (mustelid size) were recovered from Pit Structure 3 (roof fall) and Room 11 (Feature 1, bin) (Tables 8.3 and 8.4). Both were edentulous, with no definite evidence of cultural use. Several small carnivores are found in the area today. Neusius (1986) recorded pine martin (*Martes americana*), mink (*Mustela vison*), and long-tailed weasel (*M. frenata*) from Dolores-area Pueblo I sites.

Medium carnivore

Three specimens—an isolated tooth from modern ground surface in the midden, a tooth from collapsed structural debris in Room 7, and an edentulous premaxilla from a pit (Feature 18) in Courtyard 3 (Tables 8.3, 8.4, and 8.5)—could be identified only as medium-size carnivore (small canid, or bobcat size). The edentulous premaxilla is comparable in size to many of the canid specimens recovered but morphologically was not identical to any of the canid species recorded for the site. The tooth from Room 7 had been fashioned into a pendant; the remaining two specimens showed no signs of cultural modification.

Order Artiodactyla
Family Cervidae

***Odocoileus hemionus* (mule deer):** *Odocoileus hemionus* is the only deer extant in southwestern Colorado (Armstrong 1972). Most *Odocoileus* bone recovered from Duckfoot is probably associated with the prehistoric occupation. Unmodified specimens identified as *Odocoileus* were recovered from the midden, Courtyard 1, Pit Structures 2 and 3, and Rooms 4, 6, 7, 10, 12, 14, 16, and 19 (Tables 8.2–8.5). The majority of these specimens were found in roof fall or other structural collapse, and some

were burned. Whether the burning occurred during food preparation (cooking) or was the result of bones coming into contact with fire during the destruction of some structures when the site was abandoned could not be ascertained. The unmodified specimens were primarily lower-leg bones and teeth.

Numerous tools made from *Odocoileus* bone were also recovered from Duckfoot. In fact, modified *Odocoileus* bone was more common than unmodified bones. Deer bone tools were recovered from all four pit structures, seven rooms (Rooms 1, 7, 8, 11, 15, 16, and 19), Courtyard 3, and the midden. The majority were found in roof fall deposits and on structure floors, and awls are the predominant tool type. The most unusual tool is the edentulous mandible found in Pit Structure 1 postabandonment fill. The dorsal border of this mandible had been used as an abrading edge; longitudinal striations suggestive of scraping and smoothing are present on both the lingual and buccal sides. (See Chapter 5 for a more detailed discussion of bone tools.)

Neusius (1986) found mule deer to be the most common artiodactyl at Dolores Pueblo I sites and proposed that winter hunting was the most probable method of procurement (Neusius 1988). No evidence was present on any of the Duckfoot *Odocoileus* specimens that could be used to infer season of hunting.

Artiodactyl, indeterminate

Unidentified artiodactyl material was recovered from nine surface rooms, all four pit structures, the three courtyard areas, and the midden (Tables 8.2–8.5). Most of the bones recovered from structures were found on floors and in roof or wall fall strata; several feature fill and postabandonment contexts are represented as well. A number of tools, mostly awls, are included in the assemblage.

Most of this material is probably *Odocoileus*, although it is possible that some may be *Ovis canadensis* (bighorn), another artiodactyl of the same general size that was found historically in the region (Armstrong 1972). *Ovis* was common in Dolores-area Pueblo I sites but could not be positively identified in the Duckfoot assemblage.

The presence of fetal artiodactyl material in Room 15 and Pit Structure 2 is suggestive of winter occupation. The material is the general size of late-winter fetal deer, although it could not be matched with known-age specimens available. If the postulated age of late winter for the fetal material is correct, the presence of these specimens tends to support Neusius's (1988) case for winter hunting of deer by Pueblo I inhabitants of the region.

Mammal, indeterminate

Almost half (43.5 percent) of the osteological material recovered from Duckfoot could not be identified with any degree of confidence beyond the level of mammalian. This material consists primarily of highly fragmented specimens. An attempt was made, on the basis of size and cortical thickness, to assign individual bones to either small-mammal or medium-mammal categories. When even this general assessment could not be made, the specimens were identified simply as indeterminate mammal. Most of the material is small mammal, probably rabbit or large rodent, which conforms to the trend seen in the identifiable assemblage, where large numbers of such mammals were identified. A number of tools, including eight gaming pieces (five from midden deposits), eight awls, one pendant, and one needle, were identified in the assemblage.

Discussion

Of the 5,710 individual bones or bone fragments that make up the Duckfoot faunal assemblage, 1,059 (18.6 percent) were recovered from structure floors (excluding Room 20), 730 (12.8 percent) were recovered from roof fall strata (or mixed roof and wall fall deposits), and the remainder (3,921 specimens, or 68.7 percent) were recovered from other contexts, including feature fills, postabandonment structure fills, courtyard deposits, and the midden area (Appendix D). When the animal bone assemblage from structures only is considered (4,596 bones, excluding Room 20), floor and roof fall values rise to 23.0 percent and 15.9 percent, respectively. Within structures, it appears that some vertical displacement resulted in the recovery from postabandonment deposits of bones that ordinarily would be expected to occur on living floors. For example, domestic dog remains from Room 15, a deer mandible tool from Pit Structure 1, and an awl from Pit Structure 4 were found in collapsed wall debris or in postabandonment deposits above collapsed structural debris. Possible explanations for the mixing of site deposits include animal disturbance (burrowing and digging) and the washing of materials from courtyard areas into structure depressions after abandonment.

Differential preservation does not appear to have been a factor in the recovery of faunal material from Duckfoot, nor, for the most part, does mode of recovery. However, in a few instances, the recovery of extremely large numbers of bones may have been due at least partly to excavation technique; for example, in Rooms 11 and 13 and in Pit Structure 2, sediments that were water-screened through ¹⁄₁₆-in mesh (as opposed to dry-screened through ¼-in mesh) yielded particularly large quantities of rodent and small-mammal bones.

The Duckfoot faunal assemblage is composed overwhelmingly of mammal bones (5,643 specimens, or 98.8 percent). Reptiles and birds account for only 67 items and make up the remaining 1.2 percent of the assemblage.

Within the identifiable mammal assemblage (those remains that could be identified at least to order), lagomorphs (hares and rabbits) are the most common animals represented. *Sylvilagus* was found in all 4 pit structures and 16 of the 20 rooms and accounts for 33.0 percent of the identifiable mammal assemblage. *Lepus* was recovered from all 4 pit structures and 14 rooms and constitutes 10.6 percent of the identifiable mammal assemblage. *Sylvilagus* was found in greater numbers than *Lepus* in all rooms except Rooms 2, 4, 6, 13, and 18. In the midden area, *Lepus* occurs in higher numbers and in more excavation units (42.8 percent) than does *Sylvilagus* (32.8 percent). Overall, the evidence suggests that the inhabitants of the Duckfoot site were heavily dependent on both species, most likely for food, skins, and tool manufacture. Lagomorphs have been an important subsistence item throughout western North America, both ethnographically and prehistorically (Steward 1938; Walker 1988; see also Neusius 1985b). In the unidentifiable mammal assemblage, small mammals (rodent and rabbit size) far outnumber all other size categories, making up 83.5 percent of the unidentifiable category. This pattern parallels that seen in the identifiable assemblage.

Odocoileus bones make up only 1.9 percent of the identifiable mammal assemblage; however, when indeterminate artiodactyls, most of which probably are *Odocoileus*, are included, the proportion of all artiodactyls is 10.3 percent. Most identifiable *Odocoileus* bones from Duckfoot were modified into tools, and 36 of the 92 bone tools analyzed (39.1 percent) were made of *Odocoileus* bone. This suggests a significant dependence on this animal for at least tool manufacture; whether deer were procured primarily for food or for tool manufacture or both is not known.

Only one domestic species—*Canis familiaris*, domestic dog—is represented in the Duckfoot faunal assemblage. Remains of turkey (*Meleagris gallopavo*) may also represent a domesticated form, but the characteristics necessary for such a determination were not present on the recovered specimens. The skeletal morphology of modern domestic turkey is highly distinct from that of extant wild turkeys. However, osteological differences between early domestic turkeys and wild turkeys are less clear (Olsen 1968, 1979b; McKusick 1986). The small number of bones at Duckfoot (12) suggests little cultural use of turkeys at this site.

The presence of cottontail, jackrabbit, and prairie dog remains in Duckfoot site deposits possibly supports a model of garden-hunting, that is, the opportunistic hunting of animal pests that are attracted to agricultural plots (Benz 1984; Neusius 1984; Neusius and Gould 1988). Such a strategy serves two purposes: it eliminates animal competition for food crops and provides a source of meat protein. Deer may also have been taken in this fashion, as they too are invaders of agricultural habitat and are capable of inflicting even more crop damage than the smaller animals.

9

Human Skeletal Remains

J. Michael Hoffman

Introduction

Human skeletal remains provide the only direct evidence of prehistoric people's lives. With a minimum sample of 16 individuals, including 14 discrete burials, the Duckfoot human remains assemblage is the largest from a late Pueblo I site known to this author, and it has the potential to address a number of questions relevant to the overall Duckfoot research design: Is the age-sex structure of the skeletal population consistent with what one would expect for a late Pueblo I site? Are there any appreciable differences between the skeletal remains recovered from pit structures and those recovered from the midden? Do the pit structure skeletons reflect death by natural causes or death as the result of some kind of catastrophic event? Is there evidence of trauma? Of physiological stress? Of cannibalism? How does the Duckfoot skeletal assemblage compare to other late Pueblo I assemblages in the area? In this chapter, these and related questions are addressed when the quantity and quality of the data allow; in addition, basic descriptive data and statistical summaries of measurements and observations recorded during analysis are presented for the interested reader.[1]

The approach used in the Duckfoot analysis follows the structure of Frank Saul's (1972, 1976) "osteobiographic paradigm." Central to the paradigm is the recognition that there are important lessons in both the individual burials

and the collective data they represent about the population. The primary data are the individual burials; the population data are secondary levels of abstraction and must be treated with more caution. Beyond this basic notion, more specific questions about who was there (age-sex profiles), what they were like (stature, disease, anomalies, personal characteristics), and what happened to them (during life and at death) are explored in the context of the research questions above and other issues as they arise.

Underlying many of the research questions and the analysis of the remains is a basic ecological framework that places human populations in their natural (biotic and abiotic), social, and cultural environments. All these environments provide direct or indirect stressors to which individuals and groups must respond or perish. Many of the analyses undertaken in this study are based on this framework. (See Goodman et al. [1984] for an excellent explication of this model.)

A word of caution is in order with regard to the interpretations offered in this chapter and the data sets on which they are based: problems stemming from incomplete data and small sample and subsample sizes restricted some aspects of the study and limited the extent to which analysis results could be used for comparative and interpretive purposes. It is hoped that this report strikes a balance between reasonable analysis and interpretation on the one hand and informed speculation on the other.

1. The Duckfoot human remains study would not have been possible without the assistance and support of several individuals. At Colorado College, Laura Fulginitti assisted in some of the early data collection while she was an undergraduate student; Kimberly Spurr provided later assistance in data collection and labeled virtually all of the material, mostly during her year as paraprofessional in the Department of Anthropology. I thank my good friend, professional colleague, and personal dentist, Dr. Joseph Gentile of Colorado Springs, for x-raying the fused incisors of Burial 10 and for the discussion about possible diagnoses. Finally, my deepest appreciation goes to my wife and daughter, Merla and Beth, for their patience, understanding, and support when too many evenings and weekends were spent "working with the bones" and pounding the computer keyboard. I hope this effort is worthy of their sacrifices. All errors of commission or omission in this chapter reside solely with its author.

Methods

Cleaning and Restoration

Following transfer of the material to the Biological Anthropology Research Laboratory at Colorado College, each burial was unpacked, washed under clear tap water, and left to air-dry on open tables. Stubborn deposits were loosened and removed with fingers, dental picks, and toothbrushes. Because the material arrived in generally excellent condition from the field (with some exceptions as noted below), little restoration had to take place initially. Where damage had occurred, it was mostly in the form of broken or crushed crania because of the collapse of pit structure roofs; when possible, repairs were made using Elmer's school glue.

Assessment of Age at Death

Estimations of age at death were made through traditional physical anthropological techniques, supplemented by some newer approaches when the estimations proved difficult or ambiguous. Children's ages (birth to 12 years) were assessed by dental development and eruption standards following Ubelaker (1978), because these standards are more appropriate for archaeological populations. Supplementary techniques included the use of epiphyseal development and fusion standards (Krogman and Iscan 1986; Stewart 1979), developmental standards for specific (nonepiphyseal) bones (Krogman and Iscan 1986; Redfield 1970), epiphyseal size (Sundick 1978), and diaphyseal lengths (Hoffman 1979). Priority was given to dental development and eruption, although good congruence was found between ages determined from the dentition and those determined from other methods when intraindividual comparisons could be made.

Subadult (12 to 18 years) and adult (18 years and older) ages at death were estimated by epiphyseal fusion and dental eruption standards for the former (Krogman and Iscan 1986; Ubelaker 1978) and primarily by assessment of pubic symphyseal remodeling for the latter (Suchey et al. 1986; Suchey et al. 1988; supplemented by Todd 1920; McKern and Stewart 1957; Gilbert and McKern 1973), while keeping in mind the difficulties associated with pubic symphyseal aging techniques, especially when applied to females (Suchey 1979). Additional techniques followed recent aging standards established by Webb and Suchey (1985) and Lovejoy et al. (1985) for other skeletal regions.

In recent years, researchers (Iscan et al. 1985; Iscan and Loth 1986, revised in Krogman and Iscan 1986) have developed a method of adult age estimation based on the metamorphosis of the anterior rib end. This method is employed in particularly difficult aging situations—for example, in the not uncommon situation in which a female pubic symphysis "looks old," but other skeletal and/or dental indicators suggest a younger age. The latter situation occurs quite often and may result from accelerated metamorphosis of the pubic symphysis secondary to childbirth or other unknown forces. Closure of the basilar suture—that is, the spheno-occipital synchondrosis—was taken as a demarcation of adulthood when pubic symphyseal information was not adequate. Also, because of the general unreliability and variability of cranial suture closure (Stewart 1979), this older method of estimating age at death in adults was not used except when other methods were not available or proved contradictory within an individual specimen. Age estimations made on the basis of cranial suture closure followed the recent guidelines of Meindl and Lovejoy (1985).

Determination of Sex

Determination of sex for the adult and older subadult individuals was accomplished almost exclusively by morphological assessment of the pelvis and skull, with emphasis on the former (Keen 1950; Krogman 1962; Phenice 1969; Stewart 1979). When these skeletal elements were damaged or missing, sex was assessed by visual comparison of overall bone size and robusticity of unknowns with those in the sample whose sex had previously been determined. In most cases in which the pelvis and skull were missing or were so fragmented as to be unusable, the remaining skeletal elements were also so fragmentary that the use of discriminant functions for determining sex was not possible. Also in this regard, the more complete specimens were so few that discriminant functions could not be generated internally for the sample. (See Dittrick and Suchey [1986] for very recent work on sex determination by discriminant function analysis for prehistoric California material. This approach was not used, because the Duckfoot site population was considered too different in genetic origin and environmental context.) In one case, sex was determined for an adult humerus by using metric comparisons with other sexed humeri in the sample.

Although attempts have been made to determine sex of infant, child, and younger subadult individuals (e.g., Boucher 1955, 1957; Hunt and Gleiser 1955; Weaver 1980), the methods are somewhat unreliable, population-specific, and require nearly complete remains for adequate assessment (Ubelaker 1978), or, in the case of Weaver's (1980) method, at least a recovered ilium. For these reasons, primarily the lack of adequate population-specific standards, sex was not stated for immature individuals, that is, individuals younger than 16 to 18 years of age.

Demographic Techniques

Important research questions at the Duckfoot site center on population structure—that is, its size and stability. The questions posed are important for understanding larger issues of social stability and change through time in the local sociocultural and environmental settings and region. The limited sample size from Duckfoot and the question of whether the sample is representative of a mortality cohort of the larger Duckfoot population make most demographic inferences difficult at best, if not impossible. However, some comparisons can be made using life tables and the data they provide regarding survivorship and life expectancy. The life table model used here is the composite life table developed by Swedlund and Armelagos (1969) for archaeological populations (see also Swedlund [1975], Moore et al. [1975], Swedlund and Armelagos [1976], and Hassan [1981] for a fuller discussion; and Weiss [1973, 1975] for the theoretical basis and application of life tables for demographic modeling). The composite life table is based on the ages at death of a sample of individuals from some time span, quite often several generations in prehistoric samples, and is based on the assumption that the probability of dying with respect to each age interval has not changed significantly through time. This model differs markedly from those based on living populations, which either sample the population at a point, or period, in time or follow a cohort from birth to death.

From the above literature, we can glean the following general principles and assumptions regarding composite life tables based on prehistoric samples from cemeteries or other burial contexts:

- The population is assumed to be stable (that is, the birth and death rates are fixed over time); the population is either growing, declining, or staying the same size (stationary), but at a constant rate.
- Model life tables from cemetery samples are based on reasonably large skeletal samples (most studies in the past 20 years have been based on about 100 to 300+ individuals) derived from large cemeteries or pooled from a number of similar sites in a region.
- Model life tables from cemetery samples are based on samples from reasonably long periods of time (quite often several hundred years).
- Cemetery samples are assumed to represent the average birth and death experience of the population under study, whether from a single site with a large sample or from a pooled, multisite sample.
- Over time, various demographic disturbances—wars, epidemics, migrations, etc.—are "averaged out," so that a cemetery sample is representative of the average birth and death experience of the sample, pooled or unpooled.

In the case of the Duckfoot sample of 14 (or 15) known-age individuals, we cannot be sure that any of the above assumptions are met, nor do we have adequate means of testing them. The sample is very small and comes from a relatively constricted time period, 20 to 25 years, based on archaeological reconstruction. This span is sufficiently shorter than the lifespan of a "typical" cohort of individuals whose age-at-death experience is presented in a composite life table. Therefore, we cannot assume with any degree of confidence that the skeletal sample at Duckfoot is representative of what the average death experience would have been for a cohort that lived its entire life and was buried in the same location and whose members were then all recovered during excavation. We can be sure, though, that if there were any serious demographic disturbances at Duckfoot, this temporally constrained sample did not live through a sufficient number of generations to smooth out, or average, these disturbances. In short, the small sample size and short occupation of Duckfoot seriously limit the kinds of demographic inferences that can be attempted. Despite these problems, several comparisons will be made with the results of recent regional studies.

The composite life table model has been used for prehistoric Southwest populations most recently by Stodder (1987) in her analysis of the human skeletal materials recovered during the Dolores Archaeological Program (DAP) excavations. Stodder's work is especially useful because she places the DAP results in the context of the larger Mesa Verde region. The life table procedures used in the current analysis of the Duckfoot human skeletal sample follow Stodder's (1987:414–424) approach to allow direct comparison between the two studies.

The small sample at Duckfoot, made up of the actual number of individuals whose ages could be determined, and the inability to generate meaningful statistical summaries for the female and male subsamples separately preclude many comparisons with Stodder's data. Thus, although Stodder generated survivorship and life-expectancy statistics for both sexes, meaningful statistics could not be produced here because of the small sample. Sample size also inhibited exploration of such questions as female fertility and population growth (or decline). Although we may have the largest sample of individuals from a single late Pueblo I site in the American Southwest, only 15 individuals could be aged, including 5 adult males and 4 adult females. Stodder's samples were several orders of magnitude larger, 150 and 178, for the early and late Mesa Verde–region groups, respectively.

Cranial Deformation

Descriptive categories of cranial deformation types follow Reed (1963), as more recently modified by Bennett (1973a, 1975). Basic types included are the lambdoidal and occipital (or vertico-occipital). The former type has "the flattened plane . . . centered roughly around lambda and inclined at an angle of approximately 100 degrees or more from the horizontal plane," whereas the latter type shows an "angle of inclination from the horizontal . . . approximately 90 degrees or less" (Bennett 1975:3).

Cranial and Postcranial Measurements

The measurements of the cranium and postcranium are fairly standard and traditional in physical anthropology. Major sources for the metric measurements include Bass (1979), Brothwell (1981), and Giles (1970). Other sources for the remaining measures are Comas (1960), Finnegan (1978), Howells (1970), Martin (1957), Olivier (1969), and Wilder (1920). Complete lists of all measurements and indices, as well as source references for all, are provided in Appendix E. Some readers may wonder why certain measurements, particularly of the cranium, were not taken. In most cases, these omissions resulted from the lack of certain measuring instruments, for example, coordinate calipers or head spanners.

Stature Estimation

Stature calculations follow the formulae of Genoves ([1967], corrected in Bass [1987:29]) for Mesoamerican females and males. Although it is widely recognized that these formulae give different results than formulae derived by others (e.g., Trotter and Gleser 1958; Trotter 1970), most human osteologists working with southwestern United States prehistoric populations use Genoves's formulae because they believe that southwestern populations were shorter than most native North Americans and thus more similar to the Mesoamericans in Genoves's sample. There is also a closer genetic link between prehistoric southwestern populations and Mesoamericans. Stature estimations based on incomplete long bones (Steele 1970) were not attempted, because Steele's formulae do not include Native American samples, and the fragments from Duckfoot would not have added significant information.

Cranial Nonmetric Variants

Although sample size precludes the application of nonmetric or discrete data variation in a variety of biological distance equations, these data are included for the interested reader and for the time in the future when such small data samples may be handled effectively. Because of the widely acknowledged problem of systematic errors in the observation and collection of nonmetric data by different investigators, this author was the sole collector of these data. A list of the variants and their source references can be found in Appendix E.

Dental Observations

Primary sources for the dental observations made in this study include Anderson (1962), Brothwell (1981), Scott and Symons (1971), and Ubelaker (in Bass 1979). The high degree of dental attrition in the adults precluded the collection of important kinds of data, such as molar cusp patterns and numbers, as well as several tooth measurements. In addition, many of the adult skeletons in the Duckfoot sample have extremely worn or missing teeth, which precludes meaningful statistical summaries of the timing of enamel hypoplasia event formation. This is unfortunate because Stodder (1987) uses this information for the larger Mesa Verde region in some very productive ways.

Identification of Diseases

The identification of specific pathologic processes or entities is difficult at best for most diseases that leave their impress on the skeleton. With no soft tissue available for histological examination to aid in diagnosis, specific disease identification is often impossible. More often, only general disease processes such as inflammation or tumor formation can be identified. However, several recent attempts to develop models for the identification of specific diseases have gone far to close this diagnostic gap between the living and recently deceased on the one hand and those who have died in the prehistoric past on the other. The works of Buikstra (1976), Ortner and Putschar (1981), Steinbock (1976), and Hart (1983), among others, provide a variety of methods for the solution of specific and more general diagnostic problems in paleopathology. These sources and others cited below in the discussion section serve as the basis for identifying pathologies in the Duckfoot site people.

Statistical Analysis

Statistical analysis of the Duckfoot sample is limited in scope because of the small total sample size and the incompleteness of many of the variables examined. Because the basic sample of 14 individuals (the burials) is composed of children and adults of both sexes, meaningful statistical analysis should involve the use of subsamples: children of specific ages and adult females and males considered separately. However, subdividing the total sample in this way would have produced subsamples too small to be significant for most hypothesis testing about intrasite and intersite variability, especially for the metric and nonmetric data of the crania and postcrania.

Analyses of categorical data were performed to investigate intrasite differences in burial location and disease. Chi square tests of significance of these distributions were used, many of which used Yates' correction for small cell sizes. Summary statistics for the cranial and postcranial metric data, as well as life table calculations and graphs, were calculated using a Quattro™ spreadsheet, version 1.0 (Borland International 1987). Cross-tabulations and chi square tests of categorical data were made using Statgraphics®, version 3.0 (STSC, Inc. 1988).

Description of Human Remains

The descriptions below are intended to provide a more complete picture of what the individuals at Duckfoot were like—including personal attributes, diseases, anomalies, and the condition of their remains—than mere statistical summaries could. Also included in the descriptions is a graphic depiction of the differences in completeness between the remains recovered from the pit structures and those from the midden area and roomblock. Following the descriptions of the numbered burials (Burials 1 through 14) is a list of all miscellaneous human bone found throughout Duckfoot; this list should be consulted for descriptions of isolated bones that may belong to some of the numbered burials.

Human Burials

Burial 1

LOCATION: Pit Structure 1, floor
GENERAL DESCRIPTION: This is one of two adults found in Pit Structure 1, which burned at abandonment. This individual was on his back across the hearth, left arm resting across abdomen, right arm flexed at elbow with hand near right shoulder, both hips and knees partly flexed. The head was toward the southwest; the face was turned up and slightly to the south. Burial 1 is nearly 100 percent intact, missing only some small bones of the hands and feet. Archaeological reconstruction indicates that the individual was placed on the floor before the pit structure was burned, because burned roof fall was found over the skeleton. Only the left knee region is burned; the remainder of the skeleton does not show signs of burning, charring, or even smudging, although the body was covered with dirt and ash.
AGE: 30 to 39 years
SEX: male
STATURE: 63.68 in, calculated from the right femur
CRANIAL DEFORMATION: lambdoidal, slightly asymmetrical to the left
MAJOR PATHOLOGIES OR INJURIES: fusion (congenital synostosis) of bodies of fifth and sixth cervical vertebrae (Klippel-Feil syndrome); possible old, healed depressed fracture about 3 cm above right browridge; no cribra orbitalia or porotic hyperostosis; minimal lipping of lower lumbar vertebrae indicative of osteoarthritis.
MAJOR ANOMALIES: Klippel-Feil syndrome (see above)
DENTITION, MAXILLARY: 16 teeth present; 1 carious left first molar; abscess or severe periodontal disease has exposed mesiobuccal root of right third molar; mesiolingual torsion of both central incisors.
DENTITION, MANDIBULAR: 14 teeth present; both third molars absent premortem; occlusal caries in left first and second molars; anterior tooth crowding.

DENTITION, GENERAL: mild to moderate calculus formation; moderate to severe periodontal disease with alveolar recession and root exposure; enamel hypoplasia.

Burial 2

LOCATION: Pit Structure 1, floor and fill above floor
GENERAL DESCRIPTION: This fragmented yet nearly complete and partly burned skeleton is the second of two burials found in Pit Structure 1. Charred wood found under the skeleton may indicate that the body was placed on top of the roof before it burned and collapsed or that fuel used to ignite the structure was placed under the body, on the structure floor. The burial was found on its back, slightly turned onto its right side; the head was toward the east-northeast, facing north. Both arms were flexed at the elbows, the left leg was flexed at the knee, and the right leg was straight. The left half of the cranial vault is burned and fragmented, with some charring severe enough to affect the entire thickness of the vault bones; that is, charring is more than mere surface burning. The fragmentation is consistent with the skull vault's having exploded from the "steaming" of the intracranial contents in an intensely hot fire of some duration or with damage sustained as the roof collapsed onto the skull and then continued to burn. Burned postcranial bones include the distal left humerus, proximal left ulna and radius, several right metacarpals and phalanges, and the right tibia and fibula, plus most of the adjacent tarsals and metatarsals. There is surface erosion of many bones.
AGE: 30 to 39 years
SEX: male (The sex of this individual is somewhat uncertain, but a number of pelvic and cranial measurements, plus comparison with definite females from elsewhere in the site, indicate a male. However, the postcranial skeleton in general is rather nonrobust for a male, in spite of a very robust facial skeleton.)
STATURE: 63.41 in, calculated from the right femur
CRANIAL DEFORMATION: lambdoidal, symmetrical
MAJOR PATHOLOGIES OR INJURIES: left maxillary sinusitis; no porotic hyperostosis or cribra orbitalia; small Schmorl's nodes on inferior body of seventh thoracic vertebra; mild osteoarthritic lipping of lumbar vertebrae; in general, this is a rather disease-free individual for someone in his fourth decade of life at this time in prehistory.
MAJOR ANOMALIES: two supernumerary left maxillary teeth: one has a caniniform shape and is located buccally between the incisors, the other is a "peg" found between the second and third molar space buccally.
DENTITION, MAXILLARY: 18 teeth present, including 2 supernumeraries described above; interproximal mesial caries in right first molar; anteriorly located supernumerary tooth caused crowding of lateral incisor and canine.
DENTITION, MANDIBULAR: 16 teeth present; occlusal caries in right central incisor; exposed root canal of right central

incisor led to periapical abscess formation; crowding of all anterior teeth from premolar to premolar, with some torsion of left central and lateral incisors.

DENTITION, GENERAL: mild calculus formation and periodontal disease; enamel hypoplasia; presence of anterior supernumerary caused an asymmetrical pattern of attrition with greater wear on right side of dentition than on left side, where the supernumerary is located.

Burial 3

LOCATION: Pit Structure 2, floor

GENERAL DESCRIPTION: This incomplete and fragmented burial was found face down, with the head to the south. Numerous sandstone rocks were clustered over and around the skeleton. Notable is the lack of any limb bones, with the exception of both hipbones and fragments of both scapulae; this is in marked contrast to the nearly complete remains found in Pit Structures 1 and 3. The skull is cracked in several places around the sides and posterior vault and facial regions, and it is missing some fragments; the mandible is complete. Postmortem warping prevents an accurate reconstruction of most fragments. The vertebral column is intact through the sacrum; most ribs are present, but many are broken and/or incomplete. The sternum is also present. Carnivore tooth marks, recognized as depressed punctures on the surfaces of several bones, indicate gnawing on "green" bone, or bone that was relatively fresh postmortem. The pattern of missing limb bones is typical of that associated with carnivore activity on a relatively fresh body from immediately after death until several weeks or months after, until decomposition is nearly complete.

AGE: 30 to 39 years. This was a difficult individual to age because of the disparity in age estimations by different techniques: over 18 years old on the basis of third molar eruption, iliac crest and medial clavicular epiphysis fusion, and basilar suture closure; 40.0 ± 12.2 years (33.7–46.3 years, 95 percent confidence interval) on the basis of rib phase analysis (phase 5, female); 35–39 years on the basis of auricular surface morphology (phase 4); and 60.0 ± 12.0 years (42–87 years range) on the basis of pubic symphyseal aging (Suchey-Brooks, phase VI [Suchey et al. 1988]). Estimates based on aging techniques that rely on bone surface remodeling were revised downward, especially those obtained with the pubic symphyseal technique, because of the apparent multiple childbirth episodes (evidenced by the deep dorsal pubic pitting and similar changes in the preauricular sulcus region). Also supporting the younger age estimate is the moderate amount of tooth wear present, an amount not consistent with what one would expect in a member of this population in her sixth or later decade of life.

SEX: female

STATURE: could not calculate

CRANIAL DEFORMATION: lambdoidal, asymmetrical to the right

MAJOR PATHOLOGIES OR INJURIES: probable well-healed, depressed fracture of the left occipital squama, 1 cm left of the midline and 1 cm below a bipartite ossicle at lambda, and apparently involving only the outer table of bone, as there is no disruption of the normal surface of the internal lamina; mild, healed cribra orbitalia (porotic type) of the left orbital roof, but no porotic hyperostosis.

MAJOR ANOMALIES: none

DENTITION, MAXILLARY: 12 teeth present; missing are both third molars and the right first and second premolars; no caries; lingual torsion of both central incisors.

DENTITION, MANDIBULAR: 13 teeth present; missing are both second molars and the left first premolar; occlusal caries in both third molars; lingual torsion of both central incisors.

DENTITION, GENERAL: no abscesses and only very slight calculus formation; moderate periodontal disease.

Burial 4

LOCATION: Pit Structure 3, floor

GENERAL DESCRIPTION: This adult was covered with burned and unburned roof fall, and portions of the skeleton are charred and slightly calcined, especially the right cranial vault and facial skeleton. The postcranial skeleton is virtually complete and unbroken, but the skull is broken and incomplete. Skull breakage appears to be the result of the structure roof's falling on top of the skull and not of antemortem damage or injury. The skull was lying mostly on its left side. This side is more intact than the right side, which sustained the impact of the collapsing roof. The skull's right side was crushed, flattened, and burned in situ. The burial was oriented in a north-south direction, on its back, with the head toward the north, facing east. The knees were flexed so that the feet were close to the hips. The right arm was extended along the side of the torso, and the left arm was flexed at the elbow, with the hand near the neck. Rodents had disturbed the bones of the left hand in such a way that several carpal, metacarpal, and phalangeal bones were found scattered near the pelvis.

AGE: 30 to 39 years

SEX: male

STATURE: 63.95 in, calculated from the right femur

CRANIAL DEFORMATION: lambdoidal, symmetrical

MAJOR PATHOLOGIES OR INJURIES: no cribra orbitalia or porotic hyperostosis; mild osteitis of the hard palate, a general accompaniment of poor oral health; left fourth rib shows a healed fracture 5.5 cm from the anterior end; mild to moderate osteoarthritis of the thoracic and lumbar spine, and of the hip, knee, and ankle joints.

MAJOR ANOMALIES: none

DENTITION, MAXILLARY: 13 teeth present; missing antemortem were the right third and first molars and the right second premolar; 4 molar teeth have caries and there are two abscesses on the right side of the upper jaw, the largest involving both the second premolar and first molar root areas contiguously.

DENTITION, MANDIBULAR: 12 teeth present; missing antemortem were both third and second molars; caries are present in both first molars (that on the left is massive and has nearly destroyed the entire crown); several abscesses; some crowding of the mandibular dentition has resulted in lingual torsion of both mandibular canines.

DENTITION, GENERAL: moderate to severe attrition of the teeth; minimal calculus formation; moderate periodontal disease.

Burial 5

LOCATION: midden, 60N 99E

GENERAL DESCRIPTION: This is one of seven individuals recovered from burial pits in the midden. The skeleton is very incomplete and fragmented, but it is not burned. About one-half of the skull is present. Most of the facial skeleton is missing, and most vault bones are fragmented or incomplete. The postcranium is completely missing except for the right humeral shaft, which has both ends broken off (shaft fragment length = 138 mm). Orientation appears to be east-southeast, with the face turned west-southwest. On the basis of archaeological evidence, the body is believed to have been buried intact; the missing bones are assumed to have been destroyed or carried off by animals and/or to have decomposed in the moist pit environment.

AGE: 5 to 6 years

SEX: indeterminate, because of immaturity

STATURE: could not be calculated

CRANIAL DEFORMATION: lambdoidal, asymmetrical to the right

MAJOR PATHOLOGIES OR INJURIES: mild cribra orbitalia (porotic type) bilaterally, but no porotic hyperostosis of the vault; no other pathologies present.

MAJOR ANOMALIES: none

DENTITION, MAXILLARY: nine deciduous teeth present and fully erupted; missing postmortem is the left deciduous canine; nine adult maxillary teeth are present in various stages of development, with the left first molar erupting.

DENTITION, MANDIBULAR: 6 deciduous teeth present and fully erupted; missing are the left molars and canine, plus the right canine; 11 adult mandibular teeth are present in various stages of development, with the right molar erupting.

DENTITION, GENERAL: mandible and maxillae are very fragmented and incomplete; no evidence of major dental disease, typical of this age group; dentition is especially useful here for determining age at death.

Burial 6

LOCATION: midden, 56N 109E

GENERAL DESCRIPTION: This adult came from an indistinct, shallow pit in the midden very near modern ground surface. It was lying supine, with its head toward the south-southeast. The arms were extended alongside the torso, and the legs were tightly flexed with both knees to the west (left side of body). The cranium is about 80 to 85 percent present, although it is fragmented and missing some parts of the vault and nasal region, particularly the left side and posterior vault. The mandible is complete. The postcranium is fairly complete but fragmented, and it shows evidence of much carnivore activity and damage, especially to epiphyses, vertebral bodies and spines, foot bones, and the posterior innominates. A spiral fracture is present in the upper half of the right femur, but it cannot be determined whether the fracture occurred before or after death, while the bone was still "green." The break may be the result of carnivore chewing, although canine tooth marks are not seen in the immediate region of the fracture. Some of the damage is probably the result of foot and wheelbarrow traffic over the shallow, undiscovered skeleton during earlier excavation. The skeleton is unburned.

AGE: 40 to 49 years

SEX: male

STATURE: could not be calculated

CRANIAL DEFORMATION: lambdoidal, asymmetrical to the right

MAJOR PATHOLOGIES OR INJURIES: no cribra orbitalia or porotic hyperostosis; very mild periostitis of both femoral and tibial diaphyses; Schmorl's nodes on twelfth thoracic through fifth lumbar vertebral bodies range from slight to moderate indentations; mild degenerative changes of most joint surfaces, manifested as slight erosion and pitting, indicative of osteoarthritis.

Especially interesting is the possible combination of two left lower-limb pathologies:

- An exostosis of the medial lip of the left femoral linea aspera in the region of attachment of a hip adductor.
- The presence, about midshaft, of 6- to 7-mm circular facets on opposing surfaces of the left fourth and fifth metatarsals. The facets have the appearance of the pitted, roughened surface of a pseudoarthrosis and could be related to a possible fracture of the fifth metatarsal, since the metatarsal's distal half is angled slightly laterally.

These two pathologies may be the result of a single traumatic episode in which the individual broke his foot and, as he fell, tore the insertion of the hip adductor, which eventually healed and ossified as an exostosis.

There is also an old, healed, subperiosteal hematoma of the left anterior tibial shaft—an old "shin bruise."

MAJOR ANOMALIES: none

DENTITION, MAXILLARY: 11 teeth present; missing antemortem are both third molars and the right second incisor, missing postmortem are both central incisors; 4 carious teeth are present, as is a large abscess at the right first molar, which consists only of two root stumps.

DENTITION, MANDIBULAR: 13 teeth present; missing antemortem are both third molars and the left second molar; 3 carious teeth; abscess at the left central incisor.

DENTITION, GENERAL: mild calculus and moderate to severe periodontitis in both jaws; attrition is generally severe, with some anterior teeth functioning with root stumps only.

Burial 7

LOCATION: midden, 56N 109E

GENERAL DESCRIPTION: A shallow burial pit contained the remains of this child, found in a fully flexed position on its left side, with the head to the east, facing southwest. Some disturbance to the arms, portions of the spinal column, and the feet was noted, probably the result of rodent activity. The skull is about 80 percent complete, but there has been damage to the left temporal region and adjacent cranial base (involving the temporal, occipital, sphenoid, zygomatic, and posterior maxilla). The postcranium is only about one-third complete, and most elements present are incomplete. Missing are most thoracic and lumbar vertebrae, sternum, both scapulae, left humerus, both radii and ulnae, all carpals and metacarpals, 90 percent of the ribs, both patellae, all right tarsals and metatarsals, most tarsals and metatarsals of the left foot, and all phalanges (pedal and manual). The skeleton is unburned.

AGE: 6 to 7 years

SEX: indeterminate, because of immaturity

STATURE: not determined, because of immaturity

CRANIAL DEFORMATION: lambdoidal, asymmetrical to the right

MAJOR PATHOLOGIES OR INJURIES: mild, healed, bilateral cribra orbitalia, but no porotic hyperostosis; no other pathologies noted, including dental diseases.

MAJOR ANOMALIES: none

DENTITION, MAXILLARY: 4 deciduous teeth are present (first and second premolars bilaterally)—other deciduous teeth were lost postmortem or were exfoliated by erupting permanent teeth; 10 permanent teeth are present—both first molars are fully erupted, both central incisors are erupting (rotated mesiolingually), and both permanent lateral incisors and both permanent second and third molars are unerupted.

DENTITION, MANDIBULAR: four deciduous teeth present (first and second premolars bilaterally)—others lost postmortem or exfoliated; nine permanent teeth present—left and right first molars are fully erupted, both central incisors and the left lateral incisor are erupting, and the left and right second and third molars have not erupted.

DENTITION, GENERAL: no evidence of caries, abscesses, or periodontal disease.

Burial 8

LOCATION: midden, 56N 107E

GENERAL DESCRIPTION: Burial 8 was recovered from a shallow pit in the midden. The skull and postcranium are each 75 to 80 percent present; most elements present are complete. The cranium shows damage to the left temporal and facial areas, but the clean surfaces of the broken bone indicate that at least some of it is the result of very recent damage (some may be damage sustained by the skeleton before its excavation because of traffic related to nearby excavation work). The body was partly flexed and slightly twisted, with the head to the east-northeast, facing southeast. The knees were to the south, the legs and pelvis were on their left side, the shoulders were nearly flat, and the left humerus was under the chest. All metatarsals and pedal phalanges are missing, as are the sternum, left clavicle, right innominate, hand bones, and most of the right femur. The disappearance of these bones is possibly the result of animal disturbance.

AGE: 50+ years

SEX: male

STATURE: 64.30 in, calculated from the left femur

CRANIAL DEFORMATION: lambdoidal, asymmetrical to the right

MAJOR PATHOLOGIES OR INJURIES: no cribra orbitalia or porotic hyperostosis; mild osteitis of the hard palate; mild periostitis of both tibial and fibular shafts; one Schmorl's node present on the superior surface of the body of the fourth lumbar vertebra; moderate to severe osteoarthritis throughout the vertebral column but most noticeable in the lower cervical and upper lumbar regions; both sacroiliac joints also show moderate degenerative joint disease.

MAJOR ANOMALIES: none

DENTITION, MAXILLARY: eight teeth present; missing are the right third molar through second premolar, left lateral incisor, canine, and first and second molars; caries in six teeth and abscesses in four.

DENTITION, MANDIBULAR: 10 teeth present; missing are the right second molar, the left canine through first molar (bone missing in this region), and the left third molar; 4 teeth are carious, and 2 have abscesses.

DENTITION, GENERAL: minimal calculus and moderate to severe periodontitis throughout the jaws; attrition is moderate to severe, especially in the anterior dentition where there are several functioning root stumps.

Burial 9

LOCATION: midden, 54N 109E

GENERAL DESCRIPTION: A shallow pit in the midden contained the fully flexed skeleton of this child. The upper torso was in a supine position, with the head to the southwest and the hands folded onto the upper abdomen and lower chest; it is relatively complete. The leg bones are missing except the right tibia and a fragment of the left fibula; some bones of both feet are present as well. The arrangement of the leg bones that were present suggests that the legs were fully flexed. The right side of the face was damaged, and there is evidence of a great deal of carnivore chewing throughout the skeleton, especially in areas of cancellous bone, such as joint ends.

AGE: 10 to 11 years

SEX: possibly male, although the immaturity of the specimen makes this identification tentative at best; the most reliable indicator of male sex is the presence of an already squared-off mental eminence of the mandible, a characteristic not seen in the females of this sample.

STATURE: could not calculate, because of immaturity

CRANIAL DEFORMATION: lambdoidal, asymmetrical to the left

MAJOR PATHOLOGIES OR INJURIES: none

MAJOR ANOMALIES: none

DENTITION, MAXILLARY: because the right maxilla is missing, the only tooth from that side is the first permanent premolar; on the left side, only six permanent teeth are present: the central incisor, first premolar, and first molar are fully erupted; the canine is erupting; and the second premolar and second molar are not yet erupted.

DENTITION, MANDIBULAR: the left deciduous second molar is present; the 14 other teeth present are all permanent, with both third molars missing postmortem; the left and right first molars, first premolars, and all 4 incisors are fully erupted; both canines are erupting; the second premolars and second molars bilaterally are not erupted; the right central incisor is rotated mesiolingually, and both canines are rotated mesiobuccally.

DENTITION, GENERAL: minimal dental calculus and periodontitis; attrition is minimal on the erupted permanent teeth.

Burial 10

LOCATION: midden, 52N 107E

GENERAL DESCRIPTION: This burial is the only one of the seven recovered from the midden to be covered by sandstone slabs. Found in a pit, the body was supine and partly flexed, with the head toward the northeast. The arms were extended alongside the torso, and the legs were slightly flexed, with the knees to the right (north). Skull and postcranium are each 90 to 95 percent complete. The cranial vault sustained some damage to the right posterior side. Most vertebrae are missing their spinous processes; missing postcranial bones include the left ulna, all bones of the left wrist and hand, the right patella, and most bones of the right ankle and foot. There is evidence of carnivore chewing, especially on vertebral bodies, anterior rib ends, and the epiphyseal ends of long bones. The body is unburned.

AGE: 18 to 20 years

SEX: female

STATURE: 58.33 in, calculated from the right femur

CRANIAL DEFORMATION: lambdoidal, symmetrical

MAJOR PATHOLOGIES OR INJURIES: mild, healed cribra orbitalia (porotic type) present bilaterally, but no porotic hyperostosis; mild osteitis of the hard palate; mild periostitis of the shafts of the femora, tibiae, and fibulae.

MAJOR ANOMALIES: bilateral septal apertures in the distal humeri; prominent costoclavicular ligament attachment sites in both clavicles (especially notable given the relatively young age of this female); see also mandibular dentition below.

DENTITION, MAXILLARY: 14 teeth present, missing only the third molars postmortem; four occlusal caries and one abscess; both lateral incisors rotated mesiobuccally.

DENTITION, MANDIBULAR: 13 teeth present, missing only the right second molar and left first molar antemortem; the count of 13 teeth results from the fact that, because of congenital fusion of both crown and root, the right central and lateral incisors were counted as 1 tooth (14 present if counted as 2). Because of the smaller space taken by the fused incisors, the left central incisor is slightly over the midline on the right side and is rotated mesiobuccally; the left lateral incisor and both canines are rotated slightly mesiolingually. There are four occlusal caries and no abscesses.

DENTITION, GENERAL: calculus virtually absent, and periodontal disease only mildly expressed; attrition very modest, as befits the young adult age.

Burial 11

LOCATION: Pit Structure 4, floor

GENERAL DESCRIPTION: This is a severely burned, fragmented, and incomplete infant burial, one of three individuals found on the floor of the smallest pit structure at Duckfoot. The head was to the southeast, but little else regarding body orientation could be determined because of the extremely fragmented condition of the burial. All bone fragments, of which there are several hundred, are thoroughly charred or calcined, the latter indicating high burning temperatures over a prolonged period of time. Identifiable fragments include the following: cranium, right hemi-mandible, left proximal femur, left ischium, right proximal femur, ribs, vertebrae, and several fragments of tibia and fibula. There also are numerous small, unidentifiable fragments.

AGE: 3 to 4 years

SEX: indeterminate, because of immaturity

STATURE: could not be calculated, because of immaturity

CRANIAL DEFORMATION: indeterminate because of fragmented and incomplete posterior vault

MAJOR PATHOLOGIES OR INJURIES: none

MAJOR ANOMALIES: none

DENTITION: very fragmented and incomplete, although the right hemi-mandible is present, with several teeth and/or fragments in situ, both deciduous and unerupted permanent teeth.

Burial 12

LOCATION: Pit Structure 4, floor

GENERAL DESCRIPTION: This adult burial is more complete and less thoroughly burned than the two infant burials in Pit Structure 4. The fully extended body was in a supine position, with the head and shoulders on top of the hearth, pointed southeast. The arms were flexed over the chest. The body was surrounded and partly covered by several large, unmodified pieces of sandstone that appear to have been deliberately and carefully arranged. The majority of the skeleton is burned. Portions not burned include most of the skull except for the lower face, the upper vertebral column, most of both shoulders (including scapulae and proximal humeri), and some fingers of the left hand. Missing cranial elements are mostly from the midface; missing postcranial bones include the sternum, several carpals, the left patella, and the right calcaneus. The partly burned bones are consistently burned on the top rather than on the bottom, and the cranium, which rested on the ash layer covering the hearth, is unburned. These observations clearly demonstrate that the hearth could not have been the cause of the conflagration that destroyed the structure roof.

AGE: 50+ years

SEX: female

STATURE: could not be calculated

CRANIAL DEFORMATION: lambdoidal, symmetrical

MAJOR PATHOLOGIES OR INJURIES: osteoarthritic changes of the right sacroiliac joint and some lipping of the rim of the sacral promontory; old, healed cribra orbitalia, but no porotic hyperostosis.

MAJOR ANOMALIES: none

DENTITION, MAXILLARY: nine teeth present, some burned; teeth missing antemortem include the right first premolar and lateral incisor and the left second premolar and first molar; teeth missing postmortem include the left lateral incisor and left second and third molars; there are four carious teeth and four abscesses, the latter all associated with carious teeth.

DENTITION, MANDIBULAR: all mandibular teeth lost antemortem except for the left third molar; most alveolar bone completely or partly resorbed; evidence of several abscesses, but the resorption makes it difficult to provide an exact count.

DENTITION, GENERAL: one unburned tooth may be a mandibular incisor, but it is severely worn, especially on its lingual surface; minimal dental calculus and evidence of periodontitis on bone still present.

Burial 13

LOCATION: midden, 54N 99E

GENERAL DESCRIPTION: This adolescent skeleton was recovered from a relatively deep pit in the midden; the tight flexure of the body was necessitated by the small size, in plan, of the burial pit. The body lay on its right side, with the head to the east, facing north. The skeleton is only one-quarter to one-third present, and those skeletal elements that are present are fragmented to varying degrees. The bone is in a generally rotted condition indicative of constant soil moisture, a result of the deepness of this particular pit whose base is bedrock. There is evidence of a great deal of carnivore activity, especially in the canine and carnassial tooth puncture marks on vertebral bodies and epiphyseal ends of long bones. The skeleton is unburned.

AGE: 12 to 15 years

SEX: possibly female, on the basis of the fairly broad sciatic notch and rounded mentum of the mandible; given the adolescent age of this individual, however, the sex estimation should be viewed with caution.

STATURE: could not be calculated

CRANIAL DEFORMATION: lambdoidal, asymmetrical to the right

MAJOR PATHOLOGIES OR INJURIES: none

MAJOR ANOMALIES: none

DENTITION, MAXILLARY: 12 teeth present; missing are both central incisors and the third molars; 3 carious teeth, but no abscesses in the fragments of maxillae.

DENTITION, MANDIBULAR: 13 teeth present; missing are the right canine and both third molars; 3 teeth are carious, but there are no abscesses in the portions of the mandible present; all 4 incisors and the left second premolar are rotated mesiolingually.

DENTITION, GENERAL: calculus is minimally present; periodontitis is present on the fragment of left mandible remaining; attrition is relatively minor.

Burial 14

LOCATION: Pit Structure 4, floor

GENERAL DESCRIPTION: This child, the third of the burials in Pit Structure 4, is very fragmented, incomplete, and severely burned, with virtually all bones and teeth charred through or calcined. The head was oriented to the northwest, but, because of the fragmentation of the skeleton, body orientation could not be determined.

AGE: 5 to 6 years

SEX: indeterminate, because of immaturity

STATURE: could not be calculated, because of immaturity

CRANIAL DEFORMATION: lambdoidal, although the symmetry or asymmetry of the deformation could not be assessed, because of fragmented and incomplete posterior vault.

MAJOR PATHOLOGIES OR INJURIES: porotic hyperostosis moderately expressed on the occipital squama may be active but is difficult to evaluate because of the condition of the cranial vault bones; insufficient evidence to evaluate for cribra orbitalia; no other pathologies or injuries.

MAJOR ANOMALIES: none

DENTITION, MAXILLARY: left and right maxillae only partly intact; left—unerupted permanent central incisor, deciduous first molar with crown broken off, and four other deciduous teeth not in place; right—unerupted permanent canine and first premolar, both deciduous molar crowns broken off, and deciduous incisors and canine not in place. Isolated maxillary teeth include two permanent incisors, one permanent canine, and two permanent premolar crowns.

DENTITION, MANDIBULAR: fragment of the left anterior mandible with an unerupted permanent central incisor in situ; isolated mandibular teeth include a deciduous canine, first and second deciduous molars, and permanent first and second molar crowns.

DENTITION, GENERAL: other tooth fragments include an incisor and molars.

Miscellaneous Human Bone

Except where noted below, none of the following miscellaneous skeletal elements could be directly associated with any of the numbered burials at the site.

Pit Structure 1

Above wall/roof fall: Left proximal femur fragment, adult; includes head, neck, and shaft to approximately 6 cm below lesser trochanter; matches right femur from Burial 2. *Roof fall:* Left fifth metacarpal, adult, unburned; left capitate, pisiform, hamate (two fragments), and triquetral fragment; consistent with missing bones of Burial 2; also some small unidentifiable fragments. *Floor:* Additional bones from Burial 1: right posterior rib fragment; right distal phalanx, first digit, foot; right distal phalanx of foot, not first digit; right permanent maxillary third molar with large mesial cervical caries, two small occlusal caries.

Pit Structure 2

Floor: Two adult premolars; one left rib midshaft fragment, probably adult, unburned, consistent with missing Burial 3 skeletal element; three right rib midshaft fragments, probably adult, unburned, consistent with missing Burial 3 skeletal elements; one right midshaft rib fragment, probably adult, unburned, consistent with missing Burial 3 skeletal element.

Pit Structure 3

Roof fall: Right proximal tibial epiphysis—unfused but near adult size, unburned, probably the "mate" of the left epiphysis of Burial 13, as they match exactly in size; adult molar without crown, probably associated with Burial 4; fragment of basi-occipital from left anterior border of foramen magnum, probably adult, does *not* belong to Burial 4, nor can it be associated with any other numbered burial at the site.

Pit Structure 4

Roof fall: Left rib, posterior half, adult, mild arthritic changes to head, several canid puncture marks, consistent with Burial 12. *Floor:* Two immature cranial vault fragments (one burned, one unburned), consistent with either Burial 11 or Burial 14; two foot phalanges (first digit plus second, third, or fourth), adult, unburned, consistent with Burial 12; two tooth fragments (possible canine and premolar), burned, immature roots, consistent with Burial 14.

Room 1

Wall fall: Three rib fragments (one left, one right, and an anterior rib end fragment that appears to have been fractured or injured antemortem, as there is evidence of healing), adult, unburned; second cervical vertebra (axis) and a fragmented third cervical vertebra, adult, unburned, compatible with remains described above from Room 1.

Room 2

Mixed fill: Left talus, adult, moderately eroded, unburned, possibly Burial 3.

Room 3

Wall fall: Sacrum, probably adult, very fragmented and incomplete. *Roof fall:* Fragments of at least five immature thoracic vertebrae; neural rings fused to bodies, but vertebral rims not fused; unburned; age compatible with that of remains described below; could be associated with Burial 5, but not definitive. Two unfused epiphyses (right distal femur and right proximal tibia); greatest width of each is 43 mm and 33 mm, respectively, consistent with a child aged 5.0 ± .5 years, on the basis of Sundick (1978, 1979); unburned; could be associated with Burial 5, but not definitive.

Room 4

Wall fall: Right clavicular midshaft, possibly male; very small fragments—clavicular(?); left rib, adult, incomplete (two fragments glued); left patella, adult, multiple bone spurs at site of quadriceps tendon insertion; right clavicular midshaft, possibly female; vertebra fragments; left second rib, adult, missing anterior tip. *Floor:* Left first rib, adult, unburned.

Room 5

Floor: Four very small fragments of immature bone, possibly epiphyses or carpals, all damaged and/or eroded.

Room 6

Wall fall: Left talus, adult, unburned, recent postmortem abrasion.

Room 12

Wall/roof fall: Four thoracic vertebra fragments, adult, unburned.

Room 14

Wall fall: Left first rib, anterior end broken off, adult; left clavicle, adult, both ends damaged, trowel mark near lateral end; several small scapular fragments (including part of a glenoid cavity), possibly immature.

Room 15

Wall fall: Anterior half of vertebral body, immature, lower cervical or upper thoracic; nearly complete cervical vertebra, adult, unburned; thoracic vertebra arch fragment, adult, unburned; left innominate bone (six fragments glued), adult female, 50+ years old, deeply pitted preauricular sulcus generally indicating at least one pregnancy (estimated pubis length = 89 mm, estimated ischium length = 78 mm), represents the fourth identifiable adult female at the site, since definitely not part of Burials 3, 10, or 12.

Room 19

Wall/roof fall: Small cranial and other cancellous fragments; immature cranial fragment.

Courtyard 1

Modern ground surface: Lower thoracic vertebra, adult, incomplete, canid tooth marks.

Courtyard 3

Feature 19: Cranial vault fragment (probable left posterior parietal, with portions of sagittal, lambdoidal, and squamosal sutures), immature, unburned.

Midden

52N 109E: Right second metacarpal, adult, unburned.
54N 99E: Right hamate, adult, unburned.
54N 103E: Left second metacarpal, adult, unburned; right clavicle shaft, missing both ends, adult(?).
54N 109E: Immature right talus, unburned, possible "mate" for Burial 7; immature metacarpal or metatarsal fragment; small fragment, possibly from left temporal, near external auditory meatus.

56N 103E: Immature thoracic vertebra, same stage of development as thoracic vertebrae of Burial 7.
56N 105E: Immature left ischium, no fusion, unburned, 49 mm long, probably the "mate" to the right ischium of Burial 7.
56N 107E: Right first and second metatarsal fragments, adult, unburned; other small unidentifiable fragments.
56N 109E: Immature metatarsal diaphysis, maximum length = 37 mm, possibly from Burial 7; left proximal phalanx, adult; many small fragments (long bone, rib, vertebra, tarsal[?]).
58N 99E: Fragment of immature pars lateralis of the occipital (with fused segment of squamous portion), including right occipital condyle and foramen magnum rim segment; unfused anteriorly; four to six years old (from Redfield [1970] in Krogman and Iscan [1986:110]); could be from Burial 5.
58N 101E: Left calcaneus, adult, canid puncture marks, maximum length = 72 mm; right humerus, adult, male (represents an additional adult male besides those of Burials 1, 2, 4, 6, and 8), canid puncture marks at both ends, prominent attachment site for teres major or latissimus dorsi muscles:

maximum length = 321 mm
maximum diameter of head = 45 mm
proximal end breadth = 49 mm
distal end breadth = 59 mm (estimate)
anterior-posterior midshaft diameter = 20 mm
medial-lateral midshaft diameter = 20 mm

Lumbar vertebra with broken arch, adult; rib fragments, probably adult; midcervical vertebra, adult, unburned; two vertebral body fragments, probably lumbar; other fragments.
58N 107E: Manubrium fragment, adult, unburned, arthritic changes at costal articulations.
58N 109E: Right first distal phalanx, foot, adult; posterior calcaneal fragment; many long-bone fragments; unidentifiable fragments.
60N 97E: Very immature right clavicle, diaphyseal length less than 75 mm, five to six years old (following Sundick 1978, 1979), could be from Burial 5, as it was recovered from the same general vicinity.
60N 101E: Probably left fourth metatarsal fragment, adult, charcoal-stained.
60N 103E: Left fourth metatarsal, adult; right third or fourth metacarpal, adult, proximal end missing.
65N 94E: Right navicular (hand), adult, unburned; body of rib, 32 mm long, possibly immature human; thin, slender fragment, 11 mm long, unknown.
68N 113E: Bone fragments—mostly, if not all, adult mandible, possibly female; six adult teeth (four molars, second premolar, and an incisor or canine root stump), most crowns broken off postmortem.
Mixed proveniences: Left fourth metatarsal fragment, adult, charcoal-stained.

Analysis

The Duckfoot Skeletal Sample

The Duckfoot skeletal sample is composed of a minimum of 16 individuals. This number is not derived from a minimum bone element count, which is less than 16 for all individual bones and teeth present, but is based on the count of individually identified burials plus additional unique bones. The sex and age profiles, as well as the burial locations and relative completeness of the remains, leave little doubt that there are 14 individual burials from the pit structures and midden. In addition, there is one adult female innominate (Room 15) and one adult male humerus (midden, 58N 101E) that, because of duplication of skeletal elements, cannot belong to any of the numbered adult burials. The 14 discrete burials are summarized in Table 9.1.

There are potentially more individuals present at the site, particularly in the unexcavated portion of the midden, but the inability to identify them and to match isolated bones with specific additional individuals precludes a reasonable estimate of total number of individuals. Surely some of the isolated bones recovered from the midden belong to identified midden burials, given the substantial postburial animal disturbance. In some cases, as noted in the burial descriptions and list of miscellaneous bone, matches of isolated bone with specific burials were very reasonable and probable; for example, several immature bones are probably the mates of those belonging to Burial 7. In other cases, tentative associations could only be suggested; for instance, an immature clavicular diaphysis is consistent with Burial 5, but a positive association cannot be proved. In still other contexts—for example, Room 4—it could be demonstrated that, because of duplication of bones or age incompatibility, isolated remains definitely

were *not* associated with any of the numbered burials. However, because such bones could have belonged to the additional female and male noted above, statements about additional individuals beyond the definite 16 cannot be made.

The vast majority of the human skeletal material, then, is associated with the 14 numbered burials from pit structures and the midden. Different numbers of individuals are used in the different analyses described below. In the examination of mortuary practice, only the 14 numbered burials are used, because the additional male and female isolated elements come from contexts that cannot be specifically traced to a primary burial location. In the various demographic analyses, both 15 and 16 individuals are used at various times (the 14 numbered burials plus the adult female, 50+ years old, represented by a single innominate, and the adult male, age unknown, represented by a single humerus). Sometimes the same analysis is repeated using either 15 or 16 individuals. These separate analyses are presented so the reader can make appropriate inferences depending on whether it is assumed that the isolated female innominate and the male humerus originate from individuals in the Duckfoot site or were transported from somewhere else; the former is more likely, however. The statistical summary of cranial and postcranial metric variation includes all identifiable individuals (the 14 burials plus the isolated male humerus and female innominate), which helps in some minimal way to increase the sample sizes for these statistical analyses.

Mortuary Practice

Table 9.2 cross-tabulates the number of burials found in the pit structures and in the midden by age category, either adult or child (less than 18 years of age). A chi square test of this distribution indicates that there is no significant difference from a random distribution. Table 9.3 presents the same information broken down by sex, with unknown sexes included; Table 9.4 includes only adults of known sex. As with the distribution by age, that for sex is not significantly different from a random distribution as tested by chi square.

An additional analysis comparing the metric or nonmetric variation of the sexes within and between the two primary burial areas was not attempted because of the small adult sample sizes when broken down by sex and the incompleteness of the data sets for the individuals from Pit Structure 4 and most of the midden burials. Such an analysis might have been informative with regard to the question of whether there are any differences between the individuals buried in the pit structures and those buried in the midden, and it may have been useful in exploring questions of biological distance or even familial relationships. See the "Discussion" section later in the chapter for speculation based on degree of

Table 9.1. Summary of Primary Burials

Burial No.	Sex	Age (years)	Location	Condition
1	male	30–39	Pit Structure 1	burned
2	male	30–39	Pit Structure 1	burned
3	female	30–39	Pit Structure 2	unburned
4	male	30–39	Pit Structure 3	burned
5	unknown	5–6	midden	unburned
6	male	40–49	midden	unburned
7	unknown	6–7	midden	unburned
8	male	50+	midden	unburned
9	unknown	10–11	midden	unburned
10	female	18–20	midden	unburned
11	unknown	3–4	Pit Structure 4	burned
12	female	50+	Pit Structure 4	burned
13	unknown	12–15	midden	unburned
14	unknown	5–6	Pit Structure 4	burned

Table 9.2. Cross-tabulation of Age Status by
Burial Location

| Age Status | | Location | | Total |
		Midden	Pit Structure	
Adult	N	3	5	8
	%	42.9	71.4	57.1
Child	N	4	2	6
	%	57.1	28.6	42.9
TOTAL	N	7	7	14
	%	50.0	50.0	100.0

Table 9.3. Cross-tabulation of Sex by Burial Location,
Unknown Sexes Included

| Sex | | Location | | Total |
		Midden	Pit Structure	
Female	N	1	2	3
	%	14.3	28.6	21.4
Male	N	2	3	5
	%	28.6	42.9	35.7
Unknown	N	4	2	6
	%	57.1	28.6	42.9
TOTAL	N	7	7	14
	%	50.0	50.0	100.0

Table 9.4. Cross-tabulation of Sex by Burial Location,
Known Sexes Only

| Sex | | Location | | Total |
		Midden	Pit Structure	
Female	N	1	2	3
	%	33.3	40.0	37.5
Male	N	2	3	5
	%	66.7	60.0	62.5
TOTAL	N	3	5	8
	%	37.5	62.5	100.0

metric variation within Duckfoot and between it and Dolores skeletal material.

One obvious difference between the burials in the midden area and those in the pit structures is the burning of structures and bodies in the latter burial sites. Table 9.5 illustrates this dramatic difference in the treatment of the burials, whether intentional or not. This is a highly significant distribution ($\chi^2 = 7.29$, $df = 1$, p = 0.0069 with Yates' correction). Chi square tests of the distribution of burned/unburned burials by age status (adult versus child) and by sex were not significant. These distributions are shown in Tables 9.6, 9.7, and 9.8. Table 9.7 includes individuals of known sex only; Table 9.8 includes those of unknown sex as well.

None of the burials show evidence of interpersonal trauma. Regardless of burial site, then, the condition of all broken and/or incomplete skeletons can be best explained by natural or human-caused circumstances other than trauma.

Midden burials (Burials 5, 6, 7, 8, 9, 10, and 13), virtually without exception, show varying degrees of disturbance and direct alteration by carnivore gnawing. Evidence of canid tooth marks—canine and carnassial tooth punctures—is widespread; gnaw marks are particularly common on the epiphyseal ends of long bones, source of red marrow. This evidence points to disturbance and destruction of the relatively shallow burials in the midden. Archaeological evidence also indicates some rodent disturbance of burials. Many bones from the midden, particularly those not directly associated with the midden burials, also show surface erosion and weathering, which indicates some degree of movement by natural processes, possibly soil slumping and water transport.

Within the pit structures, about half the burials are relatively well preserved. Those from Pit Structure 1 (Burials 1 and 2) and Pit Structure 3 (Burial 4) are nearly complete and, with some exceptions, show little damage to the skeleton. Burial 1 is intact and nearly complete, missing only some of the small bones of the hands and feet, primarily phalanges; bones not present are probably missing as the result of rodent activity. This burial is minimally burned in terms of duration of exposure to the fire and extent of the body involved. The only burned portion of Burial 1 is the left knee region, which was near a burned roof timber. The position of Burial 1 does not appear to be that of an individual who struggled to escape a burning structure; this, coupled with the fact that this burial position is relatively common in the region, argues for death before the pit structure burned.

Burial 2, found mostly in contact with the floor but also with some burned material *under* the body, is nearly complete but fragmented and is more thoroughly burned than Burial 1. The left half of the cranial vault and adjacent facial region is broken, but, as noted above, this does not appear to be the result of antemortem trauma. The damage probably was caused by the impact of the collapsing roof or by the skull's "exploding" as a result of heating of the intracranial contents. Because the cranium is primarily scorched and charred but not calcined, indicating a relatively low burning temperature, the former explanation is more likely. The apparently random distribution of the burned areas of Burial 2 is consistent with the scenario of various parts of the body being in contact with burning or smoldering timbers just long enough to burn away soft tissues and char the bone but not long enough, or at a high-enough temperature, to bring about the recrystalization changes of the inorganic bone salts seen in calcination (Stewart 1979). These changes are accompanied by a "greying" of the bone and a distinctive metallic "clink"

Table 9.5. Cross-tabulation of Burial Location by Condition

Location		Condition		Total
		Burned	Unburned	
Midden	N	0	7	7
	%	0.0	87.5	50.0
Pit structure	N	6	1	7
	%	100.0	12.5	50.0
TOTAL	N	6	8	14
	%	42.9	57.1	100.0

Table 9.6. Cross-tabulation of Age Status by Condition

Age Status		Condition		Total
		Burned	Unburned	
Adult	N	4	4	8
	%	66.7	50.0	57.1
Child	N	2	4	6
	%	33.3	50.0	42.9
TOTAL	N	6	8	14
	%	42.9	57.1	100.0

Table 9.7. Cross-tabulation of Sex by Condition, Known Sexes Only

Sex		Condition		Total
		Burned	Unburned	
Female	N	1	2	3
	%	25.0	50.0	37.5
Male	N	3	2	5
	%	75.0	50.0	62.5
TOTAL	N	4	4	8
	%	50.0	50.0	100.0

Table 9.8. Cross-tabulation of Sex by Condition, Unknown Sexes Included

Sex		Condition		Total
		Burned	Unburned	
Female	N	1	2	3
	%	16.7	25.0	21.4
Male	N	3	2	5
	%	50.0	25.0	35.7
Unknown	N	2	4	6
	%	33.3	50.0	42.9
TOTAL	N	6	8	14
	%	42.9	57.1	100.0

when the bone is lightly percussed. Although the evidence does not conclusively support either archaeological interpretation of this burial—on the roof when it collapsed or on the floor, with the roof collapsing over it—the latter scenario seems more reasonable, in part because this is the pattern present elsewhere at Duckfoot and in the Dolores valley (Kane 1988; Stodder 1987). If this is true, then the body may have been on its side on the floor and may have subsequently rolled onto some burned roof debris or onto fuelwood used to ignite the roof.

Burial 4 from Pit Structure 3 is similar to Burial 1 in that it was placed on the floor of the pit structure before the structure burned and probably after death. The very localized damage and burning of the skeleton, basically confined to the right half of the cranial vault and adjacent face, coupled with the presence of burned roof fall overlying the right side of the skull, argues for the roof as the source of the damage. Partial reconstruction of the cranial vault and lack of evidence of a depressed skull fracture support the argument for a nontraumatic source of the damage.

In contrast to the burials just described, Burials 11, 14, and 12, two young children and an older female from Pit Structure 4, are more fragmented and more thoroughly burned, especially the children. The two children, Burials 11 and 14, are so thoroughly burned that many skeletal elements are calcined and do not show the pattern of scorching noted in the previous descriptions; the majority of the bones are completely charred. This more-thorough burning, compared with that of the adult female in the pit structure, is probably best explained by the much thinner soft tissue layers in the children, which would have burned away much more quickly, exposing the underlying skeleton more directly and for a longer period of time. Compared with the children, Burial 12, the adult female from Pit Structure 4, shows more scorching and charring than calcination, which indicates less exposure time and/or lower temperature of exposure. That the body is burned unevenly is best explained by the fact that the unburned body parts were protected by sandstone slabs or did not come in contact with the burning roof debris. As is the case with Burials 1 and 4, it is likely that Burial 12 was deliberately interred before the roof was burned.

Finally, Burial 3 from Pit Structure 2 is the only unburned skeleton recovered from a pit structure. It is also the most incomplete, especially of the adult skeletons from these structures. The limb bones, both arms and legs with their distal appendages, are missing, probably the result of carnivore activity, as evidenced by numerous canine tooth puncture marks on recovered skeletal parts. The loss of limb bones is not uncommon in contemporary forensic contexts in which bodies are disturbed by dogs or coyotes. After gnawing through major joint areas adjacent to the marrow-rich epiphyseal bone ends, canids typically carry limbs away from the corpse for further consumption or transport to a den in the area (Willey and Snyder 1989). Damage to the extant skeleton (cracked cranium and numerous broken ribs) is consistent with damage resulting from the collapse of the roof and does not appear to be the result of interpersonal trauma. Archaeological evidence suggests that this burial was a secondary interment, with the body already disarticulated and incomplete when it was placed on the floor and partly covered with rocks. There was little disarticulation of the extant skeletal parts and little vertical displacement to indicate carnivores' digging down to the body through roof fall.

Table 9.9 lists the individual burials by body position, degree of flexure, and head orientation. Table 9.10 presents the frequencies with which each of the specific positions and orientations occurs.

Cannibalism

None of the skeletons, regardless of place of interment, whether burned or not and whether complete or not, show evidence of butchering that might indicate cannibalism. Turner's (1988) three signs of cannibalism—butchered, burned, and broken bones—are not all present here. Although most of the pit structure burials are burned and several of these are broken, there is no evidence of cut marks on any of the bones to support the notion of dismemberment before burning (i.e., cooking). The burned and broken condition of the bones can be explained more simply as the natural result of cremation in a burning structure, with fragmentation due mostly to falling roof timbers and/or animal disturbance.

Demography

In the following demographic summaries, most analyses use 15 or 16 individuals, depending on whether a particular age estimation is necessary or mere status as an adult is acceptable. The two additional individuals added to the various demographic analyses are the adult female (50+ years old) represented by a single innominate bone and the adult male (age estimation not possible) represented by a single humerus.

Table 9.11 cross-tabulates sex by general age category, adult versus child (under 18 years of age). Table 9.12 presents the same data by age range in years. A graphical representation of Table 9.12 is seen in Figure 9.1, except that the adult male of unknown age is not plotted (note that the one female in the 10–20 age group is the Burial 10 individual, whose age is estimated at 18 to 20 years and who therefore is considered an adult).

The most notable features of Figure 9.1 are the expected large number of subadults, six total, and the unexpected large number of adults aged 50+ years, three individuals in the total sample. The percentage of children under 18 years of age, 37.5 percent (based on a total sample of 16),

Table 9.9. Burial Positions and Orientations

Burial No.	Position	Flexure	Orientation (in degrees)[a]	Facing
1	supine	partial	SW 221	up
2	right side	partial	ENE 75	N
3	prone	unk	SE 143	down
4	supine	partial	N 353	E
5	unk	unk	ESE 123	WSW
6	supine	partial	SSE 149	up
7	left side	full	E 100	SW
8	left side	partial	ENE 68	SE
9	supine	full	SW 211	NE
10	supine	partial	NE 51	up
11	unk	unk	ESE 116?	unk
12	supine	extended	SE 139	up
13	right side	full	NE 45	N
14	unk	unk	NW 70?	unk

unk = unknown, because of poor preservation or missing skeletal elements.
[a] Orientation is the angle toward the head from true north.

Table 9.10. Frequency Tabulations of Body Position, Degree of Flexure, and Head Orientation

	N	%
Body Position (N = 11)		
Prone	1	9.1
Supine	6	54.5
Left side	2	18.2
Right side	2	18.2
Degree of Flexure (N = 10)		
Fully flexed	3	30.0
Partially flexed	6	60.0
Extended	1	10.0
Head Orientation (N = 14)		
North	1	7.1
Northeast	2	14.3
East-northeast	2	14.3
East	1	7.1
East-southeast	2	14.3
Southeast	2	14.3
South-southeast	1	7.1
Southwest	2	14.3
Northwest	1	7.1

in this mortality sample is not an unusual number for prehistoric populations or even for many modern, underdeveloped countries (Hinkes 1983:17). At the other end of the age distribution, 20.0 percent of the sample that could be aged (N = 15) is at least 50 years old, a very unusual figure for any population not from current developed countries.

To further explore the mortality data from the Duckfoot site, the burial data are processed in life tables (Tables 9.13–9.18). It is important to note, however, that the

Table 9.11. Cross-tabulation of Age Status by Sex

| Age Status | | Sex | | | Total |
		Female	Male	Unknown	
Adult	N	4	6	0	10
	%	100.0	100.0	0.0	62.5
Child	N	0	0	6	6
	%	0.0	0.0	100.0	37.5
TOTAL	N	4	6	6	16
	%	25.0	37.5	37.5	100.0

NOTE: Includes 14 burials, plus one adult female, represented by a single innominate, and one adult male, represented by a single humerus.

Table 9.12. Cross-tabulation of Age Range by Sex

| Age Range (years) | | Sex | | | Total |
		Female	Male	Unknown	
3–4	N	0	0	1	1
	%	0.0	0.0	16.7	6.3
5–6	N	0	0	2	2
	%	0.0	0.0	33.3	12.5
6–7	N	0	0	1	1
	%	0.0	0.0	16.7	6.3
10–11	N	0	0	1	1
	%	0.0	0.0	16.7	6.3
12–15	N	0	0	1	1
	%	0.0	0.0	16.7	6.3
18–20	N	1	0	0	1
	%	25.0	0.0	0.0	6.3
30–39	N	1	3	0	4
	%	25.0	50.0	0.0	25.0
40–49	N	0	1	0	1
	%	0.0	16.7	0.0	6.3
50+	N	2	1	0	3
	%	50.0	16.7	0.0	18.8
Unknown	N	0	1	0	1
	%	0.0	16.7	0.0	6.3
TOTAL	N	4	6	6	16
	%	25.0	37.5	37.5	100.0

NOTE: Includes 14 burials, plus one adult female (50+ years), represented by a single innominate, and one adult male (age unknown), represented by a single humerus. There were no individuals in the 20–29 age interval.

skeletons represent the local people who died during the habitation of the site and do not necessarily provide a profile of those alive at any point in time or during the entire span of site occupation. The composite life table analysis presented here uses mortality data, not data of individuals alive at some point in time or during a cohort's lifespan, as other kinds of life tables do. Generally, the last category (percent surviving) is not included in life tables,

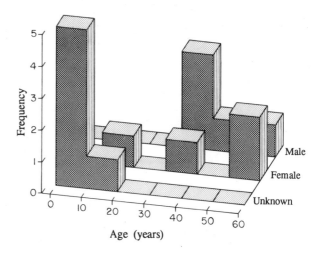

Figure 9.1. Age-sex distribution, the Duckfoot site. Figure does not include the one male of unknown age.

although the data are easily calculated from life table data because "percent surviving" is l_x expressed as a percentage of the total population. The category *is* presented here, however, because it is also used in the comparative graphs of the percentage of individuals surviving each age interval for Duckfoot and the Mesa Verde–region populations.

Two slightly different life tables are presented: (1) an expanded version that matches the age categories of the life tables in Stodder's (1987) analysis and (2) a condensed version of the former, which collapses the adult age categories into 10-year age ranges instead of 5-year intervals. The expanded life table is used to directly compare the Duckfoot mortality data with Stodder's data from the larger Mesa Verde region. The collapsed version of the life table, and its comparison with similar collapsed tables of Stodder's data, is used because this author believes that it is not possible with current methods to estimate adult ages at death in categories of five years' breadth. Numerous contemporary studies of samples in which ages at death were known (see several examples in Krogman and Iscan [1986], Suchey et al. [1986], and Suchey et al. [1988]) clearly indicate that age-estimation techniques have too high a standard error to warrant placing all adult individuals into five-year categories, especially beyond the fourth decade of life.

Stodder's life table data (1987:416–417, Tables 20.40, 20.41) have been recalculated for the current analysis because different computer programs have different numerical rounding techniques and numbers of significant digits used in calculations. However, the calculation formulae follow Stodder's and are presented in the discussion of methods above. Stodder's life table results, for both the early (A.D. 600–975) and the late (A.D. 975–1300) Mesa Verde samples, are presented and compared with

Table 9.13. Composite Life Table, Duckfoot

x	d'_x	d_x	l_x	q_x	L_x	e^o_x*	e^o_x	% Surviving
0–2	0	.0	1000	.0000	1000.0	6.17	12.33	100.00
3–7	4	266.7	1000	.2667	866.7	5.17	25.83	100.00
8–11	1	66.7	733	.0909	700.0	5.86	23.45	73.33
12–15	1	66.7	667	.1000	633.3	5.40	21.60	66.67
16–20	1	66.7	600	.1111	566.7	4.94	24.72	60.00
21–25	0	.0	533	.0000	533.3	4.50	22.50	53.33
26–30	0	.0	533	.0000	533.3	3.50	17.50	53.33
31–35	2	133.3	533	.2500	466.7	2.50	12.50	53.33
36–40	2	133.3	400	.3333	333.3	2.17	10.83	40.00
41–45	1	66.7	267	.2500	233.3	2.00	10.00	26.67
46–50	0	.0	200	.0000	200.0	1.50	7.50	20.00
50+	3	200.0	200	1.0000	100.0	.50	2.50	20.00
								.00
	15							

NOTE: The life table categories, or functions, are defined below, following Stodder's (1987:416) conventions for the function abbreviations. (These definitions, and their abbreviations, differ somewhat from those used by Swedlund and Armelagos [1976], Ubelaker [1978], and most other workers in paleodemographic research. The major differences are Stodder's L_x and e^o_x* functions, the latter of which is usually replaced by the more standard T_x. The resulting values of e^o_x, then, differ when calculated by the two techniques—Stodder's method giving lower values, especially in the early years.)

x = the age interval

d'_x = number of deaths in each age interval

d_x = number of deaths in each age interval, based on a population of 1000 = $\frac{d'_x}{n}$ (1000)

l_x = number of individuals surviving at beginning of each age interval = l_x minus previous d_x, where l_x for the first age category is 1000

q_x = the probability of dying in each age interval = $\frac{d_x}{l_x}$

L_x = number of individuals surviving between each age interval and the following age interval = $\frac{(l_x) + (l_x + 1)}{2}$

e^o_x* = mean life expectancy at beginning of each age interval, expressed in age interval = $\frac{\sum L_x}{l_x}$

e^o_x = life expectancy in years at end of each age interval = (e^o_x*) (age interval size, in years)

the Duckfoot mortality data, even though Duckfoot falls into the chronological range of Stodder's early regional sample.

The recalculation of Stodder's data generally confirms her results for the early sample, but an error was detected in her late-sample life table. Her function that this author has designated l_x is a constantly decreasing function of the number of individuals, based on a radix of 1,000 individuals, who have survived to the beginning of a particular age category. In her late sample, age category 31–35 years, she presents a value of 474 for the function l_x. This value should be 247. This error affects both the life-expectancy and percentage-surviving calculations and, therefore, graphic comparisons between the Duckfoot assemblage and Stodder's data. Also, the final values of q_x in Stodder's life tables for both the early and the late samples should be 1.0000, not .1000, as presented by her (Stodder 1987:416, 417).

The composite life table for Duckfoot, using the expanded age categories of Stodder's original tables, is presented in Table 9.13. The recalculated values for Stodder's early and late samples are in Tables 9.14 and 9.15, respectively. The life tables based on the collapsed age categories are provided in Tables 9.16, 9.17, and 9.18 for Duckfoot and the early and late Mesa Verde–region samples, respectively.

The three life tables with expanded adult age categories (Tables 9.13–9.15) allow direct comparison of Duckfoot and Mesa Verde–region life-expectancy values, e^o_x. These comparisons are presented in Figure 9.2. A similar graph using collapsed adult age categories is shown in Figure 9.3. Figure 9.2 reveals gross differences between the Duckfoot

Table 9.14. Composite Life Table, Mesa Verde Region, Early Sample

x	d'_x	d_x	l_x	q_x	L_x	$e^o_x *$	e^o_x	% Surviving
0–2	16	107	1000	.1067	946.7	5.10	10.20	100.00
3–7	21	140	893	.1567	823.3	4.65	23.25	89.33
8–11	6	40	753	.0531	733.3	4.42	17.68	75.33
12–15	10	67	713	.0935	680.0	3.64	14.56	71.33
16–20	11	73	647	.1134	610.0	2.96	14.82	64.67
21–25	26	173	573	.3023	486.7	2.28	11.40	57.33
26–30	17	113	400	.2833	343.3	2.05	10.25	40.00
31–35	15	100	287	.3488	236.7	1.66	8.31	28.67
36–40	15	100	187	.5357	136.7	1.29	6.43	18.67
41–45	6	40	87	.4615	66.7	1.19	5.96	8.67
46–50	5	33	47	.7143	30.0	.79	3.93	4.67
50+	2	13	13	1.0000	6.7	.50	2.50	1.33
								.00
	150							

SOURCE: Adapted and recalculated from Stodder (1987:416).
NOTE: Refer to Table 9.13 for an explanation of life table abbreviations.

Table 9.15. Composite Life Table, Mesa Verde Region, Late Sample

x	d'_x	d_x	l_x	q_x	L_x	$e^o_x *$	e^o_x	% Surviving
0–2	30	168.5	1000	.1685	915.7	4.48	8.96	100.00
3–7	22	123.6	831	.1486	769.7	4.28	21.42	83.15
8–11	11	61.8	708	.0873	677.0	3.94	15.78	70.79
12–15	13	73.0	646	.1130	609.6	3.27	13.10	64.61
16–20	29	162.9	573	.2843	491.6	2.63	13.14	57.30
21–25	21	118.0	410	.2877	351.1	2.47	12.36	41.01
26–30	8	44.9	292	.1538	269.7	2.27	11.35	29.21
31–35	16	89.9	247	.3636	202.2	1.59	7.95	24.72
36–40	14	78.7	157	.5000	118.0	1.21	6.07	15.73
41–45	9	50.6	79	.6429	53.4	.93	4.64	7.87
46–50	4	22.5	28	.8000	16.9	.70	3.50	2.81
50+	1	5.6	6	1.0000	2.8	.50	2.50	.56
								.00
	178							

SOURCE: Adapted and recalculated from Stodder (1987:417).
NOTE: Refer to Table 9.13 for an explanation of life table abbreviations.

and Mesa Verde–region life-expectancy data sets after the second age interval. However, this is due partly to the problem of age intervals used in composite life tables. In order to fit the adult Duckfoot data into intervals of five years, individual burials had to be arbitrarily assigned to five-year age groups—for example, to either the 41–45 or the 46–50 group. This investigator believes that it is impossible to estimate adult ages this closely and that using five-year age intervals introduces a bias that can influence the calculations in either a positive or a negative direction, depending on whether the individual is placed in the older or the younger age interval, respectively. More serious is the addition of age intervals, particularly with such a small data set, that have no individuals in them.

Ultimately, life-expectancy calculations are based on the cumulative experience of individuals who survive to a particular age interval, expressed as l_x. As the number of empty age categories increases, more total individuals accumulate throughout the column of l_x values. This can easily be seen by comparing the Duckfoot l_x values in Tables 9.13 and 9.16. The l_x values are grossly inflated because the same individuals are carried into several successive, but empty, age intervals, which then influences subsequent calculations that are based on l_x values—for example, L_x and e^o_x. Primarily for this reason, the best comparison between the Duckfoot life expectancies and the Mesa Verde–region data is seen in Figure 9.3, which uses collapsed age intervals.

Table 9.16. Composite Life Table, Duckfoot, Collapsed Adult Age Intervals

x	d'_x	d_x	l_x	q_x	L_x	e^o_x*	e^o_x	% Surviving
0–2	0	.0	1000	.0000	1000.0	5.03	10.07	100.00
3–7	4	266.7	1000	.2667	866.7	4.03	20.17	100.00
8–11	1	66.7	733	.0909	700.0	4.32	17.27	73.33
12–15	1	66.7	667	.1000	633.3	3.70	14.80	66.67
16–20	1	66.7	600	.1111	566.7	3.06	15.28	60.00
21–30	0	.0	533	.0000	533.3	2.38	11.88	53.33
31–40	4	266.7	533	.5000	400.0	1.38	6.88	53.33
41–50	1	66.7	267	.2500	233.3	1.25	6.25	26.67
50+	3	200.0	200	1.0000	100.0	.50	2.50	20.00
								.00
	15							

NOTE: Refer to Table 9.13 for an explanation of life table abbreviations.

Table 9.17. Composite Life Table, Mesa Verde Region, Early Sample, Collapsed Adult Age Intervals

x	d'_x	d_x	l_x	q_x	L_x	e^o_x*	e^o_x	% Surviving
0–2	16	107	1000	.1067	946.7	4.47	8.93	100.00
3–7	21	140	893	.1567	823.3	3.94	19.70	89.33
8–11	6	40	753	.0531	733.3	3.58	14.32	75.33
12–15	10	67	713	.0935	680.0	2.75	11.01	71.33
16–20	11	73	647	.1134	610.0	1.98	9.92	64.67
21–30	43	287	573	.5000	430.0	1.17	5.87	57.33
31–40	30	200	287	.6977	186.7	.85	4.24	28.67
41–50	11	73	87	.8462	50.0	.65	3.27	8.67
50+	2	13	13	1.0000	6.7	.50	2.50	1.33
								.00
	150							

SOURCE: Adapted and recalculated from Stodder (1987:416).
NOTE: Refer to Table 9.13 for an explanation of life table abbreviations.

Table 9.18. Composite Life Table, Mesa Verde Region, Late Sample, Collapsed Adult Age Intervals

x	d'_x	d_x	l_x	q_x	L_x	e^o_x*	e^o_x	% Surviving
0–2	30	168.5	1000	.1685	915.7	4.00	8.00	100.00
3–7	22	123.6	831	.1486	769.7	3.71	18.55	83.15
8–11	11	61.8	708	.0873	677.0	3.27	13.08	70.79
12–15	13	73.0	646	.1130	609.6	2.53	10.14	64.61
16–20	29	162.9	573	.2843	491.6	1.79	8.97	57.30
21–30	29	162.9	410	.3973	328.7	1.31	6.54	41.01
31–40	30	168.5	247	.6818	162.9	.84	4.20	24.72
41–50	13	73.0	79	.9286	42.1	.57	2.86	7.87
50+	1	5.6	6	1.0000	2.8	.50	2.50	.56
								.00
	178							

SOURCE: Adapted and recalculated from Stodder (1987:417).
NOTE: Refer to Table 9.13 for an explanation of life table abbreviations.

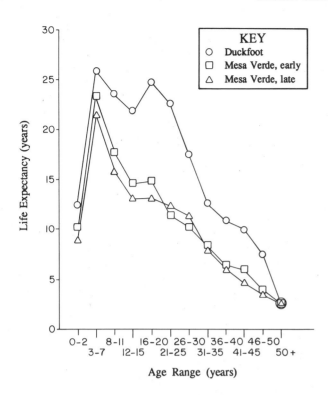

Figure 9.2. Life expectancy, 5-year adult age intervals, Duckfoot versus Mesa Verde–region data. The Mesa Verde data were recalculated from Stodder (1987:416, 417).

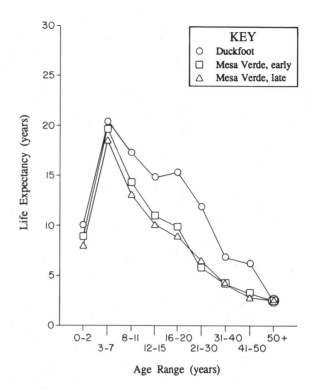

Figure 9.3. Life expectancy, 10-year adult age intervals, Duckfoot versus Mesa Verde–region data. The Mesa Verde data were recalculated from Stodder (1987:416, 417).

For all age intervals, Duckfoot life expectancies are higher than either of the two Mesa Verde–region life expectancies. Compared with the early Mesa Verde–region data, Duckfoot life expectancy is 1.14 years greater at the beginning of the 0- to 2-year interval, .47 years greater at the start of the 3- to 7-year interval, and then markedly greater for the remainder of the age intervals. The greatest difference in life expectancy occurs for the 21- to 30-year age interval, for which the Duckfoot sample has 6.01 years greater life expectancy than the early Mesa Verde–region sample (11.88 years compared with 5.87 years).

The major reason for the substantial disparities in the life-expectancy data is the large number of Duckfoot individuals who survived to reach at least the 31- to 40-year age interval—8 of 15 total inhabitants who could be aged, or 53.3 percent. For the early regional sample, only 43 of 150, or 28.7 percent, of the total sample lived to this age interval. One must be cautious, however, before assigning too much significance to these data. Compared with other prehistoric populations in the American Southwest (e.g., Stodder 1987; Bennett 1975; Palkovich 1980; Akins 1986), the Duckfoot sample has a dearth of very young individuals and a larger proportion of adults, especially in the older age categories. This is probably the result of two factors: (1) the sample from Duckfoot may not be the complete sample of those who died during the site's occupation

(how many individuals are still buried in the midden?), and (2) the sample could easily reflect a random demographic sampling, especially if one considers the small sample size and short duration of the site's occupation. It is likely that both types of sampling error are factors here.

The percentage of individuals who survived to enter a particular age interval is portrayed in Figures 9.4 and 9.5 for the 5-year adult age intervals and 10-year adult age intervals, respectively. Figure 9.4 provides the most direct comparison of the Duckfoot data with the Mesa Verde–region data because it parallels exactly the age intervals in Stodder's (1987) presentation of the latter data set. However, the same criticism applies here as to the life-expectancy data; namely, that the empty age-interval categories influence the calculation of the percentage of individuals surviving into each age interval. The effect here, however, is not as noticeable as in the life-expectancy data, because of the way this calculation is made. Nonetheless, Figure 9.5 provides the most reasonable comparison of the three data sets.

The major differences between Duckfoot and Mesa Verde survivorship—that is, the percentage who survived to the beginning of an age interval—are in the second age interval (3–7 years) and for all adult intervals from 31–40 years onward. The first difference is due to the fact that no Duckfoot individuals died between birth and two years of age; all survived to at least three years of age. From 8

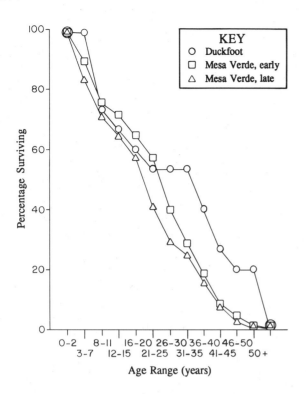

Figure 9.4. Survivorship curves, 5-year adult age intervals, Duckfoot versus Mesa Verde–region data. The Mesa Verde data were recalculated from Stodder (1987:416, 417).

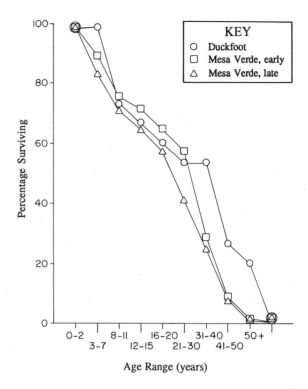

Figure 9.5. Survivorship curves, 10-year adult age intervals, Duckfoot versus Mesa Verde–region data. The Mesa Verde data were recalculated from Stodder (1987:416, 417).

through 30 years of age, Duckfoot survivorship data closely parallel the early regional sample data; in fact, the Duckfoot values are somewhat lower than the Mesa Verde values for these years. However, since no one from Duckfoot died in his or her twenties and this was a large death interval for the early regional sample, a much larger percentage of individuals survived to reach at least 30 years of age at Duckfoot. Also, three Duckfoot individuals, or 20 percent of the sample, survived to at least 50, whereas only 1.33 percent of the early Mesa Verde sample survived 50 years or more.

Cranial Deformation

The only summary statement that can be made about the form of cranial deformation in the Duckfoot sample is that all individuals for whom sufficient cranial material existed showed the lambdoidal type of deformation. Most of the deformities were asymmetrical to one side or the other, but this is to be expected (Bennett 1975; Stodder 1987). The lambdoidal pattern seen here is, by far, the most prevalent of the cranial deformation patterns found in the early Pueblo periods, including those reported by Stodder (1987:368). The vertical-occipital type did not become widespread until the Pueblo III period (Reed 1949). Specific information for each Duckfoot cranium can be found in the description of individual burials above.

Cranial and Postcranial Metric Variation

Metric variation in the cranial skeleton is summarized separately for the adult males and females in Table 9.19. Similar tabulations and summaries for the postcranial skeleton follow in Table 9.20. A modest amount of sexual dimorphism is present. However, the incompleteness of many of the adult skeletons, the small size of the total Duckfoot sample, and the even smaller subsample sizes when the assemblage is broken down by sex make meaningful comparisons with other data sets difficult. Measurements and indices for individual burials are on file at the Anasazi Heritage Center and are also available from the author.

Stature

Stature could be calculated for only five adults, four males and one female (Table 9.21). Using the formulae of Genoves (1967), these estimates are presented for calculations based on a multiple-bone formula (designated "All bones" in Table 9.21) and on formulae for individual femora and tibiae. Stodder's calculations (1987:372, Table 20.17) also use Genoves's formula for the femur and are compared with the Duckfoot data in Table 9.22. Estimates of stature for Duckfoot males are virtually identical to those in Stodder's data, whereas the single female is about 3 inches

Table 9.19. Summary of Adult Cranial Measurements (in mm) and Indices

	Male Summary Statistics						Female Summary Statistics					
	N	Min.	Max.	Mean	SD	Var.	N	Min.	Max.	Mean	SD	Var.
MEASUREMENTS												
Cranium												
Maximum length	3	161	177	169.00	6.53	42.67	1	154	154	154.00	—	—
Maximum breadth	4	137	154	147.00	6.20	38.50	2	131	148	139.50	8.50	72.25
Biasterionic breadth	3	100	130	115.67	12.28	150.89	2	99	101	100.00	1.00	1.00
Minimum frontal breadth	2	92	98	95.00	3.00	9.00	3	90	96	92.67	2.49	6.22
Bizygomatic breadth	3	133	147	141.67	6.18	38.22	2	129	136	132.50	3.50	12.25
Basion-bregma height	4	140	146	143.50	2.29	5.25	2	131	133	132.00	1.00	1.00
Basion-nasion length	4	100	103	101.25	1.30	1.69	2	91	93	92.00	1.00	1.00
Basion-prosthion length	5	91	101	94.40	3.56	12.64	2	86	89	87.50	1.50	2.25
Upper facial height	4	72	83	76.50	4.39	19.25	2	69	70	69.50	.50	.25
Total facial height	4	117	128	122.00	5.05	25.50	2	114	115	114.50	.50	.25
Nasal height	4	50	56	53.50	2.60	6.75	2	47	52	49.50	2.50	6.25
Nasal breadth	3	26	26	26.00	.00	.00	2	26	26	26.00	.00	.00
Left orbit height	2	36	37	36.50	.50	.25	2	32	36	34.00	2.00	4.00
Left orbit breadth	2	37	43	40.00	3.00	9.00	2	36	36	36.00	.00	.00
Right orbit height	4	34	37	35.25	1.30	1.69	2	33	36	34.50	1.50	2.25
Right orbit breadth	4	39	44	41.00	1.87	3.50	2	34	35	34.50	.50	.25
Interorbital breadth	3	22	24	23.33	.94	.89	2	24	24	24.00	.00	.00
Biorbital breadth	3	95	106	100.00	4.55	20.67	2	94	95	94.50	.50	.25
Palate external length	5	53	57	54.60	1.62	2.64	2	50	52	51.00	1.00	1.00
Palate external breadth	4	67	71	68.75	1.48	2.19	2	61	67	64.00	3.00	9.00
Palate internal length	5	46	49	47.60	1.02	1.04	2	43	45	44.00	1.00	1.00
Palate internal br. at M2	5	37	46	39.40	3.38	11.44	2	34	40	37.00	3.00	9.00
Left mastoid length	4	31	33	32.00	.71	.50	2	26,	26	26.00	.00	.00
Right mastoid length	4	31	36	33.00	1.87	3.50	2	24	26	25.00	1.00	1.00
Foramen magnum length	5	34	41	36.40	2.42	5.84	2	32	34	33.00	1.00	1.00
Foramen magnum breadth	4	29	32	30.75	1.09	1.19	3	24	27	25.67	1.25	1.56
Maximum circumference	2	484	505	494.50	10.50	110.25	1	454	454	454.00	—	—
Frontal arc	3	122	145	131.67	9.74	94.89	2	111	126	118.50	7.50	56.25
Parietal arc	4	104	131	118.50	11.72	137.25	2	90	102	96.00	6.00	36.00
Occipital arc	4	110	121	113.25	4.55	20.69	2	114	129	121.50	7.50	56.25
Frontal chord	4	108	119	113.25	4.15	17.19	2	98	107	102.50	4.50	20.25
Parietal chord	4	94	117	105.50	10.11	102.25	2	85	99	92.00	7.00	49.00
Occipital chord	4	90	104	95.75	5.21	27.19	2	93	105	99.00	6.00	36.00
Simotic chord	3	8	14	10.67	2.49	6.22	2	9	9	9.00	.00	0.00
Mandible												
Symphyseal height	5	33	42	35.80	3.25	10.56	3	30	36	32.67	2.49	6.22
Min. breadth ramus, left	5	32	37	34.80	1.72	2.96	3	30	34	32.33	1.70	2.89
Max. breadth ramus, left	5	44	45	44.40	.49	.24	3	40	44	42.33	1.70	2.89
Bigonial breadth	4	95	114	103.00	7.11	50.50	3	81	96	90.67	6.85	46.89
Bicondylar breadth	3	125	130	127.00	2.16	4.67	3	112	118	115.00	2.45	6.00
Body height at M1-2	5	25	31	28.80	2.40	5.76	3	22	29	25.33	2.87	8.22
Body thickness at M2	5	17	25	19.60	3.07	9.44	3	14	18	16.33	1.70	2.89
Interforaminal breadth	5	43	50	46.40	2.33	5.44	3	43	48	46.00	2.16	4.67
Body (corpal) length	5	77	83	78.60	2.33	5.44	3	69	81	73.67	5.25	27.56
Ramus height, left	5	53	65	60.40	4.08	16.64	3	54	61	57.33	2.87	8.22
Coronoid height, left	5	58	71	64.20	4.45	19.76	3	58	63	60.33	2.05	4.22
Total mandibular length	5	102	115	106.80	4.40	19.36	3	93	107	99.00	5.89	34.67
Gonial angle	5	115	124	119.40	2.87	8.24	3	117	123	119.67	2.49	6.22
INDICES												
Cranial	3	83.62	88.17	85.62	1.90	3.59	1	85.06	85.06	85.06	—	—
Cranial module	3	146.00	157.00	152.44	4.69	21.95	1	138.67	138.67	138.67	—	—
Mean height	3	89.85	93.96	91.67	1.71	2.93	1	91.93	91.93	91.93	—	—
Length-height	3	82.49	86.96	85.08	1.89	3.59	1	85.06	85.06	85.06	—	—

Table 9.19. Summary of Adult Cranial Measurements (in mm) and Indices *(continued)*

	Male Summary Statistics						Female Summary Statistics					
	N	Min.	Max.	Mean	*SD*	Var.	N	Min.	Max.	Mean	*SD*	Var.
Breadth-height	4	92.86	102.19	97.75	3.34	11.16	2	89.86	100.00	94.93	5.07	25.68
Upper facial	3	49.66	54.14	52.53	2.03	4.14	2	50.74	54.26	52.50	1.76	3.11
Total facial	3	79.59	88.28	85.28	4.02	16.19	2	83.82	89.15	86.49	2.66	7.09
Nasal	3	46.43	52.00	49.48	2.30	5.31	2	50.00	55.32	52.66	2.66	7.07
Left orbital	1	97.30	97.30	97.30	—	—	2	88.89	100.00	94.44	5.56	30.86
Right orbital	4	82.93	90.00	86.05	2.76	7.62	2	94.29	105.88	100.08	5.80	33.62
External palatal	4	124.07	130.19	126.78	2.56	6.54	2	117.31	134.00	125.65	8.35	69.66
Internal palatal	5	77.08	93.88	82.70	5.78	33.43	2	75.56	93.02	84.29	8.73	76.28
Fronto-parietal	2	65.77	67.15	66.46	.69	.48	2	64.86	68.70	66.78	1.92	3.68
Cranio-facial	2	94.16	97.08	95.62	1.46	2.14	2	91.89	98.47	95.18	3.29	10.83
Gnathic	4	92.00	99.02	94.06	2.89	8.33	2	92.47	97.80	95.14	2.66	7.10

SD = standard deviation; var. = variance; min. = minimum; max. = maximum; br. = breadth.

Table 9.20. Summary of Adult Postcranial Measurements (in mm) and Indices

		Male Summary Statistics						Female Summary Statistics					
		N	Min.	Max.	Mean	*SD*	Var.	N	Min.	Max.	Mean	*SD*	Var.
MEASUREMENTS													
Humerus													
Maximum length	L	3	295	316	304.67	8.65	74.89	1	273	273	273.00	—	—
	R	6	297	321	308.83	9.97	99.47	1	277	277	277.00	—	—
Maximum diam. head	L	4	41	44	41.75	1.30	1.69	1	35	35	35.00	—	—
	R	6	40	45	43.00	2.08	4.33	1	36	36	36.00	—	—
Proximal end breadth	L	3	43	49	45.67	2.49	6.22	1	41	41	41.00	—	—
	R	6	41	51	47.00	3.65	13.33	1	41	41	41.00	—	—
Distal end breadth	L	2	57	57	57.00	.00	.00						
	R	5	55	61	57.80	2.04	4.16	2	53	55	54.00	1.00	1.00
a-p midshaft diam.	L	3	10	19	15.33	3.86	14.89	1	15	15	15.00	—	—
	R	6	15	21	18.17	2.34	5.47	1	17	17	17.00	—	—
m-l midshaft diam.	L	3	19	23	21.00	1.63	2.67	1	17	17	17.00	—	—
	R	6	20	24	21.33	1.37	1.89	1	19	19	19.00	—	—
Radius													
Maximum length	L	4	237	258	247.25	7.46	55.69						
	R	4	234	257	244.00	9.46	89.50	1	218	218	218.00	—	—
Maximum diam. head	L	3	20	22	21.00	.82	.67	1	19	19	19.00	—	—
	R	4	20	23	21.00	1.22	1.50	1	19	19	19.00	—	—
Distal end breadth	L	3	29	33	30.67	1.70	2.89	1	29	29	29.00	—	—
	R	4	20	34	28.75	5.26	27.69	1	27	27	27.00	—	—
a-p midshaft diam.	L	4	10	12	10.50	.87	.75						
	R	4	10	13	10.75	1.30	1.69	1	11	11	11.00	—	—
m-l midshaft diam.	L	4	12	15	13.75	1.30	1.69						
	R	4	13	16	14.25	1.09	1.19	1	12	12	12.00	—	—
Ulna													
Maximum length	L	2	253	261	257.00	4.00	16.00						
	R	3	248	263	256.00	6.16	38.00	1	234	234	234.00	—	—
Shaft length	L	3	226	244	234.67	7.36	54.22						
	R	5	221	242	230.80	6.94	48.16	1	209	209	209.00	—	—
Distal end breadth	L	3	16	17	16.67	.47	.22						
	R	4	15	18	16.50	1.12	1.25	1	15	15	15.00	—	—
Trochlear notch height	L	2	22	24	23.00	1.00	1.00						
	R	4	21	25	23.25	1.48	2.19	2	20	22	21.00	1.00	1.00
a-p midshaft diam.	L	2	12	13	12.50	.50	.25						
	R	4	13	16	14.00	1.22	1.50	1	13	13	13.00	—	—

Table 9.20. Summary of Adult Postcranial Measurements (in mm) and Indices *(continued)*

			Male Summary Statistics						Female Summary Statistics				
		N	Min.	Max.	Mean	*SD*	Var.	N	Min.	Max.	Mean	*SD*	Var.
m-l midshaft diam.	L	2	10	11	10.50	.50	.25						
	R	4	10	12	11.50	.87	.75	1	11	11	11.00	—	—
Femur													
Maximum length	L	2	429	431	430.00	1.00	1.00	1	381	381	381.00	—	—
	R	3	419	425	422.00	2.45	6.00	1	380	380	380.00	—	—
Bicondylar length	L	2	426	427	426.50	.50	.25	1	378	378	378.00	—	—
	R	3	418	422	419.33	1.89	3.56	1	378	378	378.00	—	—
Trochanteric length	L	1	412	412	412.00	—	—	1	359	359	359.00	—	—
	R	3	396	404	400.33	3.30	10.89	1	360	360	360.00	—	—
Maximum diam. head	L	4	41	47	43.25	2.28	5.19	1	37	37	37.00	—	—
	R	4	41	46	42.75	1.92	3.69	2	38	39	38.50	.50	.25
a-p subtroch. diam.	L	5	23	25	23.80	.98	.96	1	21	21	21.00	—	—
	R	4	22	26	23.75	1.48	2.19	2	20	21	20.50	.50	.25
m-l subtroch. diam.	L	5	31	32	31.80	.40	.16	1	27	27	27.00	—	—
	R	4	30	32	31.25	.83	.69	2	27	32	29.50	2.50	6.25
a-p midshaft diam.	L	2	27	32	29.50	2.50	6.25	1	24	24	24.00	—	—
	R	3	26	30	27.67	1.70	2.89	1	24	24	24.00	—	—
m-l midshaft diam.	L	2	24	25	24.50	.50	.25	1	22	22	22.00	—	—
	R	3	23	25	23.67	.94	.89	1	22	22	22.00	—	—
Epicondylar breadth	L	1	75	75	75.00	—	—	1	66	66	66.00	—	—
	R	2	73	76	74.50	1.50	2.25						
Fibula													
Maximum length	L	2	339	353	346.00	7.00	49.00	1	310	310	310.00	—	—
	R	4	337	354	348.50	6.95	48.25	1	312	312	312.00	—	—
Tibia													
Maximum length	L	1	352	352	352.00	—	—	1	313	313	313.00	—	—
	R	3	350	361	356.33	4.64	21.56	1	319	319	319.00	—	—
Physiological length	L	2	340	350	345.00	5.00	25.00	1	309	309	309.00	—	—
	R	3	339	353	347.33	6.02	36.22	1	312	312	312.00	—	—
Max. diam. prox. end	L	1	70	70	70.00	—	—	1	62	62	62.00	—	—
	R	2	70	71	70.50	.50	.25	1	64	64	64.00	—	—
Nut. for. a-p diam.	L	3	37	38	37.33	.47	.22	1	26	26	26.00	—	—
	R	4	32	38	35.75	2.28	5.19	1	28	28	28.00	—	—
Nut. for. m-l diam.	L	3	17	26	22.00	3.74	14.00	1	20	20	20.00	—	—
	R	4	19	26	23.00	2.74	7.50	1	20	20	20.00	—	—
Calcaneus													
Maximum length	L	2	76	78	77.00	1.00	1.00	2	66	67	66.50	.50	.25
	R	4	76	78	77.00	.71	.50						
Clavicle													
Maximum length	L	4	150	155	152.25	1.79	3.19	2	128	143	135.50	7.50	56.25
	R	5	146	153	150.40	2.42	5.84	2	131	142	136.50	5.50	30.25
Scapula													
Maximum length	L	3	147	151	148.33	1.89	3.56						
	R	4	142	148	145.50	2.18	4.75						
Maximum breadth	L	3	102	108	105.33	2.49	6.22						
	R	4	97	104	100.75	2.59	6.69	1	93	93	93.00	—	—
Spine length	L	2	141	143	142.00	1.00	1.00						
	R	4	133	140	137.75	2.77	7.69	1	124	124	124.00	—	—
Supraspinous line length	L	3	48	53	51.00	2.16	4.67	1	37	37	37.00	—	—
	R	4	44	54	49.50	3.64	13.25						
Infraspinous line length	L	3	107	118	111.67	4.64	21.56						
	R	4	108	115	110.75	2.68	7.19	1	113	113	113.00	—	—
Glenoid cavity length	L	5	35	43	37.20	2.93	8.56	2	32	34	33.00	1.00	1.00
	R	5	36	40	37.20	1.47	2.16	3	31	34	32.33	1.25	1.56
Glenoid cavity breadth	L	4	23	27	26.00	1.73	3.00	1	23	23	23.00	—	—
	R	5	24	28	26.60	1.50	2.24	1	23	23	23.00	—	—

Table 9.20.   Summary of Adult Postcranial Measurements (in mm) and Indices *(continued)*

		Male Summary Statistics						Female Summary Statistics					
		N	Min.	Max.	Mean	*SD*	Var.	N	Min.	Max.	Mean	*SD*	Var.
Pelvis													
Max. innominate height	L	3	197	208	204.00	4.97	24.67						
	R	3	195	208	200.00	5.72	32.67	2	188	198	193.00	5.00	25.00
Max. innominate breadth	L	2	147	159	153.00	6.00	36.00						
	R	3	142	158	149.33	6.60	43.56	1	144	144	144.00	—	—
Pubis length	L	3	79	89	83.00	4.32	18.67	3	89	94	91.67	2.05	4.22
	R	4	67	92	79.00	9.41	88.50	1	92	92	92.00	—	—
Ischium length	L	3	84	86	85.33	.94	.89	1	78	78	78.00	—	—
	R	3	85	89	86.67	1.70	2.89	2	82	82	82.00	.00	.00
Maximum pelvic breadth		2	264	283	273.50	9.50	90.25	1	248	248	248.00	—	—
Sagittal diameter inlet		2	92	103	97.50	5.50	30.25	1	107	107	107.00	—	—
Transverse diameter inlet		2	120	121	120.50	.50	.25	1	125	125	125.00	—	—
Transverse diameter outlet		1	105	105	105.00	—	—						
Interspinous breadth		1	90	90	90.00	—	—						
Sacrum													
Maximum anterior breadth		5	113	123	117.00	3.74	14.00	2	115	121	118.00	3.00	9.00
Maximum anterior height		3	105	117	112.00	5.10	26.00	2	102	106	104.00	2.00	4.00
S-1 breadth		5	50	55	52.20	1.72	2.96	2	44	48	46.00	2.00	4.00
S-1 length		5	31	36	33.40	2.24	5.04	2	27	29	28.00	1.00	1.00
INDICES													
Platymeric	L	5	71.88	78.13	74.84	2.81	7.92	1	77.78	77.78	77.78	—	—
	R	4	70.97	81.25	75.97	3.69	13.58	2	65.63	74.07	69.85	4.22	17.85
Platycnemic	L	3	45.95	70.27	58.91	10.00	99.91	1	76.92	76.92	76.92	—	—
	R	4	52.78	70.27	64.40	6.90	47.60	1	71.43	71.43	71.43	—	—
Robusticity, femur	L	2	11.94	13.38	12.66	.72	.52	1	12.17	12.17	12.17	—	—
	R	3	11.96	12.68	12.24	.31	.10	1	12.17	12.17	12.17	—	—
Radio-humeral	L	3	80.34	81.85	81.28	.67	.45						
	R	4	78.26	82.72	80.32	1.66	2.75	1	78.70	78.70	78.70	—	—
Humero-femoral	L	1	68.45	68.45	68.45	—	—	1	71.65	71.65	71.65	—	—
	R	3	69.88	71.36	70.86	.69	.47	1	72.89	72.89	72.89	—	—
Tibio-femoral	L	1	82.44	82.44	82.44	—	—	1	82.80	82.80	82.80	—	—
	R	2	82.94	86.36	84.65	1.71	2.93	1	84.39	84.39	84.39	—	—
Ischium-pubis	L	2	94.05	94.19	94.12	.07	.00	1	114.10	114.10	114.10	—	—
	R	3	77.91	93.26	86.07	6.31	39.76	1	112.20	112.20	112.20	—	—
Scapular	L	3	69.39	72.11	71.01	1.17	1.37						
	R	4	68.31	71.23	69.24	1.17	1.37						
Sacral		3	99.12	107.62	102.53	3.67	13.44	2	112.75	114.15	113.45	.70	.49

SD = standard deviation; var. = variance; min. = minimum; max. = maximum; diam. = diameter; a-p = anterior-posterior; m-l = medial-lateral; subtroch. = subtrochanteric; prox. = proximal; nut. for. = nutrient foramen; L = left; R = right.

shorter than Stodder's means, but still within the range of variation she reports, except for the early sample with the DAP sample removed.

Cranial Nonmetric Variation

The same cautionary note made above for the metric variation tabulations and summaries applies here for nonmetric cranial variation: the data set is simply too small to allow meaningful comparisons at the present time. Data were collected only for the eight adult crania and are available at the Anasazi Heritage Center and from the

Table 9.21.   Adult Stature Estimations, in Inches

Burial No.:		1	2	4	8	10
Sex:		Male	Male	Male	Male	Female
Location:		PS 1	PS 1	PS 3	Midden	Midden
All bones	L			62.87		
	R	62.51		62.63		61.24
Femur	L			64.48	64.30	58.43
	R	63.68	63.41	63.95		58.33
Tibia	L			64.07		58.63
	R	64.77		63.92	64.54	59.27

SOURCE: After Genoves (1967) in Bass (1987:29).
PS = pit structure; L = left; R = right.

Table 9.22. Comparative Mean Stature Estimates, in Inches, for Duckfoot, DAP, and Mesa Verde–Region Early Sample

Sex	Duckfoot			DAP			Mesa Verde Region 1[a]			Mesa Verde Region 2[b]		
	n	Mean	SD	n	Mean	SD	n	Mean	SD	n	Mean	SD
Male	4	63.83	.62	9	63.90	1.87	17	63.76	1.11	26	63.81	1.38
Female	1	58.33	—	12	61.40	1.96	12	61.15	.96	24	61.28	1.52

SOURCE: Mesa Verde data from Stodder (1987:372, Table 20.17).
NOTE: All stature estimates are based on femur length.
SD = standard deviation.
[a] Does not include the DAP (Dolores Archaeological Program) sample.
[b] Includes the DAP sample.

author. Immature crania are not included in most analyses using nonmetric cranial data because of variation resulting from incomplete development. No summary statistics are presented, because these would serve no useful purpose in the current study.

Diseases, Injuries, and Anomalies

In the following discussion of paleopathologies, most entities are merely described and listed by affected burial; major pathologies are listed in tabular form in Appendix F. Too many burials are fragmented and incomplete to allow meaningful statistical summaries; on the other hand, some processes are so pervasive—for example, periodontal disease in the adults—that one can simply state that the process is present in everyone for whom adequate skeletal material exists. Also, as with other data sets, the small total sample size and the even smaller subsamples by sex and/or age status give unreasonably small subsets of data for statistical analysis in most cases. Dental pathologies and anomalies are described in the section on dentition.

Systemic Processes—Manifestations of Anemia

Cribra orbitalia and porotic hyperostosis are generally considered manifestations of anemia, regardless of cause (Carlson et al. 1974; Cockburn 1977; El-Najjar et al. 1975; El-Najjar et al. 1976; Hengen 1971; Mensforth et al. 1978; Moseley 1965; Ortner and Putschar 1981; Steinbock 1976; Stuart-Macadam 1987; and others). Anemia is defined as an abnormally low red blood cell count and/or abnormally low amounts of hemoglobin in the red blood cells. The most common causes of anemia are iron deficiency (resulting from dietary deficiencies, chronic blood loss from intestinal parasitism or gastrointestinal disease of other kinds, malaria, and other etiologies) and a variety of congenital hemolytic anemias, including sickle-cell disease and thalassemia.

In the American Southwest, paleoepidemiological research in the past 15 to 25 years (Armelagos 1967; El-Najjar et al. 1975; El-Najjar et al. 1976; Palkovich 1987; Zaino 1967; and others) has indicated that both porotic hyperostosis and cribra orbitalia are manifestations of iron-deficiency anemia, often coupled with protein malnutrition, probably resulting from diet and/or gastrointestinal infections. In their recent reviews of the literature concerning these two manifestations of anemia, Hinkes (1983:47–54) and Palkovich (1987) have cogently argued that they should be treated separately in osteological analyses because they probably represent different etiologies. This approach will be followed in the analysis below.

Cribra orbitalia. Eleven of the Duckfoot crania are sufficiently complete to be examined for the presence of cribra orbitalia. Five of 11, or 45.5 percent, show some degree of development of this skeletal sign of anemia; these 5 include 2 children and 3 adults, the latter all females (an example of healed cribra orbitalia is shown in Figure 9.6). Table 9.23 presents the distribution of cribra orbitalia by age status, child or adult, for the 11 skulls; Table 9.24 shows the distribution by sex for those individuals whose sex could be determined. The distribution by age status is not significantly different from a random distribution as tested by chi square. However, the distribution of cribra orbitalia by sex is statistically significant ($\chi^2 = 4.30$ with Yates' correction, $df = 1$, $p = 0.038$). Table 9.25 illustrates the distribution of cribra orbitalia by burial location. This distribution, too, is not significant.

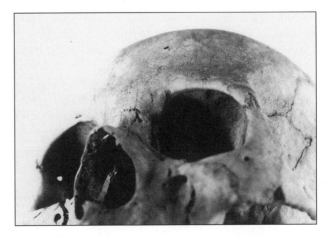

Figure 9.6. Healed cribra orbitalia, left supero-lateral orbit, Burial 3.

Table 9.23. Cross-tabulation of Cribra Orbitalia by
Age Status

Age Status		Cribra Orbitalia		Total
		Absent	Present	
Adult	N	5	3	8
	%	83.3	60.0	72.7
Child	N	1	2	3
	%	16.7	40.0	27.3
TOTAL	N	6	5	11
	%	54.5	45.5	100.0

Table 9.24. Cross-tabulation of Cribra Orbitalia by Sex

Sex		Cribra Orbitalia		Total
		Absent	Present	
Female	N	0	3	3
	%	0.0	100.0	37.5
Male	N	5	0	5
	%	100.0	0.0	62.5
TOTAL	N	5	3	8
	%	62.5	37.5	100.0

Table 9.25. Cross-tabulation of Cribra Orbitalia by
Burial Location

Location		Cribra Orbitalia		Total
		Absent	Present	
Midden	N	3	3	6
	%	50.0	60.0	54.5
Pit structure	N	3	2	5
	%	50.0	40.0	45.5
TOTAL	N	6	5	11
	%	54.5	45.5	100.0

Table 9.26. Cross-tabulation of Porotic Hyperostosis by
Age Status

Age Status		Porotic Hyperostosis		Total
		Absent	Present	
Adult	N	8	0	8
	%	66.7	0.0	61.5
Child	N	4	1	5
	%	33.3	100.0	38.5
TOTAL	N	12	1	13
	%	92.3	7.7	100.0

Table 9.27. Cross-tabulation of Porotic Hyperostosis by Sex

Sex		Porotic Hyperostosis		Total
		Absent	Present	
Female	N	3	0	3
	%	37.5	0.0	37.5
Male	N	5	0	5
	%	62.5	0.0	62.5
TOTAL	N	8	0	8
	%	100.0	0.0	100.0

Table 9.28. Cross-tabulation of Porotic Hyperostosis by
Burial Location

Location		Porotic Hyperostosis		Total
		Absent	Present	
Midden	N	7	0	7
	%	58.3	0.0	53.8
Pit structure	N	5	1	6
	%	41.7	100.0	46.2
TOTAL	N	12	1	13
	%	92.3	7.7	100.0

Although a moderate percentage of individuals show evidence of anemia in the form of cribra orbitalia, none of the manifestations of this condition are more than mild or barely modest, which indicates that the individual experiences probably were not very severe or long-standing. The Duckfoot specimens certainly do not manifest the more pronounced characteristics of cribra orbitalia seen in other prehistoric Southwestern groups (e.g., El-Najjar et al. 1975).

Porotic hyperostosis. Of the 13 skulls examined for porotic hyperostosis, only 1, or 7.7 percent, shows signs of the condition. The distributions of porotic hyperostosis by age status, sex, and burial location are presented in Tables 9.26, 9.27, and 9.28. None of these distributions are statistically significant. More skulls were examined for porotic hyperostosis than for cribra orbitalia because of

differential preservation of the two areas of the skull in which the respective conditions are manifested. Porotic hyperostosis is evident in the skull vault, of which there are more whole or fragmented examples; cribra orbitalia is observed on the upper wall of the eye orbit, a more fragile area of bone that does not preserve as well or as frequently as does the cranial vault.

Is there an association between the two manifestations of anemia? Table 9.29 presents the cross-tabulation of porotic hyperostosis by cribra orbitalia. Eleven individuals have sufficient cranial vault and orbital remains to be examined for both conditions. The distribution is not significant; in fact, because of the distribution of empty cells, the chi square statistic could not be calculated. It is interesting that the individual who has porotic hyperostosis does *not* have cribra orbitalia.

Table 9.29. Cross-tabulation of Porotic Hyperostosis by
Cribra Orbitalia

Cribra Orbitalia		Porotic Hyperostosis		Total
		Absent	Present	
Absent	N	6	0	6
	%	54.5	0.0	54.5
Present	N	5	0	5
	%	45.5	0.0	45.5
TOTAL	N	11	0	11
	%	100.0	0.0	100.0

Table 9.30. Cross-tabulation of Periostitis
by Age Status

Age Status		Periostitis		Total
		Absent	Present	
Adult	N	5	3	8
	%	50.0	100.0	61.5
Child	N	5	0	5
	%	50.0	0.0	38.5
TOTAL	N	10	3	13
	%	76.9	23.1	100.0

Compared with other southwestern populations in general (Akins 1986; El-Najjar et al. 1975; El-Najjar et al. 1976; Hinkes 1983; Palkovich 1980) and with populations in the greater Mesa Verde region, including Dolores (Miles 1975; Stodder 1987), the Duckfoot sample shows relatively mild expressions of these manifestations of anemia and fewer examples of porotic hyperostosis than of cribra orbitalia. These indicators of nutritional and/or infectious stress in the Duckfoot population are at relatively low levels of intensity and frequency.

Systemic Processes—Inflammation, Including Periostitis

Inflammatory processes, usually but not always indicating infection in prehistoric populations, are relatively infrequent in the people of Duckfoot. Three adults show an inflammation (osteitis) of the hard palate, each case being associated with generally poor dental hygiene indicated by numerous carious teeth and several abscesses per mouth. Burial 2 has a mild inflammation of the left maxillary sinus. This was detected only because of damage to the enclosing maxillary bone, and it cannot be stated how much more prevalent sinus infections may have been in the population.

The ubiquitous, but etiologically poorly understood (cf. Birkett 1983:102; Ortner and Putschar 1981:129–138), prehistoric manifestation of periostitis is noted in only three adults, for a total of 23.1 percent of the 13 skeletons having sufficient postcranial material for examination. These manifestations are mild and, at least among the bones present in the sample, are limited to the bones of the lower limb, more frequently the tibia (three cases) and less frequently the femur and fibula (two cases each). Tables 9.30, 9.31, and 9.32 show the distributions of periostitis by age status, sex, and burial location, respectively. None of the distributions are statistically significant.

Systemic Processes—Osteoarthritis

Osteoarthritis, a progressive, degenerative disease of the joints, does not show up in any of the Duckfoot adults until

Table 9.31. Cross-tabulation of Periostitis by Sex

Sex		Periostitis		Total
		Absent	Present	
Female	N	2	1	3
	%	40.0	33.3	37.5
Male	N	3	2	5
	%	60.0	66.7	62.5
TOTAL	N	5	3	8
	%	62.5	37.5	100.0

Table 9.32. Cross-tabulation of Periostitis by
Burial Location

Location		Periostitis		Total
		Absent	Present	
Midden	N	3	3	6
	%	30.0	100.0	46.2
Pit structure	N	7	0	7
	%	70.0	0.0	53.8
TOTAL	N	10	3	13
	%	76.9	23.1	100.0

the 30–39 and older age intervals and is seen in a total of six adults (all those over 30 years of age, except Burial 3, which is too incomplete for specific age evaluation). In general, in the older individuals more joints are affected and the involvement of the affected joints is more serious. The vertebral column is the most frequently affected part of the skeleton; five of the six adults with osteoarthritis show involvement of the spine. The most common site in the spine is the lumbar area. All vertebral cases of osteoarthritis show mild to moderate manifestations of the degenerative disease; there are none of the severe cases of vertebral body lipping, with or without ankylosis, that one might expect in a sample with several people over 50 years of age. Other affected joints are randomly distributed among four of the adult primary burials. Burial 4

shows mild involvement of the hip, knee, and ankle; Burial 6 has mild erosion, pitting, and some lipping of most major joints in the upper and lower limbs; Burials 8 and 12 have involvement of the sacroiliac joints. One miscellaneous bone, a manubrium fragment from the midden, shows degenerative changes at the costal cartilage articulations.

Statistical summaries are not provided here, because so many of the burials, especially those from the midden, are fragmented and incomplete, and to give frequencies based on involved bones of those present would not be procedurally sound given the small size of the sample. However, one general comment can be made: considering the age structure of the Duckfoot sample, it is surprising that more degenerative joint changes are not evident.

Systemic Processes—Other

Several of the older individuals at Duckfoot—specifically, Burials 6, 8, and 12—may show signs of osteoporosis associated with age, also known as endocrine osteopenia. The postcranial bones of these people are lighter, show a sparser and more open network of exposed cancellous bone, and exhibit some degree of fine porosity of the bone surfaces. However, without complete radiographic survey of the postcranial skeletons and comparisons with known standards, this is a difficult diagnosis to make. Caution must be applied here because local soil conditions can result in the leaching of postcranial bones to the point where they become as brittle and weak as those affected by osteoporosis (Steinbock 1976:260). There are no indications of neoplasms, rampant nutritional deficiencies, abnormal metabolic processes (other than the aforementioned ones), or other systemic diseases in the Duckfoot inhabitants.

Injuries, Fractures

Injuries and fractures are relatively few in the Duckfoot assemblage. There is a possible old, well-healed, depressed fracture of the right forehead in Burial 1 (Figure 9.7 *left*). Alternatively, this small (less than 1 cm), round depression about 3 cm above the browridge could be the result of a scalp cyst or similar entity, which remodeled the external vault surface into a depression. Burial 3, an adult female, also has what may be a depressed skull fracture (Figure 9.7 *right*). It is located in the occipital squama, about 1 cm left of the midline and 1 cm below the lambdoidal suture, is nearly circular in outline (about 18 mm in diameter), and is depressed below the surrounding outer table by several millimeters. If this is a fracture, it involved only the outer lamina of the skull, since the inner lamina shows no abnormal features, not even remodeling. As in Burial 1, this moderately large cavity could have been caused by a sebaceous cyst of the scalp or similar structure. The left fourth rib of Burial 4, an adult male, has a well-healed, well-aligned fracture of the anterior segment. One rib from the miscellaneous collection of bones also shows a well-healed fracture.

One of the more interesting injuries noted involves the possible association of two left lower-limb pathologies in Burial 6, an adult male. The features of interest are (1) an exostosis of the medial lip of the linea aspera at the level of the nutrient foramen in the region of the adductor longus muscle and (2) circular facets on the opposing surfaces of the fourth and fifth metatarsals about midshaft. The facets are raised above the diaphyseal surface, are rounded in outline, and have pitted, roughened surfaces most commonly seen in a pseudoarthrosis (see Stewart [1974] and Hoffman [1976] for descriptions and illustrations of the

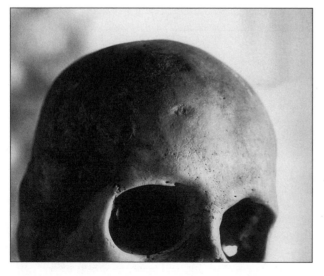 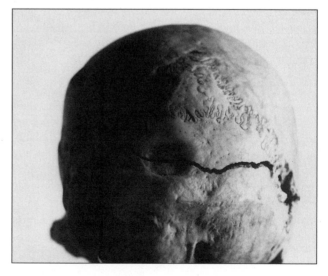

Figure 9.7. Remodeling of the cranium as the result of healed cranial fractures or cystic structures of the scalp: *left*, Burial 1, right central frontal region; *right*, Burial 3, left occipital squama (note: fracture line is postmortem).

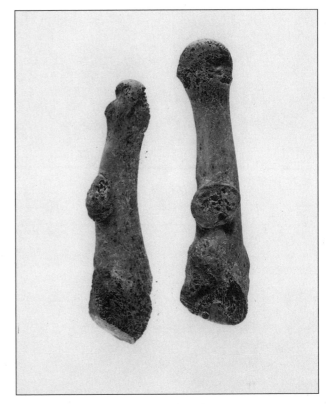

Figure 9.8. Femoral exostosis and metatarsal facets, Burial 6: *left,* left posterior femur with exostosis, left fourth and fifth metatarsals with midshaft elevated facets (note angulation of fifth metatarsal shaft, which indicates possible old healed fracture); *right,* metatarsals rotated to show osteoarthritic-like facet surfaces.

latter). The involved fifth metatarsal probably has a well-healed midshaft fracture, because the distal half of the bone is angled slightly laterally. These may be related or separate injuries; the "ages" of the injuries are similar enough for a single episode. If these injuries represent a single traumatic episode, it could be one in which the foot (fifth metatarsal) broke and the hip adductor tore as the individual fell; or, depending on the circumstances, the adductor may have been torn at the same time the metatarsal fractured. The torn muscle insertion then healed and ossified as an exostosis; the metatarsal healed with enough angulation to cause the pseudoarthrosis-like facets to form (Figure 9.8). This individual also has an old, healed, subperiosteal hematoma of the left anterior tibial crest, a "shin bruise," which may also be part of the injury complex described above.

The only other injuries identified in the Duckfoot people are indicated by the presence of Schmorl's nodes in several adults (Burials 2, 6, and 8). Schmorl's nodes are depressions of various sizes in either the superior or the inferior surfaces of the vertebral bodies caused by disc herniations, with subsequent erosion of the body surfaces. These are variously expressed in the three individuals: Burial 2 has a single node on the inferior surface of the seventh thoracic vertebra, Burial 6 has them on most surfaces of the twelfth

thoracic through fifth lumbar vertebrae, and Burial 8 has a single small depression on the superior surface of the fourth lumbar vertebra. These lesions are generally the result of compressive trauma to the spine or are seen in older individuals with generalized osteoporosis. In the latter instance, loss of substantive bone can cause some degree of disc herniation without much compression of the intervertebral disc; this may be the reason for their presence in Burials 6 and 8, since each is an older individual and shows early signs of generalized bone loss.

Anomalies, Including Congenital Malformations

There are three cases of major anomalous skeletal or dental variations in the Duckfoot sample. Two of these, supernumerary teeth and fused teeth, are described below in the section on dentition. The remaining anomaly is a case of Klippel-Feil syndrome (see Warkany [1971:909–914] for complete description) in Burial 1, which is manifested as a "fusion" of the fifth and sixth cervical vertebrae (see Figure 9.9). The vertebral bodies form a single mass without an intervertebral disc space; the synovial articular joints are narrowed, but still open, to slits less than 1 mm high. The condition is not a true fusion of the bodies but, rather, the result of incomplete segmentation of the meso-

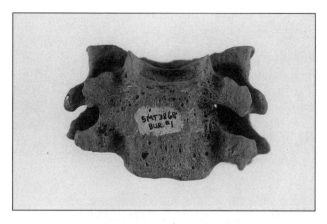

Figure 9.9. Klippel-Feil syndrome, congenital fusion of fifth and sixth vertebral bodies, Burial 1: *top,* anterior; *bottom,* lateral.

dermal somites early in embryonic life. The type of cervical synostosis seen here is the most common. The etiology of this condition, as described by Warkany (1971:909–910), is rather heterogeneous, with hereditary and environmental causes acting separately or in tandem. It is often associated with other malformations of the spine and other organ systems. Depending on the extent of involved vertebrae, there may be some neuromuscular limitations especially of the upper limbs. Given the very slender and nonrobust clavicles and upper limb bones of Burial 1, it is tempting to infer that such limitations existed; however, similar nonrobusticity of the clavicles and upper limb bones also is present in Burial 2, which does not have the syndrome.

The only other anomalous condition observed in the Duckfoot assemblage is the occurrence of bilateral septal apertures of the humeri in Burial 10, an adult female.

Dentition

The dentitions of the Duckfoot inhabitants range from completely intact through partly present to absent. As expected, the jaws and teeth of the pit structure burials are more complete than those of the midden burials. Overall,

dental health is poor, especially in the older adults, which is to be expected of a prehistoric horticultural people with a high-carbohydrate diet. However, the children, at least those under 12 years of age, are generally free of major dental diseases. None of them show periodontal disease, calculus formation, carious teeth, or abscesses. One older child, Burial 13, does have three carious teeth in both the upper and the lower jaws, all of which are permanent teeth.

Information regarding basic dental morphology, such as cusp pattern and number, was not recorded in this study because so many of the adults were missing teeth, and the teeth that were present were so worn that much of the information from the crown was missing. To statistically summarize information from such a small database—and an incomplete one at that—would distort the actual population values (if they could be known). Information on how many and which teeth are present in each individual and on general condition is presented above in the description of individual burials.

Periodontitis, Calculus Formation

Among the adults, periodontal disease and calculus formation are universally present, as one would expect in a population with a high-carbohydrate diet and poor dental care. Periodontal disease, an inflammation primarily of the gums and underlying alveolar bone, is manifested as a reduction in, and rounding off of, the edges of alveolar bone, a retreating of alveolar bone from the necks of the teeth, and a fine porosity of the bone surface. Periodontal disease in the Duckfoot population ranges from mild to severe, with the most serious manifestations in the older adults who still have sufficient alveolar bone to examine. Calculus is identified from the hard, calcarious deposits on the tooth surfaces. It is only mildly expressed in most cases and moderately expressed in a few.

Caries

Table 9.33 presents the frequency and percentage of carious teeth per mouth, both jaws combined. It should be kept in mind that most likely this is an underestimate of the actual carious teeth because some of the individuals included in the tabulation were missing teeth postmortem, either because the teeth had dropped out of their sockets or because the jaws were incomplete.

In the sample of 13 individuals who had dentitions sufficiently complete to be tabulated, there is an average of 3.7 carious teeth per mouth. Four children had no caries, and one, the oldest (12 to 15 years old), had six carious teeth. The eight adults for whom dental material exists have a minimum of 2 carious teeth per mouth; the highest number is 10, in Burial 8, one of the older males. The average is 5.3 carious teeth per adult. The actual frequency

Table 9.33. Frequency Tabulation of Number of Carious
Teeth per Mouth (N = 13)

No. of Carious Teeth per Mouth	N	%
0	4	30.8
1	0	0.0
2	2	15.4
3	1	7.7
4	1	7.7
5	0	0.0
6	2	15.4
7	1	7.7
8	1	7.7
9	0	0.0
10	1	7.7

Table 9.34. Cross-tabulation of Caries
by Age Status

Age Status		Number of Caries				Total
		0	1–3	4–6	7–10	
Adult	N	0	3	2	3	8
	%	0.0	100.0	66.7	100.0	61.5
Child	N	4	0	1	0	5
	%	100.0	0.0	33.3	0.0	38.5
TOTAL	N	4	3	3	3	13
	%	30.8	23.1	23.1	23.1	100.0

Table 9.35. Cross-tabulation of Caries by Sex

Sex		Number of Caries			Total
		1–3	4–6	7–10	
Female	N	1	1	1	3
	%	33.3	50.0	33.3	37.5
Male	N	2	1	2	5
	%	66.7	50.0	66.7	62.5
TOTAL	N	3	2	3	8
	%	37.5	25.0	37.5	100.0

Table 9.36. Cross-tabulation of Caries by Burial Location

Location		Number of Caries				Total
		0	1–3	4–6	7–10	
Midden	N	3	0	1	3	7
	%	75.0	0.0	33.3	100.0	53.8
Pit structure	N	1	3	2	0	6
	%	25.0	100.0	66.7	0.0	46.2
TOTAL	N	4	3	3	3	13
	%	30.8	23.1	23.1	23.1	100.0

of caries in the living individuals was probably higher, since many of the adults are missing one or more teeth, possibly because of caries leading to abscess formation, with consequent loss of the tooth and remodeling of the alveolar bone. Caries may also be obliterated as a result of the relatively rapid rate of dental attrition seen here, which would also cause an underestimate of carious teeth.

The maxillary teeth are more frequently affected by caries than are the mandibular; 24 maxillary teeth and 18 mandibular ones are carious (some teeth have multiple caries). Most frequently affected in both jaws are the molar teeth. Caries are present both in the occlusal surfaces, especially of the molars, and interproximally in the cervical region between adjacent teeth. Some of the caries are massive and have destroyed more than half of the tooth crown.

Tables 9.34, 9.35, and 9.36 present the distribution of carious teeth by age status, sex, and burial location, respectively. The distribution by age status, as expected, is significant ($\chi^2 = 10.18$, $df = 3$, $p = 0.017$); in this sample, children younger than about 10 years of age have no cavities, as opposed to the adults and the one older child of 12 to 15 years. The distributions of caries by sex and by location are not significant. However, if one were to examine only the individuals who have caries, the cross-tabulation of caries by burial location would show a statistically significant distribution ($\chi^2 = 6.430$, $df = 2$, $p = 0.043$).

Because occlusal caries penetrate the enamel and the underlying dentine, they can eventually involve the pulp chamber and result in the formation of periapical abscesses at the root tip. Another cause of abscess formation, perhaps more common than that related to carious teeth, is dental wear, or attrition. As the enamel and the dentine wear away, the pulp cavity is exposed, with the probable result of infection traveling down the root to its apex, where an abscess cavity forms (see Figure 9.10).

Abscesses

Table 9.37 shows the frequency and percentage of individuals by the number of abscesses they have in both jaws combined. Again, most likely this is an underestimate of the abscesses present. All the adults except Burial 3 show visual evidence of abscess formation; none of the children do. The number of abscesses in adults ranges from one, the most frequent number, to six, in Burial 8, the adult male afflicted with the largest number of carious teeth. Among the eight adults, there is an average of 2.4 abscesses per mouth. It is notable that 3 adults, 23.1 percent of the total sample of 13, had four or more abscesses.

Distributions of abscesses by age status, sex, and burial location are presented in Tables 9.38, 9.39, and 9.40, respectively. As with caries, the distribution of abscesses by age status is statistically significant ($\chi^2 = 9.48$, $df = 2$,

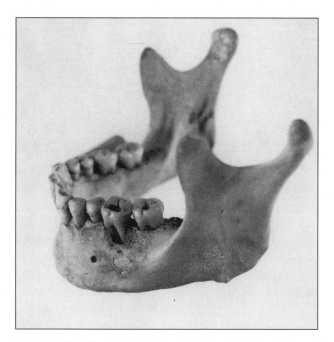

Figure 9.10. General dental pathologies: *left,* Burial 1, moderate occlusal caries (left first molar, second molar), periapical abscess (left first molar); *right,* Burial 4, large occlusal caries (left first-second molars) and abscess cavity (right second premolar-first molar).

Table 9.37. Frequency Tabulation of Number of Abscesses per Mouth (N = 13)

No. of Abscesses per Mouth	N	%
0	6	46.2
1	3	23.1
2	1	7.7
3	0	0.0
4	2	15.4
5	0	0.0
6	1	7.7

p = 0.017). The distributions by sex and by burial location are not significant, even when the latter is examined only for individuals who have abscesses.

Given the relationship between caries and periapical abscesses, the latter, as expected, are also more frequent in the upper jaws than in the lower. Of a total of 19 abscesses, 13 are maxillary in origin and 6 are mandibular. Then, too, these data probably do not reflect the total frequency of abscesses in the living, given the loss of teeth, drainage of abscess pockets, and subsequent healing and remodeling of the alveolar bone.

Dental Anomalies

Different dental anomalies were noted in two individuals, Burials 2 and 10. Two supernumerary teeth are present in the left maxilla of Burial 2; the first has a caniniform shape and is located buccally between the left incisors. The second is a typical peg-shaped tooth of rather modest size located buccally between the second and third molars. The anterior supernumerary had caused some crowding and displacement of the lateral incisor and canine. Because of its relatively small size, the peg-shaped supernumerary did not seem to have caused any displacement or crowding of the molars. In addition, the anterior extra tooth apparently had allowed less attrition to take place in the left side of the mouth than in the right, probably because of the extra buttressing and enamel present on the left side. The peg tooth is below the occlusal plane of the molars and played no role in this differential attrition. See Figure 9.11.

What appears to be a fusion of the right central and lateral mandibular incisors in Burial 10 is actually a dental anomaly difficult to diagnose (cf. Goaz and White 1982; Spouge 1973). There is a single incisor on the right side of the mandible, with a large double crown and a centrally located ridge running vertically along the labial surface of the tooth (Figure 9.12). There is a double, fused root but only a single pulp chamber and root canal; the latter can be seen in X rays of the tooth. (It should be noted that, on the basis of crown and root morphology, the two remaining mandibular incisors appear to be normal left-sided teeth.) The most likely explanation for the anomaly is either congenital fusion of the incisors (the result of partial fusion of the developing tooth germs) or a phenomenon known as gemination or twinning (the result of a dividing single tooth germ). Goaz and White (1982:371) would use the presence

Table 9.38. Cross-tabulation of Abscesses by Age Status

Age Status		Number of Abscesses			Total
		0	1–3	4–6	
Adult	N	1	4	3	8
	%	16.7	100.0	100.0	61.5
Child	N	5	0	0	5
	%	83.3	0.0	0.0	38.5
TOTAL	N	6	4	3	13
	%	46.2	30.8	23.1	100.0

Table 9.39. Cross-tabulation of Abscesses by Sex

Sex		Number of Abscesses			Total
		0	1–3	4–6	
Female	N	1	1	1	3
	%	100.0	25.0	33.3	37.5
Male	N	0	3	2	5
	%	0.0	75.0	66.7	62.5
TOTAL	N	1	4	3	8
	%	12.5	50.0	37.5	100.0

Table 9.40. Cross-tabulation of Abscesses by Burial Location

Location		Number of Abscesses			Total
		0	1–3	4–6	
Midden	N	4	2	1	7
	%	66.7	50.0	33.3	53.8
Pit structure	N	2	2	2	6
	%	33.3	50.0	66.7	46.2
TOTAL	N	6	4	3	13
	%	46.2	30.8	23.1	100.0

of separate root canals to argue for a diagnosis of congenital fusion rather than gemination; Spouge (1973:135) would also argue for congenital fusion, but on the basis of the lack of a second incisor on the same side of the jaw. Given that the Duckfoot specimen shows characteristics that could be used to support either interpretation, the author of this chapter declines to make a diagnosis. Regardless of the diagnosis, the reduced space normally taken up by two teeth has resulted in the mesial drift of the left central incisor across the midline by several millimeters. No abnormal patterns of attrition are associated with this anomaly.

Discussion

Mortuary Practice

The Duckfoot data support some of Stodder's (1987:363) arguments, based on Dolores data, that several common assumptions about early Anasazi mortuary behavior patterns are fallacious. Three of Stodder's five contentions are supported at Duckfoot: (1) midden deposits are *not* the most common burial locations, (2) extended burials are found in the Southwest before Pueblo III times, and (3) cremations are present in the Anasazi record. Stodder's two remaining contentions—that there is evidence of status differentiation and that ritually oriented cannibalism and violence occurred—are not sustained by the Duckfoot data.

The 14 principal burials at Duckfoot are evenly divided between pit structures and the midden, with 7 burials each; there were no discrete burials in the room block. No significant differences in age at death or in sex distribution were noted between the two burial locations. However, six of seven pit structure burials, from three of four pit structures, are burned; these include three males, one female, and two children. The presence of burned burials *only* in the pit structures is highly significant and indicates

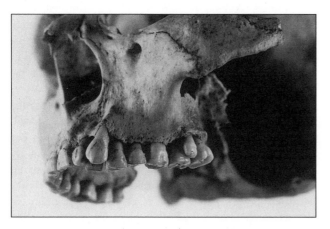 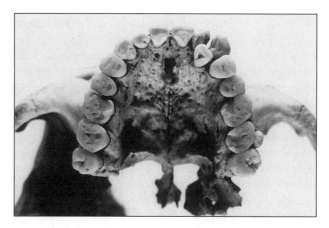

Figure 9.11. Burial 2, caniniform supernumerary tooth between left incisors (lateral incisor displaced lingually) and small peg tooth between left second and third molars: *left*, left antero-lateral view; *right*, occlusal view. Both extra teeth are positioned buccally.

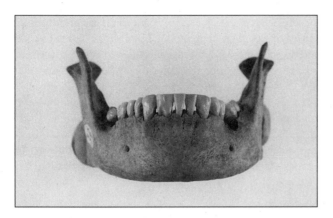 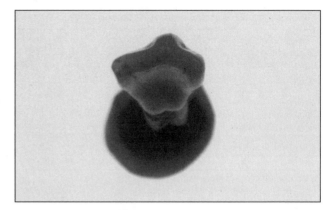

Figure 9.12. Congenital fusion, involving both crowns and roots, of right central and lateral mandibular incisors, Burial 10: *left,* **anterior view of mandible;** *right,* **occlusal view of fused incisors.**

very strongly that burning of the dead is associated in a nonrandom way with burning of pit structures. If it were not, one would expect to see some burned individuals from the midden as well. The one extended burial at Duckfoot is Burial 12, the burned adult female from Pit Structure 4. She is also one of only three burials to be covered or partly covered by sandstone rocks, the other two cases being Burial 10, a female in the midden, and Burial 3, a female in Pit Structure 2. The covering of bodies in this manner was also noted at Dolores (Stodder 1987:351, Table 20.3).

The one unburned burial from a pit structure at Duckfoot is the incomplete skeleton of an adult female on the floor of Pit Structure 2. Numerous canid tooth marks are present on most of the remaining bones, indicating that animal disturbance is the most likely reason for the missing elements. The other pit structure burials are virtually intact except for the two infants from Pit Structure 4, whose remains probably were consumed more thoroughly by fire than those of the adults.

A variety of burial positions and orientations is noted in the Duckfoot burials; supine, partly flexed, and head toward the east are the most common. With the exception of Burial 2 in Pit Structure 1, all burned burials intact enough to be evaluated indicate that the dead were deliberately placed on the floor of the pit structures before the structures were burned and that all burial positions are consistent with known body positions from a wide area of the Colorado Plateau. The exception, an adult male (Burial 2) found partly on top of burned timbers, is not easily explained. However, the damaged portion of the Burial 2 cranial vault is also burned, which suggests that the damage is fire related and not the result of antemortem injury or trauma. The burned vault fragments are charred thoroughly enough to indicate that the fragmentation is fire related— either the normal breakage that occurs when bone is burned or the fragmentation that results from the steaming of the intracranial contents and subsequent explosion of the vault from built-up intracranial pressure. It is possible that this

individual was killed before being placed in the pit structure, but there is no skeletal evidence of a cause of death unless the cranial damage is related to death, which does not appear to be the case. The same may be stated for the other pit structure burials, whose careful positioning argues against the individuals having been deliberately killed before interment.

Cannibalism

Although there are reports of cannibalism from the greater Anasazi region (e.g., Flinn et al. 1976; Nickens 1974) and even the local region (Turner 1988, for Dolores), there is no evidence of it at Duckfoot. Turner's three criteria for cannibalism—butchered, burned, and broken—are not all met at Duckfoot. This, of course, begs the question of whether these criteria are sufficient for identifying cannibalism from skeletal remains.

Demography

The human skeletal remains recovered from Duckfoot include children and adults of both sexes. Of principal interest are the 14 discrete burials, plus 2 additional individuals who cannot be accounted for on the basis of missing elements of the main 14. The structure of the Duckfoot population, or at least that of the sample recovered, is unlike that of other prehistoric groups from the local area or the larger Southwest. It contains no infants 0 to 2 years old and no young adults 20 to 29 years old, and it is marked by a higher proportion of adults at least 50 years of age than one would expect on the basis of contemporary demographic profiles of underdeveloped countries or compared with other large archaeological skeletal samples from the Southwest or elsewhere. Each of these aspects of the Duckfoot assemblage will be considered in turn, and possible explanations for the observed patterns will be explored.

Hinkes's (1983:17) survey of percentage of subadults (less than 21 years old) in large archaeological skeletal samples indicates a range of 30 to 67 percent for southwestern populations: 41 percent for Point of Pines (Bennett 1973b), 30 percent for Pecos Pueblo (Goldstein 1963), 40 percent for Mesa Verde (Bennett 1975), and 67 percent for Grasshopper Pueblo (Hinkes 1983). Stodder (1987) indicates that 42.7 percent of the early Mesa Verde sample died before the age of 21. Furthermore, Hinkes (1983:16) states that 16.5 percent of the Grasshopper sample consisted of fetal and neonate remains, and an additional 21.7 percent consisted of individuals from birth to two years of age. Although Stodder (1987) did not report any fetal remains from the early Mesa Verde sample, her data indicate that at least 11 percent of the early regional sample and 4.7 percent of the Dolores sample are under two years of age. The total subadult population at Duckfoot, including the one 18- to 20-year-old female for comparative purposes, is 43.8 percent of the minimum total of 16, which places the site well within the expected overall range. However, the earliest years of life are underrepresented, and in fact there are no infants or children under three years of age in the assemblage. Thus, although the total percentage of subadults at Duckfoot is not unusual for a prehistoric horticultural group, the absence of these early-age individuals is perplexing.

Hinkes's survey of large archaeological skeletal samples also includes a calculation of the percentage of subadults who are less than 5 years of age in the total group of subadults less than 21 years of age (Hinkes 1983:17):

$$\frac{N < 5 \text{ years old}}{N < 21 \text{ years old}}$$

The two lowest percentages, 12 percent (Pecos Pueblo) and 16 percent (Mesa Verde), are calculated for two of the oldest samples in terms of when the excavations took place. More-recently excavated large samples have higher percentages of individuals under five years of age, with Point of Pines at 31 percent and Grasshopper at an incredible 75 percent. It is possible that excavation techniques and a tendency to collect only the remains of older, adult individuals resulted in an underrepresentation of young children in the earliest excavated samples. The Duckfoot assemblage, with only one out of seven subadults (14.3 percent) less than five years old, is similar to these early collections. However, if we assume that the excavations at Duckfoot recovered all human remains, then we might infer that factors other than archaeological recovery technique and bias are playing a role here.

Another unusual characteristic of the Duckfoot skeletal assemblage is the absence of young adults in their twenties and the apparent high percentage of individuals at least 50 years of age. Recalculation of Stodder's data from the early and late Mesa Verde–region samples indicates that only 1.33 percent and .56 percent, respectively, survived to

reach the 50+ age interval. Other populations from the Southwest are similar to the Mesa Verde–region populations for adults aged 50 years or more: 2.6 percent for Arroyo Hondo (Palkovich 1980:32), .0 percent for Mancos Canyon (Nickens 1974:31), .0 percent for Yellow Jacket (Swedlund 1969:100), and 9.9 percent for Grasshopper, including individuals 45 years and older (Hinkes 1983:16). With 20 percent of the assemblage composed of individuals 50 years or older (3 of 15 individuals who could be aged), Duckfoot clearly represents a departure from the established trend. However, the high proportion of individuals over 50 years of age in the Duckfoot sample should be viewed with some caution. A reasonable explanation for this high proportion is the likely underrepresentation of infants, particularly in the 0–2 age interval. For many prehistoric populations, this is the age group with the largest percentage of dead individuals in the subadult age groups, if not all age groups. Fewer young will automatically result in a higher percentage of older individuals. Then, too, the lack of any individuals dying in their third decade of life (20 to 29 years old), often the highest death-age interval for adults, means that there were more individuals who survived to the later age intervals.

A number of factors may be responsible for the unusual mortality patterns noted for the Duckfoot site, including excavation and recovery methods, poor preservation, postmortem disturbance of graves, specific living conditions that may have created an unusual population structure, and sampling bias. For instance, it is possible that not all infant skeletons present at Duckfoot were recovered from the excavated areas or that additional skeletons remain in those areas of the site that were not excavated, primarily the midden. The former possibility is unlikely, given the excellent recovery of other very fragile materials from various parts of the site; however, especially in the case of newborns, the fragility of the skeletons could result in complete decomposition in local soils or complete consumption of the body by carnivores after burial. All the midden burials and the adult female from Pit Structure 2 show evidence of postmortem disturbance by animals. Given the shallowness of the midden burials, it is also conceivable that heavy rains and runoff could have washed away burial pit overlays and uncovered bodies, which were then exposed to disruption by animal, geological, and meteorological forces. Again, this would have affected the very young infants more than the older individuals.

The overall health of the Duckfoot people appears to have been good, at least as indicated by the skeletal evidence. Although bones do not reveal all possible diseases and stress-related conditions, there is nothing in the skeletons to indicate current or chronic processes that would have seriously threatened anyone at the site, nor is there anything unusual in the disease profile that would have resulted in the unique population structure observed.

Sampling bias may well account for much of the apparent discrepancy between the Duckfoot and the regional mortality profiles. Two possible sources of sampling bias have already been discussed: failure to identify human remains in excavated areas and incomplete excavation of the site. The former seems unlikely; the latter may account for some missing individuals, although it probably cannot account for all, unless all the young adults happened to be in the relatively small portion of the midden not excavated. It is possible that additional burials are located in neither the pit structures nor the midden but elsewhere. Certainly the dearth of Anasazi burials from the entire Colorado Plateau makes this not too unreasonable to consider.

The source of bias that potentially has had the greatest effect on the Duckfoot population structure is the duration of site occupation, estimated at only 20 to 25 years. This short span did not allow even a single generation of individuals to experience completely their mortality expectations, much less encompass the several generations needed to smooth out occasional random fluctuations in the overall pattern. This factor, combined with the small sample size and perhaps one or more of the other problems discussed above, dictates that the Duckfoot population be viewed, not as an average or representative sample, but as a unique cohort of people who happened to die at Duckfoot during its brief occupation.

One of the difficulties in interpreting the mortality profile at Duckfoot lies in not knowing whether the individuals died throughout the occupation or whether they died within a short period of time, perhaps in the final days or weeks before abandonment. Dolores data from McPhee Village indicate that there was gradual abandonment by household or interhousehold groups rather than complete abandonment of the site in toto (Kane 1988:42). If this type of abandonment applies to the Duckfoot site, what other lines of archaeological evidence might be brought to bear on the issue? Are there climatological or other data to suggest successive subnormal crop harvests, which might have led to gradual abandonment? Is there evidence of chronic nutritional stress, violence, or cannibalism? On the basis of the Duckfoot skeletal remains, the answer to the last question is no. If one assumes that most of the individuals in the Duckfoot assemblage died during a brief period of time, what could have caused the mass mortality? One possibility is plague, especially if it rapidly fulminated to the pneumonic form. This is a highly contagious and virulent form of the disease, which could have easily spread throughout the population and killed a large proportion of those present at the site in a short period of days or weeks. Plague is endemic in the contemporary Southwest and may have been present prehistorically. Rodents, especially prairie dogs and ground squirrels, are secondary hosts for the bacterium-infested fleas and serve as the major reservoir today. That plague killed some of the Duckfoot inhabitants is a possibility, but one that cannot be proved, because the disease leaves no trace in the skeletal system. It is possible that the Duckfoot data represent several kinds of abandonment and related mortality behavior, both gradual and abrupt. If the pit structure burials represent ritual abandonment, perhaps they also represent a brief but intense episode of mortality. Conversely, the midden burials might represent the occasional death of individuals under more normal circumstances. However, skeletal evidence neither proves nor disproves any of these scenarios, which for want of additional evidence must remain speculative only.

A final question regarding the makeup of the Duckfoot population is related to the foregoing discussion: If the total Duckfoot population was between 25 and 40 during the 20- to 25-year occupation of the site, is the actual mortality cohort recovered an expected one? The simple answer is yes, but it is not the only possible cohort: any number of other mortality profiles would be equally plausible. The duration of occupation at Duckfoot precludes adequate assessment because, in such a short period of time, the actual cohort of individuals living at Duckfoot could have experienced almost any random pattern of deaths.

Paleopathology

Skeletons do not reveal most causes of death or most of the acute diseases that afflict people. Without an adequate sample to be examined from an age/sex perspective, discussions of an epidemiological nature are not reasonable, except in the simplest of terms; therefore, only a very general overview is provided here. The patterns of disease at Duckfoot are unusual in that they present a picture of a fairly healthy population, except for the inevitable oral-cavity disease seen throughout prehistoric southwestern populations. Manifestations of nutritional stress, specifically anemia, are mild and represent old episodes and not active disease at the time of death, even in the children, with the exception of the possible active porotic hyperostosis in Burial 14. Periostitis, sometimes linked to manifestations of anemia, is infrequent and mild. Some authors (Birkett 1983) do not even think that periostitis, especially the mild form on lower limb bone diaphyses, is infectious in origin. Most evidence of osteoarthritis is in the vertebral column; surprisingly little is in the other major joints. All of these manifestations are what one would expect for an age-related, chronic, progressive, degenerative disease.

Injuries are few and all of them can be explained in the context of an active population involved in hard physical labor and as the consequences of the inevitable accidents that befall everyone. These include two rib fractures, Schmorl's depressions in several vertebral columns, the interesting femoral exostosis and fractured metatarsal in one individual, and a probable depressed fracture of a posterior cranial vault. The last can be explained as accidental—the result of a fall or of sharply raising the head under an immovable object.

Dental diseases, especially caries and abscesses, occur in frequencies commonly seen in prehistoric populations with a high-carbohydrate diet. The only significant observation about the distribution of these diseases is that the three individuals with the greatest number of carious teeth (7 to 10) are all from midden burial pits. This difference between the two burial locations, however, is not sustained for the number of abscesses per individual. Given the small sample size, the difference in caries prevalence should be viewed with caution and not overemphasized.

Stress

The discussion of diseases and injuries at Duckfoot leads to the conclusion that stress, as inferred from several skeletal indicators, was not very high. As said earlier, the most serious diseases are associated with poor dental health. Although these diseases can be debilitating, they are so pervasive in the prehistoric record of horticulturists that it is difficult to pin down whether lack of oral health played a role in the mortality of individuals. Certainly dental caries and periodontal disease can lead to abscesses in the supporting dental structures. These abscess pockets can then be the origin of hematogenous infections which, especially without antibiotic treatment, can kill the victim by a generalized septicemia or a more localized abscess in a vital organ. The bones in these cases may never reveal the systemic infection but only the site of a potential locus of infection, the abscess cavity itself (Ortner and Putschar 1981:442). Because all the adults except Burial 3 had evidence of at least one dental abscess, this source of morbidity and possibly mortality cannot be discounted. It just cannot be proved.

Indicators of dietary stress—cribra orbitalia and porotic hyperostosis—are at low levels of prevalence and degrees of expression. With the one exception noted above, none of the cases are active, all having been present earlier in the lifetime of the individuals, who now show remnants of the formerly active process of marrow expansion and cortical bone erosion.

Periostitis of the long-bone diaphyses, another potential indicator of acute and chronic stress and indicating both dietary and infectious synergisms, is present in only 3 of the 14 principal burials. In all cases, it is only mildly expressed. Because the etiology of this entity is apparently both varied (Ortner and Putschar 1981) and controversial (Birkett 1983), it is difficult to evaluate its role in stress at Duckfoot.

Intrasite Variation

Most of the comments one can make about intrasite variation at Duckfoot have already been made in one context or another above. One of the more frustrating aspects of this study was the inability to tease out sufficiently large subsamples of data—subsamples broken down by burial location and by sex—to test hypotheses about differences between the burial locations. With only eight adults, nearly evenly distributed by sex in the two primary burial locations (see Table 9.4), subsamples range in size from one to three. These samples are simply too small to be evaluated from a metric perspective, although several categorical data analyses were attempted.

The most significant aspect of intrasite variation among the human remains at Duckfoot is the presence of burned burials only in the pit structures. This is not related to age or sex or other identifiable differences when compared with the midden burials. There is nothing in the human osteological data to suggest possible reasons why certain individuals were buried in pit structures and why most of them (six of seven individuals) were burned to some extent in the apparently deliberate burning and collapse of structure roofs. Studies elsewhere (Glennie 1983; Wilshusen 1985, 1986c) indicate that it is very difficult to burn down a pit structure accidentally. As Kane (1988) reports for McPhee Village at Dolores, pit structures were collapsed before and during abandonment, but only a small number of them were burned. The combination of so many burned, collapsed pit structures at Duckfoot makes it an interesting variation on a practice known from several scattered locations but first systematically noted at Dolores. (See Stodder [1987] for a review of these other locales of pit structure collapse with burials.)

What other patterns, if any, indicate possible differences between pit structure and midden burials? Other than the burned versus unburned distinction, only the number of caries is different in the two burial groups: those from the midden have more caries than those from the pit structures. This may reflect nothing more than age distribution, with the midden yielding a few more older individuals with more carious teeth. However, other possibilities warrant consideration. Does the observed pattern reflect differential access to dietary resources? Do the midden burials with more caries actually reflect a diet higher in carbohydrates? Or, conversely, did the pit structure sample have greater access to animal protein? Is this an indication of status differentiation? The data are suggestive, but the sample size limits interpretation.

An additional observation relates to the fact that Pit Structure 4, the last pit structure built at the site, is the only one that contained remains of children. In this respect, Pit Structure 4 also differs from any of the Dolores pit structures containing burials, all of which had only adults interred in them (Stodder 1987:358, Table 20.9).

The other intrasite difference in skeletal remains relates to the distribution of cribra orbitalia in the Duckfoot population: among the adults, only females had evidence of this condition, presumably from childhood episodes of anemia. Does this reflect differential access to dietary resources? Even though it appears that local water and maize may have had sufficient iron content, chelating agents such as phytate in maize inhibit the absorption of adequate amounts. If the female children were getting more maize-based gruel and less animal protein, this difference in diet may account for the sex-related difference in cribra orbitalia. Did the male offspring receive more iron-rich food or food that provided iron in a more available form; that is, did they have greater access to animal protein? Unfortunately, the fact that the sex of the children cannot be determined makes this a tantalizing speculation only. However, two of the cases of cribra orbitalia are in older females, one 30 to 39 years old and the other 50+ years old. If Duckfoot was inhabited for only 20 to 25 years, then these two females must have lived elsewhere during the years when they most likely had active anemia, their childhood years. So, if there was differential access to animal protein, it probably did not occur at Duckfoot for these two women.

The People of Duckfoot in Regional Perspective

The perception of Duckfoot's uniqueness is based on comparisons with other archaeological skeletal data from the region, primarily data reported by Stodder (1987) for Dolores and Mesa Verde–region samples. These data are briefly recounted below, following several comparative statements about the Duckfoot metric data, including stature estimations. The ever-present cautionary note about the sample and subsample sizes must be kept in mind. Duckfoot is intriguing because of its very circumscribed occupation span, but this militates against a good mortality profile that could reasonably represent the pattern of deaths for one or more complete generational cohorts of people.

Tables 9.41 and 9.42, which compare Duckfoot and Dolores metric data, use Stodder's data as published in several tables; the data were not recalculated as was done for the life table data. Stodder's published data, however, contain some errors, of which the reader should be aware. For instance, she reports male and female epicondylar breadths of the femur range from 42 to 46 mm, which is much too small for any adult population. In general, though, her data appear reasonable.

Table 9.41 presents summary statistics for cranial measurements and indices of the Duckfoot and Dolores male and female data sets. Table 9.42 presents comparable statistics for the postcranial data sets. Because the Dolores data sets do not include all the measurements taken in the current study, only those used in both studies are presented.

Male crania from Duckfoot, in general, have larger vaults and somewhat smaller faces than those from Dolores. In contrast, Duckfoot female crania are generally smaller than their Dolores counterparts in nearly all dimensions. Some of these differences may be due to the random

effects of small sample sizes. However, it cannot be discounted that the smaller Duckfoot female crania may reflect some differential access to food, especially protein; if so, given the ages of two of the females (see discussion above on intrasite variation), this probably occurred before these women lived at Duckfoot. It may also suggest that the Duckfoot females, who originate from a single site, are more genetically related, with a tendency toward smaller skulls, whereas Dolores females are pooled from numerous sites and reflect more genetic heterogeneity. Perhaps a combination of genetic and environmental circumstances are at work, producing the smaller Duckfoot female crania. Although sample-size differences between Duckfoot and Dolores may be a factor, it is interesting that, for both males and females, the Duckfoot cranial data are much less variable than the Dolores cranial data. Surely this is due in great part to the origin of the samples—a single site's data versus data pooled from a number of sites. A single site's population is likely to be more homogeneous in both genetic and environmental components of variation, which would result in less metric variation.

Postcranial data from the two assemblages are more difficult to interpret because the Duckfoot sample sizes for postcranial data are even smaller than those for cranial data. With some exceptions, male postcranial bones from Duckfoot are longer but have smaller diameters than those from Dolores. Given the complex interactions between genetics and environment, this observation is difficult to explain. The postcranial bones of Duckfoot females, however, are generally smaller in both length and diameter than those of their counterparts at Dolores. Again, this may reflect genetic homogeneity, genes for smaller bones in general, or dietary differences. These results must be viewed with caution, however, because most Duckfoot female postcranial bones are represented by only a single bone. As with the cranial measurements, in general the variation in postcranial measurements at Duckfoot is much lower than at Dolores, at least for the males. Given the single bones for the female Duckfoot data set, this observation must be limited to the males. As is the case with the cranial measurements, the single-site versus multisite database problem is certainly playing a role here.

The comparative data from Duckfoot and Dolores, especially the cranial data, highlight another difference, this one in the degree of metric variation in females as compared to males, regardless of site origin. With a few exceptions, the female cranial data show less metric variation than do the male data, with Duckfoot showing this trend to a greater degree than Dolores. It is possible that the smaller variation reflects a basic pattern of matrilineal social organization with matrilocality. If the females at Duckfoot are more closely related genetically than are the males, who may originate from several sites, it is possible that the greater genetic homogeneity among the females is reflected in the smaller metric variation within the site.

Table 9.41. Adult Cranial Measurements (in mm) and Indices, Duckfoot versus DAP Data

		Male Summary Statistics						Female Summary Statistics					
		Duckfoot			DAP			Duckfoot			DAP		
	N	Mean	SD	N	Mean	SD	N	Mean	SD	N	Mean	SD	
MEASUREMENTS													
Cranium													
Maximum length	3	169.00	6.53	7	150.6	15.1	1	154.00	—	6	139.0	9.2	
Maximum breadth	4	147.00	6.20	6	144.0	14.1	2	139.50	8.50	4	144.5	9.6	
Minimum frontal breadth	2	95.00	3.00	6	104.7	5.9	3	92.67	2.49	6	97.2	5.1	
Bizygomatic breadth	3	141.67	6.18	5	132.4	6.3	2	132.50	3.50	4	126.3	8.5	
Basion-bregma height	4	143.50	2.29	4	141.3	4.0	2	132.00	1.00	5	137.4	6.4	
Basion-nasion length	4	101.25	1.30	4	97.0	5.6	2	92.00	1.00	5	99.8	17.1	
Basion-prosthion length	5	94.40	3.56	4	91.3	4.3	2	87.50	1.50	5	95.0	18.2	
Upper facial height	4	76.50	4.39	6	77.8	8.1	2	69.50	.50	6	70.3	6.2	
Total facial height	4	122.00	5.05	3	126.3	4.7	2	114.50	.50	3	141.3	9.6	
Nasal height	4	53.50	2.60	6	49.3	9.8	2	49.50	2.50	6	37.7	8.7	
Nasal breadth	3	26.00	.00	5	30.4	2.4	2	26.00	.00	5	26.4	1.5	
Left orbit height	2	36.50	0.50	5	40.0	3.8	2	34.00	2.00	5	39.4	3.2	
Left orbit breadth	2	40.00	3.00	5	41.6	4.4	2	36.00	.00	5	40.8	4.3	
Right orbit height	4	35.25	1.30	6	39.0	5.3	2	34.50	1.50	6	39.7	3.7	
Right orbit breadth	4	41.00	1.87	6	40.3	2.3	2	34.50	.50	6	40.8	4.3	
Interorbital breadth	3	23.33	.94	2	29.3	6.7	2	24.00	.00	5	27.4	3.8	
Palate external length	5	54.60	1.62	6	57.5	3.3	2	51.00	1.00	5	55.8	1.8	
Palate external breadth	4	68.75	1.48	4	69.8	4.1	2	64.00	3.00	5	63.4	2.1	
Mandible													
Symphyseal height	5	35.80	3.25	10	34.9	6.1	3	32.67	2.49	14	33.8	3.6	
Min. breadth ramus, left	5	34.80	1.72	9	36.4	7.8	3	32.33	1.70	13	36.6	3.4	
Bigonial breadth	4	103.00	7.11	4	94.8	3.8	3	90.67	6.85	6	94.0	4.7	
Bicondylar breadth	3	127.00	2.16	2	122.0	2.8	3	115.00	2.45	6	112.8	6.0	
Body thickness at M2	5	19.60	3.07	11	18.4	4.2	3	16.33	1.70	14	17.6	3.5	
Interforaminal breadth	5	46.40	2.33	6	49.5	3.4	3	46.00	2.16	11	49.5	4.8	
Ramus height, left	5	60.40	4.08	8	66.4	6.6	3	57.33	2.87	13	63.0	5.3	
Coronoid height, left	5	64.20	4.45	12	61.4	5.5	3	60.33	2.05	12	57.3	4.4	
Total mandibular length	5	106.80	4.40	4	111.3	8.2	3	99.00	5.89	8	102.1	10.0	
INDICES													
Cranial	3	85.62	1.90	4	103.38	12.40	1	85.06	—	5	96.52	10.14	
Cranial module	3	152.44	4.69	4	147.60	4.70	1	138.67	—	5	14.02	42.60	
Length-height	3	85.08	1.89	4	93.11	10.72	1	85.06	—	5	99.32	7.07	
Breadth-height	4	97.75	3.34	4	95.29	6.20	2	94.93	5.07	5	95.31	4.60	
Upper facial	3	52.53	2.03	5	62.10	6.23	2	52.50	1.76	4	54.85	6.36	
Total facial	3	85.28	4.02	3	100.13	2.45	2	86.49	2.66	3	116.14	46.13	
Nasal	3	49.48	2.30	5	63.37	15.17	2	52.66	2.66	5	73.57	22.61	
Left orbital	1	97.30	—	5	96.51	12.08	2	94.44	5.56	5	97.19	6.99	
External palatal	4	126.78	2.56	4	118.32	4.96	2	125.65	8.35	5	113.65	1.58	
Gnathic	4	94.06	2.89	2	92.63	11.24	2	95.14	2.66	6	91.51	6.57	

SOURCE: DAP (Dolores Archaeological Program) data after Stodder (1987:369, 370, Tables 20.13, 20.14).

SD = standard deviation; min. = minimum.

However, the sample sizes are insufficient to test this idea more rigorously at the present time.

Compared with the Dolores data (from Stodder 1987:351–353, Tables 20.4–20.6), Duckfoot burial patterns show less variation and more intensification of some patterns seen occasionally at Dolores. Whereas burials at Dolores included primary and secondary burials—single, double, or multiple for each—as well as cremations, it appears that all Duckfoot burials, except for Burial 3, were primary (discounting the miscellaneous bone, for which burial type is not known). Burial locations, too, are more restricted at Duckfoot, where burials were found only in pit structures (floor and fill) and the midden, whereas at Dolores burials included not just these locations, but also surface structures (fill, floor, and subfloor), pit structure ventilator shafts, and extramural activity areas. When all

Table 9.42. Adult Postcranial Measurements (in mm) and Indices, Duckfoot versus DAP Data

| | | Male Summary Statistics | | | | | | Female Summary Statistics | | | | | |
| | | Duckfoot | | | DAP | | | Duckfoot | | | DAP | | |
		N	Mean	SD	N	Mean	SD	N	Mean	SD	N	Mean	SD
MEASUREMENTS													
Humerus													
Maximum length	L	3	304.67	8.65	9	305.4	12.9	1	273.00	—	8	289.5	21.2
Maximum diameter head	L	4	41.75	1.30	11	44.4	5.5	1	35.00	—	9	38.8	3.8
Distal end breadth	L	2	57.00	.00	10	50.7	12.7				11	48.7	11.4
a-p midshaft diameter	L	3	15.33	3.86	9	19.4	3.3	1	15.00	—	8	18.5	3.8
m-l midshaft diameter	L	3	21.00	1.63	9	21.2	1.4	1	17.00	—	8	21.3	4.5
Radius													
Maximum length	L	4	247.25	7.46	9	234.2	21.1				13	221.8	15.4
Maximum diameter head	L	3	21.00	.82	7	24.3	3.7	1	19.00	—	13	21.8	2.2
Distal end breadth	L	3	30.67	1.70	10	29.9	8.2	1	29.00	—	14	28.3	6.5
Ulna													
Maximum length	L	2	257.00	4.00	8	244.3	16.2				11	235.2	16.2
Distal end breadth	L	3	16.67	.47	9	18.0	7.1				10	18.5	5.2
Trochlear notch height	L	2	23.00	1.00	10	23.2	4.0				15	22.6	1.6
Femur													
Maximum length	L	2	430.00	1.00	9	425.2	21.2	1	381.00	—	11	411.2	19.7
Bicondylar length	L	2	426.50	.50	9	421.2	19.9	1	378.00	—	10	410.1	19.9
Maximum diameter head	L	4	43.25	2.28	12	48.1	4.1	1	37.00	—	11	42.5	3.9
a-p subtroch. diameter	L	5	23.80	.98	14	28.4	4.7	1	21.00	—	14	26.7	2.8
m-l subtroch. diameter	L	5	31.80	.40	14	33.3	5.9	1	27.00	—	14	32.7	19.4
a-p midshaft diameter	L	2	29.50	2.50	8	28.1	5.8	1	24.00	—	11	27.4	3.7
m-l midshaft diameter	L	2	24.50	.50	8	28.0	4.4	1	22.00	—	11	26.6	3.5
Epicondylar breadth	L	1	75.00	—	10	45.2	3.9	1	66.00	—	10	42.4	4.9
Fibula													
Maximum length	L	2	346.00	7.00	6	334.8	18.5	1	310.00	—	10	330.8	19.9
Tibia													
Maximum length	L	1	352.00	—	7	355.6	11.7	1	313.00	—	10	344.4	26.7
Max. diam. proximal end	L	1	70.00	—	9	72.3	5.9	1	62.00	—	9	65.1	3.4
Nut. for. a-p diameter	L	3	37.33	.47	10	37.8	5.0	1	26.00	—	15	33.3	3.7
Nut. for. m-l diameter	L	3	22.00	3.74	9	26.6	4.4	1	20.00	—	17	22.8	3.8
Clavicle													
Maximum length	L	4	152.25	1.79	7	147.4	8.7	2	135.50	7.50	10	134.3	7.5
Pelvis													
Max. innominate height	L	3	204.00	4.97	5	202.2	11.2				6	187.5	10.1
Max. innominate breadth	L	2	153.00	6.00	5	138.6	8.3				7	141.0	8.6
Sacrum													
Maximum anterior breadth		5	117.00	3.74	6	107.5	8.0	2	118.00	3.00	7	114.7	4.6
Maximum anterior height		3	112.00	5.10	4	124.8	14.1	2	104.00	2.00	4	112.8	6.4
INDICES													
Platymeric	L	5	74.84	2.81	9	90.87	20.32	1	77.78	—	11	82.78	20.04
Platycnemic	L	3	58.91	10.00	7	72.51	10.36	1	76.92	—	11	65.81	10.95
Robusticity, femur	L	2	12.66	.72	6	13.22	.98	1	12.17	—	9	13.13	1.64
Radio-humeral	L	3	81.28	.67	6	77.28	.96				7	76.94	2.27
Humero-femoral	L	1	68.45	—	6	71.59	2.16	1	71.65	—	6	74.89	2.20
Tibio-femoral	L	1	82.44	—	7	82.55	28.28	1	82.80	—	9	84.63	3.14
Sacral		3	102.53	3.67	4	84.12	9.42	2	113.45	.70	4	104.17	6.74

SOURCE: DAP (Dolores Archaeological Program) data from Stodder (1987:370, 371, Tables 20.15, 20.16).

SD = standard deviation; diam = diameter; a-p = anterior-posterior; m-l = medial-lateral; subtroch. = subtrochanteric; prox. = proximal; nut. for. = nutrient foramen; max. = maximum; L = left.

time periods are considered, it is seen that 55 to 70 percent of Dolores burials came from pit structure contexts, versus 50 percent at Duckfoot; 20 to 29 percent came from middens, versus 50 percent at Duckfoot; and 5 to 20 percent came from all other contexts, whereas none can be definitely identified as originating elsewhere at Duckfoot. With regard to pit structure burials, even though these burials were documented for both Duckfoot and Dolores-area sites, differences were noted between the two. At Duckfoot, pit structure interments consisted of a single adult male (Burial 4 from Pit Structure 3), two adult males (Burials 1 and 2 from Pit Structure 1), a single adult female (Burial 3 from Pit Structure 2), and an adult female with two children (Burials 12, 11, and 14 from Pit Structure 4). At Dolores, pit structure burials included only single adult females or adult female/male double interments; no children were interred in pit structures (Stodder 1987:358, Table 20.9).

Cremations are rare at Dolores, common at Duckfoot. At Dolores, one double interment is interpreted as a cremation (Lightfoot et al. 1988), and two burned, collapsed pit structures contained burials (one double and one single interment) that Stodder (1987:359) believes should be identified as cremations, too. These three burials account for less than 10 percent of all Dolores burials. In contrast, burned burials, or cremations, involved 6 of 14, or 43 percent, of the burials at Duckfoot. Does this difference represent an intensification of ritual behavior associated with pit structure abandonment and/or with disposal of the dead at Duckfoot? Since burned or cremated burials are relatively rare throughout the Anasazi area, including those associated with pit structures, why this particular pattern at Duckfoot? Does it reflect an attempt at more social control? Why the difference between burials from pit structures and those from the midden at Duckfoot? Was the midden the common burial ground used by everyone until the period of final abandonment, during which the pit structures were used exclusively?

The Duckfoot population shows fewer and less intense expressions of disease than does the Dolores sample. At Dolores (Stodder 1987:438–442), 75 percent of 191 crania sufficiently complete for examination showed cribra orbitalia and/or porotic hyperostosis. At Duckfoot, these frequencies are 45.5 percent for cribra orbitalia and 7.7 percent for porotic hyperostosis. Certainly for the childhood and early teen years—years of rapid growth and potential nutritional stress, when these manifestations of anemia are most active—the sample at Duckfoot shows less frequent and less intense skeletal signs of anemia. In addition, several of the Duckfoot adults had healed cribra orbitalia, which suggests that the stress occurred when the individuals were children living elsewhere and lowers even further the actual percentage of anemic indicators active during the Duckfoot occupation. The apparent general good health of the population indicated by the skeletal remains might strengthen the speculation that some specific

acute occurrence—plague?—was responsible for many of the deaths at Duckfoot. Plague is a natural disaster that would leave no evidence in the osseous tissues.

Stodder (1987:385) indicates that three individuals at Dolores had evidence of primary periostitis, but because she does not report the number of individuals for whom there was sufficient skeletal material for an examination, it is not known what percentage the three individuals represent. Four of 13 individuals at Duckfoot, or 30.8 percent, had periostitis, all mildly expressed.

When one compares the oral cavities of Duckfoot and Dolores people, a major difference is immediately apparent. Stodder (1987:397, Figure 20.26) indicates an average of $2 \pm .5$ caries per mouth over most adult age intervals. The data at Duckfoot are 3.7 caries per mouth, all ages included, and 5.3 caries per mouth if only adults are counted, including one individual with 10 carious teeth! Dolores data for abscesses indicate about 1 abscess per adult mouth, as compared with 2.4 for the Duckfoot adults. If one keeps in mind that the abscess rate is intimately tied to caries rate plus degree of attrition, the most reasonable explanation for this disparity in dental disease is that the Duckfoot people had a more cariogenic diet—a diet that probably included less animal protein and wild vegetable resources and more carbohydrates, especially maize. If this inference is correct, it is an additional indication that the basic maize-based horticulture of the Duckfoot population was not failing, necessitating a shift to alternative food-procurement techniques. In other words, the data concerning caries further strengthen the case for nonnutritional stress leading to the abandonment of Duckfoot—a sharp contrast with the indications from Dolores. It should be kept in mind, however, that the Duckfoot data are much more temporally restricted than those provided by Stodder for the Dolores area and therefore may represent a rather brief period of intensive agriculture and ingestion of carbohydrates.

Summary and Conclusions

Small sample and subsample sizes, coupled with the short period of occupation, have made many analyses of the Duckfoot human skeletal material difficult at best, if not impossible. So, what can we conclude at this point?

The people of Duckfoot appear to be a group that was not under active stress of nutritional or morbid origin at the various times of their deaths. Most skeletal signs of stress—cribra orbitalia, porotic hyperostosis, periostitis, personal injury indicative of interpersonal strife, and cannibalism—were either absent or at very low levels of expression, and almost all of them indicate old, inactive processes. The exception is the very poor oral health as documented by the number of caries and abscesses per mouth. Except for their oral health, the Duckfoot inhabitants *appear* to have been reasonably healthy people.

However, because skeletal remains do not provide information about many acute stressors, especially for adults past their growth and development years, a variety of these more acute stressors could not be examined. These include acute gastrointestinal diseases with attendant dehydration and electrolyte imbalance, contaminated water supplies, infectious processes such as plague, and acute poisonings. The human skeletal evidence, therefore, does not provide an answer to the question of why the site was abandoned or whether it was abandoned gradually or quickly, following some unique episodic event that may have killed a number of occupants.

There are some subtle indications that *perhaps* there was status differentiation between the individuals interred in the pit structures and those buried in the midden. A higher caries rate in the latter might indicate less access to animal protein and a greater reliance on carbohydrates. Similarly, there are indications of possible differential access to animal protein by sex, since the females from the site were smaller than those at Dolores, whereas the Duckfoot and Dolores males were the same size. However, small sample sizes and the difficulties of comparing a single site's data with a pooled multisite database make this inference rather tenuous. In addition, two of the three adult females were old enough for it to be assumed that they were not born at Duckfoot and that they had reached adulthood elsewhere. Although the problem of small subsample sizes precludes adequate hypothesis testing, there are indications that the Duckfoot inhabitants followed the general western Pueblo pattern of matrilocality and matrilineality, since the variation in cranial metric measurements among the females is very small compared with that of the males at the site and the pooled female data at Dolores.

The inhabitants of Duckfoot engaged in patterns of mortuary behavior similar to those that had been system-atically noted at Dolores, but they appear to have intensified a few of them. A higher proportion of individuals, partly or completely burned, was from pit structures that apparently were deliberately set afire and collapsed onto the bodies. The total context of these burials, including habitus and lack of indicators of personal violence, is a strong indication that the pit structures, or at least three of the four, were ritually abandoned tombs. Another difference between Duckfoot and Dolores is the presence at Duckfoot of some male-only and child interments in pit structures. No primary burials were recovered from areas other than the pit structures and midden.

It was not possible to explore whether the demographic profile of Duckfoot fit into the general regional pattern from Dolores or the larger Mesa Verde region, mostly because of the short occupation span of the site, which made it impossible to build an adequate mortality profile reflecting all the inhabitants. What Duckfoot presents in this regard is a record of the random deaths of some of the site inhabitants. These deaths include children and adults of both sexes but are notably lacking in very young children up to three years of age and young adults in their twenties.

Archaeological sites of such brief occupation as Duckfoot offer only a glimpse into the daily lives of the inhabitants, their activities, their aches and pains, and the accidents that befell them; they provide a record of who died but, in most cases, not why. What is unusual in the evidence provided by Duckfoot—an apparently typical hamlet—is the intensified ritual behavior associated with death and structure abandonment. In this respect, Duckfoot reveals a particular pattern of variation that is often lost in multicomponent sites of long occupation. The latter, where data sets tend to be larger, provide more of an "average" view of what life was like, rather than the unique pattern seen at Duckfoot.

10

Synthesis

Ricky R. Lightfoot

Excavations at the Duckfoot site have added much to our understanding of the late Pueblo I period in the Mesa Verde region. Investigations of this period have taken place throughout much of the middle half of this century and include the work of Morris (1919, 1939), Roberts (1930), Martin (1938, 1939), Brew (1946), and Hayes and Lancaster (1975). More recently, the Dolores Archaeological Program (DAP) has conducted investigations of numerous Pueblo I sites in the Dolores River valley, the results of which are described in a number of volumes (e.g., Kane 1986; Petersen and Orcutt 1987; Kane and Robinson 1988; Lipe et al. 1988; Petersen 1988). Building on his own studies of Pueblo I architecture and social organization in the Dolores River valley, Wilshusen (1991) has synthesized data on Pueblo I villages in southwestern Colorado, including documentation of several Pueblo I villages that had not been previously recorded or reported in published literature.

The layout of Duckfoot is similar to that of other late Pueblo I settlements in the northern Southwest and consists of an above-ground roomblock with 19 contiguous rooms in two rows, 4 pit structures, and an isolated room southwest of the roomblock. Construction details and the overall architectural plan are used to define three architectural suites, each consisting of back rooms, front rooms, an associated pit structure, and a courtyard area. On the basis of 375 dated tree-ring samples from the site, the roomblock (with the possible exception of Room 17) and Pit Structures 1, 2, and 3 are interpreted as having been built in the mid- to late A.D. 850s (Volume 2; Lightfoot 1992b). Pit Structure 4 was built in about A.D. 872 or 873. The latest tree-ring date from the site is A.D. 876, and the site is believed to have been abandoned by about A.D. 880. Three pit structures and seven rooms were burned, and the fourth pit structure was intentionally dismantled, at abandonment.

Site-Formation Processes

Through excavation and recording, archaeologists use physical traces of the past to make inferences about the culture that left them behind. We can document the spatial context of materials in the modern archaeological record, but the association with past cultural behavior is entirely inferential: the link between the artifacts, features, and structures observed in the present and the prehistoric behaviors that produced them are based on theory-dependent arguments. The content and structure of the archaeological record is a result of the activities of a prehistoric cultural system, the activities associated with the abandonment of sites and structures, and the natural and cultural processes that have acted on the site since its abandonment. These topics are addressed in detail in Volume 2, but several aspects of the use and abandonment of structures are reviewed here.

Structure Use

Seven of the 19 rooms in the Duckfoot roomblock are front rooms (excluding Rooms 14 and 17) that originally may have functioned as domestic, or living, rooms. Rooms 12, 13, 16, and 19 contained hearths, storage cists or bins, manos, metates, and debris (for example, sherds and flakes) related to the use of these structures for a variety of domestic activities throughout the site's occupation. Room 11 also had a hearth and storage features that were in use until abandonment, but it lacked manos and metates. The absence of food-processing tools may be explained by the presence of several manos and metates in the fill of the adjoining back room (Room 7): the grinding tools in this back room may have been moved there from Room 11 in the final days of occupation. Therefore, Room 11, too, is interpreted to have been a domestic room. Evidence of remodeling in Rooms 10 and 15 suggests that these two front rooms changed from living rooms used for diverse domestic activities to rooms used for more limited or specialized purposes. In both rooms, the hearths had been truncated or destroyed by the construction of storage features before abandonment. In Room 10, a subfloor cist had been dug through the hearth, and three large bins were probably late additions to the room. The construction of a cist in Room 15 obliterated any sign of a hearth, except for some ash on the floor. In addition, both rooms lacked full complements of grinding tools on their floors. The change in the use of Rooms 10 and 15 apparently involved installation of internal storage facilities and cessation of domestic activities such as cooking and food preparation.

Back rooms at Duckfoot, as at other Pueblo I sites, are believed to have been used predominantly for storage (e.g., Gross 1987). This interpretation is based on the absence of hearths, the relative paucity of artifacts on floors, and the recognition that food storage is a necessary part of a successful agricultural strategy because it provides a buffer against fluctuations in food availability (e.g., Burns 1983). At several sites, large quantities of corn have been recovered from back rooms that burned at abandonment (e.g., Brew 1946; Gross 1987:65; Hayes and Lancaster 1975:34; Morris 1939), which further supports the interpretation that back rooms, in general, were used for food storage. Given the ethnographic and archaeological evidence, it is likely that corn was stacked on the cob inside storage rooms rather than being shelled and stored in containers. However, liquids or granular solids, such as seeds, might have been stored in pottery containers; large clusters of jar sherds in Room 2 fill may be the remnants of such storage containers. Several back rooms at Duckfoot also yielded evidence that they were used to store tools (for example, a cluster of manos in the fill of Room 1, manos and peckingstones in Rooms 6 and 8, and weathered igneous cobbles in Room 8). These items would have taken up little space, and their presence would not have precluded the use of the rooms for food storage as well. Several back rooms show evidence of a change in use late in the occupation. Room 5 is a back room that, on the basis of the presence of a small corner bin and a shallow, informal fire pit, is interpreted to have been used, at least temporarily, as a living room. Room 4 had a burned spot and ashes on the floor in addition to hundreds of splinters of small-mammal bones, many of which were burned. It appears that one or several meals were prepared in Room 4 late in the occupation of the site, although the debris could also be related to a brief post-abandonment reuse of the site.

Pit structures at Duckfoot show evidence of having been used for domestic, storage, and ritual activities. The presence of hearths, manos, metates, cooking and serving vessels, and a variety of other tools on the floors of pit structures supports the argument that these structures were used for domestic activities. Pit Structures 1, 2, and 4 also have wing walls that separate an area of floor along the southern wall that was predominantly used for storage. One of the main differences between the pit structure and room feature assemblages is that the former include many sand-filled pits (Varien and Lightfoot 1989). Wilshusen (1989) presented data from a broad range of temporal and spatial contexts to support his argument that sand-filled pits in pit structures were used in ritual activities, either physically (for example, altar sockets) or symbolically (for example, sipapus).

Abandonment

Three of the 4 pit structures, 6 of 7 front rooms (not counting Rooms 14 and 17), and 1 of 10 back rooms were burned at the time of abandonment. The burned roof fall lay directly on the floors of these structures or on top of fuel used to ignite the roofs, with no intervening layer of naturally deposited sediment. This stratigraphic sequence indicates that burning of the structures was an abandonment, not a postabandonment, event. In all three burned pit structures, human skeletons covered or overlapped the hearths, yet the bones were burned only on the top, where they were exposed to the heat of the burning roof or of fires set inside the structure to ignite the roof. Although it is not clear why so many bodies were deposited in structures at abandonment, it appears that abandonment was rapid, with no intent to return. Structures were destroyed with usable tools and containers left inside. These details of abandonment suggest that the site may have been abandoned rapidly as the result of some catastrophe that caused the death of six or more individuals, including men, women, and children, and that the structures were destroyed as part of a funerary and abandonment ritual. If there had been any intent to return, it is likely that the usable artifacts would have been cached for future recovery and the buildings left standing for future repair, reuse, or salvage. It is possible that some tools and containers were cached in unburned storage rooms and later retrieved.

The abundance of secondary refuse on the floors of Pit Structure 2 and Room 15 suggests that these structures were abandoned slightly before the other structures. Stratigraphic evidence from Pit Structure 2 also supports an argument that the roof of this structure was intentionally dismantled and the earthen portion of the roof dumped into the structure. This argument is further supported by the fact that two of the main post holes were filled with postabandonment sediment, and the other two were filled with clean sand, capped with silt loam, and covered by piles of clean sand on the floor. However, the results of sherd-refitting analysis showed numerous refits between sherds on the floor of Pit Structure 2 and sherd containers on the floor of other structures that did not have abundant trash on their floors.

Botanical evidence suggests that at least some, if not all, of the burned structures were abandoned in the fall. The burned material covering the floors of Pit Structure 1, Pit Structure 3, and Room 12 contained the charred remains of immature sagebrush (*Artemisia tridentata*-type) flowering heads. Although some might infer that these remains were part of the structure roof, several arguments suggest otherwise. First, flowers would not have served much purpose structurally. Second, if the flowering heads had been part of the roof, it is unlikely that they would have been so well preserved; the fire that destroyed the structure almost certainly would have completely consumed the dry, fragile plant material. Finally, if flowering sagebrush had been used as roof-closing material, one would expect impressions of it to have preserved in the charred building mud. Instead, hardened remnants consistently preserved the impressions of smooth, round beams. Given these arguments, it seems more probable that the sagebrush flowering heads found on the structure floors and in roof fall are the remains of fuel that was brought in to deliberately destroy the structures at abandonment. Modern sagebrush flowering heads at a comparable stage of maturity were observed in the area around Duckfoot in September, which leads to the inference that at least the burned structures, and possibly the entire site, were abandoned in the fall.

Tree-ring dates from pit structures at Duckfoot support the architectural interpretation that Pit Structure 4 was built well after the other structures. This small pit structure was added in about A.D. 873 between two existing pit structures. In the Dolores River valley, a similar phenomenon occurred in the A.D. 870s and 880s in at least two large villages, Grass Mesa (Lipe et al. 1988) and Rio Vista (Wilshusen 1986b): small pit structures were added to preexisting roomblock and pit structure complexes shortly before the villages were abandoned. The construction of these small pit structures without the addition of associated rooms may be the result of immigration of new residents to the site, but in a situation in which the occupation was expected to be of short duration. In other words, the addition of these structures might signal a relatively long term (eight years?) anticipation that the sites would be abandoned in the not-to-distant future. The latest dates from the Duckfoot site are two cutting dates of A.D. 876, and these are used as a basis for arguing that the site was abandoned by about A.D. 880.

Economy

Subsistence

The subsistence economy at Duckfoot appears to have been similar to that of other sites occupied during the late A.D. 800s in the region, consisting of a balance of cultigens, wild plant foods, and wild animals, mostly small mammals. In flotation (macrobotanical) samples (see Chapter 6), cheno-am seeds, corn (*Zea mays*) kernels, and groundcherry (*Physalis longifolia*) seeds were the most common plant food remains recovered from food-processing, cooking, or storage features. Cheno-am seeds and groundcherry fruit would have been available for harvest in mid- to late summer, along with seeds of prickly pear (*Opuntia*), purslane (*Portulaca*), and stickleaf (*Mentzelia albicaulis*), which also were identified in some flotation samples from Duckfoot. These foods could have been consumed primarily at harvest time, or they could have been stored for use during the winter and spring as well. During the early summer, Duckfoot residents probably collected Indian rice grass (*Stipa hymenoides*) florets, tansymustard (*Descurainia*) seeds, and squawbush (*Rhus aromatica*) fruit, which ripen around mid-June in the region. During the fall, the subsistence economy probably focused on harvesting and storing the cultivated crops of corn, beans, and squash. Corn kernels and cobs are abundant in the macrobotanical remains. Evidence of beans and squash is less abundant, but the recovery of a *Phaseolus* cotyledon from Pit Structure 1 and the presence of *Cucurbita* pollen in several structures attest to the prehistoric use of these plants at Duckfoot.

Protein from meat seems to have been derived mostly from small mammals (cottontail, jackrabbit, and prairie dog) that probably were procured using an opportunistic "garden-hunting" strategy. The bones of deer and other artiodactyls (*Odocoileus hemionus* and artiodactyl, indeterminate) are common but not abundant at the site. Most of the artiodactyl bones recovered are limb bones, and most of those identifiable to species are tools. If deer were hunted regularly, they apparently were butchered offsite, and only the meat and limb bones were brought back to Duckfoot. The presence of fetal artiodactyl bones supports Neusius's (1988:1212) proposition that deer were hunted mostly in the winter.

On the basis of his analysis of human remains from Duckfoot, Hoffman concludes that the health and nutrition of the Duckfoot inhabitants were quite good. In fact, dental caries and periodontal abscesses are the most common health problems. The incidence of dental caries and abscesses is typical of agricultural populations with a high-carbohydrate diet. This suggests that agricultural productivity was good during the Duckfoot occupation. Further support for this argument is provided by the low incidence and mild expression of nutritional-stress indicators, such as cribra orbitalia and porotic hyperostosis. With only one exception, manifestations of these conditions in the Duckfoot population represent old episodes and not conditions that were active at the time of death. In fact, two of the cases are in older females who probably suffered from the anemia that caused their cribra orbitalia in their younger years, before they lived at Duckfoot.

Seasonality of Site and Structure Use

The question of seasonality of site and structure use is important to understanding Pueblo I economy and social organization. This is a long-standing problem in southwestern archaeology (Gillespie 1976; Gilman 1987; Martin 1939; Morris 1939; Powell 1983; Schlanger 1987). Martin (1939) suggested that when both surface and pit structures were present on the same site, surface structures were summer dwellings and pit structures were winter dwellings. Morris (1939) proposed that in Pueblo I sites the rooms were primarily residential, whereas the pit structures were "protokivas," that is, structures that no longer were being used primarily for general domestic purposes but that served increasingly as specialized religious structures. Gilman (1987) argued that the use of pit structures correlated with winter sedentism and reliance on stored food. On the basis of a world-wide ethnographic survey, she proposed that pit structures were winter residences for groups with at least biseasonal residential mobility. Gilman (1987) argued that as dependence on agriculture increased, groups became more sedentary and began to rely more on above-ground architecture. She did not address the Pueblo I situation, in which there were both pit structures and above-ground residential rooms at the same site. Schlanger (1987) compared floor areas of pit structures and associated rooms at Dolores to evaluate the possibility that surface rooms were occupied in summer and pit structures in winter by the same group of people. She found that, beginning about A.D. 800, the floor areas of pit structures no longer would have provided adequate space for all the residents of a pueblo.

In Volume 2, architectural evidence is used to argue that, at Duckfoot, front rooms and pit structures were almost interchangeable in terms of their domestic use. Comparisons of the floor areas of pit structures and domestic rooms in each architectural suite demonstrate that there is considerable variability between the two types of structures. In two of the three architectural suites, there would not have been enough room for the pit structures to accommodate all the residents of the rooms associated with them. This argues against a seasonal consolidation of several domestic-room-based households into a pit structure, as might be implied by some models of seasonality (e.g., Gilman 1987).

One reason for relying on agriculture is to produce a food surplus that may be consumed during winter and that provides a buffer against food shortages in subsequent years. The presence of ample storage facilities in Pueblo I roomblocks and the absence of these facilities elsewhere on the landscape suggest that these residential sites were occupied year-round. Macrobotanical remains and fossil pollen indicate that wild and cultivated plants were collected throughout the spring, summer, and fall and brought into structures at Duckfoot. The botanical record is of limited value in trying to determine winter use, because plants in this region are dormant during the winter. However, faunal remains, in the form of fetal artiodactyl bones that were recovered from Room 15 and Pit Structure 2, provide some indication of winter hunting. Walker (Chapter 8) notes that these bones are the general size of late-winter fetal deer. In general, the hearths of surface rooms contain a greater diversity of seeds, kernels, and achenes of food taxa than do the hearths in pit structures. This could be interpreted as supporting the argument that, during the summer, surface rooms were used more intensively for domestic activities than were pit structures. This inference assumes that there was greater dietary diversity in the summer, when wild plant foods could be harvested and consumed rather than being stored for winter consumption. For example, Indian rice grass (*Stipa hymenoides*) is present in the hearths of surface rooms but is absent in those of pit structures. Indian rice grass bears achenes (seeds) in early summer, and its absence from pit structures could indicate that pit structures were used less intensively in the summer than were domestic rooms. On the other hand, cactus (cholla) pollen is more abundant in pit structures than in surface rooms. The presence of cactus pollen suggests that the flowering parts of cactus were introduced into pit structures during the summer flowering season, although it is also possible that cacti were collected, dried, and stored for later use.

In summary, there seems little reason to believe that the Duckfoot site was not occupied throughout the year. There are strong theoretical reasons to expect that the diversity of structures (domestic rooms, storage rooms, and pit structures) accommodated the diversity of year-round activities and year-round storage. Botanical and faunal remains support the argument that the site was occupied throughout the year. Although there may have been seasonal differences in the way structures were used, there is no evidence of a rigid dichotomy between use in one season and non-use in another.

Technology

The pottery vessels used, broken, and discarded at Duckfoot were predominantly gray ware cooking vessels; smaller numbers of gray ware storage jars (ollas) were recovered. A typical assemblage of cooking vessels for an architectural suite probably would have included two small jars (0.25 to 2.5 liters), two to three medium cooking jars (2.5 to 5.0 liters), two large cooking jars (more than 5.0 liters), three ollas (narrow-neck jar with capacity of more than 6 liters), one or two bowls, and two or three other vessel forms (such as miniatures, dippers, seed jars, and effigies) (see Volume 2). The cooking jars recovered from Duckfoot are wide-mouth gray ware (Chapin, Moccasin, and Mancos) jars that are heavily sooted, probably because they were placed directly in the hearth for cooking. Many sherd containers were also used in food preparation, as evidenced by their contexts and use-wear traces (sooting and stains). The storage vessels recovered are narrow-mouth gray ware (Chapin) jars, or ollas, that could have been used to store water or small seeds or both. Corn generally is thought to have been stored on the cob with the husks on, as it was in historic pueblos (Gross 1987). Bowls and "other" vessels are composed largely of decorated white ware (predominantly Piedra Black-on-white) and red ware (predominantly Bluff Black-on-red) vessels.

Hegmon's (Chapter 3) analysis of white ware pottery from Duckfoot indicates that white ware vessels were made on a very localized basis. This is indicated by petrographic analysis of white ware temper, which showed significant differences in temper type between Duckfoot and Dolores-area pottery. Hegmon's compositional analysis shows that Duckfoot white wares were chemically very heterogeneous, which indicates that the white wares were produced at a very small scale (such as that of a household). Small-scale production could indicate that there were several production units at Duckfoot or that white wares were produced at several locations in the Duckfoot area and then distributed through exchange.

Corn kernels and probably other seeds were ground in trough metates with two-hand manos. However, several slab metates and manos suitable for use with slab metates were recovered from structure floors at Duckfoot. Generally, slab metates are not thought to have been used until the Pueblo II period, but apparently there was some experimentation with this type of grinding tool late in the Pueblo I period. There are also a few one-hand, or "biscuit," manos in the Duckfoot site assemblage. One-hand manos are generally believed to have been used with basin metates, but basin metates were not found at Duckfoot. The one-hand manos could have been used for miscellaneous grinding on a variety of surfaces, without metates to be used specifically with them. The surfaces of these grinding tools apparently were refurbished by pecking with pecking-stones that were made predominantly from nodular cores. Peckingstones are the most numerous stone tools in the Duckfoot assemblage.

There are a variety of projectile point types in the Duckfoot assemblage, ranging from Archaic dart points to Pueblo III arrow points. Over half of the 65 projectile points found at Duckfoot are dart points of styles typically identified with the Archaic (20) and Basketmaker (14) periods; the remaining 31 points are arrow points. The substantial proportion of earlier styles of dart points in the Duckfoot collection may be interpreted as indicating that the use of darts propelled by atlatls remained popular into Pueblo I times. On the other hand, these points may have been collected as curios or as objects that were thought to possess magical power.

Worked vegetal materials were not well preserved at Duckfoot, the only pieces in the collection being several basket fragments, probably from a single basket, on the floor of Pit Structure 3; two pieces of twine from the roof fall of Room 3; and six cordage fragments on the floor of Pit Structure 4. Bone awls are common, and a single needle was recovered; both are frequently interpreted as tools for weaving or sewing. Ornaments in the assemblage include beads, pendants, shell bracelets, and possibly bone tubes.

Social Organization

On the basis of construction details, the roomblock at Duckfoot is believed to have been built as three distinct room suites, each consisting of two to three front rooms (excluding Rooms 14 and 17) and three to four back rooms. Each room suite is associated with a pit structure and an area of intervening courtyard to form an architectural suite interpreted to be the facilities used by a single household. In Volume 2, arguments supporting this interpretation are further developed and tested, with the results pointing to a model of Pueblo I households resembling extended families.

Throughout the Mesa Verde area, the Anasazi of the late ninth century were living predominantly in aggregated settlements composed of clusters of roomblocks. Some settlements consisted of relatively dense clusters of hundreds of rooms (for example, Grass Mesa Village); others consisted of clusters of smaller roomblocks that accounted for a comparable number of structures but that were more widely dispersed across the landscape (for example, McPhee Village) (Lipe et al. 1988; Kane and Robinson 1988; Wilshusen 1991). In the Crow Canyon drainage area to the east of Duckfoot, there are as many as 12 settlements with Pueblo I components, and many of them are habitation sites comparable to Duckfoot. In addition, there is a small, aggregated village composed of four roomblock clusters (the Cirque site) about 3 km east of Duckfoot. The smaller, dispersed hamlets and the Cirque site cluster may have been part of a single community that occupied and farmed the land in the upper Crow Canyon watershed. Great kivas may have been important in the ritual integration of Pueblo I communities (Adler 1989; Lightfoot 1988; Morris 1939:84; Wilshusen 1991:228–232), and surface topography suggests that such a structure may be present at the Cirque site. Cross-cultural studies (e.g., Adler 1989; Wilshusen 1991; Adler and Wilshusen 1990) indicate that specialized integrative facilities were likely to exist in communities with a population of about 200 or more. Wilshusen (1991:206) estimates the population of the Cirque site to have been about 25 households, which would represent approximately 125 to 175 individuals. The presence of a great kiva would suggest that the Cirque site was the center of a larger dispersed community that may well have included Duckfoot.

Even within the larger aggregated roomblocks in the region, there are suites of rooms and associated pit structures comparable to the units defined as architectural suites at Duckfoot. In the DAP model of household organization, these suites were interpreted as being occupied by two or more cooperating households. In Volume 2, it is argued that each of these suites was occupied by a single household, comparable in size to an extended family.

Educational Achievements of the Duckfoot Project

As a training laboratory, the Duckfoot site served its purpose well: over 4,000 students, of all ages and from all walks of life, worked alongside archaeologists committed to quality research. For many laypeople, the attraction of archaeology is fueled by glossy magazine photographs showing riches recovered from Egyptian tombs or Mayan temples. The Duckfoot site, on the other hand, consisted of the dwellings of a small group of ordinary farmers who possessed little that was ornate or extravagant. The Duckfoot project emphasized the importance of gathering information through careful, planned study focused on context rather than on objects, and program participants responded with interest and enthusiasm, regardless of how difficult or mundane the day-to-day work. They assisted us in almost all facets of field and laboratory research, everyone doing his or her best with the training we had provided. Individually and collectively, participants in Crow Canyon educational programs made a tremendous contribution to our research, and we believe that they left with a greater appreciation of what archaeology is and how, and why, it is done.

Appendixes

Table A.1. Sherd Data Summary, by Study Unit

WARE AND TYPE	Room 1 Other Fill N	%	Room 2 Surface 1 N	%	Room 2 Other Fill N	%	Room 2 Total N	%	Room 3 Surface 1 N	%	Room 3 Roof N	%	Room 3 Other Fill N	%
Mesa Verde														
Mudware														
Gray Ware														
Chapin Gray	2	1.0									4	1.6	3	.8
Moccasin Gray	8	3.8			10	.9	10	.9	1	8.3	14	5.5	26	6.7
Mancos Gray	2	1.0			51	4.7	51	4.7			7	2.7	1	.3
Indeterminate Plain Gray	180	86.1	7	100.0	1,007	93.5	1,014	93.5	10	83.3	224	87.5	334	86.5
Mancos Corrugated														
Mesa Verde Corrugated														
Indeterminate Corrugated	3	1.4											1	.3
White Ware														
Chapin Black-on-white														
Piedra Black-on-white	3	1.4			5	.5	5	.5					1	.3
Cortez Black-on-white	2	1.0			1	.1	1	.1					3	.8
Mancos Black-on-white														
McElmo Black-on-white														
Mesa Verde Black-on-white														
Indeterminate White	8	3.8			3	.3	3	.3			1	.4	6	1.6
Nonlocal														
San Juan Red Ware														
Abajo Red-on-orange											1	.4	1	.3
Bluff Black-on-red											3	1.2	5	1.3
Deadmans Black-on-red									1	8.3			1	.3
Indeterminate San Juan Red	1	.5									2	.8	4	1.0
Other Nonlocal Red														
Other Nonlocal White														
FORM														
Jar	199	95.2	6	85.7	1,060	98.4	1,066	98.3	9	75.0	238	93.0	367	95.1
Bowl	10	4.8	1	14.3	7	.6	7	.6	3	25.0	17	6.6	11	2.8
Other					10	.9	11	1.0			1	.4	8	2.1
TOTAL	209	100.0	7	.6	1,077	99.4	1,084	100.0	12	1.8	256	39.1	386	59.0

NOTE: Structure surfaces include surfaces and surface-associated features containing cultural or constructional fill. Roofs include roof fall and mixed roof and wall fall deposits. Other structure fill includes all other structure deposits exclusive of modern ground surface.

Table A.1. Sherd Data Summary, by Study Unit (*continued*)

WARE AND TYPE	Room 3 Total N	%	Room 4 Surface 1 N	%	Room 4 Other Fill N	%	Room 4 Total N	%	Room 5 Surface 1 N	%	Room 5 Other Fill N	%	Room 5 Total N	%
Mesa Verde														
Mudware														
Gray Ware														
Chapin Gray	7	1.1			2	1.5	2	1.1	1	2.1	1	.8	2	1.1
Moccasin Gray	41	6.3	3	6.0	9	6.9	12	6.7			4	3.1	4	2.2
Mancos Gray	8	1.2			6	4.6	6	3.3			1	.8	1	.6
Indeterminate Plain Gray	568	86.9	37	74.0	83	63.8	120	66.7	45	93.8	93	71.0	138	77.1
Mancos Corrugated														
Mesa Verde Corrugated														
Indeterminate Corrugated	1	.2			2	1.5	2	1.1			2	1.5	2	1.1
White Ware														
Chapin Black-on-white	1	.2									3	2.3	3	1.7
Piedra Black-on-white	3	.5	1	2.0			1	.6			1	.8	1	.6
Cortez Black-on-white														
Mancos Black-on-white														
McElmo Black-on-white														
Mesa Verde Black-on-white														
Indeterminate White	7	1.1	3	6.0	8	6.2	11	6.1			16	12.2	16	8.9
Nonlocal														
San Juan Red Ware														
Abajo Red-on-orange	2	.3												
Bluff Black-on-red	8	1.2	4	8.0	14	10.8	18	10.0			5	3.8	5	2.8
Deadmans Black-on-red	1	.2												
Indeterminate San Juan Red	7	1.1	2	4.0	6	4.6	8	4.4	.2	4.2	5	3.8	7	3.9
Other Nonlocal Red														
Other Nonlocal White														
TOTAL	654	100.0	50	27.8	130	72.2	180	100.0	48	26.8	131	73.2	179	100.0
FORM														
Jar	614	93.9	39	78.0	105	80.8	144	80.0	43	89.6	111	84.7	154	86.0
Bowl	31	4.7	10	20.0	25	19.2	35	19.4	3	6.3	20	15.3	23	12.8
Other	9	1.4	1	2.0			1	.6	2	4.2			2	1.1
TOTAL	654	100.0	50	27.8	130	72.2	180	100.0	48	26.8	131	73.2	179	100.0

Table A.1. Sherd Data Summary, by Study Unit (*continued*)

WARE AND TYPE	Room 6								Room 7					
	Surface 1		Roof		Other Fill		Total		Surface 1		Roof		Other Fill	
	N	%	N	%	N	%	N	%	N	%	N	%	N	%
Mesa Verde														
Mudware														
Gray Ware														
Chapin Gray	1	.8	1	.8	8	2.4	10	1.7			3	1.7	1	.8
Moccasin Gray	8	6.0	13	10.1	14	4.2	35	5.9			7	4.0	4	3.3
Mancos Gray					8	2.4	8	1.3			1	.6	1	.8
Indeterminate Plain Gray	114	85.7	110	85.3	270	81.1	494	83.0	20	100.0	134	77.5	102	84.3
Mancos Corrugated														
Mesa Verde Corrugated					1	.3	1	.2					1	.8
Indeterminate Corrugated														
White Ware														
Chapin Black-on-white	2	1.5												
Piedra Black-on-white			1	.8	9	2.7	12	2.0			2	1.2		
Cortez Black-on-white											4	2.3	1	.8
Mancos Black-on-white														
McElmo Black-on-white														
Mesa Verde Black-on-white														
Indeterminate White	1	.8			6	1.8	7	1.2			3	1.7	1	.8
Nonlocal														
San Juan Red Ware														
Abajo Red-on-orange														
Bluff Black-on-red	5	3.8	2	1.6	4	1.2	11	1.8			13	7.5	1	.8
Deadmans Black-on-red														
Indeterminate San Juan Red	2	1.5	2	1.6	13	3.9	17	2.9			6	3.5	8	6.6
Other Nonlocal Red														
Other Nonlocal White														
FORM														
Jar	124	93.2	124	96.1	288	86.5	536	90.1	19	95.0	150	86.7	109	90.1
Bowl	9	6.8	5	3.9	24	7.2	38	6.4	1	5.0	19	11.0	11	9.1
Other					21	6.3	21	3.5			4	2.3	1	.8
TOTAL	133	22.4	129	21.7	333	56.0	595	100.0	20	6.4	173	55.1	121	38.5

Table A.1. Sherd Data Summary, by Study Unit (continued)

WARE AND TYPE	Room 7 Total N	%	Room 8 Surface 1 N	%	Room 8 Other Fill N	%	Room 8 Total N	%	Room 9 Surface 1 N	%	Room 9 Roof N	%	Room 9 Other Fill N	%
Mesa Verde														
Mudware														
Gray Ware														
Chapin Gray	4	1.3			5	3.3	5	3.2					3	1.0
Mocasin Gray	11	3.5			11	7.2	11	7.1					13	4.5
Mancos Gray	2	.6			1	.7	1	.6						
Indeterminate Plain Gray	256	81.5	2	100.0	116	76.3	118	76.6	8	100.0	6	60.0	225	78.7
Mancos Corrugated														
Mesa Verde Corrugated														
Indeterminate Corrugated	1	.3												
White Ware														
Chapin Black-on-white	2	.6												
Piedra Black-on-white	4	1.3												
Cortez Black-on-white	1	.3												
Mancos Black-on-white														
McElmo Black-on-white														
Mesa Verde Black-on-white														
Indeterminate White	4	1.3			7	4.6	7	4.5					20	7.0
Nonlocal														
San Juan Red Ware														
Abajo Red-on-orange	1	.3												
Bluff Black-on-red	14	4.5			3	2.0	3	1.9			2	20.0	1	.3
Deadmans Black-on-red													7	2.4
Indeterminate San Juan Red	14	4.5			9	5.9	9	5.8			2	20.0	17	5.9
Other Nonlocal Red														
Other Nonlocal White														
FORM														
Jar	278	88.5	2	100.0	137	90.1	139	90.3	8	100.0	6	60.0	257	89.9
Bowl	30	9.6			14	9.2	14	9.1			4	40.0	27	9.4
Other	6	1.9			1	.7	1	.6					2	.7
TOTAL	314	100.0	2	1.3	152	98.7	154	100.0	8	2.6	10	3.3	286	94.1

Table A.1. Sherd Data Summary, by Study Unit (*continued*)

	Room 9		Room 10								Room 11			
	Total		Surface 1		Roof		Other Fill		Total		Surface 1		Roof	
WARE AND TYPE	N	%	N	%	N	%	N	%	N	%	N	%	N	%
Mesa Verde														
Mudware														
Gray Ware														
Chapin Gray	3	1.0			4	1.2	1	.4	5	.7			7	1.7
Moccasin Gray	13	4.3	5	4.5	23	6.7	8	3.0	36	5.0	5	5.6	11	2.7
Mancos Gray							2	.8	2	.3	1	1.1	2	.5
Indeterminate Plain Gray	239	78.6	100	90.9	296	86.0	227	85.7	623	86.6	82	91.1	352	86.7
Mancos Corrugated														
Mesa Verde Corrugated														
Indeterminate Corrugated							3	1.1	3	.4				
White Ware														
Chapin Black-on-white					2	.6	1	.4	3	.4			7	1.7
Piedra Black-on-white					3	.9	1	.4	4	.6			7	1.7
Cortez Black-on-white														
Mancos Black-on-white														
McElmo Black-on-white														
Mesa Verde Black-on-white														
Indeterminate White	20	6.6	2	1.8	4	1.2	13	4.9	19	2.6	1	1.1	10	2.5
Nonlocal														
San Juan Red Ware														
Abajo Red-on-orange	1	.3												
Bluff Black-on-red	9	3.0	2	1.8	7	2.0	5	1.9	14	1.9	1	1.1	5	1.2
Deadmans Black-on-red														
Indeterminate San Juan Red	19	6.3	1	.9	5	1.5	4	1.5	10	1.4			5	1.2
Other Nonlocal Red														
Other Nonlocal White														
FORM														
Jar	271	89.1	104	94.5	323	93.9	244	92.1	671	93.3	88	97.8	374	92.1
Bowl	31	10.2	5	4.5	21	6.1	20	7.5	46	6.4	2	2.2	27	6.7
Other	2	.7	1	.9			1	.4	2	.3			5	1.2
TOTAL	304	100.0	110	15.3	344	47.8	265	36.9	719	100.0	90	12.0	406	54.0

Table A.1. Sherd Data Summary, by Study Unit (continued)

WARE AND TYPE	Room 11 Other Fill N	%	Room 11 Total N	%	Room 12 Surface 1 N	%	Room 12 Roof N	%	Room 12 Other Fill N	%	Room 12 Total N	%	Room 13 Surface 1 N	%
Mesa Verde														
Mudware														
Gray Ware														
Chapin Gray	5	2.0	12	1.6			10	1.1			10	1.0	1	.3
Moccasin Gray	6	2.3	22	2.9	11	40.7	54	5.8	3	6.0	68	6.7	24	6.6
Mancos Gray	6	2.3	9	1.2			18	1.9	1	2.0	19	1.9	2	.5
Indeterminate Plain Gray	205	80.1	639	85.0	16	59.3	688	73.5	43	86.0	747	73.7	330	90.4
Mancos Corrugated														
Mesa Verde Corrugated														
Indeterminate Corrugated	2	.8	2	.3			4	.4			4	.4		
White Ware														
Chapin Black-on-white	3	1.2	10	1.3			2	.2	1	2.0	3	.3		
Piedra Black-on-white	4	1.6	11	1.5			29	3.1			29	2.9	4	1.1
Cortez Black-on-white	1	.4	1	.1			1	.1			1	.1		
Mancos Black-on-white	1	.4	1	.1			4	.4			4	.4		
McElmo Black-on-white														
Mesa Verde Black-on-white														
Indeterminate White	14	5.5	25	3.3			93	9.9			93	9.2	1	.3
Nonlocal														
San Juan Red Ware														
Abajo Red-on-orange	1	.4	1	.1			2	.2			2	.2		
Bluff Black-on-red	6	2.3	12	1.6			18	1.9	2	4.0	20	2.0	3	.8
Deadmans Black-on-red														
Indeterminate San Juan Red	2	.8	7	.9			13	1.4			13	1.3		
Other Nonlocal Red														
Other Nonlocal White														
FORM														
Jar	230	89.8	692	92.0	24	88.9	873	93.3	47	94.0	944	93.2	355	97.3
Bowl	24	9.4	53	7.0	3	11.1	52	5.6	3	6.0	58	5.7	6	1.6
Other	2	.8	7	.9			11	1.2			11	1.1	4	1.1
TOTAL	256	34.0	752	100.0	27	2.7	936	92.4	50	4.9	1,013	100.0	365	37.9

Table A.1. Sherd Data Summary, by Study Unit (continued)

WARE AND TYPE	Room 13 Roof N	%	Other Fill N	%	Total N	%	Room 14 Surface 1 N	%	Other Fill N	%	Total N	%	Room 15 Surface 1 N	%
Mesa Verde														
Mudware														
Gray Ware														
Chapin Gray	1	.7	3	.7	5	.5			6	2.8	6	2.4	6	1.0
Moccasin Gray	8	5.6	22	4.8	54	5.6	2	6.3	4	1.9	6	2.4	42	6.7
Mancos Gray	1	.7	11	2.4	14	1.5			1	.5	1	.4	1	.2
Indeterminate Plain Gray	126	88.1	383	84.2	839	87.1	30	93.8	184	85.6	214	86.6	540	86.5
Mancos Corrugated									2	.9	2	.8		
Mesa Verde Corrugated														
Indeterminate Corrugated			5	1.1	5	.5			1	.5	1	.4	1	.2
White Ware														
Chapin Black-on-white			2	.4	2	.2								
Piedra Black-on-white			3	.7	7	.7								
Cortez Black-on-white														
Mancos Black-on-white			5	1.1	5	.5								
McElmo Black-on-white														
Mesa Verde Black-on-white														
Indeterminate White	2	1.4	10	2.2	13	1.3			7	3.3	7	2.8	11	1.8
Nonlocal														
San Juan Red Ware														
Abajo Red-on-orange														
Bluff Black-on-red	1	.7	5	1.1	9	.9			4	1.9	4	1.6	15	2.4
Deadmans Black-on-red														
Indeterminate San Juan Red	4	2.8	6	1.3	10	1.0			6	2.8	6	2.4	8	1.3
Other Nonlocal Red														
Other Nonlocal White														
FORM														
Jar	138	96.5	434	95.4	927	96.3	32	100.0	203	94.4	235	95.1	587	94.1
Bowl	3	2.1	19	4.2	28	2.9			12	5.6	12	4.9	32	5.1
Other	2	1.4	2	.4	8	.8							5	.8
TOTAL	143	14.8	455	47.2	963	100.0	32	13.0	215	87.0	247	100.0	624	21.3

Table A.1. Sherd Data Summary, by Study Unit (continued)

| | Room 15 | | | | Room 16 | | | | | | | | Room 17 | |
| | Other Fill | | Total | | Surface 1 | | Surface 2 | | Other Fill | | Total | | Surface 1 | |
WARE AND TYPE	N	%	N	%	N	%	N	%	N	%	N	%	N	%
Mesa Verde														
Mudware														
Gray Ware														
Chapin Gray	56	2.4	62	2.1			3	.7	3	.9	6	.7	3	5.2
Moccasin Gray	147	6.4	189	6.5	2	5.7	30	6.7	23	7.0	55	6.8	3	5.2
Mancos Gray	2	.1	3	.1			36	8.1	6	1.8	42	5.2		
Indeterminate Plain Gray	1,912	83.0	2,452	83.8	29	82.9	342	76.9	272	82.9	643	79.6	48	82.8
Mancos Corrugated	2	.1	2	.1					1	.3	1	.1		
Mesa Verde Corrugated														
Indeterminate Corrugated	5	.2	5	.2					3	.9	3	.4		
White Ware														
Chapin Black-on-white	2	.1	3	.1										
Piedra Black-on-white	13	.6	13	.4			2	.4	3	.9	5	.6		
Cortez Black-on-white	1	.0	1	.0										
Mancos Black-on-white	1	.0	1	.0					1	.3	1	.1		
McElmo Black-on-white														
Mesa Verde Black-on-white														
Indeterminate White	63	2.7	74	2.5			8	1.8	10	3.0	18	2.2	3	5.2
Nonlocal														
San Juan Red Ware														
Abajo Red-on-orange														
Bluff Black-on-red	71	3.1	86	2.9	4	11.4	13	2.9	4	1.2	21	2.6	1	1.7
Deadmans Black-on-red	9	.4	9	.3										
Indeterminate San Juan Red	19	.8	27	.9			11	2.5	2	.6	13	1.6		
Other Nonlocal Red														
Other Nonlocal White														
FORM														
Jar	2,071	89.9	2,658	90.8	28	80.0	403	90.6	308	93.9	739	91.5	54	93.1
Bowl	193	8.4	225	7.7	4	11.4	41	9.2	19	5.8	64	7.9	4	6.9
Other	39	1.7	44	1.5	3	8.6	1	.2	1	.3	5	.6		
TOTAL	2,303	78.7	2,927	100.0	35	4.3	445	55.1	328	40.6	808	100.0	58	26.9

Table A.1. Sherd Data Summary, by Study Unit (continued)

WARE AND TYPE	Room 17 Other Fill N	%	Room 17 Total N	%	Room 18 Surface 1 N	%	Room 18 Roof N	%	Room 18 Other Fill N	%	Room 18 Total N	%	Room 19 Surface 1 N	%
Mesa Verde														
Mudware														
Gray Ware														
Chapin Gray	1	.6	4	1.9					1	1.4	1	.8	10	1.6
Moccasin Gray	11	7.0	14	6.5	1	4.2	1	3.4			2	1.6	50	7.8
Mancos Gray	2	1.3	2	.9									2	.3
Indeterminate Plain Gray	138	87.3	186	86.1	19	79.2	26	89.7	68	91.9	113	89.0	538	83.9
Mancos Corrugated														
Mesa Verde Corrugated														
Indeterminate Corrugated														
White Ware														
Chapin Black-on-white									1	1.4	1	.8	5	.8
Piedra Black-on-white														
Cortez Black-on-white														
Mancos Black-on-white														
McElmo Black-on-white														
Mesa Verde Black-on-white														
Indeterminate White	3	1.9	6	2.8	1	4.2			1	1.4	2	1.6	6	.9
Nonlocal														
San Juan Red Ware														
Abajo Red-on-orange														
Bluff Black-on-red			1	.5	1	4.2			1	1.4	2	1.6	11	1.7
Deadmans Black-on-red														
Indeterminate San Juan Red	3	1.9	3	1.4	2	8.3	2	6.9	2	2.7	6	4.7	19	3.0
Other Nonlocal Red														
Other Nonlocal White														
FORM														
Jar	153	96.8	207	95.8	21	87.5	27	93.1	70	94.6	118	92.9	627	97.8
Bowl	5	3.2	9	4.2	3	12.5	2	6.9	3	4.1	8	6.3	11	1.7
Other									1	1.4	1	.8	3	.5
TOTAL	158	73.1	216	100.0	24	18.9	29	22.8	74	58.3	127	100.0	641	33.5

Table A.1. Sherd Data Summary, by Study Unit (continued)

WARE AND TYPE	Room 19 Roof N	Room 19 Roof %	Other Fill N	Other Fill %	Total N	Total %	Room 20 Surface 1 N	Surface 1 %	Room 20 Roof N	Roof %	Total N	Total %	Pit Structure 1 Surface 1 N	Surface 1 %
Mesa Verde														
Mudware														
Gray Ware														
Chapin Gray	7	1.3	5	.7	22	1.1	4	1.0	12	1.0	16	1.0	17	1.2
Moccasin Gray	40	7.5	60	8.1	150	7.8	20	5.2	59	5.1	79	5.1	135	9.3
Mancos Gray	2	.4	5	.7	9	.5	8	2.1	13	1.1	21	1.4	41	2.8
Indeterminate Plain Gray	421	79.4	616	82.7	1,575	82.2	334	87.7	1,040	89.0	1,374	88.7	1,167	80.3
Mancos Corrugated														
Mesa Verde Corrugated														
Indeterminate Corrugated	1	.2	1	.1	2	.1								
White Ware														
Chapin Black-on-white	1	.2	2	.3	8	.4			3	.3	3	.2	13	.9
Piedra Black-on-white	1	.2	2	.3	3	.2	5	1.3	6	.5	11	.7		
Cortez Black-on-white														
Mancos Black-on-white														
McElmo Black-on-white														
Mesa Verde Black-on-white														
Indeterminate White	15	2.8	19	2.6	40	2.1	5	1.3	25	2.1	30	1.9	48	3.3
Nonlocal														
San Juan Red Ware														
Abajo Red-on-orange														
Bluff Black-on-red	13	2.5	9	1.2	33	1.7	2	.5	3	.3	5	.3	12	.8
Deadmans Black-on-red														
Indeterminate San Juan Red	29	5.5	26	3.5	74	3.9	3	.8	7	.6	10	.6	21	1.4
Other Nonlocal Red														
Other Nonlocal White														
FORM														
Jar	500	94.3	698	93.7	1,825	95.3	365	95.8	1,134	97.1	1,499	96.8	1,377	94.7
Bowl	28	5.3	45	6.0	84	4.4	15	3.9	31	2.7	46	3.0	57	3.9
Other	2	.4	2	.3	7	.4	1	.3	3	.3	4	.3	20	1.4
TOTAL	530	27.7	745	38.9	1,916	100.0	381	24.6	1,168	75.4	1,549	100.0	1,454	49.5

Table A.1. Sherd Data Summary, by Study Unit (*continued*)

| | Pit Structure 1 | | | | | | | | Pit Structure 2 | | | | | |
| | Surface 2 | | Roof | | Other Fill | | Total | | Surface 1 | | Surface 2 | | Roof | |
WARE AND TYPE	N	%	N	%	N	%	N	%	N	%	N	%	N	%
Mesa Verde														
Mudware														
Gray Ware														
Chapin Gray			6	7.0	37	2.7	54	1.8	13	.9	1	1.5	9	2.0
Moccasin Gray					60	4.3	201	6.8	99	6.6	4	5.9	30	6.5
Mancos Gray			4	4.7	36	2.6	81	2.8	60	4.0	1	1.5	6	1.3
Indeterminate Plain Gray	3	100.0	67	77.9	1,163	83.4	2,400	81.7	1,147	76.6	55	80.9	361	78.6
Mancos Corrugated														
Mesa Verde Corrugated									2	.1				
Indeterminate Corrugated					5	.4	5	.2						
White Ware														
Chapin Black-on-white			2	2.3	1	.1	3	.1	4	.3			1	.2
Piedra Black-on-white					6	.4	19	.6	29	1.9	2	2.9	5	1.1
Cortez Black-on-white					1	.1	1	.0	8	.5				
Mancos Black-on-white														
McElmo Black-on-white														
Mesa Verde Black-on-white														
Indeterminate White			2	2.3	30	2.2	80	2.7	46	3.1	5	7.4	20	4.4
Nonlocal														
San Juan Red Ware														
Abajo Red-on-orange					1	.1	1	.0					1	.2
Bluff Black-on-red			2	2.3	19	1.4	33	1.1	19	1.3			9	2.0
Deadmans Black-on-red					3	.2	3	.1						
Indeterminate San Juan Red			3	3.5	32	2.3	56	1.9	71	4.7			17	3.7
Other Nonlocal Red														
Other Nonlocal White														
FORM														
Jar	3	100.0	78	90.7	1,152	82.6	2,610	88.9	1,395	93.1	66	97.1	424	92.4
Bowl			6	7.0	236	16.9	299	10.2	90	6.0	1	1.5	35	7.6
Other			2	2.3	6	.4	28	1.0	13	.9	1	1.5		
TOTAL	3	.1	86	2.9	1,394	47.5	2,937	100.0	1,498	58.2	68	2.6	459	17.8

Table A.1. Sherd Data Summary, by Study Unit (*continued*)

| WARE AND TYPE | Pit Structure 2 | | | | Pit Structure 3 | | | | | | | | | |
| | Other Fill | | Total | | Surface 1 | | Surface 2 | | Roof | | Other Fill | | Total | |
	N	%	N	%	N	%	N	%	N	%	N	%	N	%
Mesa Verde														
Mudware														
Gray Ware														
Chapin Gray	6	1.1	29	1.1	4	.6			17	2.4	13	2.5	34	1.8
Moccasin Gray	31	5.6	164	6.4	52	7.8			41	5.9	26	5.1	120	6.4
Mancos Gray	15	2.7	82	3.2	17	2.6			7	1.0	1	.2	26	1.4
Indeterminate Plain Gray	434	78.8	1,997	77.5	559	84.1			551	79.2	405	79.3	1,515	80.8
Mancos Corrugated							1	50.0						
Mesa Verde Corrugated							1	50.0						
Indeterminate Corrugated			2	.1							1	.2	1	.1
White Ware														
Chapin Black-on-white	2	.4	7	.3	2	.3			10	1.4	4	.8	16	.9
Piedra Black-on-white	7	1.3	43	1.7	2	.3			6	.9	8	1.6	16	.9
Cortez Black-on-white			8	.3										
Mancos Black-on-white	1	.2	1	.0							1	.2	1	.1
McElmo Black-on-white														
Mesa Verde Black-on-white														
Indeterminate White	25	4.5	96	3.7	26	3.9			39	5.6	21	4.1	86	4.6
Nonlocal														
San Juan Red Ware														
Abajo Red-on-orange	1	.2	2	.1					1	.1			1	.1
Bluff Black-on-red	9	1.6	37	1.4	1	.2			10	1.4	17	3.3	28	1.5
Deadmans Black-on-red	1	.2	1	.0										
Indeterminate San Juan Red	18	3.3	106	4.1	2	.3			14	2.0	14	2.7	30	1.6
Other Nonlocal Red	1	.2	1	.0										
Other Nonlocal White														
FORM														
Jar	506	91.8	2,391	92.8	650	97.7	2	100.0	631	90.7	456	89.2	1,739	92.8
Bowl	37	6.7	163	6.3	7	1.1			56	8.0	51	10.0	114	6.1
Other	8	1.5	22	.9	8	1.2			9	1.3	4	.8	21	1.1
TOTAL	551	21.4	2,576	100.0	665	35.5	2	.1	696	37.1	511	27.3	1,874	100.0

Table A.1. Sherd Data Summary, by Study Unit (continued)

	Pit Structure 4								Courtyards 1-3		Midden			
	Surface 1		Roof		Other Fill		Total				MGS		Other	
WARE AND TYPE	N	%	N	%	N	%	N	%	N	%	N	%	N	%
Mesa Verde														
Mudware									3	.0				
Gray Ware														
Chapin Gray	14	1.8	12	3.8	2	1.6	28	2.3	118	1.0	69	.8	589	1.0
Moccasin Gray	49	6.5	35	11.0	4	3.2	88	7.3	540	4.4	334	3.6	2,203	3.8
Mancos Gray	14	1.8	9	2.8			23	1.9	106	.9	67	.7	488	.8
Indeterminate Plain Gray	632	83.4	249	78.1	100	80.6	981	81.7	10,386	85.0	7,773	84.6	47,883	82.4
Mancos Corrugated									1	.0	1	.0	1	.0
Mesa Verde Corrugated														
Indeterminate Corrugated									43	.4	21	.2	46	.1
White Ware														
Chapin Black-on-white	4	.5							32	.3	17	.2	197	.3
Piedra Black-on-white			1	.3	3	2.4	8	.7	73	.6	86	.9	756	1.3
Cortez Black-on-white													18	.0
Mancos Black-on-white									7	.1	6	.1	15	.0
McElmo Black-on-white														
Mesa Verde Black-on-white									1	.0				
Indeterminate White	6	.8	3	.9	7	5.6	16	1.3	340	2.8	388	4.2	2,323	4.0
Nonlocal														
San Juan Red Ware														
Abajo Red-on-orange									5	.0	1	.0	25	.0
Bluff Black-on-red	21	2.8	7	2.2	5	4.0	33	2.7	187	1.5	125	1.4	1,205	2.1
Deadmans Black-on-red											14	.2	6	.0
Indeterminate San Juan Red	18	2.4	3	.9	3	2.4	24	2.0	368	3.0	281	3.1	2,356	4.1
Other Nonlocal Red									2	.0	1	.0	3	.0
Other Nonlocal White													2	.0
TOTAL	758	63.1	319	26.6	124	10.3	1,201	100.0	12,212	100.0	9,184	13.6	58,116	86.4
FORM														
Jar	707	93.3	308	96.6	107	86.3	1,122	93.4	11,145	91.3	8,494	92.5	52,771	90.8
Bowl	48	6.3	11	3.4	16	12.9	75	6.2	733	6.0	655	7.1	5,058	8.7
Other	3	.4			1	.8	4	.3	334	2.7	35	.4	287	.5
TOTAL	758		319		124		1,201	100.0	12,212	100.0	9,184	13.6	58,116	86.4

Table A.1. Sherd Data Summary, by Study Unit (continued)

WARE AND TYPE	Midden Total N	%	MGS[a] N	%	Other[b] N	%	Site Total N	%
Mesa Verde								
Mudware							3	.0
Gray Ware								
Chapin Gray	658	1.0	45	.9	65	1.5	1,225	1.1
Moccasin Gray	2,537	3.8	192	4.0	180	4.1	4,883	4.4
Mancos Gray	555	.8	24	.5	39	.9	1,086	1.0
Indeterminate Plain Gray	55,656	82.7	4,195	87.3	3,674	83.9	93,336	83.2
Mancos Corrugated	2	.0	2	.0			6	.0
Mesa Verde Corrugated							4	.0
Indeterminate Corrugated	67	.1	43	.9	32	.7	228	.2
White Ware								
Chapin Black-on-white	214	.3	15	.3	22	.5	348	.3
Piedra Black-on-white	842	1.3	23	.5	39	.9	1,175	1.0
Cortez Black-on-white	18	.0	3	.1	3	.1	37	.0
Mancos Black-on-white	21	.0	3	.1	3	.1	51	.0
McElmo Black-on-white			1	.0	3	.1	4	.0
Mesa Verde Black-on-white			1	.0			2	.0
Indeterminate White	2,711	4.0	121	2.5	144	3.3	4,004	3.6
Nonlocal								
San Juan Red Ware								
Abajo Red-on-orange	26	.0	1	.0	2	.0	45	.0
Bluff Black-on-red	1,330	2.0	34	.7	66	1.5	2,023	1.8
Deadmans Black-on-red	20	.0			5	.1	39	.0
Indeterminate San Juan Red	2,637	3.9	104	2.2	102	2.3	3,688	3.3
Other Nonlocal Red	3	.0					5	.0
Other Nonlocal White	3	.0	1	.0			5	.0
FORM								
Jar	61,265	91.0	4,519	94.0	3,895	88.9	102,602	91.4
Bowl	5,713	8.5	253	5.3	372	8.5	8,574	7.6
Other	322	.5	36	.7	112	2.6	1,021	.9
TOTAL	67,300	100.0	4,808	100.0	4,379	100.0	112,197	100.0

MGS = modern ground surface.
[a] All modern ground surface proveniences except for the midden.
[b] All site deposits not included elsewhere; includes arbitrary units and general site.

Table A.2. Stone/Mineral Tool Data Summary, by Study Unit

Artifact Type	Room 1 Other Fill N	%	Room 2 Surface 1 N	%	Room 2 Other Fill N	%	Room 2 Total N	%	Room 3 Roof N	%	Room 3 Other Fill N	%	Room 3 Total N	%	Room 4 Surface 1 N	%
Grinding Tools																
Metate (all types)											1	25.0	1	7.1		
Metate rest																
Mano (all types)	8	61.5			6	100.0	6	85.7	4	40.0	2	50.0	6	42.9	3	60.0
Abrader									1	10.0			1	7.1	1	20.0
Pounding Tools																
Hammerstone			1	100.0			1	14.3								
Peckingstone	3	23.1							3	30.0			3	21.4		
Maul	1	7.7														
Chopping Tools																
Axe (all types)																
Axe/maul																
Cutting, Piercing, Scraping Tools																
Biface	1	7.7														
Projectile point											1	25.0	1	7.1		
Drill																
Modified flake																
Modified core																
Other chipped-stone tool																
Ornaments																
Bead																
Pendant																
Miscellaneous																
Modified cobble									2	20.0			2	14.3	1	20.0
Polishing stone																
Polishing/hammerstone																
Stone disk																
Other modified stone/mineral																
TOTAL	13	100.0	1	14.3	6	85.7	7	100.0	10	71.4	4	28.6	14	100.0	5	31.3

NOTE: Structure surfaces include surfaces and surface-associated features containing cultural or constructional fill. Roofs include roof fall and mixed roof and wall fall deposits. Other structure fill includes all other structure deposits exclusive of modern ground surface.

Table A.2. Stone/Mineral Tool Data Summary, by Study Unit (continued)

Artifact Type	Room 4				Room 5						Room 6					
	Other Fill		Total		Surface 1		Other Fill		Total		Surface 1		Roof		Other Fill	
	N	%	N	%	N	%	N	%	N	%	N	%	N	%	N	%
Grinding Tools																
Metate (all types)	1	9.1	1	6.3											2	9.1
Metate rest																
Mano (all types)	1	9.1	4	25.0	5	45.5	3	50.0	8	47.1	3	27.3			3	13.6
Abrader	1	9.1	2	12.5			1	16.7	1	5.9					4	18.2
Pounding Tools																
Hammerstone					1	9.1			1	5.9					2	9.1
Peckingstone	3	27.3	3	18.8	2	18.2	1	16.7	3	17.6	6	54.5	1	33.3	4	18.2
Maul											1	9.1			1	4.5
Chopping Tools																
Axe (all types)																
Axe/maul																
Cutting, Piercing, Scraping Tools																
Biface	2	18.2	2	12.5												
Projectile point					1	9.1			1	5.9			1	33.3	1	4.5
Drill																
Modified flake					1	9.1	1	16.7	2	11.8					1	4.5
Modified core																
Other chipped-stone tool															1	4.5
Ornaments																
Bead																
Pendant																
Miscellaneous																
Modified cobble	1	9.1	2	12.5	1	9.1			1	5.9					1	4.5
Polishing stone													1	33.3	1	4.5
Polishing/hammerstone																
Stone disk																
Other modified stone/mineral	2	18.2	2	12.5							1	9.1			1	4.5
TOTAL	11	68.8	16	100.0	11	64.7	6	35.3	17	100.0	11	30.6	3	8.3	22	61.1

Table A.2. Stone/Mineral Tool Data Summary, by Study Unit (continued)

Artifact Type	Room 6 Total N	Room 6 Total %	Room 7 Surface 1 N	Room 7 Surface 1 %	Room 7 Roof N	Room 7 Roof %	Room 7 Other Fill N	Room 7 Other Fill %	Room 7 Total N	Room 7 Total %	Room 8 Surface 1 N	Room 8 Surface 1 %	Room 8 Other Fill N	Room 8 Other Fill %	Room 8 Total N	Room 8 Total %
Grinding Tools																
Metate (all types)	2	5.6			4	13.8			4	12.1						
Metate rest																
Mano (all types)	6	16.7			5	17.2			5	15.2	2	66.7			2	28.6
Abrader	4	11.1			2	6.9			2	6.1						
Pounding Tools																
Hammerstone	2	5.6														
Peckingstone	11	30.6	2	66.7	4	13.8	1	100.0	7	21.2	1	33.3			1	14.3
Maul	2	5.6														
Chopping Tools																
Axe (all types)																
Axe/maul																
Cutting, Piercing, Scraping Tools																
Biface	2	5.6														
Projectile point	1	2.8			2	6.9			2	6.1			1	25.0	1	14.3
Drill	1	2.8														
Modified flake					1	3.4			1	3.0						
Modified core													1	25.0	1	14.3
Other chipped-stone tool					1	3.4			1	3.0						
Ornaments																
Bead																
Pendant																
Miscellaneous																
Modified cobble	1	2.8														
Polishing stone	2	5.6			5	17.2			5	15.2			1	25.0	1	14.3
Polishing/hammerstone																
Stone disk																
Other modified stone/mineral	2	5.6	1	33.3	5	17.2			6	18.2			1	25.0	1	14.3
TOTAL	36	100.0	3	9.1	29	87.9	1	3.0	33	100.0	3	42.9	4	57.1	7	100.0

Table A.2. Stone/Mineral Tool Data Summary, by Study Unit (continued)

| | Room 9 | | | | | | Room 10 | | | | | | | | Room 11 | |
| | Roof | | Other Fill | | Total | | Surface 1 | | Roof | | Other Fill | | Total | | Surface 1 | |
Artifact Type	N	%	N	%	N	%	N	%	N	%	N	%	N	%	N	%
Grinding Tools																
Metate (all types)	1	12.5	1	25.0	2	16.7			1	9.1			1	6.3		
Metate rest																
Mano (all types)	3	37.5	1	25.0	4	33.3			2	18.2	1	25.0	3	18.8	1	11.1
Abrader	1	12.5			1	8.3			2	18.2			2	12.5		
Pounding Tools																
Hammerstone	1	12.5			1	8.3										
Peckingstone			1	25.0	1	8.3	1	100.0	3	27.3	1	25.0	5	31.3	5	55.6
Maul									1	9.1			1	6.3		
Chopping Tools																
Axe (all types)																
Axe/maul																
Cutting, Piercing, Scraping Tools																
Biface			1	25.0	1	8.3										
Projectile point											1	25.0	1	6.3		
Drill																
Modified flake	2	25.0			2	16.7										
Modified core																
Other chipped-stone tool																
Ornaments																
Bead																
Pendant																
Miscellaneous																
Modified cobble															1	11.1
Polishing stone																
Polishing/hammerstone																
Stone disk																
Other modified stone/mineral									2	18.2	1	25.0	3	18.8	2	22.2
TOTAL	8	66.7	4	33.3	12	100.0	1	6.3	11	68.8	4	25.0	16	100.0	9	34.6

322

Table A.2. Stone/Mineral Tool Data Summary, by Study Unit (continued)

Artifact Type	Room 11 Roof N	%	Other Fill N	%	Total N	%	Room 12 Surface 1 N	%	Roof N	%	Other Fill N	%	Total N	%	Room 13 Surface 1 N	%
Grinding Tools																
Metate (all types)							3	13.6	1	5.6			4	9.3	2	9.1
Metate rest																
Mano (all types)	4	36.4			5	19.2	5	22.7	2	11.1			7	16.3	7	31.8
Abrader							1	4.5	1	5.6			2	4.7	1	4.5
Pounding Tools																
Hammerstone									1	5.6			1	2.3		
Peckingstone	4	36.4	3	50.0	12	46.2	10	45.5	5	27.8	2	66.7	17	39.5	7	31.8
Maul																
Chopping Tools																
Axe (all types)	1	9.1	1		1	3.8										
Axe/maul							1	4.5					1	2.3		
Cutting, Piercing, Scraping Tools																
Biface									1	5.6			1	2.3		
Projectile point																
Drill									1	5.6			1	2.3		
Modified flake	1	9.1	1	16.7	2	7.7			1	5.6			1	2.3		
Modified core																
Other chipped-stone tool																
Ornaments																
Bead																
Pendant									1	5.6			1	2.3		
Miscellaneous																
Modified cobble					1	3.8	1	4.5	1	5.6			2	4.7	5	22.7
Polishing stone											1	33.3	1	2.3		
Polishing/hammerstone									1	5.6			1	2.3		
Stone disk									1	5.6			1	2.3		
Other modified stone/mineral	1	9.1	2	33.3	5	19.2	1	4.5	1	5.6			2	4.7		
TOTAL	11	42.3	6	23.1	26	100.0	22	51.2	18	41.9	3	7.0	43	100.0	22	55.0

Table A.2. Stone/Mineral Tool Data Summary, by Study Unit (continued)

	Room 13						Room 14						Room 15			
	Roof		Other Fill		Total		Surface 1		Other Fill		Total		Surface 1		Other Fill	
Artifact Type	N	%	N	%	N	%	N	%	N	%	N	%	N	%	N	%
Grinding Tools																
Metate (all types)					2	5.0									1	3.4
Metate rest																
Mano (all types)	4	36.4	1	14.3	12	30.0	1	50.0			1	20.0	2	33.3	11	37.9
Abrader					1	2.5									2	6.9
Pounding Tools																
Hammerstone	3	27.3	1	14.3	11	27.5	1	50.0	1	33.3	1	20.0				
Peckingstone	1	9.1			1	2.5			1	33.3	2	40.0	2	33.3	4	13.8
Maul																
Chopping Tools																
Axe (all types)			1	14.3	1	2.5									1	3.4
Axe/maul																
Cutting, Piercing, Scraping Tools																
Biface	1	9.1			1	2.5									1	3.4
Projectile point	1	9.1			1	2.5										
Drill																
Modified flake			1	14.3	1	2.5										
Modified core															2	6.9
Other chipped-stone tool																
Ornaments																
Bead																
Pendant																
Miscellaneous																
Modified cobble			2	28.6	7	17.5							1	16.7	3	10.3
Polishing stone															1	3.4
Polishing/hammerstone																
Stone disk	1	9.1			2	5.0							1	16.7	1	3.4
Other modified stone/mineral			1	14.3	2	5.0			1	33.3	1	20.0			2	6.9
TOTAL	11	27.5	7	17.5	40	100.0	2	40.0	3	60.0	5	100.0	6	17.1	29	82.9

Table A.2. Stone/Mineral Tool Data Summary, by Study Unit (continued)

Artifact Type	Room 15 Total		Room 16 Surface 1		Surface 2		Other Fill		Total		Room 17 Other Fill		Room 18 Surface 1		Other Fill	
	N	%	N	%	N	%	N	%	N	%	N	%	N	%	N	%
Grinding Tools																
Metate (all types)	1	2.9	1	16.7	3	25.0			4	15.4	2	50.0				
Metate rest																
Mano (all types)	13	37.1	3	50.0	4	33.3	1	12.5	8	30.8			1	100.0	1	50.0
Abrader	2	5.7														
Pounding Tools																
Hammerstone																
Peckingstone	6	17.1			3	25.0	2	25.0	5	19.2	2	50.0			1	50.0
Maul																
Chopping Tools																
Axe (all types)	1	2.9														
Axe/maul																
Cutting, Piercing, Scraping Tools																
Biface							1	12.5	1	3.8						
Projectile point	1	2.9														
Drill					1	8.3			1	3.8						
Modified flake	2	5.7														
Modified core																
Other chipped-stone tool																
Ornaments																
Bead																
Pendant							1	12.5	1	3.8						
Miscellaneous																
Modified cobble	4	11.4	1	16.7					1	3.8						
Polishing stone	1	2.9					1	12.5	1	3.8						
Polishing/hammerstone							1	12.5	1	3.8						
Stone disk	2	5.7														
Other modified stone/mineral	2	5.7	1	16.7	1	8.3	1	12.5	3	11.5						
TOTAL	35	100.0	6	23.1	12	46.2	8	30.8	26	100.0	4	100.0	1	33.3	2	66.7

Table A.2. Stone/Mineral Tool Data Summary, by Study Unit (continued)

Artifact Type	Room 18 Total		Room 19 Surface 1		Roof		Other Fill		Total		Room 20 Surface 1		Roof		Total	
	N	%	N	%	N	%	N	%	N	%	N	%	N	%	N	%
Grinding Tools																
Metate (all types)					2	11.1			2	4.3			1	33.3	1	16.7
Metate rest																
Mano (all types)	2	66.7	7	35.0	7	38.9	1	12.5	15	32.6						
Abrader			2	10.0			2	25.0	4	8.7						
Pounding Tools																
Hammerstone																
Peckingstone	1	33.3	4	20.0	2	11.1	2	25.0	8	17.4	2	66.7	1	33.3	3	50.0
Maul																
Chopping Tools																
Axe (all types)			1	5.0			1	12.5	2	4.3						
Axe/maul																
Cutting, Piercing, Scraping Tools																
Biface																
Projectile point					2	11.1			2	4.3						
Drill																
Modified flake			3	15.0	1	5.6	2	25.0	6	13.0						
Modified core																
Other chipped-stone tool																
Ornaments																
Bead																
Pendant																
Miscellaneous																
Modified cobble			2	10.0	2	11.1			4	8.7			1	33.3	1	16.7
Polishing stone																
Polishing/hammerstone											1	33.3			1	16.7
Stone disk																
Other modified stone/mineral			1	5.0	2	11.1			3	6.5						
TOTAL	3	100.0	20	43.5	18	39.1	8	17.4	46	100.0	3	50.0	3	50.0	6	100.0

Table A.2. Stone/Mineral Tool Data Summary, by Study Unit (continued)

| | Pit Structure 1 | | | | | | | | Pit Structure 2 | | | | | | | |
| | Surface 1 | | Roof | | Other Fill | | Total | | Surface 1 | | Surface 2 | | Roof | | Other Fill | |
Artifact Type	N	%	N	%	N	%	N	%	N	%	N	%	N	%	N	%
Grinding Tools																
Metate (all types)	4	11.4					4	7.0	3	5.3	1	50.0	4	22.2	1	5.0
Metate rest																
Mano (all types)	10	28.6	2	100.0	3	15.0	15	26.3	4	7.0			2	11.1	7	35.0
Abrader	1	2.9			1	5.0	2	3.5	4	7.0	1	50.0	3	16.7		
Pounding Tools																
Hammerstone	2	5.7			1	5.0	3	5.3	1	1.8					1	5.0
Peckingstone	4	11.4			6	30.0	10	17.5	29	50.9			3	16.7	8	40.0
Maul	2	5.7					2	3.5								
Chopping Tools																
Axe (all types)	1	2.9			1	5.0	2	3.5	2	3.5					1	5.0
Axe/maul																
Cutting, Piercing, Scraping Tools																
Biface	3	8.6														
Projectile point					1	5.0	4	7.0								
Drill																
Modified flake					4	20.0	4	7.0	6	10.5			2	11.1	2	10.0
Modified core																
Other chipped-stone tool									2	3.5						
Ornaments																
Bead																
Pendant													1	5.6		
Miscellaneous																
Modified cobble	6	17.1			1	5.0	7	12.3	4	7.0			1	5.6		
Polishing stone																
Polishing/hammerstone																
Stone disk					1	5.0	1	1.8								
Other modified stone/mineral	2	5.7			1	5.0	3	5.3	2	3.5			2	11.1		
TOTAL	35	61.4	2	3.5	20	35.1	57	100.0	57	58.8	2	2.1	18	18.6	20	20.6

Table A.2. Stone/Mineral Tool Data Summary, by Study Unit (continued)

Artifact Type	Pit Structure 2 Total		Pit Structure 3 Surface 1		Pit Structure 3 Roof		Pit Structure 3 Other Fill		Pit Structure 3 Total		Pit Structure 4 Surface 1		Pit Structure 4 Roof		Pit Structure 4 Other Fill	
	N	%	N	%	N	%	N	%	N	%	N	%	N	%	N	%
Grinding Tools																
Metate (all types)	9	9.3	8	17.4	1	33.3	1	11.1	20	22.7	5	10.6				
Metate rest			5	10.9					5	5.7						
Mano (all types)	13	13.4	13	28.3	5	15.2			18	20.5	14	29.8	1	50.0		
Abrader	8	8.2	2	4.3			1	11.1	3	3.4	1	2.1				
Pounding Tools																
Hammerstone	2	2.1	2	4.3					2	2.3	3	6.4				
Peckingstone	40	41.2	1	2.2	4	12.1	1	11.1	6	6.8	8	17.0	1	50.0		
Maul																
Chopping Tools																
Axe (all types)	3	3.1	1	2.2					1	1.1	3	6.4				
Axe/maul					1	3.0			1	1.1						
Cutting, Piercing, Scraping Tools																
Biface							1	11.1	1	1.1						
Projectile point			1	2.2					1	1.1	1	2.1				
Drill			1	2.2					1	1.1						
Modified flake	10	10.3			4	12.1	2	22.2	6	6.8	2	4.3			1	100.0
Modified core	2	2.1	1	2.2					1	1.1						
Other chipped-stone tool																
Ornaments																
Bead																
Pendant	1	1.0	1	2.2					1	1.1	1	2.1				
Miscellaneous																
Modified cobble	5	5.2	7	15.2	3	9.1			10	11.4	7	14.9				
Polishing stone					1	3.0			1	1.1						
Polishing/hammerstone											1	2.1				
Stone disk			1	2.2					1	1.1						
Other modified stone/mineral	4	4.1	2	4.3	4	12.1	3	33.3	9	10.2	1	2.1				
TOTAL	97	100.0	46	52.3	33	37.5	9	10.2	88	100.0	47	94.0	2	4.0	1	2.0

Table A.2. Stone/Mineral Tool Data Summary, by Study Unit (continued)

Artifact Type	Pit Structure 4 Total		Courtyards 1–3		Midden MGS		Midden Other		Midden Total		MGS[a]		Other[b]		Site Total	
	N	%	N	%	N	%	N	%	N	%	N	%	N	%	N	%
Grinding Tools																
Metate (all types)	5	10.0	6	5.6	3	2.5	25	3.9	28	3.7	1	2.4	5	11.4	71	4.3
Metate rest															5	.3
Mano (all types)	15	30.0	9	8.3							1	2.4	1	2.3	212	12.9
Abrader	1	2.0			2	1.6	25	3.9	27	3.6	3	7.3	2	4.5	77	4.7
Pounding Tools																
Hammerstone	3	6.0	4	3.7	4	3.3	7	1.1	11	1.5	1	2.4			33	2.0
Peckingstone	9	18.0	27	25.0	25	20.5	225	35.5	250	33.1	7	17.1	16	36.4	469	28.5
Maul							1	.2	1	.1					8	.5
Chopping Tools																
Axe (all types)	3	6.0													15	.9
Axe/maul							1	.2	1	.1					2	.1
Cutting, Piercing, Scraping Tools																
Biface	1	2.0	1	.9	4	3.3	18	2.8	22	2.9			1	2.3	30	1.8
Projectile point			3	2.8	9	7.4	29	4.6	38	5.0	2	4.9	2	4.5	65	4.0
Drill					3	2.5	4	.6	7	.9	1	2.4			11	.7
Modified flake	3	6.0	35	32.4	60	49.2	214	33.8	274	36.3	23	56.1	15	34.1	388	23.6
Modified core			2	1.9			2	.3	2	.3					6	.4
Other chipped-stone tool			5	4.6	2	1.6	16	2.5	18	2.4					27	1.6
Ornaments																
Bead			1	.9			1	.2	1	.1					2	.1
Pendant	1	2.0					2	.3	2	.3	1	2.4			8	.5
Miscellaneous																
Modified cobble	7	14.0	4	3.7	1	.8	20	3.2	21	2.8					85	5.2
Polishing stone			1	.9	2	1.6	11	1.7	13	1.7					20	1.2
Polishing/hammerstone	1	2.0	3	2.8			3	.5	3	.4	1	2.4			11	.7
Stone disk					2	1.6	5	.8	7	.9			2	4.5	15	.9
Other modified stone/mineral	1	2.0	7	6.5	5	4.1	24	3.8	29	3.8					85	5.2
TOTAL	50	100.0	108	100.0	122	16.2	633	83.8	755	100.0	41	100.0	44	100.0	1,645	100.0

MGS = modern ground surface.
[a] All modern ground surface proveniences except for the midden.
[b] All site deposits not included elsewhere; includes arbitrary units and general site.

330

Table A.3. Chipped-Stone Debris Data Summary, by Study Unit

| | Room 1 | | Room 2 | | | | | | Room 3 | | | | | |
| | Other Fill | | Surface 1 | | Other Fill | | Total | | Roof | | Other Fill | | Total | |
	N	%	N	%	N	%	N	%	N	%	N	%	N	%
MATERIAL														
Local														
Dakota	4	21.0			3	30.0	3	23.1			1	2.7	1	1.3
Morrison	13	68.4	3	100.0	6	60.0	9	69.2	38	100.0	35	94.6	73	97.3
Igneous														
Nonlocal	2	10.5												
Unknown					1	10.0	1	7.7			1	2.7	1	1.3
TYPE														
Core	1	5.3							3	7.9			3	4.0
Flake with platform	13	68.4	3	100.0	8	80.0	11	84.6	26	68.4	24	64.9	50	66.7
Flake without platform	4	21.0			2	20.0	2	15.4	8	21.0	12	32.4	20	26.7
Edge-damaged flake	1	5.3									1	2.7	1	1.3
Other									1	2.6			1	1.3
TOTAL	19	100.0	3	23.1	10	76.9	13	100.0	38	50.7	37	49.3	75	100.0

NOTE: Structure surfaces include surfaces and surface-associated features containing cultural or constructional fill. Roofs include roof fall and mixed roof and wall fall deposits. Other structure fill includes all other structure deposits exclusive of modern ground surface.

Table A.3. Chipped-Stone Debris Data Summary, by Study Unit *(continued)*

| | Room 4 | | | | | | Room 5 | | | | | | Room 6 | |
| | Surface 1 | | Other Fill | | Total | | Surface 1 | | Other Fill | | Total | | Surface 1 | |
	N	%	N	%	N	%	N	%	N	%	N	%	N	%
MATERIAL														
Local														
Dakota			8	15.7	8	14.0	18	45.0	8	18.2	26	30.9	4	21.0
Morrison	5	83.3	29	56.9	34	59.6	10	25.0	32	72.7	42	50.0	12	63.1
Igneous							3	7.5	1	2.3	4	4.8		
Nonlocal			11	21.6	11	19.3	8	20.0	2	4.5	10	11.9	1	5.3
Unknown	1	16.7	3	5.9	4	7.0	1	2.5	1	2.3	2	2.4	2	10.5
TYPE														
Core			3	5.9	3	5.3			1	2.3	1	1.2	1	5.3
Flake with platform	4	66.7	25	49.0	29	50.9	27	67.5	30	68.2	57	67.8	12	63.1
Flake without platform	2	33.3	19	37.2	21	36.8	12	30.0	13	29.5	25	29.8	5	26.3
Edge-damaged flake			4	7.8	4	7.0							1	5.3
Other							1	2.5			1	1.2		
TOTAL	6	10.5	51	89.5	57	100.0	40	47.6	44	52.4	84	100.0	19	13.6

Table A.3. Chipped-Stone Debris Data Summary, by Study Unit (continued)

| | Room 6 | | | | | | Room 7 | | | | | | | |
| | Roof | | Other Fill | | Total | | Surface 1 | | Roof | | Other Fill | | Total | |
	N	%	N	%	N	%	N	%	N	%	N	%	N	%
MATERIAL														
Local														
Dakota	1	6.7	11	10.4	16	11.4	1	33.3	3	7.9	2	11.8	6	10.3
Morrison	11	73.3	78	73.6	101	72.1	2	66.7	28	73.7	15	88.2	45	77.6
Igneous									1	2.6			1	1.7
Nonlocal	1	6.7	9	8.5	11	7.8								
Unknown	2	13.3	8	7.5	12	8.6			6	15.8			6	10.3
TYPE														
Core			2	1.9	3	2.1			3	7.9	2	11.8	5	8.6
Flake with platform	11	73.3	76	71.7	99	70.7	3	100.0	22	57.9	8	47.0	33	56.9
Flake without platform	3	20.0	24	22.6	32	22.8			9	23.7	7	41.2	16	27.6
Edge-damaged flake	1	6.7	3	2.8	5	3.6			2	5.3			2	3.4
Other			1	.9	1	.7			2	5.3			2	3.4
TOTAL	15	10.7	106	75.7	140	100.0	3	5.2	38	65.5	17	29.3	58	100.0

Table A.3. Chipped-Stone Debris Data Summary, by Study Unit (continued)

| | Room 8 | | Room 9 | | | | | | | | Room 10 | | | |
| | Other Fill | | Surface 1 | | Roof | | Other Fill | | Total | | Surface 1 | | Roof | |
	N	%	N	%	N	%	N	%	N	%	N	%	N	%
MATERIAL														
Local														
Dakota			2	12.5			8	14.3	10	13.1	1	11.1	3	7.7
Morrison	21	77.8	13	81.2	3	75.0	32	57.1	48	63.1	5	55.5	32	82.0
Igneous							6	10.7	6	7.9			1	2.6
Nonlocal							8	14.3	8	10.5	2	22.2	3	7.7
Unknown	6	22.2	1	6.2	1	25.0	2	3.6	4	5.3	1	11.1		
TYPE														
Core			1	6.2			1	1.8	2	2.6			1	2.6
Flake with platform	16	59.2	11	68.7	2	50.0	32	57.1	45	59.2	4	44.4	26	66.7
Flake without platform	9	33.3	2	12.5	1	25.0	22	39.3	25	32.9	3	33.3	10	25.6
Edge-damaged flake	2	7.4	1	6.2			1	1.8	2	2.6			2	5.1
Other			1	6.2	1	25.0			2	2.6	2	22.2		
TOTAL	27	100.0	16	21.0	4	5.3	56	73.7	76	100.0	9	11.5	39	50.0

Table A.3. Chipped-Stone Debris Data Summary, by Study Unit (continued)

| | Room 10 | | | | Room 11 | | | | | | | | Room 12 | |
| | Other Fill | | Total | | Surface 1 | | Roof | | Other Fill | | Total | | Surface 1 | |
	N	%	N	%	N	%	N	%	N	%	N	%	N	%
MATERIAL														
Local														
Dakota	4	13.3	8	10.2	55	24.8	4	11.8	5	10.0	64	20.9		
Morrison	19	63.3	56	71.8	166	74.8	27	79.4	40	80.0	233	76.1	6	85.7
Igneous	1	3.3	2	2.6			1	2.9	1	2.0	2	.6		
Nonlocal	6	20.0	11	14.1			2	5.9	2	4.0	4	1.3	1	14.3
Unknown			1	1.3	1	.4			2	4.0	3	1.0		
TYPE														
Core			1	1.3			3	8.8	2	4.0	5	1.6	2	28.6
Flake with platform	12	40.0	42	53.8	45	20.3	22	64.7	24	48.0	91	29.7	2	28.6
Flake without platform	16	53.3	29	37.2	175	78.8	6	17.6	22	44.0	203	66.3	2	28.6
Edge-damaged flake	2	6.7	4	5.1	2	.9	3	8.8	1	2.0	4	1.3	1	14.3
Other			2	2.6					1	2.0	3	1.0		
TOTAL	30	38.5	78	100.0	222	72.5	34	11.1	50	16.3	306	100.0	7	4.6

Table A.3. Chipped-Stone Debris Data Summary, by Study Unit (continued)

| | Room 12 | | | | | | Room 13 | | | | | | | |
| | Roof | | Other Fill | | Total | | Surface 1 | | Roof | | Other Fill | | Total | |
	N	%	N	%	N	%	N	%	N	%	N	%	N	%
MATERIAL														
Local														
Dakota	13	9.5			13	8.5	3	14.3	3	8.8	6	6.2	12	7.9
Morrison	109	80.1	8	80.0	123	80.4	16	76.2	23	67.6	68	70.8	107	70.9
Igneous	1	.7			1	.6	1	4.8	2	5.9	7	7.3	10	6.6
Nonlocal	9	6.6	2	20.0	12	7.8			5	14.7	15	15.6	20	13.2
Unknown	4	2.9			4	2.6	1	4.8	1	2.9			2	1.3
TYPE														
Core	2	1.5			4	2.6	1	4.8	2	5.9	2	2.1	5	3.3
Flake with platform	92	67.6	7	70.0	101	66.0	15	71.4	19	55.9	58	60.4	92	60.9
Flake without platform	31	22.8	1	10.0	34	22.2	3	14.3	9	26.5	29	30.2	41	27.1
Edge-damaged flake	8	5.9	2	20.0	11	7.2	2	9.5	3	8.8	6	6.2	11	7.3
Other	3	2.2			3	2.0			1	2.9	1	1.0	2	1.3
TOTAL	136	88.9	10	6.5	153	100.0	21	13.9	34	22.5	96	63.6	151	100.0

Table A.3. Chipped-Stone Debris Data Summary, by Study Unit (continued)

| | Room 14 | | Room 15 | | | | | | Room 16 | | | | | |
| | Other Fill | | Surface 1 | | Other Fill | | Total | | Surface 1 | | Surface 2 | | Other Fill | |
	N	%	N	%	N	%	N	%	N	%	N	%	N	%
MATERIAL														
Local														
Dakota	4	11.1	13	52.0	6	8.6	19	20.0			19	20.2	5	12.2
Morrison	29	80.5	10	40.0	55	78.6	65	68.4	4	100.0	69	73.4	29	70.7
Igneous											4	4.2	4	9.7
Nonlocal	1	2.8	2	8.0	3	4.3	5	5.3					2	4.9
Unknown	2	5.5			6	8.6	6	6.3			2	2.1	1	2.4
TYPE														
Core	1	2.8	1	4.0	1	1.4	2	2.1	2	50.0			1	2.4
Flake with platform	24	66.7	17	68.0	50	71.4	67	70.5			43	45.7	18	43.9
Flake without platform	11	30.5	6	24.0	13	18.6	19	20.0	2	50.0	48	51.1	19	46.3
Edge-damaged flake			1	4.0	3	4.3	4	4.2			3	3.2	1	2.4
Other					3	4.3	3	3.1					2	4.9
TOTAL	36	100.0	25	26.3	70	73.7	95	100.0	4	2.9	94	67.6	41	29.5

Table A.3. Chipped-Stone Debris Data Summary, by Study Unit (continued)

| | Room 16 | | Room 17 | | | | | | Room 18 | | | | | |
| | Total | | Surface 1 | | Other Fill | | Total | | Surface 1 | | Roof | | Other Fill | |
	N	%	N	%	N	%	N	%	N	%	N	%	N	%
MATERIAL														
Local														
Dakota	24	17.3	1	11.1	6	19.3	7	17.5			1	7.7	2	7.1
Morrison	102	73.4	5	55.5	19	61.3	24	60.0	1	100.0	7	53.8	21	75.0
Igneous	8	5.7	1	11.1	3	9.7	4	10.0			3	23.1		
Nonlocal	4	2.9			3	9.7	3	7.5					3	10.7
Unknown	1	.7	2	22.2			2	5.0			2	15.4	2	7.1
TYPE														
Core	1	.7			1	3.2	1	2.5						
Flake with platform	63	45.3	6	66.7	21	67.7	27	67.5	1	100.0	6	46.1	15	53.6
Flake without platform	67	48.2	3	33.3	9	29.0	12	30.0			5	38.5	9	32.1
Edge-damaged flake	6	4.3											2	7.1
Other	2	1.4									2	15.4	2	7.1
TOTAL	139	100.0	9	22.5	31	77.5	40	100.0	1	2.4	13	30.9	28	66.7

Table A.3. Chipped-Stone Debris Data Summary, by Study Unit (continued)

| | Room 18 | | Room 19 | | | | | | | | Room 20 | | | |
| | Total | | Surface 1 | | Roof | | Other Fill | | Total | | Surface 1 | | Roof | |
	N	%	N	%	N	%	N	%	N	%	N	%	N	%
MATERIAL														
Local														
Dakota	3	7.1	2	9.5	6	10.5	8	9.9	16	10.1	9	23.1	26	32.5
Morrison	29	69.0	18	85.7	45	78.9	66	81.5	129	81.1	30	76.9	53	66.2
Igneous	3	7.1			1	1.7	2	2.5	3	1.9			1	1.2
Nonlocal	5	11.9	1	4.8	5	8.8	5	6.2	11	6.9				
Unknown	2	4.8												
TYPE														
Core	22	52.4	1	4.8	1	1.7			2	1.2	25	64.1	43	53.7
Flake with platform	14	33.3	13	61.9	38	66.7	56	69.1	107	67.3	13	33.3	32	40.0
Flake without platform	2	4.8	5	23.8	16	28.1	20	24.7	41	25.8	1	2.6	2	2.5
Edge-damaged flake	4	9.5	2	9.5	2	3.5	3	3.7	7	4.4			3	3.7
Other							2	2.5	2	1.2				
TOTAL	42	100.0	21	13.2	57	35.8	81	50.9	159	100.0	39	32.8	80	67.2

Table A.3. Chipped-Stone Debris Data Summary, by Study Unit (continued)

| | Room 20 | | Pit Structure 1 | | | | | | | | | | Pit Structure 2 | |
| | Total | | Surface 1 | | Surface 2 | | Roof | | Other Fill | | Total | | Surface 1 | |
	N	%	N	%	N	%	N	%	N	%	N	%	N	%
MATERIAL														
Local														
Dakota	35	29.4	8	14.0			1	16.7	17	12.5	26	12.9	168	36.4
Morrison	83	69.7	48	84.2	2	100.0	4	66.7	102	75.0	156	77.6	254	55.1
Igneous	1	.8					1	16.7	3	2.2	4	2.0	4	.9
Nonlocal			1	1.7					11	8.1	12	6.0	14	3.0
Unknown									3	2.2	3	1.5	21	4.5
TYPE														
Core	68	57.1	3	5.3			1	16.7	2	1.5	6	3.0	4	.9
Flake with platform	45	37.8	25	43.8	1	50.0	3	50.0	98	72.0	127	63.2	90	19.5
Flake without platform	3	2.5	19	33.3	1	50.0	1	16.7	31	22.8	52	25.9	353	76.6
Edge-damaged flake	3	2.5	4	7.0			1	16.7	5	3.7	10	5.0	6	1.3
Other			6	10.5							6	3.0	8	1.7
TOTAL	119	100.0	57	28.3	2	1.0	6	3.0	136	67.7	201	100.0	461	81.0

Table A.3. Chipped-Stone Debris Data Summary, by Study Unit (continued)

| | Pit Structure 2 | | | | | | | | Pit Structure 3 | | | | | |
| | Surface 2 | | Roof | | Other Fill | | Total | | Surface 1 | | Roof | | Other Fill | |
	N	%	N	%	N	%	N	%	N	%	N	%	N	%
MATERIAL														
Local														
Dakota			4	19.0	7	8.1	179	31.4	59	56.2	10	7.7	3	6.5
Morrison	1	100.0	16	76.2	72	83.7	343	60.3	23	21.9	96	74.4	34	73.9
Igneous			1	4.8			5	.9	13	12.4	13	10.1	4	8.7
Nonlocal					5	5.8	19	3.3	6	5.7	5	3.9	3	6.5
Unknown					2	2.3	23	4.0	4	3.8	5	3.9	2	4.3
TYPE														
Core			1	4.8	1	1.2	6	1.0	1	.9	7	5.4	1	2.2
Flake with platform	1	100.0	14	66.7	57	66.3	162	28.5	24	22.8	52	40.3	31	67.4
Flake without platform			4	19.0	25	29.1	382	67.1	27	25.7	58	45.0	9	19.6
Edge-damaged flake			2	9.5	2	2.3	10	1.7	2	1.9	8	6.2	4	8.7
Other					1	1.2	9	1.6	51	48.6	4	3.1	1	2.2
TOTAL	1	.2	21	3.7	86	15.1	569	100.0	105	37.5	129	46.1	46	16.4

Table A.3. Chipped-Stone Debris Data Summary, by Study Unit (continued)

| | Pit Structure 3 | | Pit Structure 4 | | | | | | | | | | Courtyards 1–3 | | Midden MGS | |
| | Total | | Surface 1 | | Roof | | Other Fill | | Total | | | | | | | |
	N	%	N	%	N	%	N	%	N	%	N	%	N	%
MATERIAL														
Local														
Dakota	72	25.7	22	40.0	2	28.6			24	32.0	188	10.7	580	14.0
Morrison	153	54.6	30	54.5	3	42.8	11	84.6	44	58.7	1,401	80.0	3,117	75.4
Igneous	30	10.7	1	1.8					1	1.3	29	1.6	123	3.0
Nonlocal	14	5.0									73	4.2	197	4.8
Unknown	11	3.9	2	3.6	2	28.6	2	15.4	6	8.0	61	3.5	118	2.8
TYPE														
Core	9	3.2	2	3.6					2	2.7	27	1.5	40	1.0
Flake with platform	107	38.2	37	67.3	5	71.4	8	61.5	50	66.7	1,091	62.3	2,358	57.0
Flake without platform	94	33.6	12	21.8	2	28.6	5	38.5	19	25.3	537	30.6	1,505	36.4
Edge-damaged flake	14	5.0	4	7.3					4	5.3	63	3.6	140	3.4
Other	56	20.0									34	1.9	92	2.2
TOTAL	280	100.0	55	73.3	7	9.3	13	17.3	75	100.0	1,752	100.0	4,135	24.3

Table A.3. Chipped-Stone Debris Data Summary, by Study Unit (*continued*)

| | Midden | | | | MGS[a] | | Other[b] | | Site Total | |
| | Other | | Total | | | | | | | |
	N	%	N	%	N	%	N	%	N	%
MATERIAL										
Local										
Dakota	1,745	13.5	2,325	13.7	122	13.0	93	13.0	3,308	14.1
Morrison	9,921	77.0	13,038	76.6	728	77.8	552	77.2	17,781	75.9
Igneous	395	3.1	518	3.0	21	2.2	17	2.4	670	2.9
Nonlocal	521	4.0	718	4.2	21	2.2	28	3.9	1,022	4.4
Unknown	300	2.3	418	2.4	44	4.7	25	3.5	631	2.7
TYPE										
Core	198	1.5	238	1.4	7	.7	5	.7	340	1.4
Flake with platform	7,923	61.5	10,281	60.4	573	61.2	445	62.2	13,893	59.3
Flake without platform	4,064	31.5	5,569	32.7	307	32.8	213	29.8	7,843	33.5
Edge-damaged flake	468	3.6	608	3.6	32	3.4	33	4.6	843	3.6
Other	229	1.8	321	1.9	17	1.8	19	2.6	493	2.1
TOTAL	12,882	75.7	17,017	100.0	936	100.0	715	100.0	23,412	100.0

MGS = modern ground surface.

[a] All modern ground surface proveniences except for the midden.

[b] All site deposits not included elsewhere; includes arbitrary units and general site.

Table B.1. Duckfoot Site Tree-Ring Dates

Study Unit	Tree-Ring Lab No.	Species	Date	Symbol	Study Unit	Tree-Ring Lab No.	Species	Date	Symbol
General Site	568	JUN	833	r		122	PP	800	vv
	569	JUN	862	r		283	JUN	811	vv
Courtyard	107	PNN	828	+vv		246	JUN	825	+vv
	95	PNN	852	vv		243	JUN	827	vv
	673	JUN	860	vv		250	JUN	827	+vv
	675	JUN	862	vv		254	JUN	828	vv
	672	JUN	864	vv		247	PNN	847	r
Midden	89	PNN	803	vv		253	JUN	848	+vv
	260	JUN	836	vv		286	JUN	848	+vv
	259	PNN	850	vv		445	PNN	849	+r
	93	JUN	856	vv		120	PNN	850	vv
	91	JUN	858	+B		290	JUN	850	+v
	94	JUN	858	r		279	JUN	852	vv
Room 2	98	PNN	822	vv		282	JUN	852	+vv
	108	PNN	828	vv		287	JUN	852	vv
	99	PNN	830	vv		292	JUN	852	+v
	106	PNN	847	vv		124	PNN	854	r
	97	PNN	852	vv		281	JUN	854	vv
Room 3	105	PNN	826	+vv		251	PNN	855	r
	104	JUN	842	vv		280	JUN	855	v
	103	JUN	843	vv		284	JUN	855	r
	102	PP	854	r		289	PNN	855	r
	100	PNN	856	vv		119	PNN	856	r
	101	PNN	858	r		121	PNN	856	r
Room 4	444	JUN	789	vv		123	PNN	856	r
	443	JUN	828	vv		288	PNN	856	r
Room 6	278	JUN	873	+vv		447	PNN	856	vv
Room 10	677	JUN	825	+vv		125	JUN	857	+r
	448	JUN	841	+vv		249	PNN	857	v
	679	PNN	850	vv		61	JUN	866	+rB
	450	JUN	861	v	Room 15	96	PNN	857	vv
	449	PNN	873	vv		109	PNN	858	vv
Room 11	430	JUN	796	+vv	Room 16	126	JUN	828	r
	428	JUN	799	vv	Room 18	442	JUN	853	vv
	435	JUN	828	vv	Room 19	678	JUN	800	vv
	438	JUN	839	r		686	PNN	806	vv
	429	JUN	844	+v		683	JUN	828	vv
	432	JUN	846	++vv		684	JUN	828	rB
	258	PNN	853	vv		687	PNN	849	r
	434	PNN	865	+vv		689	JUN	858	vv
	440	JUN	871	vv		681	PNN	859	r
	439	JUN	872	+vv		688	PNN	864	vv
Room 12	113	JUN	776	vv		680	JUN	868	++r
	118	JUN	790	vv	Room 20	676	JUN	874	vv
	110	JUN	800	vv	Pit Structure 1	33	PP	788	r
	272	JUN	801	+vv		5	PP	789	v
	115	JUN	807	vv		22	PNN	831	+r
	117	JUN	814	vv		51	PNN	831	r
	111	JUN	820	+vv		2	JUN	832	r
	116	JUN	832	vv		42	JUN	832	vv
	114	JUN	851	vv		14	PNN	841	++v
	266	JUN	854	vv		81	JUN	843	vv
Room 13	446	JUN	774	vv		66	PNN	844	++r
	244	JUN	782	+vv		67	PNN	845	++r
						52	JUN	846	vv

Table B.1. Duckfoot Site Tree-Ring Dates *(continued)*

Study Unit	Tree-Ring Lab No.	Species	Date	Symbol	Study Unit	Tree-Ring Lab No.	Species	Date	Symbol
Pit Structure 1	69	PNN	848	v		65	JUN	866	+r
(continued)	85	JUN	851	++B		68	JUN	866	+rB
	13	PNN	853	r		73	JUN	866	+rB
	24	PNN	853	r		75	JUN	866	+rB
	54	PNN	853	v		77	JUN	866	+rB
	63	PNN	853	r		4	JUN	872	rB
	56	JUN	854	vv		31	JUN	873	v
	72	JUN	854	vv	Pit Structure 2	426	JUN	822	vv
	76	JUN	854	v	Pit Structure 3	462	JUN	766	vv
	19	PNN	856	vv		452	JUN	787	+r
	7	PNN	859	vv		461	JUN	799	r
	20	JUN	859	vv		495	JUN	805	vv
	1	JUN	862	+r		575	JUN	813	vv
	10	PNN	863	+r		471	PNN	814	vv
	11	JUN	863	vv		556	JUN	816	vv
	40	JUN	863	+r		529	DF	817	vv
	44	PNN	863	+r		540	PNN	822	vv
	83	JUN	863	vv		467	PNN	824	+r
	112	JUN	864	vv		552	PNN	824	+r
	3	JUN	865	+r		584	JUN	828	vv
	9	JUN	865	+rB		581	PNN	829	vv
	12	JUN	865	+v		483	PNN	830	r
	15	JUN	865	+rB		521	PNN	830	rB
	23	PNN	865	+r		546	PNN	830	r
	25	PNN	865	+r		532	JUN	831	r
	27	JUN	865	+r		458	JUN	832	rB
	32	PP	865	+r		497	PP	832	r
	34	JUN	865	+rB		501	PP	834	vv
	36	PNN	865	+r		509	JUN	834	r
	37	JUN	865	+r		453	JUN	835	vv
	38	PNN	865	+r		473	JUN	835	r
	41	PNN	865	+r		489	JUN	835	r
	45	PNN	865	+r		527	JUN	835	r
	46	PNN	865	+r		539	PP	835	r
	50	PNN	865	+r		545	JUN	836	rB
	55	PP	865	+r		492	JUN	837	v
	58	PNN	865	+r		557	JUN	837	vv
	59	PNN	865	+r		567	JUN	838	vv
	62	PNN	865	+r		578	PNN	838	vv
	70	PNN	865	+r		502	PNN	842	vv
	71	JUN	865	+r		533	JUN	843	++B
	74	JUN	865	+B		500	JUN	848	r
	78	JUN	865	+r		511	JUN	848	r
	80	JUN	865	vv		537	JUN	848	r
	82	PNN	865	+r		490	PNN	849	vv
	86	PNN	865	+r		498	JUN	851	vv
	87	PNN	865	+r		518	JUN	851	r
	6	JUN	866	+r		459	PNN	852	rB
	17	JUN	866	+rB		472	JUN	852	vv
	21	JUN	866	+r		526	JUN	852	vv
	26	JUN	866	+r		541	PNN	852	r
	30	JUN	866	+rB		474	JUN	853	r
	35	JUN	866	+r		494	FIR	853	vv
	39	JUN	866	+r		561	PNN	854	r
	48	JUN	866	+rB		469	JUN	855	vv
	60	JUN	866	+r					

Table B.1. Duckfoot Site Tree-Ring Dates *(continued)*

Study Unit	Tree-Ring Lab No.	Species	Date	Symbol	Study Unit	Tree-Ring Lab No.	Species	Date	Symbol
Pit Structure 3 *(continued)*	508	DF	855	r		530	JUN	872	+r
	522	DF	855	r		542	JUN	872	vv
	525	DF	855	r		586	JUN	872	+r
	579	DF	855	r		455	JUN	873	vv
	520	JUN	856	+r		456	JUN	873	rB
	559	PNN	857	r		577	JUN	873	vv
	451	JUN	858	v		503	PNN	874	r
	485	JUN	858	r		585	JUN	874	+vv
	517	JUN	858	v		454	JUN	876	+r
	457	JUN	859	rB		481	PNN	876	r
	482	JUN	859	r		564*	PNN	802	vv
	487	JUN	859	r		563*	JUN	803	vv
	499	JUN	859	r		574*	JUN	818	vv
	505	PNN	859	r		565*	JUN	826	vv
	510	JUN	859	r		572*	JUN	829	vv
	514	JUN	859	r		571*	FIR	854	vv
	515	JUN	859	r		573*	JUN	854	vv
	523	PNN	859	r		566*	JUN	861	vv
	524	JUN	859	r	Pit Structure 4	649	JUN	689	vv
	538	JUN	859	r		644	JUN	804	vv
	543	PNN	859	r		631	JUN	811	vv
	547	PNN	859	r		611	JUN	812	vv
	550	JUN	859	r		591	JUN	822	vv
	562	JUN	859	rB		620	JUN	826	vv
	582	PNN	859	r		614	JUN	829	vv
	554	PP	860	r		606	PNN	832	vv
	519	PNN	861	r		627	JUN	838	vv
	580	PNN	861	v		669	JUN	841	vv
	496	PP	862	v		607	JUN	845	vv
	507	JUN	862	r		632	JUN	852	vv
	512	JUN	862	r		633	JUN	852	vv
	536	JUN	862	v		589	JUN	857	+vv
	553	PNN	862	r		637	JUN	858	vv
	470	JUN	863	r		646	JUN	858	vv
	476	JUN	863	+r		592	PNN	861	vv
	478	JUN	863	+r		651	PNN	861	vv
	486	JUN	863	+r		662	PNN	861	vv
	531	JUN	863	r		600	JUN	862	vv
	548	JUN	863	+r		608	JUN	862	vv
	477	PNN	865	+r		594	PNN	864	++rB
	551	PNN	865	r		603	JUN	865	vv
	555	JUN	865	+vv		642	JUN	866	++vv
	463	PNN	866	vv		658	JUN	866	vv
	466	JUN	866	+r		621	JUN	868	++r
	491	JUN	866	r		636	PNN	868	+vv
	513	JUN	866	+r		612	JUN	869	+vv
	516	JUN	866	r		638	PNN	869	vv
	528	PNN	866	r		667	JUN	869	+rB
	544	JUN	866	r		619	PNN	870	+vv
	576	JUN	866	r		653	JUN	870	+vv
	587	JUN	866	r		599	PNN	871	vv
	465	PNN	867	+r		609	PNN	871	r
	479	PNN	867	+r		613	PNN	871	B
	506	PNN	867	+r		617	JUN	871	vv
	460	JUN	869	+rB		622	JUN	871	+v
	560	JUN	870	+vv					

Table B.1. Duckfoot Site Tree-Ring Dates *(continued)*

Study Unit	Tree-Ring Lab No.	Species	Date	Symbol	Study Unit	Tree-Ring Lab No.	Species	Date	Symbol
Pit Structure 4	626	JUN	871	+r		628	JUN	872	r
(continued)	629	PNN	871	rB		634	JUN	872	rB
	640	JUN	871	+v		639	JUN	872	rB
	643	PNN	871	r		641	JUN	872	r
	648	PNN	871	vv		645	JUN	872	+r
	659	PNN	871	r		647	PNN	872	r
	660	PNN	871	vv		650	PNN	872	r
	661	JUN	871	vv		652	JUN	872	r
	588	JUN	872	r		657	JUN	872	+r
	593	PNN	872	r		663	JUN	872	r
	595	JUN	872	r		665	JUN	872	r
	596	JUN	872	r		666	JUN	872	rB
	597	JUN	872	r		668	JUN	872	rB
	598	JUN	872	rB		671	JUN	872	rB
	601	JUN	872	rB		590	JUN	873	r
	602	JUN	872	+rB		605	JUN	873	r
	610	JUN	872	rB		630	JUN	873	r
	615	JUN	872	r		656	JUN	873	r
	618	JUN	872	r		664	JUN	873	r
	623	PNN	872	r		670	JUN	873	rB
	624	JUN	872	+r		616	JUN	874	vv
	625	PNN	872	r		654	JUN	875	r

JUN = *Juniperus* spp. (juniper); PNN = *Pinus edulis* (pinyon); PP = *Pinus ponderosa* (ponderosa pine); FIR = *Abies* cf. *concolor* (white fir); DF = *Pseudotsuga menziesii* (Douglas fir).

Date symbol (B, r, and v dates are treated as cutting dates in this report):

    B = bark is present.

    r = less than a full section is present, but the outermost ring is continuous around the available circumference.

    v = a subjective judgment that, although there is no direct evidence of the true outside on the sample, the date is within a very few years of being a cutting date.

    vv = there is know way of estimating how far the last ring is from the true outside; many rings may be lost.

    + = one or a few rings may be missing near the outside whose presence or absence cannot be determined because the series does not extend far enough to provide adequate crossdating.

    ++ = a ring count is necessary beyond a certain point in the series because crossdating ceases.

*Samples collected from Feature 2, a postabandonment feature in Pit Structure 3.

Appendix C

Identification Criteria for Plant Remains

Karen R. Adams

The following observations and criteria were used in the identification of plant remains from the Duckfoot site, including wood charcoal. Taxa are listed alphabetically by taxon.

Artemisia tridentata–type

Flowering head. Duckfoot specimens consist of small, oval structures up to 3 mm long, occasionally occurring two to three structures per cluster linked by short (less than 2 mm) segments of stem. Each oval structure is composed of two or three overlapping series of rounded chaffy bracts, with what appear to be tubular flowers (some only partly open) just emerging from the tops. In specimens that are broken open, up to four inverted, tear-drop-shaped items are contained inside. Some of these tear-drop-shaped items have come loose in general flotation sample debris. These structures compare well to the small flowering heads (oval structures) and immature flowers (tear-drop-shaped items) of modern *Artemisia tridentata.*

Leaf and leaf segment. These specimens are small (up to 12 mm long) and narrow (ranging from 1.5 to 3.0 mm), with one to three lobes at one end. They are relatively thin, not over 1 mm thick. In general, they resemble leaves of both *Artemisia tridentata* and *Purshia tridentata;* however, when examined in detail, the Duckfoot specimens compare more favorably with the leaves of sagebrush (*Artemisia*).

The ancient materials and modern *Artemisia* share shallow lobing, lack of leaf tip prickles, absence to slight inrolling of leaf edges, and lack of prominent lower leaf veins.

There is additional evidence linking the *Artemisia tridentata*–type flowering heads to the leaves just described. In Pit Structure 1, flotation samples from two pottery vessels yielded *Artemisia tridentata*–type leaves attached to short stem sections that still retained *Artemisia*-type flowering heads.

Wood charcoal. Vessels are present in irregular aggregates, with a flame-like dendritic arrangement; rays are obvious, up to three or more cells wide; a distinct band of cork tissue initiates each growth ring; rings often fracture along cork tissue, causing "ring shake" (a tendency to break at annual ring boundaries); many lenticular pores are visible in tangential view.

Atriplex canescens–type

Fruit core. Duckfoot specimens are quadrangular in shape, up to 6 mm long and 3 mm wide. Four rows of jagged vein stubs indicate where the papery fruit bracts were once attached.

Wood charcoal. Vessels are present; rays are absent or indistinguishable; rings display a distinct ripple-like banding pattern made up of a series of alternating arcs of vessel and nonvessel areas.

Capparaceae-type

Seed. The Duckfoot seed measures 3 × 3 mm, is ovate in face view, and displays a strongly curved embryo characteristic of members of the caper family.

Cercocarpus/Artemisia-type

Axillary bud. These units are round, pithy-centered cylinders up to 1 cm in length, covered with six or more overlapping bracts attached at intervals. In longitudinal view, the cylinders are of two types, those with one pointed end and one broken end and those with two broken ends.

These structures were compared with charred axillary buds of several local taxa, and some compared favorably with modern *Cercocarpus* axillary buds. Unfortunately, many of the Duckfoot specimens are poorly preserved. Because one other taxon in the area (*Artemisia*) has axillary buds that are similar in general appearance, the conservative approach has been to label the items *Cercocarpus/Artemisia*-type axillary buds.

Cercocarpus-type

Wood charcoal. Vessels are present, solitary, and rather widely spaced; vessel size is variable throughout ring; rays are one to two cells wide and abundant; ring boundaries are faint, demarcated by widely spaced earlywood vessels.

Cheno-am

Seed. The Duckfoot cheno-am seeds are compressed, circular to obovate in face view, and lenticular to plano-convex in cross section. Carbonization has generally obliterated any surface identifying characteristics. When the embryo can be seen in broken specimens, it appears curved into a ring.

Compositae-type

Achene. This longitudinally elongate specimen is slightly compressed into an oval shape and has one truncate end and one narrow end. Although the item is broken, it is at least 2.2 mm long. No evidence of attachments can be seen on the truncate end. The basic morphology of the specimen matches criteria given by Isely (1947:364) for a pappus-free achene of the Compositae family.

Descurainia-type

Seed. Seeds average .75 mm × .50 mm, and each has an embryo obviously bent back on itself. Surface texture is a pattern of irregularly shaped cells stacked in neat vertical rows.

Echinocereus fendleri-type

Seed. This seed is a compressed oval, with a cavity near the base and a surface coarsely roughened by rounded protuberances. Seed measures approximately 2 mm in longest dimension.

Galium-type

Fruit. The single Duckfoot specimen identified as *Galium*-type fruit is a spherical object measuring 4 mm in diameter. It is almost completely covered with closely spaced tubercles, although there is one section where the epidermis appears roughened and the tubercles are missing. No clear point of attachment or other features are present, possibly because of the slightly degraded condition of the specimen.

Galium fruit may be fleshy or dry, and its epidermis is either smooth-coated, hairy, or covered with tubercles. Modern *Galium* fruit is spherical and generally occurs in pairs (didymous) that separate at maturity; when fruits are paired, at least one flat facet generally develops where the two fruits pressed against each other. However, there is some variability in fruit arrangement, and single globose fruits have been observed in modern herbarium specimens. *Galium* fruit size varies considerably, and although many species have relatively small fruits, fruit of *Galium aparine* can measure up to 5 mm in diameter (Harrington 1964:515–516; McDougal 1973:461–462; Munz and Keck 1959:1040; Kearney and Peebles 1960:810–811).

The Duckfoot specimen compares well with modern *Galium tricorne* and *G. aparine* fruit. However, the fruit of these two species displays a very obvious point of attachment, a feature not visible on the charred Duckfoot specimen. *Galium tricorne* provides the best morphological match of all species examined, but it is not listed as growing in Colorado (Harrington 1964:515). *G. aparine* is thought to have been introduced into the United States in historic times.

Gramineae-type

Caryopsis. Gramineae (grass) caryopses, sometimes referred to as seeds or grains, are generally identified on the basis of an embryo depression on one side, which is elliptical to oblong in outline and usually lowered in relation to the remainder of the surface. All the charred grass specimens in Duckfoot flotation samples appear to have an embryo depression on one side. They vary in the relative proportion of embryo to endosperm, which suggests that members of two major grass groups (the "festucoid" grasses and the "panicoid" grasses) are present (Reeder 1957:758). The grains also vary considerably in overall dimensions, ranging from 2 × 1 mm to half that size.

Helianthus/Viguiera-type

Achene. The single Duckfoot specimen is oblong, with a truncate distal end and a rounded proximal end. The rounded proximal end has a short knob. The item is bifacial, with one convex facet and one planar facet, and measures 4 × 2.2 × 2 mm. It appears to be missing most of its outer epidermal surface, which is present only toward the proximal end. A comparison of this specimen with modern *Helianthus* achenes provides a moderately good match, although *Viguiera* is another genus of the sunflower family whose achenes are similar to those of *Helianthus*, especially if carbonization has obliterated features.

Juniperus-type

Seed. The prehistoric seeds are ovoid, hard, and thick-walled, with no evidence of facets. They have an average length × width index of 26.3 (N = 6), which places them well within the range of modern *J. osteosperma* seeds (Lentz 1984:193).

Twig/scale leaf. Individual scale leaves are triangular and usually less than 3 mm long. When still attached to a twig, they appear in whorls of two to three. All leaves were scrutinized at 30× magnification for the presence of minute teeth along the margin, which, if present, would suggest that they came from either *Juniperus monosperma* or *J. osteosperma* (Adams 1980a:380). Although some specimens are too degraded to be seen clearly, the rest have minutely denticulate margins.

Wood charcoal. An identification of *Juniperus*-type is used because it is difficult to distinguish species within this genus (Minnis 1987:121) and because both *Juniperus osteosperma* and *J. scopulorum* grow in the region today. Wood contains no vessels; resin canals are absent, but specimens occasionally may contain trauma-induced resin ducts or holes that mimic resin canals; latewood zone is narrow to very narrow; earlywood zone is wide, occupying most of the ring; boundary between latewood of one ring and earlywood of next ring is sharply demarcated.

Leguminosae-type

Seed. The single seed recovered is kidney-bean shaped and quite small, measuring 1.7 × 1.3 mm. It appears to have a suture that divides the seed into two halves, or cotyledons, characteristic of the legume family.

Lycium-type

Wood charcoal. Vessels are present, abundant, some in vessel pairs or multiples; vessels are not evenly spaced throughout ring; earlywood is demarcated by a row of small vessels, with other vessels in ring larger; often a narrow band of dense latewood displays no vessels or only solitary vessels; occasionally a ring may exhibit a dendritic arrangement of vessels in latewood; rays are one to two cells wide, abundant.

Malvaceae-type

Seed. Seeds are compressed and reniform, each with a notch near the point of attachment. The margin opposite the point of attachment is broad and composed of two ridges with an intervening shallow trough or groove. The seeds vary in size, up to 2 mm in the longest dimension. Modern *Sphaeralcea digitata* seeds (ARIZ 271461) collected in the area today are generally larger than the prehistoric specimens.

Mentzelia albicaulis-type

Seed. The prehistoric seeds vary in number and arrangement of facets, giving them the appearance of a three-to seven-sided box that has been modified by pressure. The pebbled or beaded surface texture and small size of the items (less than 1 mm in any dimension) compare well with modern *Mentzelia albicaulis* seeds.

Monocotyledon Fibrovascular Bundles

Bundles are circular to elliptical in shape, and in cross section (at 50× magnification) they appear to be composed of 30 or more individual fibers, each with a lumen in the center. Such bundles are characteristically found in the leaves of *Yucca* and *Agave* species (Bell and King 1944).

Opuntia sp. (prickly pear)-type

Seed. Seeds are subcircular disks, somewhat flattened and with a distinct groove parallel to the margin. They compare well in size (3.2 to 4.5 mm in diameter) and shape with modern prickly pear seeds. The smoothly contoured edges of holes in some of the seeds from Pit Structure 1 appear to represent teeth marks of small rodents, which suggests that the seeds were gnawed prior to carbonization.

Peraphyllum ramosissimum-type

Seed. The Duckfoot seeds are elliptic-ovoid in face view, round at the apex, widest in the middle, and taper to a narrow base. In the same view, one lateral edge is convex and the other fairly flat. When viewed in cross section, one face is convex and the other flat to concave. A flat attachment scar is visible along the margin at the base; in one prehistoric specimen the base is curved into a slight hook. The sizes of the four specimens are quite variable. The largest seed is 5 mm long, 2 mm (apex) × 3 mm (middle) × .5 mm (base) wide, and nearly 2 mm thick. The proportionately similar smallest seed is about half this size.

These prehistoric seeds were also compared with the seeds of another local shrubby rose, *Amelanchier utahensis*. This comparison was less satisfactory, as *Amelanchier* seeds appeared larger, had a more pronounced hook at the base near the point of attachment, and tended to have planar to concave faces. However, the overall similarities between *Peraphyllum* and *Amelanchier* seeds require caution in identification when only a few ancient seeds have been recovered; use of the word "type" expresses this caution.

Peraphyllum/Amelanchier-type

Wood charcoal. Vessels are present, solitary, evenly and closely spaced throughout ring, abundant, and minute; vessels are larger in earlywood, becoming smaller in latewood; rays are one to two cells wide.

Phaseolus vulgaris-type

Cotyledon. The large size (6 mm wide and longer than 16 mm) of this single, reniform cotyledon suggests that the specimen may represent domesticated *Phaseolus vulgaris* (Kaplan 1956:205). Distinguishing characteristics such as hilum and caruncle are not visible.

Phragmites australis-type

Stem. This robust stem fragment 4 mm in diameter and 1 cm long has the parallel veining indicative of a monocotyledon. The stem has one solid node. In the internodal area, a thin epidermal layer surrounds a cylinder of pith that occupies up to one-third of the area of the cross section; the remainder of the internodal area is hollow. Comparison with modern *Phragmites australis* stem segments was favorable. Although domesticated corn (*Zea mays*) is another grass with robust stems, corn stems are not hollow, but rather are filled with pith.

Physalis longifolia-type

Seed. The prehistoric seeds are flat and oval or broadly elliptic to circular, with an obscure notch and marginal scar. Surface texture is cellular reticulate, with cells arranged in a pattern parallel to the seed's margin; a number of specimens have lost a good portion of this seed coat. In specimens that are broken open, a strongly curved embryo, typical of the genus (Isely 1947), can be observed.

Modern *Physalis longifolia* seeds vary in size, with some of them twice the size of others. The charred Duckfoot specimens are 1.5 to 2.0 mm long and compare best with the smallest modern *Physalis longifolia* in the author's collection.

Pinus edulis-type

Needle fragment. Needle fragments identified as pinyon-type are semi-terete, channeled in cross section, and come to a sharp point. They appear to be from a two-needled pine.

Pinus-type

Bark scale. *Pinus*-type bark scales generally are small (less than 1 cm, largest dimension) and flat. One surface may be quite smooth, with occasional holes. In many Duckfoot specimens this smooth surface is thickened toward the center and slopes toward a thinner margin area. This contoured appearance is a common characteristic of both *Pinus edulis* and *P. ponderosa* bark (Adams 1980a:396).

Wood charcoal. *Pinus*-type wood could represent *Pinus*, *Picea*, or *Pseudotsuga* wood. Resin duct criteria that distinguish these major genera (Minnis 1987:125) were difficult to see at the magnifications employed in this study. *Pinus* wood is recognized by lack of vessels; resin canals occur throughout ring; latewood zone is usually wide; boundary between latewood of one ring and earlywood of next ring is distinct; transition from earlywood to latewood is abrupt; small lenticular pores are sometimes visible in tangential view.

Populus angustifolia-type

Wood charcoal. Same description as *Populus/Salix*-type wood (below), with these exceptions: *P. angustifolia*-type wood has many more vessel pairs and radial multiples; wood is not as semi-ring porous (rather, there is much less distinction between earlywood and latewood vessel size within a single ring); usually there is a single vessel only between two rays; vessels at the ring boundary tend to be in pairs or multiples rather than solitary.

Populus/Salix

Wood charcoal. Vessels are present, abundant, and closely spaced; first row of solitary earlywood vessels makes a noticeable ring; wood is semi-ring porous (i.e., there is a gradual transition in size from larger earlywood vessels to smaller latewood vessels); rays are one to two cells wide and abundant; the number of vessels between any two rays ranges from one to three.

Portulaca retusa-type

Seed. This seed type is represented in Duckfoot flotation samples by circular to ovate, compressed seeds that are generally less than 1 mm in diameter. They are recognized by the concentric rows of low, knobby tubercles visible on the surface.

Purshia-type

Wood charcoal. Vessels are present and solitary; a ring of closely spaced vessels is present in the earlywood at the clearly demarcated ring boundary; vessels become minute and widely spaced in latewood; rays are one to two cells wide; rings are occasionally lobate in form.

Rhus aromatica–type

Seed. The prehistoric seeds are compressed to flat and are ovoid to nearly circular in face view. They retain a slight indentation along the margin, where an irregularity suggests the presence of an attachment scar. Along the margin directly opposite the attachment scar, is a minute, lens-shaped opening with a number of longitudinal lines radiating away from it toward the main body of the seed. The specimens have a fairly smooth surface texture and measure 4.5 mm in length, 3.5 mm in width, and 2.5 mm in thickness.

Wood charcoal. Vessels are present, solitary or in pairs; vessels in earlywood appear as a distinct ring of aggregated vessels, but the remaining rings have vessels of various sizes scattered widely throughout; there are clusters of minute vessels in latewood; rays are one to two cells wide; a distinctly different pith is present.

Scirpus-type

Achene. The seed from Duckfoot is triangular, with a broken attachment (style base) at one end. It measures 1.5 × 1.0 mm.

Stipa hymenoides–type

Floret, palea, lemma, caryopsis. The florets are shaped somewhat like a light bulb, bulbous at the proximal end and pointed at the distal end. Swollen florets measure up to 4.0 × 1.5 mm, though most are smaller than this. Single caryopses (grains) are ovoid to globose in shape and have a relatively large embryo depression. They vary in diameter from 1.25 to 1.75 mm. A fusiform appearance in face view (wide in the middle and tapering to both ends), terete shape, and distinct longitudinal opening identify a number of hollow lemmas. These range up to 4.0 mm in length and 1.25 mm in width. Individual paleas are similar to the lemmas in morphology but are slightly shorter and considerably narrower in the middle. Both lemmas and paleas reveal a short, blunt callus where the floret was attached to the rachilla.

Yucca baccata–type

Seed. Seeds are flat, with a thickened rim edge and a wavy surface. They measure up to 8 mm long × 1.5 mm thick, and in width taper from 8 mm (distal end) to 4 mm (proximal end).

Type 1 Unknown

Specimens from Duckfoot classified as "Type 1" have two distinct faces, one being generally spherical and the other generally planar. The planar face has two to three indentations or rough areas that appear jagged and irregular, possibly representing the location where the item was attached to a larger entity. These items are generally small, averaging 2 to 2.5 mm in diameter.

Type 2 Unknown

This bifacial item is nearly circular in face view, averaging about 2.5 mm in diameter. Its two distinct surfaces are not at all similar. The upper surface is slightly convex and is composed of eight 1-mm-tall, globe-like protuberances radiating from the surface. Two additional protuberances have broken off, leaving concave scars. The lower concave surface appears to have been attached to another structure, on the basis of its rough and jagged texture and the presence of several small possible attachment scars. Along the circumference margin that connects the two major surfaces just described, both small globe-like structures and concave scars are visible.

Table D.1. Faunal Remains, Front Rooms, Floor and Roof/Wall Fall Contexts

Taxon	Room 10 Floor N	%	Room 10 Roof/Wall N	%	Room 11 Floor N	%	Room 11 Roof/Wall N	%	Room 12 Floor N	%	Room 12 Roof/Wall N	%	Room 13 Floor N	%
Class Reptilia														
cf. Colubridae														
Class Aves														
Large raptor, eagle (?)														
Meleagris gallopavo														
Grouse														
Zenaida macroura														
Thrush														
Avian, indeterminate														
Class Mammalia														
Order Lagomorpha														
Sylvilagus sp.	1	100.0	6	28.6			3	6.5	1	100.0	17	14.2	1	25.0
Lepus cf. *californicus*			1	4.8			2	4.3			6	5.0	2	50.0
Leporid, indeterminate			1	4.8			25	54.3			9	7.5		
Order Rodentia														
Spermophilus cf. *lateralis*														
Spermophilus sp.														
Cynomys gunnisoni			1	4.8			3	6.5			2	1.7		
Sciurid, indeterminate											1	.8		
Thomomys bottae			1	4.8										
Dipodomys ordii														
Peromyscus sp.														
Neotoma albigula											3	2.5		
Neotoma cf. *mexicana*														
Neotoma sp.							1	2.2						
Cricetid, indeterminate														
Rodent, indeterminate											3	2.5	1	25.0
Order Carnivora														
Canis familiaris														
Canis sp.			1	4.8			1	2.2						
Vulpes cf. *macrotis*														
Canid, indeterminate							4	8.7			1	.8		
Felis rufus														
Small carnivore														
Medium carnivore														
Order Artiodactyla														
Odocoileus hemionus			1	4.8			1	2.2			1	.8		
Artiodactyl, indeterminate											3	2.5		
Small mammal, indeterminate			8	38.1			6	13.0			73	60.8		
Medium mammal, indeterminate			1	4.8							1	.8		
Mammal, indeterminate														
TOTAL	1	100.0	21	100.0	0	.0	46	100.0	1	100.0	120	100.0	4	100.0

NOTE: Includes all animal bone (modified and unmodified) from floor, roof fall, and mixed roof and wall fall contexts. Floor contexts do not include features.

Table D.1. Faunal Remains, Front Rooms, Floor and Roof/Wall Fall Contexts (*continued*)

Taxon	Room 13 Roof/Wall N	Room 13 Roof/Wall %	Room 14 Floor N	Room 14 Floor %	Room 14 Roof/Wall N	Room 14 Roof/Wall %	Room 15 Floor N	Room 15 Floor %	Room 15 Roof/Wall N	Room 15 Roof/Wall %	Room 16 Floor N	Room 16 Floor %	Room 16 Roof/Wall N	Room 16 Roof/Wall %
Class Reptilia														
cf. Colubridae														
Class Aves														
Large raptor, eagle (?)														
Meleagris gallopavo														
Grouse														
Zenaida macroura														
Thrush														
Avian, indeterminate														
Class Mammalia														
Order Lagomorpha														
Sylvilagus sp.	1	.4	3	42.9			1	25.0						
Lepus cf. *californicus*	9	4.0					1	25.0			2	50.0		
Leporid, indeterminate														
Order Rodentia														
Spermophilus cf. *lateralis*											1	25.0		
Spermophilus sp.														
Cynomys gunnisoni	1	.4					2	50.0						
Sciurid, indeterminate														
Thomomys botae														
Dipodomys ordii														
Peromyscus sp.														
Neotoma albigula														
Neotoma cf. *mexicana*														
Neotoma sp.														
Cricetid, indeterminate														
Rodent, indeterminate	200	89.7												
Order Carnivora														
Canis familiaris														
Canis sp.														
Vulpes cf. *macrotis*														
Canid, indeterminate														
Felis rufus														
Small carnivore														
Medium carnivore														
Order Artiodactyla														
Odocoileus hemionus											1	25.0		
Artiodactyl, indeterminate														
Small mammal, indeterminate			4	57.1										
Medium mammal, indeterminate														
Mammal, indeterminate	11	4.9												
TOTAL	223	100.0	7	100.0	0	.0	4	100.0	0	.0	4	100.0	0	.0

Table D.1. Faunal Remains, Front Rooms, Floor and Roof/Wall Fall Contexts (*continued*)

Taxon	Room 17 Floor		Room 17 Roof/Wall		Room 19 Floor		Room 19 Roof/Wall		Total	
	N	%	N	%	N	%	N	%	N	%
Class Reptilia										
cf. Colubridae										
Class Aves										
Large raptor, eagle (?)										
Meleagris gallopavo										
Grouse										
Zenaida macroura										
Thrush										
Avian, indeterminate										
Class Mammalia										
Order Lagomorpha										
Sylvilagus sp.					1	20.0	1	50.0	35	7.9
Lepus cf. *californicus*									21	4.8
Leporid, indeterminate					2	40.0	1	50.0	41	9.3
Order Rodentia										
Spermophilus cf. *lateralis*									1	.2
Spermophilus sp.										
Cynomys gunnisoni	1	33.3							10	2.3
Sciurid, indeterminate									1	.2
Thomomys bottae									1	.2
Dipodomys ordii										
Peromyscus sp.									3	.7
Neotoma albigula										
Neotoma cf. *mexicana*										
Neotoma sp.									1	.2
Cricetid, indeterminate									1	.2
Rodent, indeterminate									204	46.3
Order Carnivora										
Canis familiaris										
Canis sp.									2	.5
Vulpes cf. *macrotis*										
Canid, indeterminate									5	1.1
Felis rufus										
Small carnivore										
Medium carnivore										
Order Artiodactyla										
Odocoileus hemionus	1	33.3			1	20.0			5	1.1
Artiodactyl, indeterminate									4	.9
Small mammal, indeterminate	1	33.3			1	20.0			93	21.1
Medium mammal, indeterminate									2	.5
Mammal, indeterminate									11	2.5
TOTAL	3	100.0	0	.0	5	100.0	2	100.0	441	100.0

Table D.2. Faunal Remains, Back Rooms, Floor and Roof/Wall Fall Contexts

Taxon	Room 1 Floor N	%	Room 1 Roof/Wall N	%	Room 2 Floor N	%	Room 2 Roof/Wall N	%	Room 3 Floor N	%	Room 3 Roof/Wall N	%	Room 4 Floor N	%
Class Reptilia														
cf. Colubridae														
Class Aves														
Large raptor, eagle (?)														
Meleagris gallopavo														
Grouse														
Zenaida macroura														
Thrush														
Avian, indeterminate											1	3.2		
Class Mammalia														
Order Lagomorpha														
Sylvilagus sp.											5	16.1	9	2.4
Lepus cf. californicus													34	9.1
Leporid, indeterminate									3	42.9	9	29.0	114	30.6
Order Rodentia														
Spermophilus cf. lateralis														
Spermophilus sp.														
Cynomys gunnisoni													3	.8
Sciurid, indeterminate														
Thomomys bottae														
Dipodomys ordii														
Peromyscus sp.											1	3.2		
Neotoma albigula														
Neotoma cf. mexicana														
Neotoma sp.														
Cricetid, indeterminate														
Rodent, indeterminate											1	3.2	1	.3
Order Carnivora														
Canis familiaris														
Canis sp.														
Vulpes cf. macrotis														
Canid, indeterminate														
Felis rufus														
Small carnivore														
Medium carnivore														
Order Artiodactyla														
Odocoileus hemionus														
Artiodactyl, indeterminate														
Small mammal, indeterminate									4	57.1	7	22.6	211	56.7
Medium mammal, indeterminate											7	22.6		
Mammal, indeterminate														
TOTAL	0	.0	0	.0	0	.0	0	.0	7	100.0	31	100.0	372	100.0

NOTE: Includes all animal bone (modified and unmodified) from floor, roof fall, and mixed roof and wall fall contexts. Floor contexts do not include features.

351

Table D.2. Faunal Remains, Back Rooms, Floor and Roof/Wall Fall Contexts (continued)

Taxon	Room 4 Roof/Wall N	%	Room 5 Floor N	%	Room 5 Roof/Wall N	%	Room 6 Floor N	%	Room 6 Roof/Wall N	%	Room 7 Floor N	%	Room 7 Roof/Wall N	%
Class Reptilia														
cf. Colubridae														
Class Aves														
Large raptor, eagle (?)														
Meleagris gallopavo														
Grouse														
Zenaida macroura														
Thrush														
Avian, indeterminate													1	4.3
Class Mammalia														
Order Lagomorpha														
Sylvilagus sp.											1	33.3	7	30.4
Lepus cf. californicus											2	66.7	4	17.4
Leporid, indeterminate													2	8.7
Order Rodentia														
Spermophilus cf. lateralis													1	4.3
Spermophilus sp.													2	8.7
Cynomys gunnisoni									2	18.2				
Sciurid, indeterminate														
Thomomys bottae														
Dipodomys ordii														
Peromyscus sp.			1	5.6										
Neotoma albigula														
Neotoma cf. mexicana														
Neotoma sp.			1	5.6										
Cricetid, indeterminate														
Rodent, indeterminate														
Order Carnivora														
Canis familiaris														
Canis sp.														
Vulpes cf. macrotis														
Canid, indeterminate														
Felis rufus														
Small carnivore														
Medium carnivore													1	4.3
Order Artiodactyla														
Odocoileus hemionus			6	33.3					6	54.5			4	17.4
Artiodactyl, indeterminate			5	27.8					3	27.3				
Small mammal, indeterminate			5	27.8										
Medium mammal, indeterminate													1	4.3
Mammal, indeterminate														
TOTAL	0	.0	18	100.0	0	.0	0	.0	11	100.0	3	100.0	23	100.0

Table D.2. Faunal Remains, Back Rooms, Floor and Roof/Wall Fall Contexts (continued)

Taxon	Room 8 Floor		Room 8 Roof/Wall		Room 9 Floor		Room 9 Roof/Wall		Room 18 Floor		Room 18 Roof/Wall		Total	
	N	%	N	%	N	%	N	%	N	%	N	%	N	%
Class Reptilia														
cf. Colubridae														
Class Aves														
Large raptor, eagle (?)														
Meleagris gallopavo														
Grouse														
Zenaida macroura														
Thrush														
Avian, indeterminate													2	.4
Class Mammalia														
Order Lagomorpha														
Sylvilagus sp.									1	50.0			36	7.4
Lepus cf. californicus									1	50.0			41	8.5
Leporid, indeterminate							14	100.0					129	26.6
Order Rodentia														
Spermophilus cf. lateralis													1	.2
Spermophilus sp.														
Cynomys gunnisoni					1	50.0							8	1.6
Sciurid, indeterminate														
Thomomys botae														
Dipodomys ordii														
Peromyscus sp.														
Neotoma albigula													1	.2
Neotoma cf. mexicana													1	.2
Neotoma sp.													1	.2
Cricetid, indeterminate													1	.2
Rodent, indeterminate													1	.2
Order Carnivora														
Canis familiaris														
Canis sp.														
Vulpes cf. macrotis														
Canid, indeterminate														
Felis rufus														
Small carnivore														
Medium carnivore													1	.2
Order Artiodactyla														
Odocoileus hemionus													4	.8
Artiodactyl, indeterminate													6	1.2
Small mammal, indeterminate					1	50.0					2	100.0	231	47.6
Medium mammal, indeterminate													21	4.3
Mammal, indeterminate														
TOTAL	0	.0	0	.0	2	100.0	14	100.0	2	100.0	2	100.0	485	100.0

Table D.3. Faunal Remains, Pit Structures, Floor and Roof Fall Contexts

Taxon	Pit Structure 1				Pit Structure 2				Pit Structure 3				Pit Structure 4				Total	
	Floor		Roof		Floor		Roof		Floor		Roof		Floor		Roof			
	N	%	N	%	N	%	N	%	N	%	N	%	N	%	N	%	N	%
Class Reptilia																		
cf. Colubridae																		
Class Aves																		
Large raptor, eagle (?)							1	4.5									1	.1
Meleagris gallopavo			1	.8													1	.1
Grouse																		
Zenaida macroura																		
Thrush																		
Avian, indeterminate	1	1.5	9	7.1	1	.4	1	4.5			1	1.2					13	1.5
Class Mammalia																		
Order Lagomorpha																		
Sylvilagus sp.	7	10.6	3	2.4	16	5.8	2	9.1	2	3.2	3	3.7	61	27.7			94	10.9
Lepus cf. californicus	1	1.5	9	7.1	8	2.9			1	1.6	4	4.9			2	25.0	25	2.9
Leporid, indeterminate			25	19.8	2	.7	1	4.5	2	3.2	5	6.2					35	4.1
Order Rodentia																		
Spermophilus cf. lateralis																		
Spermophilus sp.																		
Cynomys gunnisoni									13	20.6					1	12.5	14	1.6
Sciurid, indeterminate																		
Thomomys bottae																		
Dipodomys ordii					6	2.2	1	4.5	11	17.5	1	1.2					19	2.2
Peromyscus sp.									1	1.6							1	.1
Neotoma albigula																		
Neotoma cf. mexicana																		
Neotoma sp.																		
Cricetid, indeterminate									30	47.6							30	3.5
Rodent, indeterminate	13	19.7															13	1.5
Order Carnivora																		
Canis familiaris																		
Canis sp.											4	4.9					4	.5
Vulpes cf. macrotis							1	4.5									1	.1
Canid, indeterminate			1	.8	1	.4			3	4.8							2	.2
Felis rufus																		
Small carnivore											1	1.2					1	.1
Medium carnivore																		
Order Artiodactyla																		
Odocoileus hemionus	2	3.0	3	2.4	5	1.8					5	6.2	4	1.8	3	37.5	22	2.5
Artiodactyl, indeterminate			1	.8	3	1.1	6	27.3			45	55.6	33	15.0	2	25.0	90	10.4
Small mammal, indeterminate	41	62.1	19	15.1	234	84.5	9	40.9			7	8.6	1	.5			311	36.0
Medium mammal, indeterminate			49	38.9							5	6.2	117	53.2			174	20.2
Mammal, indeterminate	1	1.5	6	4.8	1	.4							4	1.8			12	1.4
TOTAL	66	100.0	126	100.0	277	100.0	22	100.0	63	100.0	81	100.0	220	100.0	8	100.0	863	100.0

NOTE: Includes all animal bone (modified and unmodified) from floor and roof fall contexts. Floor contexts do not include features.

Table D.4. Faunal Remains, Courtyards, Midden, Other

Taxon	Courtyards 1-3				Midden				Other[b]	
	MGS		Other[a]		MGS		Other[a]			
	N	%	N	%	N	%	N	%	N	%
Class Reptilia										
cf. Colubridae							3	.5	5	.2
Class Aves										
Large raptor, eagle (?)									1	.0
Meleagris gallopavo			1	.6			9	1.4	1	.0
Grouse									1	.0
Zenaida macroura									1	.0
Thrush									1	.0
Avian, indeterminate	1	10.0	3	1.9			9	1.4	14	.5
Class Mammalia										
Order Lagomorpha										
Sylvilagus sp.	1	10.0	13	8.2	1	2.9	31	4.7	832	27.2
Lepus cf. californicus	1	10.0	19	11.9	2	5.9	56	8.5	169	5.5
Leporid, indeterminate			21	13.2	5	14.7	90	13.7	104	3.4
Order Rodentia										
Spermophilus cf. lateralis			1	.6			2	.3	4	.1
Spermophilus sp.									5	.2
Cynomys gunnisoni			4	2.5			10	1.5	80	2.6
Sciurid, indeterminate			2	1.3	1	2.9	2	.3	4	.1
Thomomys bottae			6	3.8			3	.5	1	.0
Dipodomys ordii							3	.5	12	.4
Peromyscus sp.							1	.2	12	.4
Neotoma albigula			1	.6			1	.2	22	.7
Neotoma cf. mexicana									1	.0
Neotoma sp.							2	.3	11	.4
Cricetid, indeterminate									5	.2
Rodent, indeterminate			4	2.5	1	2.9	13	2.0	408	13.4
Order Carnivora										
Canis familiaris			6	3.8			8	1.2	21	.7
Canis sp.			4	2.5	1	2.9	18	2.7	7	.2
Vulpes cf. macrotis										
Canid, indeterminate			3	1.9			8	1.2		
Felis rufus							1	.2		
Small carnivore									1	.0
Medium carnivore			1	.6	1	2.9				
Order Artiodactyla										
Odocoileus hemionus	5	50.0	2	1.3	2	5.9	9	1.4	15	.5
Artiodactyl, indeterminate	1	10.0	7	4.4	2	5.9	118	18.0	35	1.1
Small mammal, indeterminate			35	22.0	14	41.2	178	27.1	1,211	39.6
Medium mammal, indeterminate			15	9.4			25	3.8	40	1.2
Mammal, indeterminate	1	10.0	11	6.9	4	11.8	56	8.5	37	1.2
TOTAL	10	100.0	159	100.0	34	100.0	656	100.0	3,061	100.0

NOTE: Includes all animal bone (modified and unmodified).

MGS = modern ground surface.

[a] Includes feature contents.

[b] Includes postabandonment structure fills, structure feature fills, Room 20, arbitrary units, and modern ground surface exclusive of courtyard and midden areas.

Table E.1. Source References for Cranial Measurements and Indices

No.	Name	Source	No.	Name	Source
MEASUREMENTS			CRV31	frontal chord	Brothwell 1981:83
Cranium			CRV32	parietal chord	Brothwell 1981:83
CRV1	maximum length of cranium	Bass 1979:62	CRV33	occipital chord	Brothwell 1981:83
CRV2	maximum breadth of cranium	Bass 1979:62	CRV34	simotic chord	Brothwell 1981:83
CRV3	biasterionic breadth	Howells 1970:120	Mandible		
CRV4	minimum frontal breadth	Bass 1979:67	CRV35	symphyseal height	Bass 1979:72
CRV5	bizygomatic breadth (facial width)	Bass 1979:67	CRV36	minimum breadth ramus	Bass 1979:72
			CRV37	maximum breadth ramus	Giles 1970:108
CRV6	basion-bregma height (maximum height)	Bass 1979:62	CRV38	bigonial breadth	Bass l979:72
			CRV39	bicondylar breadth	Bass 1979:72
CRV7	basion-nasion length	Giles 1970:108	CRV40	body height at M1–2	Giles 1970:108
CRV8	basion-prosthion length	Howells 1970:120	CRV41	body thickness at M2	Giles 1970:108
CRV9	upper facial height	Bass 1979:67	CRV42	interforaminal breadth	Brothwell 1981:83
CRV10	total facial height	Bass 1979:67	CRV43	body (corpal) length	Giles 1970:108
CRV11	nasal height	Bass 1979:68	CRV44	ramus height	Bass 1979:72
CRV12	nasal breadth	Bass 1979:68	CRV45	coronoid height	Brothwell 1981:83
CRV13	orbital height	Bass 1979:69	CRV46	total mandibular length	Brothwell 1981:83
CRV14	orbital breadth (width)	Bass 1979:69	CRV47	gonial angle	Olivier 1969:187
CRV15	see 13		INDICES		
CRV16	see 14		CRV48	cranial index	Bass 1979:63
CRV17	interorbital breadth (bidacryonic chord)	Brothwell 1981:83	CRV49	cranial module	Bass 1979:64
			CRV50	mean height index	Bass 1979:65
CRV18	biorbital breadth	Wilder 1920:54	CRV51	length-height index	Bass 1979:64
CRV19	palate external length	Bass 1979:70	CRV52	breadth-height index	Bass 1979:65
CRV20	palate external breadth	Bass 1979:70	CRV53	upper facial index	Bass 1979:68
CRV21	palate internal length	Bass 1979:71	CRV54	total facial index	Bass 1979:68
CRV22	palate internal breadth	Bass 1979:71	CRV55	nasal index	Bass 1979:69
CRV23	mastoid length	Giles 1970:108	CRV56	orbital index	Bass 1979:69
CRV24	see 23		CRV57	see 56	
CRV25	foramen magnum length	Comas 1960:404	CRV58	external palatal index	Bass 1979:71
CRV26	foramen magnum breadth	Comas 1960:404	CRV59	internal palatal index	Bass 1979:71
CRV27	maximum circumference	Brothwell 1981:83	CRV60	fronto-parietal index	Bass 1979:67
CRV28	frontal arc	Brothwell 1981:83	CRV61	cranio-facial index	Olivier 1969:150
CRV29	parietal arc	Brothwell 1981:83	CRV62	gnathic index	Olivier 1969:158
CRV30	occipital arc	Brothwell 1981:83			

CRV = cranial variant.

Table E.2. Source References for Postcranial Measurements and Indices

No.	Name	Source	No.	Name	Source
MEASUREMENTS			PCRV48	see 39	
PCRV1	humerus maximum length	Bass 1979:114	PCRV49	see 40	
PCRV2	humerus maximum diameter head	Bass 1979:115	PCRV50	see 41	
			PCRV51	see 42	
PCRV3	humerus proximal end breadth	Finnegan 1978:50	PCRV52	see 43	
PCRV4	humerus distal end breadth	Giles 1970:109	PCRV53	fibula maximum length	Bass 1979:187
PCRV5	humerus a-p midshaft diameter	Finnegan 1978:50	PCRV54	see 53	
PCRV6	humerus m-l midshaft diameter	Finnegan 1978:50	PCRV55	tibia maximum length	Bass 1979:187, after Trotter and Gleser 1952:473
PCRV7	see 1				
PCRV8	see 2				
PCRV9	see 3		PCRV56	tibia physiological length[a]	author's definition
PCRV10	see 4				
PCRV11	see 5		PCRV57	tibia max. diam. proximal end	Giles 1970:109
PCRV12	see 6		PCRV58	tibia nutrient foramen a-p diameter	Bass 1979:187
PCRV13	radius maximum length	Bass 1979:124			
PCRV14	radius maximum diameter head	Finnegan 1978:50	PCRV59	tibia nutrient foramen m-l diameter	Bass 1979:187
			PCRV60	see 55	
PCRV15	radius distal end breadth	Giles 1970:109	PCRV61	see 56	
PCRV16	radius a-p midshaft diameter	Martin 1957:536 #5a	PCRV62	see 57	
			PCRV63	see 58	
PCRV17	radius m-l midshaft diameter	Martin 1957:536 #4a	PCRV64	see 59	
			PCRV65	calcaneus maximum length	Olivier 1969:279
PCRV18	see 13		PCRV66	see 65	
PCRV19	see 14		PCRV67	clavicle maximum length	Bass 1979:103
PCRV20	see 15		PCRV68	see 67	
PCRV21	see 16		PCRV69	scapula maximum length (total height)	Bass 1979:94
PCRV22	see 17				
PCRV23	ulna maximum length	Bass 1979:130	PCRV70	scapula maximum breadth	Bass 1979:95
PCRV24	ulna shaft (physiological) length	Bass 1979:130	PCRV71	scapula spine length	Bass 1979:95
			PCRV72	scapula supraspinous line length	Bass 1979:95
PCRV25	ulna distal end breadth	Finnegan 1978:50			
PCRV26	ulna trochlear notch height	Finnegan 1978:50	PCRV73	scapula infraspinous line length	Bass 1979:95
PCRV27	ulna a-p midshaft diameter	Martin 1957:541 #11			
			PCRV74	scapula glenoid cavity length	Giles 1970:109
PCRV28	ulna m-l midshaft diameter	Martin 1957:541 #12	PCRV75	scapula glenoid cavity breadth	Giles 1970:109
			PCRV76	see 69	
PCRV29	see 23		PCRV77	see 70	
PCRV30	see 24		PCRV78	see 71	
PCRV31	see 25		PCRV79	see 72	
PCRV32	see 26		PCRV80	see 73	
PCRV33	see 27		PCRV81	see 74	
PCRV34	see 28		PCRV82	see 75	
PCRV35	femur maximum length	Bass 1979:168	PCRV83	innominate maximum height	Bass 1979:153
PCRV36	femur bicondylar (physiological) length	Bass 1979:168	PCRV84	innominate maximum breadth	Bass 1979:153
			PCRV85	pubis length	Bass 1979:154
PCRV37	femur trochanteric length	Olivier 1969:260	PCRV86	ischium length	Bass 1979:154
PCRV38	femur maximum diameter head	Bass 1979:168	PCRV87	see 83	
			PCRV88	see 84	
PCRV39	femur subtrochanteric a-p diameter	Bass 1979:169	PCRV89	see 85	
			PCRV90	see 86	
PCRV40	femur subtrochanteric m-l diameter	Bass 1979:169	PCRV91	maximum pelvic breadth	Olivier 1969:253
			PCRV92	sagittal diameter pelvic inlet	Olivier 1969:253
PCRV41	femur midshaft a-p diameter	Finnegan 1978:51	PCRV93	transverse diameter pelvic inlet	Olivier 1969:253
PCRV42	femur midshaft m-l diameter	Finnegan 1978:51			
PCRV43	femur epicondylar breadth	Giles 1970:109	PCRV94	transverse diameter pelvic outlet	Finnegan 1978:53
PCRV44	see 35				
PCRV45	see 36		PCRV95	interspinous pelvic breadth[b]	author's definition
PCRV46	see 37				
PCRV47	see 38				

Table E.2. Source References for Postcranial Measurements and Indices *(continued)*

No.	Name	Source	No.	Name	Source
PCRV96	sacrum maximum anterior breadth	Bass 1979:88	PCRV104	robusticity index, femur	Bass 1979:170
			PCRV105	see 104	
PCRV97	sacrum maximum anterior height	Bass 1979:88	PCRV106	radio-humeral index	Bass 1979:115
			PCRV107	see 106	
PCRV98	sacrum S-1 breadth	Martin 1957:524 #19	PCRV108	humero-femoral index	Olivier 1969:262
			PCRV109	see 108	
PCRV99	sacrum S-1 height[c]	author's definition	PCRV110	tibio-femoral index	Olivier 1969:269
			PCRV111	see 110	
INDICES			PCRV112	ischium-pubis index	Bass 1979:154
PCRV100	platymeric index, femur	Bass 1979:169	PCRV113	see 112	
PCRV101	see 100		PCRV114	scapular index	Bass 1979:95
PCRV102	platycnemic index, tibia	Bass 1979:187	PCRV115	see 114	
PCRV103	see 102		PCRV116	sacral index	Bass 1979:88

PCRV = postcranial variant; a-p = anterior-posterior; m-l = medial-lateral; max. = maximum; diam. = diameter.

[a] Tibia physiological length: taken same as tibia maximum length except that the distal measuring point is the distal edge of the talar articular surface rather than the distal end of the lateral malleolus.

[b] Interspinous pelvic breadth: distance between the spinous processes when the pelvis is articulated.

[c] Sacrum S-1 height: taken perpendicular to sacrum S-1 breadth; the antero-posterior distance of the first sacral body segment.

Table E.3. Source References for Cranial Nonmetric Variants

No.	Name	Source	No.	Name	Source
VAR1	palatine torus	Birkby 1973:24	VAR55	see 54	
VAR2	auditory torus	Birkby 1973:24	VAR56	posterior condylar canal (condyloid canal)	Birkby 1973:38
VAR3	see 2				
VAR4	mandibular torus	Birkby 1973:24	VAR57	see 56	
VAR5	see 4		VAR58	hypoglossal canal double	Birkby 1973:38
VAR6	coronal ossicle	Birkby 1973:33	VAR59	see 58	
VAR7	see 6		VAR60	tympanic dehiscence (foramen of Huschke)	Birkby 1973:39
VAR8	bregmatic ossicle	Birkby 1973:33			
VAR9	sagittal ossicle	Finnegan 1972:20	VAR61	see 60	
VAR10	ossicle at lambda	Birkby 1973:33	VAR62	pterygo-spinous foramen (Civinini)	Birkby 1973:39; Chouke 1946:203–204
VAR11	os inca	Birkby 1973:33–34			
VAR12	lambdoidal ossicle	Birkby 1973:33			
VAR13	see 12		VAR63	see 62	
VAR14	Riolan's ossicle	Birkby 1973:34	VAR64	pterygo-alar foramen (Hyrtl)	Birkby 1973:39
VAR15	see 14		VAR65	see 64	
VAR16	asterionic ossicle	Birkby 1973:34	VAR66	foramen spinosum open	Birkby 1973:40
VAR17	see 16		VAR67	see 66	
VAR18	parietal notch bone	Birkby 1973:34	VAR68	canaliculus innominatus	Birkby 1973:40
VAR19	see 18		VAR69	see 68	
VAR20	temporo-squamosal bones	Birkby 1973:34–35	VAR70	foramen ovale incomplete	Birkby 1973:40
VAR21	see 20		VAR71	see 70	
VAR22	epipteric bone	Birkby 1973:35	VAR72	posterior malar foramen	Birkby 1973:40
VAR23	see 22		VAR73	see 72	
VAR24	os japonicum	Birkby 1973:35	VAR74	accessory lesser palatine foramen	Birkby 1973:41
VAR25	see 24				
VAR26	lacrimal foramen	Birkby 1973:35	VAR75	see 74	
VAR27	see 26		VAR76	double mental foramen	Birkby 1973:42
VAR28	posterior ethmoid foramen	Birkby 1973:35	VAR77	see 76	
VAR29	see 28		VAR78	accessory mandibular foramen	Birkby 1973:42
VAR30	anterior ethmoid foramen extra-sutural	Birkby 1973:36	VAR79	see 78	
			VAR80	metopic suture	Birkby 1973:42
VAR31	see 30		VAR81	fronto-temporal articulation	Birkby 1973:43
VAR32	accessory infraorbital foramen	Birkby 1973:36	VAR82	see 81	
VAR33	see 32		VAR83	suture into infraorbital foramen	Birkby 1973:43
VAR34	zygo-facial foramen	Birkby 1973:36			
VAR35	see 34		VAR84	see 83	
VAR36	accessory zygo-facial foramen	Birkby 1973:36	VAR85	petrosquamous suture	Birkby 1973:43
VAR37	see 36		VAR86	see 85	
VAR38	supraorbital foramen	Birkby 1973:36	VAR87	spine of Henle	Birkby 1973:44
VAR39	see 38		VAR88	see 87	
VAR40	supraorbital notch	Birkby 1973:37	VAR89	double condylar facet	Birkby 1973:44
VAR41	see 40		VAR90	see 89	
VAR42	supratrochlear spur	Birkby 1973:37	VAR91	pre-condylar tubercle	Birkby 1973:44
VAR43	see 42		VAR92	see 91	
VAR44	frontal notch	Birkby 1973:37	VAR93	pharyngeal fossa	Birkby 1973:45
VAR45	see 44		VAR94	paramastoid process	Birkby 1973:45
VAR46	frontal foramen	Birkby 1973:37	VAR95	see 94	
VAR47	see 46		VAR96	mylo-hyoid bridge	Birkby 1973:45
VAR48	parietal foramen	Birkby 1973:37	VAR97	see 96	
VAR49	see 48		VAR98	external frontal sulcus	Birkby 1973:43
VAR50	mastoid foramen	Birkby 1973:38	VAR99	see 98	
VAR51	see 50		VAR100	highest nuchal line	Finnegan 1972:15
VAR52	mastoid foramen extra-sutural	Birkby 1973:38	VAR101	see 100	
VAR53	see 52		VAR102	malar tubercle	Finnegan 1972:21
VAR54	zygo-root foramen	Birkby 1973:38	VAR103	see 102	

VAR = variant.

Table F.1. Major Pathologies by Individual Burial

Burial No.	Sex	Location	Age Range	Cribra Orbitalia	Porotic Hyperostosis	Postcranial Periostitis	No. of Caries	No. of Abcesses	Other Notable Conditions
1	male	Pit Structure 1	30–39	absent	absent	absent	3	1	Klippel-Feil syndrome
2	male	Pit Structure 1	30–39	absent	absent	absent	2	1	supernumerary teeth, maxillary sinusitis
3	female	Pit Structure 2	30–39	present	absent	absent	2	0	
4	male	Pit Structure 3	30–39	absent	absent	absent	6	4	healed rib fracture
5	unknown	midden	5–6	present	absent		0	0	
6	male	midden	40–49	absent	absent	present	7	2	exostosis, left femur; fracture, left fifth metatarsal
7	unknown	midden	6–7	present	absent	absent	0	0	
8	male	midden	50+	absent	absent	present	10	6	exceptionally poor oral health
9	unknown	midden	10–11	absent	absent	absent	0	0	
10	female	midden	18–20	present	absent	present	8	1	fused incisors
11	unknown	Pit Structure 4	3–4			absent			
12	female	Pit Structure 4	50+	present	absent	absent	4	4	
13	unknown	midden	12–15		absent	absent	6	0	
14	unknown	Pit Structure 4	5–6		present	absent	0	0	

References

Adams, E. C.

1984 *Preliminary Report of Work Accomplished During 1983 Under State Permit Nos. 83-8 and 83-9.* Crow Canyon Campus, Center for American Archeology, Cortez, Colorado. Report submitted to the Bureau of Land Management, San Juan Resource Area Office, Durango, Colorado.

Adams, K. R.

1980a Pines and Other Conifers. In *Investigations at the Salmon Site: The Structure of Chacoan Society in the Northern Southwest, III,* edited by C. Irwin-Williams and P. H. Shelley, pp. 355–562. Eastern New Mexico University Anthropology Department, Portales.

1980b *Pollen, Parched Seeds and Prehistory: A Pilot Investigation of Prehistoric Plant Remains from Salmon Ruin, a Chacoan Pueblo in Northwestern New Mexico.* Eastern New Mexico University Contributions in Anthropology, no. 9. Portales.

1988 *The Ethnobotany and Phenology of Plants in and Adjacent to Two Riparian Habitats in Southeastern Arizona.* Ph.D. dissertation, University of Arizona. University Microfilms, Ann Arbor.

Adler, M. A.

1989 Ritual Facilities and Social Integration in Nonranked Societies. In *The Architecture of Social Integration in Prehistoric Pueblos,* edited by W. D. Lipe and M. Hegmon, pp. 35–52. Occasional Papers of the Crow Canyon Archaeological Center, no. 1. Cortez, Colorado.

Adler, M. A., and R. H. Wilshusen

1990 Large-Scale Integrative Facilities in Tribal Societies: Cross-Cultural and Southwestern US Examples. *World Archaeology* 22:133–145.

Adovasio, J. M.

1977 *Basketry Technology: A Guide to Identification and Analysis.* Aldine Publishing, Chicago.

Akins, N. J.

1986 *A Biocultural Approach to Human Burials from Chaco Canyon, New Mexico.* Reports of the Chaco Center, no. 9. Branch of Cultural Research, National Park Service, Santa Fe.

Allen, G. M.

1920 *Dogs of the American Aborigines.* Museum of Comparative Zoology Bulletin, no. 63(9):431–517. Harvard University, Cambridge.

Anderson, J. E.

1962 *The Human Skeleton: A Manual for Archaeologists.* National Museum of Canada, Ottawa.

Anderson, S.

1961 *Mammals of Mesa Verde National Park.* Publications of the Museum of Natural History, the University of Kansas, no. 14(3):29–67. Lawrence.

Armelagos, G. J.

1967 Future Work in Paleopathology. In *Miscellaneous Papers in Paleopathology, I,* edited by W. D. Wade, pp. 1–8. Museum of Northern Arizona Technical Series, no. 7. Flagstaff.

Armstrong, D. M.

1972 *Distribution of Mammals in Colorado.* Monograph of the Museum of Natural History, the University of Kansas, no. 3:1–415. Lawrence.

Arrhenius, G., and E. Bonatti

1965 The Mesa Verde Loess. In *Contributions of the Wetherill Mesa Archaeological Project,* assembled by D. Osborne, pp. 92–100. Memoirs of the Society for American Archaeology, no. 19.

Baker, F. C.

1939 *Fieldbook of Illinois Land Snails.* Natural History Survey Division, Urbana, Illinois.

Barefoot, A. D., and F. W. Hankins

1982 *Identification of Modern and Tertiary Woods.* Clarendon Press, Oxford.

Bass, W. M.

1979 *Human Osteology: A Laboratory and Field Manual of the Human Skeleton.* 2d ed. Missouri Archaeological Society, Columbia.

1987 *Human Osteology: A Laboratory and Field Manual.* 3d ed. Missouri Archaeological Society, Columbia.

Bell, W. H., and C. J. King

1944 Methods for the Identification of the Leaf Fibers of Mescal (*Agave*), Yucca (*Yucca*), Beargrass (*Nolina*) and Sotol (*Dasylirion*). *American Antiquity* 10:150–160.

Bennett, K. A.

1973a *The Indians of Point of Pines, Arizona: A Comparative Study of Their Physical Characteristics.* Anthropological Papers of the University of Arizona, no. 23. Tucson.

1973b On the Estimation of Some Demographic Characteristics of a Prehistoric Population from the American Southwest. *American Journal of Physical Anthropology* 39:223–232.

1975 *Skeletal Remains from Mesa Verde National Park, Colorado.* Publications in Archeology, no. 7F. National Park Service, Washington, D.C.

Benz, B. F.

1984 Appendix B: Biotic Remains. In *Dolores Archaeological Program: Synthetic Report, 1978–1981,* prepared by D. A. Breternitz, pp. 199–214. Bureau of Reclamation, Engineering and Research Center, Denver.

Birkby, W. H.

1973 *Discontinuous Morphological Traits of the Skull as Population Markers in the Prehistoric Southwest.* Ph.D. dissertation, University of Arizona. University Microfilms, Ann Arbor.

Birkett, D. A.

1983 Non-Specific Infections. In *Disease in Ancient Man: An International Symposium,* edited by G. D. Hart, pp. 99–105. Clarke Irwin, Toronto.

Bissell, S. J., and M. B. Dillon (editors)

1982 *Colorado Mammal Distribution Latilong Study.* Colorado Division of Wildlife, Denver.

Blinman, E.

1984 Dating with Neckbands: Calibration of Temporal Variation in Moccasin Gray and Mancos Gray Ceramic Types. In *Dolores Archaeological Program: Synthetic Report, 1978–1981,* prepared by D. A. Breternitz, pp. 128–138. Bureau of Reclamation, Engineering and Research Center, Denver.

1986a Additive Technologies Group Final Report. In *Dolores Archaeological Program: Final Synthetic Report,* compiled by D. A. Breternitz, C. K. Robinson, and G. T. Gross, pp. 53–101. Bureau of Reclamation, Engineering and Research Center, Denver.

1986b Exchange and Interaction in the Dolores Area. In *Dolores Archaeological Program: Final Synthetic Report,* compiled by D. A. Breternitz, C. K. Robinson, and G. T. Gross, pp. 663–701. Bureau of Reclamation, Engineering and Research Center, Denver.

1988a Ceramic Vessels and Vessel Assemblages in Dolores Archaeological Program Collections. In *Dolores Archaeological Program: Supporting Studies: Additive and Reductive Technologies,* compiled by E. Blinman, C. J. Phagan, and R. H. Wilshusen, pp. 449–482. Bureau of Reclamation, Engineering and Research Center, Denver.

1988b *The Interpretation of Ceramic Variability: A Case Study from the Dolores Anasazi.* Ph.D. dissertation, Washington State University. University Microfilms, Ann Arbor.

Bohrer, V. L.

1964 A Navajo Sweathouse. *Plateau* 36:95–99.

1975 The Prehistoric and Historic Role of the Cool-Season Grasses in the Southwest. *Economic Botany* 29:199–207.

1978 Plants That Have Become Locally Extinct in the Southwest. *New Mexico Journal of Science* 18(2):10–19.

1983 New Life from Ashes: The Tale of the Burnt Bush (*Rhus trilobata*). *Desert Plants* 5(3):122–124.

1986 Guideposts in Ethnobotany. *Journal of Ethnobiology* 6:27–43.

Bohrer, V. L., and K. R. Adams

1977 *Ethnobotanical Techniques and Approaches at Salmon Ruin, New Mexico.* Eastern New Mexico University Contributions in Anthropology, vol. 8, no. 1. Portales.

Borland International

1987 *Quattro User's Guide.* Scotts Valley, California.

Boucher, B. J.

1955 Sex Differences in the Foetal Sciatic Notch. *Journal of Forensic Medicine* 2:51–54.

1957 Sex Differences in the Foetal Pelvis. *American Journal of Physical Anthropology* 15:581–600.

Bradley, B. A.

1992 1984–1989 Excavations at Sand Canyon Pueblo: Material Culture, Subsistence, and Human Remains. Occasional Papers of the Crow Canyon Archaeological Center, Cortez, Colorado, in preparation.

Breternitz, C. D., and J. N. Morris

1988 Excavations in Area 3. In *Dolores Archaeological Program: Anasazi Communities at Dolores: Grass Mesa Village,* compiled by W. D. Lipe, J. N. Morris, and T. A. Kohler, pp. 317–439. Bureau of Reclamation, Engineering and Research Center, Denver.

Breternitz, D. A., C. K. Robinson, and G. T. Gross (compilers)

1986 *Dolores Archaeological Program: Final Synthetic Report.* Bureau of Reclamation, Engineering and Research Center, Denver.

Breternitz, D. A., A. H. Rohn, Jr., and E. A. Morris (compilers)

1974 *Prehistoric Ceramics of the Mesa Verde Region.* Museum of Northern Arizona, Ceramic Series, no. 5. Flagstaff, Arizona.

Brew, J. O.

1946 *Archaeology of Alkali Ridge, Southeastern Utah.* Papers of the Peabody Museum of Archaeology and Ethnology, vol. 21. Harvard University, Cambridge.

Brisbin, J. M.

1986 Excavations at Windy Wheat Hamlet (Site 5MT4644), a Pueblo I Habitation. In *Dolores Archaeological Program: Anasazi Communities at Dolores: Early Anasazi Sites in the Sagehen Flats Area,* compiled by A. E. Kane and G. T. Gross, pp. 638–864. Bureau of Reclamation, Engineering and Research Center, Denver.

Brisbin, J. M., A. M. Emerson, and S. H. Schlanger

1986 Excavations at Dos Casas Hamlet (Site 5MT2193), a Basketmaker III/Pueblo I Habitation Site. In *Dolores Archaeological Program: Anasazi Communities at Dolores: Early Anasazi Sites in the Sagehen Flats Area,* compiled by A. E. Kane and G. T. Gross, pp. 548–635. Bureau of Reclamation, Engineering and Research Center, Denver.

Brisbin, J. M., A. E. Kane, and J. N. Morris

1988 Excavations at McPhee Pueblo (Site 5MT4475), a Pueblo I and Early Pueblo II Multicomponent Village. In *Dolores Archaeological Program: Anasazi Communities at Dolores: McPhee Village,* compiled by A. E. Kane and C. K. Robinson, pp. 62–403. Bureau of Reclamation, Engineering and Research Center, Denver.

Brothwell, D.

1981 *Digging up Bones.* 3d ed. Cornell University Press, Ithaca.

Brown, D. E.

1982 Great Basin Conifer Woodland. In Biotic Communities of the American Southwest—United States and Mexico, edited by D. E. Brown. *Desert Plants* 4(1–4):52–57.

Buikstra, J. E.

1976 The Caribou Eskimo: General and Specific Disease. *American Journal of Physical Anthropology* 45:351–368.

Bullard, W. R., Jr.

1962 *The Cerro Colorado Site and Pithouse Architecture in the Southwestern United States Prior to A.D. 900.* Papers of the Peabody Museum of Archaeology and Ethnology, vol. 44, no. 2. Harvard University, Cambridge.

Bullock, M.

1991 It's a Wearing Process: Preliminary Observations on an Awl Use-Wear Replication Study. Paper presented at the 56th Annual Meeting of the Society for American Archaeology, New Orleans.

Burns, B. T.

1983 *Simulated Anasazi Storage Behavior Using Crop Yields Reconstructed from Tree-Rings: A.D. 652–1968.* Ph.D. dissertation, University of Arizona. University Microfilms, Ann Arbor.

Bye, R. A., Jr.

1972 Ethnobotany of the Southern Paiute Indians in the 1870's: With a Note on the Early Ethnobotanical Contributions of Dr. Edward Palmer. In *Great Basin Cultural Ecology, a Symposium,* edited by D. D. Fowler, pp. 87–104. Desert Research Institute Publications in the Social Sciences, no. 8. Reno, Nevada.

Carlson, D. S., G. J. Armelagos, and D. P. van Gerven

1974 Factors Influencing the Etiology of Cribra Orbitalia in Prehistoric Nubia. *Journal of Human Evolution* 3:405–410.

Castetter, E. F.

1935 *Ethnobiological Studies in the American Southwest, I: Uncultivated Native Plants Used as Sources of Food.* The University of New Mexico Bulletin, Biological Series, vol. 4, no. 1. Albuquerque.

Cattanach, G. S., Jr.

1980 *Long House, Mesa Verde National Park, Colorado.* Publications in Archeology, no. 7H. National Park Service, Washington, D.C.

Center for American Archeology

1983a Duckfoot. *Center for American Archeology Newsletter,* vol. 1, no. 2. Crow Canyon Archaeological Center, Cortez, Colorado.

1983b Duckfoot Site Past, Present and Future: An Excavation Update. *Center for American Archeology Newsletter,* vol. 1, no. 1. Crow Canyon Archaeological Center, Cortez, Colorado.

Chase, C. A., III, S. J. Bissell, H. E. Kingery, and W. D. Graul (editors)

1982 *Colorado Bird Distribution Latilong Study.* Colorado Division of Wildlife, Denver.

Chouke, K. S.

1946 On the Incidence of the Foramen of Civinini and the Porus Crotaphitico-Buccinatorius in American Whites and Negroes, I: Observations on 1544 Skulls. *American Journal of Physical Anthropology* 4:203–225.

Christenson, A. L.

1987 Projectile Points: Eight Millennia of Projectile Change on the Colorado Plateau. In *Prehistoric Stone Technology on Northern Black Mesa, Arizona,* authored by W. J. Parry and A. L. Christenson, pp. 143–198. Occasional Papers of the Center for Archaeological Investigations, Southern Illinois University, no. 12. Carbondale.

Clark, C., T. W. Canaday, and K. L. Petersen

1987 Domestic Dog in the Dolores Archaeological Program Faunal Assemblage. In *Dolores Archaeological Program: Supporting Studies: Settlement and Environment,* compiled by K. L. Petersen and J. D. Orcutt, pp. 276–288. Bureau of Reclamation, Engineering and Research Center, Denver.

Clary, K.

1988 Towards Modeling Prehistoric Subsistence Change by Reducing Archaeological Ambiguity. Paper presented at the 53rd Annual Meeting of the Society for American Archaeology, Phoenix.

Clay, V. L., T. W. Canaday, and S. W. Neusius

1987 A Partial Musk Ox Skeleton from Grass Mesa. In *Dolores Archaeological Program: Supporting Studies: Settlement and Environment,* compiled by K. L. Petersen and J. D. Orcutt, pp. 290–334. Bureau of Reclamation, Engineering and Research Center, Denver.

Cockburn, E. (editor)

1977 *Porotic Hyperostosis: An Enquiry.* Paleopathology Association Monograph, no. 2. Paleopathology Association, Detroit.

Colton, H. S.

1970 The Aboriginal Southwestern Indian Dog. *American Antiquity* 35:153–159.

Comas, J.

1960 *Manual of Physical Anthropology.* Rev. English ed. Charles C. Thomas, Springfield, Illinois.

Cordell, L. S., S. Upham, and S. L. Brock

1987 Obscuring Cultural Patterns in the Archaeological Record: A Discussion from Southwestern Archaeology. *American Antiquity* 52:565–577.

Cornely, J. E., and R. J. Baker

1986 *Neotoma mexicana. Mammalian Species* 262:1–7.

Costin, C. L.

1991 Craft Specialization: Issues in Defining, Documenting, and Explaining the Organization of Production. In *Archaeological Method and Theory,* vol. 3, edited by M. B. Schiffer, pp. 1–56. University of Arizona Press, Tucson.

Crabtree, D. E.

1972 *An Introduction to Flintworking.* Occasional Papers of the Idaho State University Museum, no. 28. Pocatello.

Craig, D. B.

1982 Shell Exchange Along the Middle Santa Cruz Valley During the Hohokam Pre-Classic. Paper presented at the Tucson Basin Conference, Tucson.

Culin, S.

1975 *Games of the North American Indians.* Reprinted. Dover Publications, New York. Originally published 1907, U.S. Government Printing Office, Washington, D.C.

Dale, A.

1968 *Comparative Wood Anatomy of Some Shrubs Native to the Northern Rocky Mountains.* Research Paper, no. INT-45. U.S. Forest Service, Ogden, Utah.

Deal, M.

1985 Household Pottery Disposal in the Maya Highlands: An Ethnoarchaeological Interpretation. *Journal of Anthropological Archaeology* 4:243–291.

Dittrick, J., and J. M. Suchey

1986 Sex Determination of Prehistoric Central California Skeletal Remains Using Discriminant Analysis of the Femur and Humerus. *American Journal of Physical Anthropology* 70:3–9.

Doebley, J. F.

1984 "Seeds" of Wild Grasses: A Major Food of Southwestern Indians. *Economic Botany* 38:52–64.

Doebley, J. F., and V. L. Bohrer

1983 Maize Variability and Cultural Selection at Salmon Ruin, New Mexico. *The Kiva* 49:19–37.

Drake, R. J.

1959 Nonmarine Molluscan Remains from Recent Sediments in Matty Canyon, Pima County, Arizona. *Southern California Academy of Sciences Bulletin* 58(3):146–154.

Eddy, F. W., A. E. Kane, and P. R. Nickens

1984 *Southwest Colorado Prehistoric Context: Archaeological Background and Research Directions.* Office of Archaeology and Historic Preservation, Colorado Historical Society, Denver.

Ekren, E. B., and F. N. Houser

1965 *Geology and Petrology of the Ute Mountains Area, Colorado.* Geological Survey Professional Paper, no. 481. U.S. Government Printing Office, Washington, D.C.

Elmore, F. H.

1943 *Ethnobotany of the Navajo.* The University of New Mexico Bulletin with the School of American Research Monograph Series, vol. 1, no. 7. University of New Mexico, Albuquerque.

El-Najjar, M. Y., B. Lozoff, and D. J. Ryan

1975 The Paleoepidemiology of Porotic Hyperostosis in the American Southwest: Radiological and Ecological Considerations. *American Journal of Roentgenology, Radium Therapy and Nuclear Medicine* 125:918–924.

El-Najjar, M. Y., D. J. Ryan, C. G. Turner II, and B. Lozoff

1976 The Etiology of Porotic Hyperostosis Among the Prehistoric and Historic Anasazi Indians of Southwestern United States. *American Journal of Physical Anthropology* 44:477–488.

Erdman, J. A., C. L. Douglas, and J. W. Marr

1969 *Environment of Mesa Verde, Colorado.* Archeological Research Series, no. 7-B. National Park Service, Washington, D.C.

Finnegan, M. J.

1972 *Population Definition on the Northwest Coast by Analysis of Discrete Character Variation.* Ph.D. dissertation, University of Colorado. University Microfilms, Ann Arbor.

1978 *Guide to Osteological Analysis.* 2d ed. Kansas State University, Osteology Laboratory, Manhattan.

Flinn, L., C. G. Turner II, and A. Brew

1976 Additional Evidence for Cannibalism in the Southwest: The Case of LA 4528. *American Antiquity* 41:308–318.

Flint, P. R., and S. W. Neusius

1987 Cottontail Procurement Among Dolores Anasazi. In *Dolores Archaeological Program: Supporting Studies: Settlement and Environment,* compiled by K. L. Petersen and J. D. Orcutt, pp. 256–273. Bureau of Reclamation, Engineering and Research Center, Denver.

Franciscan Fathers

1910 *An Ethnologic Dictionary of the Navaho Language.* Franciscan Fathers, Saint Michaels, Arizona.

Friedman, J.

1978 *Wood Identification by Microscopic Examination: A Guide for the Archaeologist on the Northwest Coast of North America.* Heritage Record no. 5. British Columbia Provincial Museum, Victoria.

Garrett, E. M.

1990 Petrographic Analysis of Selected Ceramics from the Northern Southwest. Ms. in possession of author, Department of Sociology and Anthropology, New Mexico State University, Las Cruces.

Genoves, S. C.

1967 Proportionality of Long Bones and Their Relation to Stature Among Mesoamericans. *American Journal of Physical Anthropology* 26:67–78.

Gilbert, B. M., and T. W. McKern

1973 A Method for Aging the Female *Os Pubis. American Journal of Physical Anthropology* 38:31–38.

Giles, E.

1970 Discriminant Function Sexing of the Human Skeleton. In *Personal Identification in Mass Disasters,* edited by T. D. Stewart, pp. 99–109. Smithsonian Institution, Washington, D.C.

Gillespie, W. B.

1976 *Culture Change at the Ute Canyon Site: A Study of the Pithouse-Kiva Transition in the Mesa Verde Region.* Unpublished Master's thesis, Department of Anthropology, University of Colorado, Boulder.

Gilman, P. A.

1987 Architecture as Artifact: Pit Structures and Pueblos in the American Southwest. *American Antiquity* 52:538–564.

Gish, J. W.

1982 Palynological Results of the Coronado Project. In *The Coronado Project Archaeological Investigations, the Specialists' Volume: Biocultural Analyses,* compiled by R. E. Gasser, pp. 96–224. Museum of Northern Arizona Research Paper, no. 23. Flagstaff.

Glennie, G. D.

1983 *Replication of an A.D. 800 Anasazi Pithouse in Southwestern Colorado.* Unpublished Master's thesis, Department of Anthropology, Washington State University, Pullman.

Goaz, P. W., and S. C. White

1982 *Oral Radiology: Principles and Interpretation.* C. V. Mosby, St. Louis.

Goldstein, M. S.

1963 Some Vital Statistics Based on Skeletal Material. *Human Biology* 35:3–12.

Goodman, A. H., D. L. Martin, G. J. Armelagos, and G. Clark

1984 Indication of Stress from Bone and Teeth. In *Paleopathology at the Origins of Agriculture,* edited by M. N. Cohen and G. J. Armelagos, pp. 13–49. Academic Press, Orlando.

Griffitts, E. A.

1987 Analysis of the *Phaseolus* Remains from the Dolores Project Area. In *Dolores Archaeological Program: Supporting Studies: Settlement and Environment,* compiled by K. L. Petersen and J. D. Orcutt, pp. 248–253. Bureau of Reclamation, Engineering and Research Center, Denver.

Gross, G. T.

1985 Unpublished analysis forms. On file, Anasazi Heritage Center, Dolores, Colorado.

1987 *Anasazi Storage Facilities in the Dolores Region: A.D. 600–920.* Ph.D. dissertation, Washington State University. University Microfilms, Ann Arbor.

Guthe, C. E.

1925 *Pueblo Pottery Making: A Study at the Village of San Ildefonso.* Papers of the Phillips Academy Southwestern Expedition, no. 2. Yale University Press, New Haven.

Hall, E. R.

1981 *The Mammals of North America.* 2 vols. John Wiley, New York.

Hally, D. J.

1981 Plant Preservation and the Content of Paleobotanical Samples: A Case Study. *American Antiquity* 46:723–742.

Hammerson, G. A., and D. Langlois (editors)

1981 *Colorado Reptile and Amphibian Distribution Latilong Study.* Colorado Division of Wildlife, Denver.

Harrington, H. D.
1964 *Manual of the Plants of Colorado.* The Swallow Press, Chicago.

Hart, G. D. (editor)
1983 *Disease in Ancient Man: An International Symposium.* Clarke Irwin, Toronto.

Hassan, F. A.
1981 *Demographic Archaeology.* Academic Press, New York.

Haury, E. W.
1937 Shell. In *Excavations at Snaketown: Material Culture,* authored by H. S. Gladwin, E. W. Haury, E. B. Sayles, and N. Gladwin, pp. 135–153. Gila Pueblo Medallion Papers, no. 25. Globe, Arizona.
1976 *The Hohokam, Desert Farmers and Craftsmen: Excavations at Snaketown, 1964–1965.* University of Arizona Press, Tucson.

Hayes, A. C., and J. A. Lancaster
1975 *Badger House Community, Mesa Verde National Park.* Publications in Archeology, no. 7E. National Park Service, Washington, D.C.

Haynes, D. D., J. D. Vogel, and D. G. Wyant (compilers)
1976 *Geology, Structure, and Uranium Deposits of the Cortez Quadrangle, Colorado and Utah.* U.S. Geological Survey Miscellaneous Investigations. Reprinted. U.S. Geological Survey. Originally published 1972, U.S. Geological Survey.

Hegmon, M.
1991 *Pueblo I Ceramic Production in Southwestern Colorado: Analyses of Igneous Rock Temper.* Report on research funded by the Karen S. Greiner Endowment for Colorado Archaeology, Colorado State University. Ms. on file, Crow Canyon Archaeological Center, Cortez, Colorado.
1992a Boundary-Making Strategies in Early Pueblo Societies: Style and Architecture in the Kayenta and Mesa Verde Regions. In *The Ancient Southwestern Community: Models and Methods for the Study of Prehistoric Social Organization,* edited by W. H. Wills and R. D. Leonard. University of New Mexico Press, Albuquerque, in press.
1992b *Style as a Social Strategy in the American Southwest.* Occasional Papers of the Crow Canyon Archaeological Center, Cortez, Colorado, in press.

Hegmon, M., and J. R. Allison
1990 The Local Economy and Regional Exchange: Early Red Ware Production in the Northern Southwest. Proposal approved for Neutron Activation Analysis at the University of Missouri Research Reactor, Columbia. Ms. on file, University of Missouri, Columbia.

Hegmon, M., W. Hurst, and J. R. Allison
1991 Production for Local Consumption and Exchange: Comparisons of Early Red and White Ware Ceramics in the San Juan Region. Paper presented at the 56th Annual Meeting of the Society for American Archaeology, New Orleans.

Hengen, O. P.
1971 Cribra Orbitalia: Pathogenesis and Probable Etiology. *Homo* 22:57–75.

Hibbard, C. W.
1958 *Summary of North American Pleistocene Mammalian Vertebrate Local Faunas.* Papers of the Michigan Academy of Science, Arts and Letters, no. 43:3–32. Ann Arbor.

Hinkes, M. J.
1983 *Skeletal Evidence of Stress in Subadults: Trying to Come of Age at Grasshopper Pueblo.* Unpublished Ph.D. dissertation, Department of Anthropology, University of Arizona, Tucson.

Hoffman, J. M.
1976 Studies in California Paleopathology, II: Comminuted Fracture of a Humerus with Pseudoarthrosis Formation. *Contributions of the University of California Archaeological Research Facility* 30:25–39. Berkeley.
1979 Age Estimations from Diaphyseal Lengths: Two Months to Twelve Years. *Journal of Forensic Sciences* 24:461–469.

[Honeycutt, L., and J. Fetterman]
1991 *Indian Camp Ranch at Cortez: Archaeology at Work, Preserving the Anasazi Legacy for the Future.* Indian Camp Ranch Report, no. 1. Archie and Mary Hanson, Templeton, California.

Howard, A. V.
1987 The La Cuidad Shell Assemblage. In *Specialized Studies in the Economy, Environment and Culture of La Cuidad,* edited by J. E. Kisselburg, G. E. Rice, and B. L. Shears, pp. 75–174. Office of Cultural Resource Management, Arizona State University, Tempe.

Howells, W. W.
1970 Multivariate Analysis for the Identification of Race from Crania. In *Personal Identification in Mass Disasters,* edited by T. D. Stewart, pp. 111–121. Smithsonian Institution, Washington, D.C.

Hunt, E. F., and I. Gleiser
1955 The Estimation of Age and Sex of Preadolescent Children from Bone and Teeth. *American Journal of Physical Anthropology* 13:479–487.

Hurst, W.
1983 The Prehistoric Peoples of San Juan County, Utah. In *San Juan County, Utah,* edited by A. K. Powell, pp. 17–44. Utah State Historical Society, Salt Lake City.

Hurst, W., M. Bond, and S. E. Schwindt
1985 Type Description: Piedra Black-on-white, White Mesa Variety. *Pottery Southwest* 12(3).

Irwin-Williams, C.
1973 *The Oshara Tradition: Origins of Anasazi Culture.* Eastern New Mexico University Contributions in Anthropology, vol. 5, no. 1. Portales.

Iscan, M. Y., and S. R. Loth
1986 Determination of Age from the Sternal Rib in White Males: A Test of the Phase Method. *Journal of Forensic Sciences* 31:122–131.

Iscan, M. Y., S. R. Loth, and R. K. Wright
1985 Age Estimation from the Rib by Phase Analysis: White Females. *Journal of Forensic Sciences* 30:853–863.

Isely, D.
1947 *Investigations in Seed Classification by Family Characteristics.* Research Bulletin, no. 351. Agricultural Experiment Station, Iowa State College of Agriculture and Mechanic Arts. Ames.

Jernigan, E. W.
1978 *Jewelry of the Prehistoric Southwest.* School of American Research and University of New Mexico Press, Santa Fe.

Jones, J. K., D. C. Carter, H. H. Genoways, R. S. Hoffmann, and D. W. Rice
1982   *Revised Checklist of North American Mammals North of Mexico, 1982.* Occasional Papers of the Museum, Texas Tech University, no. 80:1–22. Lubbock.

Jones, V. H.
1931   *The Ethnobotany of the Isleta Indians.* Unpublished Master's thesis, Department of Biology, University of New Mexico, Albuquerque.

Judd, N. M.
1981   *The Material Culture of Pueblo Bonito.* Reprinted. J & L Reprint, Lincoln, Nebraska. Originally published 1954, Smithsonian Institution, Washington, D.C.

Kane, A. E.
1986   Prehistory of the Dolores River Valley. In *Dolores Archaeological Program: Final Synthetic Report,* compiled by D. A. Breternitz, C. K. Robinson, and G. T. Gross, pp. 353–435. Bureau of Reclamation, Engineering and Research Center, Denver.

Kane, A. E., and C. K. Robinson (compilers)
1988   *Dolores Archaeological Program: Anasazi Communities at Dolores: McPhee Village.* Bureau of Reclamation, Engineering and Research Center, Denver.

Kaplan, L.
1956   The Cultivated Beans of the Prehistoric Southwest. *Annals of the Missouri Botanical Garden* 43:189–251.

Keane, S. P., and V. L. Clay
1987   Geological Sources of Unusual Minerals and Rocks of the Dolores Project Area. In *Dolores Archaeological Program: Supporting Studies: Settlement and Environment,* compiled by K. L. Petersen and J. D. Orcutt, pp. 506–546. Bureau of Reclamation, Engineering and Research Center, Denver.

Kearney, T. H., and R. H. Peebles
1960   *Arizona Flora.* University of California Press, Berkeley.

Keen, A. M.
1963   *Marine Molluscan Genera of Western North America: An Illustrated Key.* Stanford University Press, Stanford.

1971   *Sea Shells of Tropical West America: Marine Mollusks from Baja California to Peru.* 2d ed. Stanford University Press, Stanford.

Keen, J. A.
1950   Sex Differences in Skulls. *American Journal of Physical Anthropology* 8:65–79.

Kidder, A. V.
1932   *The Artifacts of Pecos.* Yale University Press, New Haven.

Kirk, D. R.
1970   *Wild Edible Plants of the Western United States.* Naturegraph Publishers, Healdsburg, California.

Kohler, T. A., and M. H. Matthews
1988   Long-Term Anasazi Land Use and Forest Reduction: A Case Study from Southwest Colorado. *American Antiquity* 53:537–564.

Krogman, W. M.
1962   *The Human Skeleton in Forensic Medicine.* Charles C. Thomas, Springfield, Illinois.

Krogman, W. M., and M. Y. Iscan
1986   *The Human Skeleton in Forensic Medicine.* 2d ed. Charles C. Thomas, Springfield, Illinois.

Kurtén, B., and E. Anderson
1980   *Pleistocene Mammals of North America.* Columbia University Press, New York.

Lancaster, J. A., J. M. Pinkley, P. F. Van Cleave, and D. Watson
1954   *Archeological Excavations in Mesa Verde National Park, Colorado, 1950.* Archeological Research Series, no. 2. National Park Service, Washington, D.C.

Lehr, J. H.
1978   *A Catalogue of the Flora of Arizona.* Desert Botanical Garden, Phoenix.

Lekson, S. H.
1986   *Great Pueblo Architecture of Chaco Canyon, New Mexico.* Reprinted. University of New Mexico Press, Albuquerque. Originally published 1984, National Park Service, Albuquerque.

Lentz, D. L.
1984   Utah Juniper (*Juniperus osteosperma*) Cones and Seeds from Salmon Ruin, New Mexico. *Journal of Ethnobiology* 4:191–200.

Leonhardy, F. C., and V. L. Clay
1985   Bedrock Geology, Quaternary Stratigraphy, and Geomorphology. In *Dolores Archaeological Program: Studies in Environmental Archaeology,* compiled by K. L. Petersen, V. L. Clay, M. H. Matthews, and S. W. Neusius, pp. 131–138. Bureau of Reclamation, Engineering and Research Center, Denver.

Lightfoot, R. R.
1985   *Report of 1984 Archeological Investigations Conducted Under State Permit No. 84-15.* Crow Canyon Center for Southwestern Archeology, Cortez, Colorado. Report submitted to the Office of the State Archaeologist, Colorado Historical Society, Denver.

1987   *Annual Report of Investigations at the Duckfoot Site (5MT3868), Montezuma County, Colorado.* Crow Canyon Archaeological Center, Cortez, Colorado. Report submitted to the Office of the State Archaeologist, Colorado Historical Society, Denver.

1988   Roofing an Early Anasazi Great Kiva: Analysis of an Architectural Model. *The Kiva* 53:253–272.

1992a   *Archaeology of the House and Household: A Case Study of Assemblage Formation and Household Organization in the American Southwest.* Ph.D. dissertation, Washington State University. University Microfilms, Ann Arbor.

1992b   Architecture and Tree-Ring Dating at the Duckfoot Site in Southwestern Colorado. *Kiva* 57:213–236.

Lightfoot, R. R., and B. A. Bradley
1986   *Field Manual.* Crow Canyon Archaeological Center, Cortez, Colorado.

Lightfoot, R. R., A. M. Emerson, and E. Blinman
1988   Excavations in Area 5, Grass Mesa Village (Site 5MT23). In *Dolores Archaeological Program: Anasazi Communities at Dolores: Grass Mesa Village,* compiled by W. D. Lipe, J. N. Morris, and T. A. Kohler, pp. 561–766. Bureau of Reclamation, Engineering and Research Center, Denver.

Lightfoot, R. R., and C. R. Van West
1986   *Excavation and Testing Conducted on Private Land Under State Permit No. 85-22.* Crow Canyon Archaeological Center, Cortez, Colorado. Report submitted to the

Office of the State Archaeologist, Colorado Historical Society, Denver.

Lightfoot, R. R., and M. D. Varien

1988 *Report of 1987 Archaeological Investigations at the Duckfoot Site (5MT3868), Montezuma County, Colorado.* Crow Canyon Archaeological Center, Cortez, Colorado. Report submitted to the Bureau of Land Management, San Juan Resource Area Office, Durango, Colorado.

Lipe, W. D., J. N. Morris, and T. A. Kohler (compilers)

1988 *Dolores Archaeological Program: Anasazi Communities at Dolores: Grass Mesa Village.* Bureau of Reclamation, Engineering and Research Center, Denver.

Lovejoy, C. O., R. S. Meindl, T. R. Pryzbeck, and R. P. Mensforth

1985 Chronological Metamorphosis of the Auricular Surface of the Ilium: A New Method for the Determination of Adult Skeletal Age at Death. *American Journal of Physical Anthropology* 68:15–28.

Lowe, C. H., and D. E. Brown

1982 Introduction. In Biotic Communities of the American Southwest—United States and Mexico, edited by D. E. Brown. *Desert Plants* 4(1–4):8–16.

Lucius, W., and D. A. Breternitz

1981 The Current Status of Red Wares in the Mesa Verde Region. In *Collected Papers in Honor of Erik Kellerman Reed,* edited by A. H. Schroeder, pp. 99–111. Papers of the Archaeological Society of New Mexico, Albuquerque.

Martin, P. S.

1936 *Lowry Ruin in Southwestern Colorado* (with reports by Lawrence Roys and Gerhardt von Bonin). Field Museum of Natural History, Anthropological Series, vol. 23, no. 1. Chicago.

1938 *Archaeological Work in the Ackmen-Lowry Area, Southwestern Colorado, 1937.* Field Museum of Natural History, Anthropological Series, vol. 23, no. 2. Chicago.

1939 *Modified Basket Maker Sites: Ackmen-Lowry Area, Southwestern Colorado, 1938.* Field Museum of Natural History, Anthropological Series, vol. 23, no. 3. Chicago.

Martin, R.

1957 *Lehrbuch der Anthropologie,* vol. 4. 3d ed. K. Saller, editor. Gustav Fischer Verlag, Stuttgart.

Matthews, M. H.

1985 Botanical Studies: Nature and Status of the Data Base. In *Dolores Archaeological Program: Studies in Environmental Archaeology,* compiled by K. L. Petersen, V. L. Clay, M. H. Matthews, and S. W. Neusius, pp. 41–60. Bureau of Reclamation, Engineering and Research Center, Denver.

1986 The Dolores Archaeological Program Macrobotanical Data Base: Resource Availability and Mix. In *Dolores Archaeological Program: Final Synthetic Report,* compiled by D. A. Breternitz, C. K. Robinson, and G. T. Gross, pp. 151–184. Bureau of Reclamation, Engineering and Research Center, Denver.

Matthews, W.

1886 Navajo Names for Plants. *American Naturalist* 20:767–777.

McDougal, W. B.

1973 *Seed Plants of Northern Arizona.* Museum of Northern Arizona, Flagstaff.

McGrew, J. C.

1979 *Vulpes macrotis. Mammalian Species* 123:1–6.

McGuire, R. H., and M. B. Schiffer

1982 *Hohokam and Patayan: Prehistory of Southwestern Arizona.* Academic Press, New York.

McKern, T. W., and T. D. Stewart

1957 *Skeletal Age Changes in Young American Males: Analysed from the Standpoint of Age Identification.* Technical Report, no. EP-45. Headquarters Quartermaster Research and Development Command, Natick, Massachusetts.

McKusick, C. R.

1986 *Southwest Indian Turkeys: Prehistory and Comparative Osteology.* Southwest Bird Laboratory, Globe, Arizona.

Meindl, R. S., and C. O. Lovejoy

1985 Ectocranial Suture Closure: A Revised Method for the Determination of Skeletal Age at Death Based on the Lateral-Anterior Sutures. *American Journal of Physical Anthropology* 68:57–66.

Mensforth, R. P., C. O. Lovejoy, J. W. Lallo, and G. J. Armelagos

1978 The Role of Constitutional Factors, Diet, and Infectious Disease in the Etiology of Porotic Hyperostosis and Periosteal Reactions in Prehistoric Infants and Children. *Medical Anthropology* 2(1), Part 2:1–59.

Miles, J. S.

1975 *Orthopedic Problems of the Mesa Verde Populations.* Publications in Archeology, no. 7G. National Park Service, Washington, D.C.

Mills, P. R.

1987 *Use-Wear Analysis of Stone Axes from Sand Canyon Pueblo Ruin (5MT765), Southwestern Colorado.* Unpublished Master's thesis, Department of Anthropology, Washington State University, Pullman.

Mindeleff, V.

1989 *A Study of Pueblo Architecture in Tusayan and Cibola.* Reprinted. Smithsonian Institution Press, Washington, D.C. Originally published 1891, U.S. Government Printing Office, Washington, D.C.

Minnis, P. E.

1987 Identification of Wood from Archaeological Sites in the American Southwest, I: Keys for Gymnosperms. *Journal of Archaeological Science* 14:121–131.

Moore, J. A., A. C. Swedlund, and G. J. Armelagos

1975 The Use of Life Tables in Paleodemography. In *Population Studies in Archaeology and Biological Anthropology: A Symposium,* edited by A. C. Swedlund, pp. 57–70. Memoirs of the Society for American Archaeology, no. 30.

Morris, D. P.

1986 *Archaeological Investigations at Antelope House.* National Park Service, Washington, D.C.

Morris, E. A.

1980 *Basketmaker Caves in the Prayer Rock District, Northeastern Arizona.* Anthropological Papers of the University of Arizona, no. 35. Tucson.

Morris, E. H.

1919 *Preliminary Account of the Antiquities of the Region Between the Mancos and La Plata Rivers in Southwestern Colorado.* Thirty-third Annual Report of the Bureau of American Ethnology for the Years 1911–1912, pp. 155–205. Washington, D.C.

1939 *Archaeological Studies in the La Plata District: Southwestern Colorado and Northwestern New Mexico.* Carnegie Institution of Washington Publication, no. 519. Washington, D.C.

Morris, E. H., and R. F. Burgh
1954 *Basket Maker II Sites Near Durango, Colorado.* Carnegie Institution of Washington Publication, no. 604. Washington, D.C.

Morris, J. N.
1988a Excavations at Weasel Pueblo (Site 5MT5106), a Pueblo I–Pueblo III Multiple-Occupation Site. In *Dolores Archaeological Program: Anasazi Communities at Dolores: McPhee Village,* compiled by A. E. Kane and C. K. Robinson, pp. 664–787. Bureau of Reclamation, Engineering and Research Center, Denver.
1988b Excavations in Area 6. In *Dolores Archaeological Program: Anasazi Communities at Dolores: Grass Mesa Village,* compiled by W. D. Lipe, J. N. Morris, and T. A. Kohler, pp. 767–846. Bureau of Reclamation, Engineering and Research Center, Denver.
1988c Excavations in Area 8. In *Dolores Archaeological Program: Anasazi Communities at Dolores: Grass Mesa Village,* compiled by W. D. Lipe, J. N. Morris, and T. A. Kohler, pp. 873–932. Bureau of Reclamation, Engineering and Research Center, Denver.

Moseley, J. E.
1965 The Paleopathological Riddle of "Symmetrical Osteoporosis." *American Journal of Roentgenology* 95:135–142.

Mueller-Dombois, D., and H. Ellenberg
1974 *Aims and Methods of Vegetation Ecology.* John Wiley and Sons, New York.

Munz, P. A., and D. D. Keck
1959 *A California Flora.* University of California Press, Berkeley.

National Climate Center, Environmental Data and Information Service, National Oceanic and Atmospheric Administration
1983 *Climate Normals for the U.S. (Base: 1951–1980).* Gale Research, Detroit.

Netting, R. M., R. R. Wilk, and E. J. Arnould (editors)
1984 *Households: Comparative and Historical Studies of the Domestic Group.* University of California Press, Berkeley.

Neusius, S. W.
1984 Garden Hunting and Anasazi Game Procurement: Perspective from Dolores. Paper presented at the 49th Annual Meeting of the Society for American Archaeology, Portland.
1985a The Dolores Archaeological Program Faunal Data Base. In *Dolores Archaeological Program: Studies in Environmental Archaeology,* compiled by K. L. Petersen, V. L. Clay, M. H. Matthews, and S. W. Neusius, pp. 117–126. Bureau of Reclamation, Engineering and Research Center, Denver.
1985b Faunal Resource Use: Perspectives from the Ethnographic Record. In *Dolores Archaeological Program: Studies in Environmental Archaeology,* compiled by K. L. Petersen, V. L. Clay, M. H. Matthews, and S. W. Neusius, pp. 101–115. Bureau of Reclamation, Engineering and Research Center, Denver.
1985c Past Faunal Distribution and Abundance Within the Escalante Sector. In *Dolores Archaeological Program: Studies in Environmental Archaeology,* compiled by K. L. Petersen, V. L. Clay, M. H. Matthews, and S. W. Neusius, pp. 65–100. Bureau of Reclamation, Engineering and Research Center, Denver.

1986 The Dolores Archaeological Program Faunal Data Base: Resource Availability and Resource Mix. In *Dolores Archaeological Program: Final Synthetic Report,* compiled by D. A. Breternitz, C. K. Robinson, and G. T. Gross, pp. 199–303. Bureau of Reclamation, Engineering and Research Center, Denver.
1988 Faunal Exploitation During the McPhee Phase: Evidence from the McPhee Community Cluster. In *Dolores Archaeological Program: Anasazi Communities at Dolores: McPhee Village,* compiled by A. E. Kane and C. K. Robinson, pp. 1208–1291. Bureau of Reclamation, Engineering and Research Center, Denver.

Neusius, S. W., and T. W. Canaday
1985 The Dolores Archaeological Program Faunal Studies Section Laboratory Manual. Ms. on file, Anasazi Heritage Center, Dolores, Colorado.

Neusius, S. W., and P. R. Flint
1985 Cottontail Species Identification: The Reliability of Mandibular Measurements. *Journal of Ethnobiology* 5:51–58.

Neusius, S. W., and M. Gould
1988 Faunal Remains: Implications for Dolores Anasazi Adaptations. In *Dolores Archaeological Program: Anasazi Communities at Dolores: Grass Mesa Village,* compiled by W. D. Lipe, J. N. Morris, and T. A. Kohler, pp. 1049–1135. Bureau of Reclamation, Engineering and Research Center, Denver.

Nickens, P. R.
1974 *Analysis of Prehistoric Human Skeletal Remains from the Mancos Canyon, Southwestern Colorado.* Ms. prepared in conjunction with Bureau of Indian Affairs, contract no. MOOC14201337.

Nickens, P. R., and D. A. Hull
1982 Archaeological Resources of Southwestern Colorado: An Overview of the Bureau of Land Management's San Juan Resource Area. In *Archaeological Resources in Southwestern Colorado,* authored by P. R. Nickens, D. A. Hull, A. D. Reed, D. D. Scott, P. J. Gleichman, and S. Eininger, pp. 1–308. Colorado Bureau of Land Management Cultural Resources Series, no. 13.

Nickerson, N. H.
1953 Variation in Cob Morphology Among Certain Archaeological and Ethnological Races of Maize. *Annals of the Missouri Botanical Garden* 40:79–111.

Olivier, G.
1969 *Practical Anthropology.* Charles C. Thomas, Springfield, Illinois.

Olsen, S. J.
1968 The Osteology of the Wild Turkey. In *Fish, Amphibian and Reptile Remains from Archaeological Sites, Part I: Southeastern and Southwestern United States,* pp. 107–133. Papers of the Peabody Museum of Archaeology and Ethnology, vol. 56, no. 2. Harvard University, Cambridge.
1979a Archaeologically, What Constitutes an Early Domestic Animal? In *Advances in Archaeological Method and Theory,* vol. 2, edited by M. B. Schiffer, pp. 175–197. Academic Press, New York.
1979b *North American Birds.* Papers of the Peabody Museum of Archaeology and Ethnology, vol. 56, no. 5. Harvard University, Cambridge.
1985 *Origins of the Domestic Dog: The Fossil Record.* University of Arizona Press, Tucson.

Ortner, D. J., and W. G. J. Putschar

1981 *Identification of Pathological Conditions in Human Skeletal Remains.* Smithsonian Contributions to Anthropology, no. 28. Smithsonian Institute Press, Washington, D.C.

Palkovich, A. M.

1980 *Pueblo Population and Society: The Arroyo Hondo Skeletal and Mortuary Remains.* Arroyo Hondo Archaeological Series, no. 3. School of American Research Press, Santa Fe.

1987 Endemic Disease Patterns in Paleopathology: Porotic Hyperostosis. *American Journal of Physical Anthropology* 74(4):527–537.

Pennak, R. W.

1978 *Fresh-Water Invertebrates of the United States.* John Wiley and Sons, New York.

Petersen, K. L.

1986 Climatic Reconstruction for the Dolores Project Area. In *Dolores Archaeological Program: Final Synthetic Report,* compiled by D. A. Breternitz, C. K. Robinson, and G. T. Gross, pp. 311–325. Bureau of Reclamation, Engineering and Research Center, Denver.

1987 Tree-Ring Transfer Functions for Estimating Corn Production. In *Dolores Archaeological Program: Supporting Studies: Settlement and Environment,* compiled by K. L. Petersen and J. D. Orcutt, pp. 216–231. Bureau of Reclamation, Engineering and Research Center, Denver.

1988 *Climate and the Dolores River Anasazi: A Paleoenvironmental Reconstruction from a 10,000-Year Pollen Record, La Plata Mountains, Southwestern Colorado.* Anthropological Papers, no. 113. University of Utah Press, Salt Lake City.

Petersen, K. L., and V. L. Clay

1987 Characteristics and Archaeological Implications of Cold Air Drainage in the Dolores Project Area. In *Dolores Archaeological Program: Supporting Studies: Settlement and Environment,* compiled by K. L. Petersen and J. D. Orcutt, pp. 186–214. Bureau of Reclamation, Engineering and Research Center, Denver.

Petersen, K. L., and J. D. Orcutt (compilers)

1987 *Dolores Archaeological Program: Supporting Studies: Settlement and Environment.* Bureau of Reclamation, Engineering and Research Center, Denver.

Phagan, C. J., and T. H. Hruby

1984 *Reductive Technologies Manual: Preliminary Analysis Systems and Procedures.* Dolores Archaeological Program Technical Reports, DAP-150. Report submitted to the Bureau of Reclamation, Salt Lake City.

Phenice, T. W.

1969 A New Developed Visual Method for Sexing the *Os Pubis. American Journal of Physical Anthropology* 30:297–301.

Pilsbry, H. A.

1939 *Land Mollusca of North America (North of Mexico).* Academy of Natural Sciences of Philadelphia Monographs, no. 3, vol. 1, part 1.

1948 *Land Mollusca of North America (North of Mexico).* Academy of Natural Sciences of Philadelphia Monographs, no. 3, vol. 2, part 2.

Pizzimenti, J. J., and R. S. Hoffmann

1973 *Cynomys gunnisoni. Mammalian Species* 25:1–4.

Powell, S. L.

1983 *Mobility and Adaptation: The Anasazi of Black Mesa, Arizona.* Southern Illinois University Press, Carbondale.

Raffensperger, C.

1988 Cooking Methods. In *Dolores Archaeological Program: Supporting Studies: Additive and Reductive Technologies,* compiled by E. Blinman, C. J. Phagan, and R. H. Wilshusen, pp. 205–208. Bureau of Reclamation, Engineering and Research Center, Denver.

Reagan, A. B.

1929 Plants Used by the White Mountain Apache Indians of Arizona. *Wisconsin Archaeologist* 8:143–161.

Redfield, A.

1970 A New Aid to Aging Immature Skeletons: Development of the Occipital Bone. *American Journal of Physical Anthropology* 33:207–220.

Reed, A. D.

1979 The Dominguez Ruin: A McElmo Phase Pueblo in Southwestern Colorado. In *The Archaeology and Stabilization of the Dominguez and Escalante Ruins,* authored by A. D. Reed, J. A. Hallasi, A. S. White, and D. A. Breternitz, pp. i–196. Colorado Bureau of Land Management Cultural Resources Series, no. 7. Bureau of Land Management, Colorado State Office, Denver.

Reed, A. D., W. K. Howell, P. R. Nickens, and J. Horn

1985 *Archaeological Investigations on the Johnson Canyon Road Project, Ute Mountain Ute Tribal Lands, Colorado.* Report prepared for the Bureau of Indian Affairs, Albuquerque.

Reed, E. K.

1949 The Significance of Skull Deformation in the Southwest. *El Palacio* 56:106–119.

1963 *Occipital Deformation in the Northern Southwest.* Regional Research Abstract, no. 310. National Park Service, Santa Fe.

Reeder, J. R.

1957 The Embryo in Grass Systematics. *American Journal of Botany* 44:756–768.

Robbins, C. S., B. Bruun, H. S. Zim, and A. Singer

1966 *Birds of North America.* Golden Press, New York.

Robbins, W. W., J. P. Harrington, and B. Freire-Marreco

1916 *Ethnobotany of the Tewa Indians.* Bureau of American Ethnology Bulletin, no. 55. Washington, D.C.

Roberts, F. H. H., Jr.

1929 *Shabik'eshchee Village—A Late Basket Maker Site in the Chaco Canyon, New Mexico.* Bureau of American Ethnology Bulletin, no. 92. Washington, D.C.

1930 *Early Pueblo Ruins in the Piedra District, Southwestern Colorado.* Bureau of American Ethnology Bulletin, no. 96. Washington, D.C.

Rohn, A. H.

1971 *Mug House, Mesa Verde National Park, Colorado.* Archeological Research Series, no. 7-D. National Park Service, Washington, D.C.

Russell, F.

1975 *The Pima Indians.* Reprinted. University of Arizona Press, Tucson. Originally published 1908, Smithsonian Institution, Washington, D.C.

Saul, F. P.

1972 *The Human Skeletal Remains of Altar de Sacrificios: An Osteobiographic Analysis.* Papers of the Peabody Museum of Archaeology and Ethnology, vol. 63, no. 2. Harvard University, Cambridge.

1976 Osteobiography: Life History Recorded in Bone. In *The Measures of Man: Methodologies in Biological Anthropology,* edited by E. Giles and J. S. Friedlaender, pp. 372–382. Peabody Museum Press, Cambridge.

Schaff, J.

1981 A Method for Reliable and Quantifiable Subsampling of Archaeological Features for Flotation. *Mid-Continental Journal of Archaeology* 6:219–248.

Schiffer, M. B.

1985 Is There a "Pompeii Premise" in Archaeology? *Journal of Anthropological Research* 41:18–41.

Schlanger, S. H.

1985 *Prehistoric Population Dynamics in the Dolores Area, Southwestern Colorado.* Ph.D. dissertation, Washington State University. University Microfilms, Ann Arbor.

1987 Population Measurement, Size, and Change, A.D. 600–1175. In *Dolores Archaeological Program: Supporting Studies: Settlement and Environment,* compiled by K. L. Petersen and J. D. Orcutt, pp. 568–613. Bureau of Reclamation, Engineering and Research Center, Denver.

Schweingruber, F. H.

1982 *Microscopic Wood Anatomy, Structural Variability of Stems and Twigs in Recent and Subfossil Woods from Central Europe.* 2d ed. F. Fluck-Wirth, Teufen, Switzerland.

Scott, J. H., and N. B. B. Symons

1971 *Introduction to Dental Anatomy.* 6th ed. E. & S. Livingstone, Edinburgh.

Shawe, D. R., G. C. Simmons, and N. L. Archbold

1968 *Stratigraphy of Slick Rock District and Vicinity, San Miguel and Dolores Counties, Colorado.* U.S. Geological Survey Professional Paper, no. 576-A. U.S. Government Printing Office, Washington, D.C.

Shepard, A. O.

1939 Technology of La Plata Pottery. In *Archaeological Studies in the La Plata District: Southwestern Colorado and Northwestern New Mexico,* authored by E. H. Morris, pp. 249–287. Carnegie Institution of Washington Publication, no. 519. Washington, D.C.

1976 *Ceramics for the Archaeologist.* Carnegie Institution of Washington Publication, no. 609. Washington, D.C.

Shifrin, L. K.

1991 Untitled draft of Master's thesis, dated May 30, 1991. Ms. on file, Crow Canyon Archaeological Center, Cortez, Colorado.

Siemer, E. G.

1977 *Colorado Climate.* Colorado Experiment Station, Colorado State University, Fort Collins.

Spouge, J. D.

1973 *Oral Pathology.* C. V. Mosby, St. Louis.

Steele, D. G.

1970 Estimation of Stature from Fragments of Long Limb Bones. In *Personal Identification in Mass Disasters,* edited by T. D. Stewart, pp. 85–97. Smithsonian Institution, Washington, D.C.

Steinbock, R. T.

1976 *Paleopathological Diagnosis and Interpretation: Bone Diseases in Ancient Human Populations.* Charles C. Thomas, Springfield, Illinois.

Stevenson, M. C.

1894 *The Sia.* In Eleventh Annual Report of the Bureau of Ethnology for the Years 1889–1890, pp. 3–157. Washington, D.C.

1915 *Ethnobotany of the Zuni Indians.* In Thirtieth Annual Report of the Bureau of American Ethnology for the Years 1908–1909, pp. 35–102. Washington, D.C.

Steward, J. H.

1938 *Basin-Plateau Aboriginal Sociopolitical Groups.* Bureau of American Ethnology Bulletin, no. 120. Washington, D.C.

Stewart, T. D.

1974 Nonunion of Fractures in Antiquity, with Descriptions of Five Cases from the New World Involving the Forearm. *Bulletin of the New York Academy of Medicine* 50:875–891.

1979 *Essentials of Forensic Anthropology, Especially as Developed in the United States.* Charles C. Thomas, Springfield, Illinois.

Stodder, A. W.

1987 The Physical Anthropology and Mortuary Practice of the Dolores Anasazi: An Early Pueblo Population in Local and Regional Context. In *Dolores Archaeological Program: Supporting Studies: Settlement and Environment,* compiled by K. L. Petersen and J. D. Orcutt, pp. 336–504. Bureau of Reclamation, Engineering and Research Center, Denver.

STSC, Inc.

1988 *Statgraphics,* Version 3.0. STSC, Rockville, Maryland.

Stuart-Macadam, P.

1987 Porotic Hyperostosis: New Evidence to Support the Anemia Theory. *American Journal of Physical Anthropology* 74:521–526.

Suchey, J. M.

1979 Problems in the Aging of Females Using the *Os Pubis. American Journal of Physical Anthropology* 51:467–470.

Suchey, J. M., S. T. Brooks, and D. Katz

1988 Instructions for use of the Suchey-Brooks system for age determination of the female os pubis. Instructional materials accompanying female pubic symphyseal models. Distributed by France Casting, Fort Collins, Colorado.

Suchey, J. M., D. V. Wiseley, and D. Katz

1986 Evaluation of the Todd and McKern-Stewart Methods of Aging the Male *Os Pubis.* In *Forensic Osteology: Advances in the Identification of Human Remains,* edited by K. Reichs, pp. 33–67. Charles C. Thomas, Springfield, Illinois.

Sundick, R. I.

1978 Human Skeletal Growth and Age Determination. *Homo* 29:228–249.

1979 Age Determination of Subadult Skeletons. Paper presented at the Annual Meeting of the American Academy of Forensic Sciences, Atlanta.

Swank, G. R.

1932 *The Ethnobotany of the Acoma and Laguna Indians.* Unpublished Master's thesis, Department of Biology, University of New Mexico, Albuquerque.

Swedlund, A. C.

1969 *Human Skeletal Material from the Yellowjacket Canyon Area, Southwestern Colorado.* Unpublished Master's thesis, Department of Anthropology, University of Colorado, Boulder.

Swedlund, A. C. (editor)

1975 *Population Studies in Archaeology and Biological Anthropology: A Symposium.* Memoirs of the Society for American Archaeology, no. 30.

Swedlund, A. C., and G. J. Armelagos

1969 Un Recherche en Paleodemographie: La Nubia Soudanaise. *Annales: Economie, Societes, Civilization* 24:1287–1298.

1976 *Demographic Anthropology.* Wm. C. Brown, Dubuque.

Todd, T. W.

1920 Age Changes in the Pubic Bone, I: The Male White Pubis. *American Journal of Physical Anthropology* 3:285–334.

Trotter, M.

1970 Estimation of Stature from Intact Long Limb Bones. In *Personal Identification in Mass Disasters,* edited by T. D. Stewart, pp. 71–83. Smithsonian Institution, Washington, D.C.

Trotter, M., and G. C. Gleser

1952 Estimation of Stature from Long Bones of American Whites and Negroes. *American Journal of Physical Anthropology* 10:463–514.

1958 A Re-Evaluation of Estimation of Stature Based on Measurements of Stature Taken During Life and of Long Bones After Death. *American Journal of Physical Anthropology* 16:79–123.

Turner, C. G., II

1988 Another Prehistoric Southwest Mass Human Burial Suggesting Violence and Cannibalism: Marshview Hamlet, Colorado. In *Dolores Archaeological Program: Aceramic and Late Occupations at Dolores,* compiled by G. T. Gross and A. E. Kane, pp. 81–83. Bureau of Reclamation, Engineering and Research Center, Denver.

Turner, R. M.

1982 Great Basin Desertscrub. In Biotic Communities of the American Southwest—United States and Mexico, edited by D. E. Brown. *Desert Plants* 4(1–4):145–155.

Ubelaker, D. H.

1978 *Human Skeletal Remains: Excavation, Analysis, Interpretation.* Aldine, Chicago.

Van West, C. R.

1986 *Cultural Resource Inventory for the 1985 Field Season in the Crow Canyon and Sand Canyon Areas, Montezuma County, Colorado.* Crow Canyon Archaeological Center, Cortez, Colorado. Report submitted to the Bureau of Land Management, San Juan Resource Area Office, Durango, Colorado.

Varien, M. D.

1988 Excavations in Areas 1 and 2. In *Dolores Archaeological Program: Anasazi Communities at Dolores: Grass Mesa Village,* compiled by W. D. Lipe, J. N. Morris, and T. A. Kohler, pp. 75–315. Bureau of Reclamation, Engineering and Research Center, Denver.

Varien, M. D., and R. R. Lightfoot

1989 Ritual and Nonritual Activities in Mesa Verde Region Pit Structures. In *The Architecture of Social Integration in Prehistoric Pueblos,* edited by W. D. Lipe and M. Hegmon,

pp. 73–87. Occasional Papers of the Crow Canyon Archaeological Center, no. 1. Cortez, Colorado.

Walker, D. N.

1988 Archaeological Evidence for the Use of Small Mammals by Prehistoric Inhabitants on the Northwestern High Plains. In *The Prairie: Roots of Our Culture; Foundation of Our Economy,* edited by A. Davis and G. Stanford, pp. 6.08.1–6. Native Prairie Association of Texas, Dallas.

Warkany, J.

1971 *Congenital Malformations: Notes and Comments.* Year Book Medical Publishers, Chicago.

Waterworth, R. M. R., and E. Blinman

1986 Modified Sherds, Unidirectional Abrasion, and Pottery Scrapers. *Pottery Southwest* 13(2):4–7.

Weaver, D. S.

1980 Sex Differences in the Ilia of a Known Sex and Age Sample of Fetal and Infant Skeletons. *American Journal of Physical Anthropology* 52:191–195.

Webb, P. A. O., and J. M. Suchey

1985 Epiphyseal Union of the Anterior Iliac Crest and Medial Clavicle in a Modern Multi-Racial Sample of American Males and Females. *American Journal of Physical Anthropology* 68:457–466.

Weiss, K. M.

1973 *Demographic Models for Anthropology.* Memoirs of the Society for American Archaeology, no. 27.

1975 Demographic Disturbance and the Use of Life Tables in Anthropology. In *Population Studies in Archaeology and Biological Anthropology: A Symposium,* edited by A. C. Swedlund, pp. 46–56. Memoirs of the Society for American Archaeology, no. 30.

Wellhausen, E. J., L. M. Roberts, and E. Hernandez X. (in collaboration with P. C. Mangelsdorf)

1952 *Races of Maize in Mexico: Their Origin, Characteristics and Distribution.* The Bussey Institution of Harvard University, Cambridge.

Wheeler, R. P.

1980a Bone and Antler Artifacts. In *Long House, Mesa Verde National Park, Colorado,* authored by G. S. Cattanach, Jr., pp. 307–315. Publications in Archeology, no. 7H. National Park Service, Washington, D.C.

1980b Stone Artifacts and Minerals. In *Long House, Mesa Verde National Park, Colorado,* authored by G. S. Cattanach, Jr., pp. 243–305. Publications in Archeology, no. 7H. National Park Service, Washington, D.C.

Whiting, A. F.

1939 *Ethnobotany of the Hopi.* Museum of Northern Arizona Bulletin, no. 15. Flagstaff.

1966 *Ethnobotany of the Hopi.* Reprinted. Northland Press, Flagstaff. Originally published 1939, Museum of Northern Arizona, Flagstaff.

Wilder, H. H.

1920 *A Laboratory Manual of Anthropometry.* P. Blakiston and Sons, Philadelphia.

Wilk, R. R., and R. M. Netting

1984 Households: Changing Forms and Functions. In *Households: Comparative and Historical Studies of the Domestic Group,* edited by R. M. Netting, R. R. Wilk, and E. J. Arnould, pp. 1–28. University of California Press, Berkeley.

Wilk, R. R., and W. L. Rathje (editors)
  1982 Archaeology of the Household: Building a Prehistory of Domestic Life. *American Behavioral Scientist* 25(6).
Willcox, G. H.
  1974 A History of Deforestation as Indicated by Charcoal Analysis of Four Sites in Eastern Anatolia. *Anatolian Studies* 24:117–133.
Willey, P., and L. M. Snyder
  1989 Canid Modification of Human Remains: Implications for Time-Since-Death Estimations. *Journal of Forensic Sciences* 34(4):894–901.
Wilshusen, R. H.
  1984 Engineering the Pithouse to Pueblo Transition. Paper presented at the 49th Annual Meeting of the Society for American Archaeology, Portland.
  1985 The Relationship Between Abandonment Mode and Feature Assemblage in Pueblo I Anasazi Protokivas. Paper presented at the 50th Annual Meeting of the Society for American Archaeology, Denver.
  1986a Excavations at Periman Hamlet (Site 5MT4671), Area 1, a Pueblo I Habitation. In *Dolores Archaeological Program: Anasazi Communities at Dolores: Middle Canyon Area,* compiled by A. E. Kane and C. K. Robinson, pp. 24–208. Bureau of Reclamation, Engineering and Research Center, Denver.
  1986b Excavations at Rio Vista Village (Site 5MT2182), a Multicomponent Pueblo I Village. In *Dolores Archaeological Program: Anasazi Communities at Dolores: Middle Canyon Area,* compiled by A. E. Kane and C. K. Robinson, pp. 210–658. Bureau of Reclamation, Engineering and Research Center, Denver.
  1986c The Relationship Between Abandonment Mode and Ritual Use in Pueblo I Anasazi Protokivas. *Journal of Field Archaeology* 13:245–254.
  1987 Architecture as Artifact—Part II: A Comment on Gilman's "Architecture as Artifact: Pit Structures and Pueblos in the American Southwest." *American Antiquity* 54:826–833.
  1988a Architectural Trends in Prehistoric Anasazi Sites During A.D. 600 to 1200. In *Dolores Archaeological Program: Supporting Studies: Additive and Reductive Technologies,* compiled by E. Blinman, C. J. Phagan, and R. H. Wilshusen, pp. 599–633. Bureau of Reclamation, Engineering and Research Center, Denver.

  1988b Household Archaeology and Social Systematics. In *Dolores Archaeological Program: Supporting Studies: Additive and Reductive Technologies,* compiled by E. Blinman, C. J. Phagan, and R. H. Wilshusen, pp. 635–647. Bureau of Reclamation, Engineering and Research Center, Denver.
  1988c Sipapus, Ceremonial Vaults, and Foot Drums (Or, a Resounding Argument for Protokivas). In *Dolores Archaeological Program: Supporting Studies: Additive and Reductive Technologies,* compiled by E. Blinman, C. J. Phagan, and R. H. Wilshusen, pp. 649–671. Bureau of Reclamation, Engineering and Research Center, Denver.
  1989 Unstuffing the Estufa: Ritual Floor Features in Anasazi Pit Structures and Pueblo Kivas. In *The Architecture of Social Integration in Prehistoric Pueblos,* edited by W. D. Lipe and M. Hegmon, pp. 89–111. Occasional Papers of the Crow Canyon Archaeological Center, no. 1. Cortez, Colorado.
  1991 *Early Villages in the American Southwest: Cross-Cultural and Archaeological Perspectives.* Ph.D. dissertation, University of Colorado. University Microfilms, Ann Arbor.
Wilson, C. D., and E. Blinman
  1991a Ceramic Types of the Mesa Verde Region. Handout prepared for the Colorado Council of Professional Archaeologists Ceramic Workshop, Boulder.
  1991b Changing Specialization of White Ware Manufacture in the Northern San Juan Region. Paper presented at the 56th Annual Meeting of the Society for American Archaeology, New Orleans.
Wood, W. R., and D. L. Johnson
  1978 A Survey of Disturbance Processes in Archaeological Site Formation. In *Advances in Archaeological Method and Theory,* vol. 1, edited by M. B. Schiffer, pp. 315–381. Academic Press, New York.
Yanovsky, E.
  1936 *Food Plants of the North American Indians.* United States Department of Agriculture Miscellaneous Publication, no. 237. Washington, D.C.
Zaino, E. C.
  1967 Symmetrical Osteoporosis, a Sign of Severe Anemia in the Prehistoric Pueblo Indians of the Southwest. In *Miscellaneous Papers in Paleopathology, I,* edited by W. D. Wade, pp. 40–47. Museum of Northern Arizona Technical Series, no. 7. Flagstaff.

Index